## ABOUT ISLAND PRESS

Since 1984, the nonprofit Island Press has been stimulating, shaping, and communicating the ideas that are essential for solving environmental problems worldwide. With more than 800 titles in print and some 40 new releases each year, we are the nation's leading publisher on environmental issues. We identify innovative thinkers and emerging trends in the environmental field. We work with world-renowned experts and authors to develop cross-disciplinary solutions to environmental challenges.

Island Press designs and implements coordinated book publication campaigns in order to communicate our critical messages, in print, in person, and online using the latest technologies, programs, and the media. Our goal: to reach targeted audiences—scientists, policymakers, environmental advocates, the media, and concerned citizens—who can and will take action to protect the plants and animals that enrich our world, the ecosystems we need to survive, the water we drink, and the air we breathe.

Island Press gratefully acknowledges the support of its work by the Agua Fund, Inc., Annenberg Foundation, The Christensen Fund, The Nathan Cummings Foundation, The Geraldine R. Dodge Foundation, Doris Duke Charitable Foundation, The Educational Foundation of America, Betsy and Jesse Fink Foundation, The William and Flora Hewlett Foundation, The Kendeda Fund, The Forrest and Frances Lattner Foundation, The Andrew W. Mellon Foundation, The Curtis and Edith Munson Foundation, Oak Foundation, The Overbrook Foundation, the David and Lucile Packard Foundation, The Summit Fund of Washington, Trust for Architectural Easements, Wallace Global Fund, The Winslow Foundation, and other generous donors.

The opinions expressed in this book are those of the author(s) and do not necessarily reflect the views of our donors.

# Noninvasive Survey Methods
# for Carnivores

# Noninvasive Survey Methods for Carnivores

EDITED BY

*Robert A. Long, Paula MacKay,*
*William J. Zielinski, Justina C. Ray*

**ISLAND**PRESS

*Washington · Covelo · London*

Noninvasive survey methods for carnivores / edited by Robert A. Long . . . [et al.].
     p. cm.
  Includes bibliographical references and index.
   ISBN-13: 978-1-59726-119-7 (cloth : alk. paper)
   ISBN-10: 1-59726-119-X (cloth : alk. paper)
   ISBN-13: 978-1-59726-120-3 (pbk. : alk. paper)
   ISBN-10: 1-59726-120-3 (pbk. : alk. paper)
   1. Carnivora—North America. 2. Mammalogy—Methodology. 3. Mammal surveys—North America. I. Long, Robert A.
   QL737.C2N47 2008
   599.7072′3—dc22                                                  2007040646

*Keywords*: carnivore, wildlife survey methods, DNA analysis for wildlife, remote cameras, scat detection dogs, noninvasive research techniques, wildlife detection devices, wildlife hair collection

# Contents

# List of North American Carnivore Species

| Common Name | Latin Name |
|---|---|
| **Canids** | |
| Coyote | *Canis latrans* |
| Gray wolf | *Canis lupus* |
| Domestic dog | *Canis lupus familiaris* |
| Red wolf | *Canis rufus* |
| Gray fox | *Urocyon cinereoargenteus* |
| Island fox | *Urocyon littoralis* |
| Arctic fox | *Vulpes lagopus* |
| Kit fox | *Vulpes macrotis* |
| Swift fox | *Vulpes velox* |
| Red fox | *Vulpes vulpes* |
| **Felids** | |
| Domestic cat | *Felis catus* |
| Ocelot | *Leopardus pardalis* |
| Margay | *Leopardus wiedii* |
| Canada lynx | *Lynx canadensis* |
| Bobcat | *Lynx rufus* |
| Jaguar | *Panthera onca* |
| Cougar, puma, or mountain lion | *Puma concolor* |
| Jaguarundi | *Puma yagouaroundi* |
| **Mephitids** | |
| American hog-nosed skunk | *Conepatus leuconotus* |
| Hooded skunk | *Mephitis macroura* |

| Common Name | Latin Name |
|---|---|
| Striped skunk | *Mephitis mephitis* |
| Western spotted skunk | *Spilogale gracilis* |
| Eastern spotted skunk | *Spilogale putorius* |
| **Mustelids** | |
| Wolverine | *Gulo gulo* |
| North American river otter | *Lontra canadensis* |
| American marten | *Martes americana* |
| Fisher | *Martes pennanti* |
| Ermine or short-tailed weasel | *Mustela ermina* |
| Long-tailed weasel | *Mustela frenata* |
| Black-footed ferret | *Mustela nigripes* |
| Least weasel | *Mustela nivalis* |
| American mink | *Neovison vison* |
| American badger | *Taxidea taxus* |
| **Procyonids** | |
| Ringtail | *Bassariscus astutus* |
| White-nosed coati | *Nasua narica* |
| Raccoon | *Procyon lotor* |
| **Ursids** | |
| American black bear | *Ursus americanus* |
| Grizzly or brown bear | *Ursus arctos* |
| Polar bear | *Ursus maritimus* |

# Acknowledgments

This volume is the result of the dedication and efforts of many people. First and foremost, we are extremely grateful to the authors who have been a part of this undertaking since our organizational workshop was held at the Essex Conference Center and Retreat in June 2005. Workshop participants were remarkably generous of their time and wisdom and played a formative role in bringing the book to fruition. Numerous authors joined the project after the original planning meeting, and to them we further extend our heartfelt gratitude. Without exception, our contributors have exhibited great knowledge, professionalism, and patience—despite our seemingly endless requests for revisions. We feel honored and fortunate to have worked with such top-notch scientists.

Each of the core chapters in the book was subjected to rigorous reviews by at least two experts in the field. These reviews were pivotal to the development and content of the individual chapters and the volume as a whole. Our reviewers included: Keith Aubry, Jerry Belant, John Boulanger, Kevin Crooks, Barbara Davenport, Terri Donovan, Matt Gompper, Robert Harrison, Aimee Hurt, Roland Kays, Marcella Kelly, Carolyn King, Tom Kucera, Kevin McKelvey, Audrey Magoun, Josh Millspaugh, Garth Mowat, Glen Sargeant, Ric Schlexer, Mike Schwartz, Rick Truex, Drew Tyre, Lisette Waits, John Weaver, Steve Weigley, Alice Whitelaw, and the late Eric York. We sincerely appreciate the assistance of these committed individuals.

We are indebted to illustrators Stephen Harrison and Patricia Kernan for their creative contributions, and to graphic artist Kevin Cross for putting his master's touch on some of our final figures.

Barbara Dean at Island Press has been supportive of this project since its early inception, and graciously encouraged us to pursue a book. Barbara's unique ability to be both friend and mentor has made her a legend in her field, and we deeply appreciate all that she has done for us. Likewise, Barbara Youngblood was a ray of sunshine even during the darker days of editing, and always provided prompt and upbeat replies to our many queries. Other colleagues at Island Press, including John Cangany, Erin Johnson, and Katherine Macdonald, were also excellent resources.

We would like to thank Jon and Kitty Harvey for their generosity on behalf of carnivore conservation—the project could not have happened without them. We gratefully acknowledge Wildlife Conservation Society Canada and the Pacific Southwest Research Station of the USDA Forest Service for their contributions as well. Finally, we thank our respective family members, whose patience and encouragement allowed us to fulfill this important vision.

# Chapter 1

# Noninvasive Research and Carnivore Conservation

*Paula MacKay, William J. Zielinski, Robert A. Long, and Justina C. Ray*

Mammalian carnivores face myriad challenges in our overcrowded world. As the all-too-familiar threats of habitat loss and fragmentation continue to increase against a backdrop of global climate change, many carnivore populations have experienced dramatic range contractions (Laliberte and Ripple 2004) and are in urgent need of protection (Ginsberg 2001; MacDonald 2001). Meanwhile, in some regions, carnivores that suffered serious declines in the past century are reclaiming lost ground as a result of natural reforestation (Foster et al. 2002; Falcucci et al. 2007) or targeted conservation inspired by their ecological importance, public appeal, or conflicts with humans (Ray 2005). Given these and other compelling scenarios, it is more critical than ever for scientists to produce relevant and sound data pertaining to carnivore distribution, habitat use, and other biological and ecological measures. While good science does not *guarantee* quality conservation, the latter will not be possible without the former.

The methods described in this book are especially important tools for those seeking to conduct surveys for members of the order Carnivora. The 230-odd species in this group exhibit remarkable diversity in body form, function, and ecology, yet they share a propensity to leave identifiable evidence of their presence in the form of tracks and droppings. Arguably more than any other animal group, carnivore foot morphology displays interspecific variation. Further, many species are characterized by territoriality, curiosity, traveling along routes, and marking behaviors that result in the prominent placement of sign—traits that lend themselves well to noninvasive survey methods. Carnivore movement patterns are also conducive to the strategic placement of devices to "capture" evidence of species presence. At the same time, the low-density populations and elusive and wide-ranging nature of most carnivores render them difficult to study with observational or traditional capture-based methods. The unique fit between noninvasive survey methods and carnivores, coupled with their importance in the conservation arena, is the impetus for this book.

## The Meaning of *Noninvasive*

The methods described in the following chapters, which we've liberally assembled under the umbrella *noninvasive*, share the common attribute that they do not require target animals to be directly observed or handled by the surveyor. Given that the term noninvasive may inherently imply judgment against methods that could be lumped together under the antonymous term *invasive*, we feel it is appropriate to discuss this apparent (and somewhat misleading) dichotomy in a bit more detail.

The word noninvasive has historically been used in a medical context, as in a diagnostic procedure that doesn't involve penetrating the skin or organism with an incision or an injection—as opposed to an invasive procedure, which does (Webster's Ninth New Collegiate Dictionary 1988). During the last fifteen years, noninvasive has been more generally applied to the remote collection of DNA samples (e.g., hair, feces) from free-ranging animals. Garshelis (2006) attributes the term's first use in this regard to Morin and Woodruff (1992). In recent applications, usage has expanded to include non-DNA-based wildlife survey techniques as well (e.g., Moruzzi et al. 2002; Gompper et al. 2006; Long et al. 2007b; Schipper 2007).

The survey methods included here are noninvasive in the broadest sense. For the reasons discussed, we grappled with other terms that might be used instead; for example, *nonintrusive* or *remote*. Numerous researchers have employed the latter term in this fashion (Sloane et al. 2000; Piggott and Taylor 2003; Frantz et al. 2004), and Garshelis (2006) suggests that this is a more exact and suitable adjective to describe sampling that occurs without human presence. We were concerned, however, that *remote survey methods* might be confused with *remote sensing* and other such technologies. Furthermore, we have observed that noninvasive has become somewhat conventional in the wildlife literature and felt that it might behoove us to adhere to an increasingly familiar term. Last, we appreciate the intention of the word noninvasive from the perspective of being minimally invasive with the animals we study.

Indeed, few would question the benefits of minimizing disturbance to target animals during wildlife surveys. We nonetheless recognize the potential risk of the term noninvasive conveying an ethical rather than a scientific basis for this publication. Further, some readers may interpret our emphasis on nonin-

vasive methods as a criticism of those that require live-capture, such as more traditional telemetry methods. Neither assessment would be accurate. First, as alluded to earlier and demonstrated throughout the book, noninvasive methods are particularly appropriate for the scientific study of carnivores given their ecology and behavior. Second, many of the contributors to this volume (including the editors) have used or continue to use telemetry methods in their work, and there is no disputing the valuable role that these methods play in wildlife research. Our goal is not to compare noninvasive techniques with capture-based methods, nor to advocate their use simply *because* they are noninvasive. Rather, we seek to provide researchers with information on the applicability of such methods for meeting survey goals. The exciting fact is that noninvasive survey methods can now yield high-quality data for modeling site occupancy, estimating population distribution and abundance, and achieving other ecological objectives. To our knowledge, no existing resource provides a comprehensive overview of contemporary noninvasive techniques. This is the gap we seek to fill.

Conventions aside, we recognize that noninvasive survey methods are not necessarily nonintrusive. While it's true that, by definition, these methods do not require physical contact between the surveyor and the surveyed, they too can have behavioral consequences. In a recent study, for example, Schipper (2007) found that camera flashes at remote camera stations resulted in trap avoidance by arboreal kinkajous (*Potos flavus*), and this effect has been observed in tigers (*Panthera tigris*) as well (Wegge et al. 2004). Further, some track stations, remote cameras, and hair collection methods utilize bait, which can result in trap avoidance or trap-happiness (see chapters 4, 5, 6, and 10), and notoriously shy species (e.g., coyotes [*Canis latrans*]) have been shown to avoid survey equipment (Harris and Knowlton 2001). It seems feasible that scat detection dog surveys could have behavioral ramifications as well—for instance, due to the presence of dogs or the removal of scats deposited for territorial marking—although no such effects have been documented to our knowledge. More generally, the mere presence of humans can clearly disturb wildlife, and injuries are not out of the question with noninvasive methods if equipment (e.g., barbed-wire hair collection devices, nails used to secure bait) is improperly deployed or interacted with by animals in unanticipated ways. Reciprocally, radio-based or global positioning system (GPS)-based telemetry may be virtually noninvasive after the initial capture (Garshelis 2006).

The bottom line is that researchers conducting *any* wildlife survey must weigh the tradeoffs associated with methods that will allow them to achieve their goals. For some surveys, the target species and primary objectives will lend themselves well to one or more affordable and effective noninvasive methods. We presume that such methods will be the obvious choice when this is the case. In other situations, the required data may only be obtainable via more traditional capture- or observation-based methods.

Even with recent advances in noninvasive methods, telemetry still offers unique advantages when attempting to address certain objectives (Mech and Barber 2002). It has only been since the advent of telemetry, particularly via satellite, that we have been able to gain a scientific appreciation of large-scale animal movement (e.g., Inman et al. 2004). If it is necessary to locate animals during key life stages (e.g., denning), or to access carcasses to document causes of mortality, telemetry may be required. Similarly, telemetry may provide more informative assessments of habitat use if both the movement and fate (e.g., survival and mortality) of individual animals can be closely tracked (Garshelis 2006). Further, the accuracy of abundance estimates for wide-ranging species may increase if information about the extent of geographic closure is available. Such data are readily accessible if a proportion of the population is telemetered. In a comparison of abundance estimators, Choate et al. (2006) found that the capture and marking of cougars (*Puma concolor*) via telemetry was the most costly of the techniques employed, but also the most sensitive for estimating population size.

## A Brief History of Noninvasive Survey Methods for Carnivores

Noninvasive methods for the study of carnivores probably date back to the origin of humans—presumably, we have always sought information on the whereabouts and habits of species that can harm us or provide us with valuable resources (e.g., hides for clothing and shelter, meat). The modern scientific study of mammalian carnivores, however, began after commercial agriculture diminished the need to hunt and gather, human social structures and technological developments (e.g., firearms, steel traps, poison) reduced the perceived threat of large carnivores, and some carnivore populations exhibited signs of decline.

In the mid- and late-twentieth century, a few field-savvy experts catalyzed public interest in tracking and other outdoor skills. These individuals captured their expertise in field guides, primarily for an audience of commercial and recreational fur trappers and naturalists striving to develop field skills as a recreational pastime. Expert naturalists such as Olaus Murie (Murie 1954), Tom Brown (Brown 1983), and Jim Halfpenny (Halfpenny 1986) provided descriptions of wildlife tracks and trails in natural substrates, and these accounts became an important foundation for the scientific study of free-ranging mammals. New descriptive accounts of animal sign continue to materialize today (e.g., Elbroch 2003; Lowery 2006), and there is a renaissance of interest in experiencing animal sign first-hand among lay persons and citizen scientists (e.g., Keeping Track, www.keepingtrack.org; CyberTracker Conservation, www.cybertracker.co.za).

In the 1970s and 1980s, the field of mammal inventory and monitoring began to emerge from the foundation established by naturalists. The close of the twentieth century found more people dwelling in cities, and fewer in rural locations where they could encounter and experience wild mammals and their sign. The number of individuals who possessed and could share skills in mammal track and trail identification dwindled.

Meanwhile, the environmental movement and associated legislation (e.g., National Environmental Protection Act [1970], Endangered Species Act of 1973), produced a political climate in which natural resource decision makers required scientifically defensible information about the status of wildlife. Scientists were being called into service to help society develop rigorous methods for detecting species and inventorying and monitoring their populations. Dependence on a declining number of experts who had developed descriptive, and often qualitative, means of identifying wildlife tracks, trails, and sign was unacceptable. Without quantitative methods to identify mammals from sign, information from traditional sources was not considered reliable enough for scientific endeavors or legal challenges. Quality control issues assumed priority, and scientists sought methods that could yield credible results when deployed and interpreted by biologists with limited field experience. This era was inaugurated in North America by efforts to determine the distribution of uncommon mustelids (i.e., Barrett 1983; Jones and Raphael 1993), and in Europe and New Zealand by stoat (*Mustela erminea*) research conducted by King and colleagues (King and Edgar 1977). Two new methods were developed during this period: (1) specially designed track-receptive surfaces enclosed in small boxes or tubes, and (2) line-triggered instamatic (110) film camera stations. These methods, along with more traditional techniques involving snow tracking and live-capture, were summarized in an important review paper on mustelids in the mid-1990s (Raphael 1994).

The field of noninvasive survey methods began to explode in the mid-1990s. In response, the USDA Forest Service sponsored and published a manual describing standardized protocols for detecting forest carnivores (i.e., American martens [*Martes americana*], fishers [*Martes pennanti*], wolverines [*Gulo gulo*], and Canada lynx [*Lynx canadensis*]) using track stations, remote cameras, and snow tracking methods (Zielinski and Kucera 1995a). This popular handbook supported a burgeoning interest, among professional biologists and amateurs alike, in detecting the presence of rare, forest-dwelling carnivores.

The most important new device presented in the Zielinski and Kucera (1995a) publication was the remote, 35mm camera triggered by either a motion sensor or the interruption of a light beam by an animal. Unlike its line-triggered predecessor, this system could be left unattended in the field for weeks and could collect up to thirty-six images (Kucera and Barrett 1993; Kucera et al. 1995a). Particularly influential during this period was the Trailmaster camera system (Goodson & Associates, Lenexa, KS). Also on the horizon, however, loomed a new and powerful technology that was destined to be introduced to the wildlife profession ever since the development of the polymerase chain reaction: DNA analysis. Zielinski and Kucera's manual foreshadowed the important role of this development, but it was not generally described to wildlife conservation practitioners until the late 1980s and early 1990s, via the papers of Kocher et al. (1989) and Morin and Woodruff (1992). Methods for the collection of genetic samples via hair snaring (Foran et al. 1997b) and scat collection via dogs (Smith et al. 2003; Wasser et al. 2004) emerged shortly thereafter.

The chapters included in this book describe the current state of the art, and their respective authors help forecast the future of noninvasive methods. Indeed, today is an exhilarating time to be a part of this field. It is now possible to identify species, sex, population, matrilines, and individuals with noninvasive methods—possibilities that our predecessors could not have imagined when they were gleaning information from tracks. With time, we presume that the methods presented here will too become outdated, and thus become part of the unfolding history of survey methods for carnivores.

## History and Scope of This Volume

This book was inspired, in part, by the Zielinski and Kucera (1995a) manual mentioned earlier. Recognizing the need for a timely and up-to-date resource for field biologists, agency personnel, graduate students, and others seeking to undertake carnivore surveys, we originally envisioned a technical report or white paper. Alas, our modest ambitions took on new fervor when we gathered together a small group of experts at the Essex Conference Center and Retreat (Essex, MA) in June 2005 (see Contributors at the end of the volume). This intensive, two-day workshop provided us with a unique opportunity to discuss noninvasive survey methods in great detail with experienced researchers. We were also able to work through our vision for a publication and to flesh out a comprehensive outline. In the end, the enthusiasm, knowledge, and dedication of the workshop participants encouraged us to pursue a book.

Although today's wildlife practitioners have an extensive toolbox of noninvasive research techniques at their disposal, they are also faced with a growing complexity of factors to consider in deciding which methods to use under which circumstances. We hope that this book will help provide direction to fellow researchers throughout the survey process, including during survey design, sample and data collection, DNA and endocrine analyses, and data analysis. Citing examples from the field, we review the suitability of various survey methods as they relate to target species, objectives, and other considerations. We also present strategies for integrating multiple noninvasive techniques into a single survey. Given the broad scope of species and topics included, this publication is, by necessity, less a "cookbook" than a comprehensive guidebook. Rather than prescribing survey protocols, it allows readers to carefully evaluate a diversity of detection methods and to develop protocols specific to their goals, region, and species of interest.

This volume focuses primarily on North American carnivores and generally follows the taxonomy of Wilson and Reeder (2005). Numerous examples and case studies originate from farther afield, however, and most of the methodological information included in the book should have global applicability. We expect that this material will also be of value to those interested in surveying other taxa with noninvasive methods. We strived to maximize consistency among chapters by employing common headings whenever possible, and to minimize redundancy by cross-referencing where appropriate. Our ultimate goal was to create a user friendly and cohesive "whole" by weaving together assorted parts representing the efforts of twenty-five experts in the field. While we furnished our contributors with a general structure, we also encouraged them to freely share their respective opinions and experiences. We were especially liberal with our guidelines for case studies, whose purpose is to illustrate real-world applications of the survey methods featured in the volume. Each case study presents a brief overview of survey objectives, methods and protocols, results, and conclusions. Contributors were invited to submit published or unpublished studies exemplifying their particular method.

## Chapters

Any successful survey begins with a solid survey design. Thus, in chapter 2, Robert Long and Bill Zielinski help set the stage for the rest of the book by providing an overview of the types of objectives that can be met using data collected via noninvasive survey methods—followed by a discussion of approaches and design considerations for surveys aimed at achieving these objectives. The material presented represents the most current thinking in survey design. For example, this chapter describes the latest thoughts on how to design detection-nondetection surveys to estimate site occupancy, as well as noninvasive capture-recapture surveys to estimate animal abundance.

The next five chapters showcase specific survey

methods. These chapters are arranged chronologically according to their emergence in the field, with natural sign surveys (chapter 3) having the longest history of use. Each of the method-specific chapters (i.e., those led by Kim Heinemeyer, Justina Ray, Roland Kays, Kate Kendall, and Paula MacKay, respectively) is roughly organized around a common framework that addresses the following topics:

- Background
- Target species
- Strengths and weaknesses
- Treatment of objectives
- Description and application of survey methods
- Practical considerations
- Survey design issues
- Sample and data collection and management
- Future directions and concluding thoughts
- Case studies

Two of these topics warrant a bit more explanation. First, in the *Treatment of Objectives* section of each chapter, the authors begin by describing how their respective methods have been used to meet four key objectives (i.e., occurrence and distribution, relative abundance, abundance and density, monitoring), and then briefly discuss other achievable objectives per a framework established in chapter 2. Second, the section entitled *Survey Design Issues* is tightly linked to the design considerations presented in chapter 2 but elaborates upon those especially relevant to the survey method at hand.

In chapter 8, Lori Campbell and coauthors discuss how multiple survey methods can be combined to meet certain objectives for one or more target species when a single method is insufficient for collecting adequate data. This chapter describes the synergy that can sometimes be achieved by integrating multiple methods and provides guidance for planning and executing multimethod surveys.

Many recent advances with noninvasive survey methods have been made possible by revolutionary developments in genetic and endocrine techniques over the last decade. Authors Mike Schwartz and Steve Monfort share their expertise with these techniques in chapter 9, respectively summarizing laboratory methods and applications for DNA and hormonal analyses, as well as pragmatic recommendations on how to collect and preserve samples to maximize survey results. The last section is dedicated to the unique objectives that can be achieved by combining molecular and endocrine approaches.

A number of the methods included in this book require attractants to draw target animals to detection devices. Decisions regarding whether or not to use attractants, which attractants to use, and how to deploy them can be pivotal to survey success. In chapter 10, Ric Schlexer thoroughly reviews the wide variety of substances and techniques available for attracting carnivores. Further, Schlexer offers helpful suggestions on how to acquire, apply, and store attractants and describes scientific efforts to test their efficacy.

Chapter 11 introduces readers to approaches that can be used to analyze and model noninvasively collected data. This chapter is not a step-by-step guide to analysis, but rather an accessible overview of the data structures and mathematical underpinnings of some of the more common analysis techniques. Andy Royle and coauthors also discuss a number of characteristics unique to noninvasively collected data, the challenges they pose, and thoughts on how to deal with them analytically.

In chapter 12, the editors attempt to tie together the multiple strands of the book. We highlight those advances we believe have contributed most to contemporary noninvasive survey methods and identify some remaining limitations. We also provide two summary tables, one designed to help readers decide which methods are most suitable for a given target species, and a second evaluating the relative strengths and weaknesses of each method. Finally, we share our vision of the future of the field, speculating on forthcoming developments and identifying other advances that we believe would further benefit carnivore research.

## A Final Note on Sharing Knowledge

The remarkable speed at which the field of noninvasive survey methods has grown is due, in part, to the tradition of sharing innovations via scientific meetings and word of mouth. It is astonishing to witness the flurry of new approaches that can be precipitated by a given workshop or key publication. Carnivore ecologists share a strong environmental ethic, and the eager transfer of ideas about new detection methods and modes of analysis is—in our opinion—a consequence of this ethic. As a group, we realize that sharing knowledge with our colleagues advances the cause of carnivore conservation. Often this means that we must cite innovations that have not yet been published or that may not be destined for peer-reviewed literature. Although the practice of citing this type of information is (understandably) discouraged by scientific journals, we have deliberately encouraged authors in this book to include important but unpublished literature or ideas. More specifically, while contributors have been instructed to cite published literature in favor of unpublished "gray" literature, they have also been supported in citing unpublished work when—in their view—an important innovation or finding is not represented in the peer-reviewed literature. In this way, the book continues the tradition that has propelled our collective work forward: an open and collegial spirit of cooperation.

# Chapter 2

# Designing Effective Noninvasive Carnivore Surveys

*Robert A. Long and William J. Zielinski*

The **methods** described throughout this book will enable researchers to noninvasively detect most, if not all, terrestrial carnivore species that occur in North America, as well as myriad carnivores worldwide. In many cases, genetic approaches will further extend the ability of these techniques to permit the identification of individuals. The **detection** of a species or individual at a **survey** location is, however, only the tip of the very large iceberg of information available to those surveying carnivores. With the appropriate approach, carnivore survey data will allow surveyors to say something about the **distribution** or **abundance** of species across extensive **survey areas**. Further, if surveys are conducted repeatedly over time, changes in carnivore **population status** can be **monitored**.

Given the remarkable breadth of potential survey objectives and **designs**, and the rapidly changing nature of the field, this chapter will necessarily be a brief introduction. Many existing resources (some of them entire books in their own right) provide comprehensive coverage of design considerations for specific survey objectives and modeling approaches—indeed, we reference many of these throughout the chapter. The existing survey and monitoring lexicon is diverse, expanding, and sometimes inconsistent.

We attempt to remedy this—at least for the application of noninvasive surveys for carnivores—by including a glossary of frequently used terms (see appendix 2.1). Terms defined in the glossary are set in boldface upon first use.

Integral to maximizing the amount of information to result from a particular carnivore survey is to carefully consider from the outset what one would like to accomplish (the objectives) and how to best go about accomplishing these objectives (the survey design). Later, appropriate analytical methods can be used to deal with data collected via designs that were suboptimal, or for which all analysis assumptions were not met. To be clear, we define a survey as one or more attempts to detect a species at either a single location or across many locations, with the intent of understanding species distribution, **occupancy**, or population size. This chapter (and indeed much of the book) focuses on four key survey objectives: assessing **occurrence** and distribution, assessing **relative abundance**, estimating abundance, and monitoring. Another class of survey objectives—those which we term secondary objectives—can be met by collecting additional data or conducting additional analyses. Examples of secondary objectives include characterizing diet or behavior, estimating survival, assessing the genetic structure of a population, or evaluating carnivore movement. Such objectives are largely beyond the scope of this chapter.

In virtually all situations involving free-ranging carnivores, it is impossible to detect or count every individual within a specified area. It is necessary, therefore, to rely on **sampling** (i.e., recording or measuring characteristics of a portion of the population) to make inferences about the actual population of interest (i.e., the statistical population). For example, if the objective is to estimate the proportion of a county occupied by gray foxes (*Urocyon cinereoargenteus*), track **station** surveys might be conducted within twenty 500 x 500 m **sites** (**sample units**) to provide an estimate of the number of sites that are occupied. We'll discuss how these sample units should be selected, but the important point here is that sampling allows researchers to infer information about a population occupying a very large area based on data collected from a much smaller, but representative, area.

# Assessing Occurrence and Distribution

The most basic carnivore survey consists of surveying a single location or area with the intent of assessing whether at least one individual of the target species occurs there. By conducting such assessments at multiple locations, it becomes possible to create simple maps documenting occurrence across a region (i.e., assessments of distribution). And by going one step further and establishing some rules about how survey locations should be selected, how far apart they should be, and how many times they should be surveyed, one can use multiple location surveys to assess the proportion of the region occupied by a species (i.e., occupancy estimation). Finally, by including site specific covariates, or using techniques that detect and model the spatial relationships between detections, it is possible to accurately estimate or predict the actual distribution of a species.

## Assessing Occurrence at a Single Location

Single-location surveys require sufficient effort to detect at least a *single individual*, and a survey is concluded when the target species is detected. Such surveys are most commonly mandated by land management agencies for detecting rare species when a proposed management activity may affect its habitat. (Note: Some land management agencies in the United States refer to these management activities as projects, and the surveys that precede them are referred to as **pre-project surveys**.) Single-location surveys can also be initiated as part of a simple inventory to determine if a species of interest occurs in an area of interest (e.g., Zielinski et al. 1995b; Resource Inventory Committee 1999), and are often conducted using unpublished **protocols**. Despite their use for a variety of taxa (e.g., USDA and USDI 2000), the value of single-location surveys as part of conservation strategies for vulnerable species is rarely stipulated.

Due to the inconsistent application of effort characterizing single-location surveys, they often produce more reliable inferences about **presence** than absence. Moreover, surveys conducted within such small areas and resulting in a single detection or a nondetection typically reveal little about the status of a carnivore population. In our opinion, such characteristics render single-location surveys the *least* useful survey application. Although data pertaining to the whereabouts of rare carnivores are always of some value, single-location surveys are usually conducted with the assumption that information regarding the presence of a species in one area, and at one point in time, will influence whether or not a habitat-altering land management project should proceed. We believe this to be an inappropriate use of carnivore surveys. As probability of occurrence cannot be easily distinguished from **probability of detection** under the above circumstances, single-location surveys are inadequate for resolving critical questions about habitat alteration. Decisions regarding the extent and intensity of habitat-altering activities should, instead, be based on a more comprehensive assessment of habitat and populations at larger and multiple scales.

Importantly, the results of single-location surveys are often compiled by researchers to determine current distributions and identify isolated populations (Temple and Temple 1986; Kucera et al. 1995b; Zielinski et al. 1995a; Aubry and Lewis 2003; Matos and Santos-Reis 2006). Thus, a retrospective summary of the results of many single-location surveys can provide a crude assessment of distribution and may be of considerable value. Because single-location surveys continue to be routinely conducted, we will briefly highlight some essential considerations for designing such surveys.

Statistically speaking, this particular survey objective does not require sampling because there is no population parameter to estimate. Although the term *sample unit* has been used to describe separate components of a survey when the goal is to detect a single individual in a survey area (e.g., Zielinski et al. 1995b), this application is not technically appropriate because statistical sampling is not actually taking place. Hence, for this objective, we simply refer to locations where one or more **devices** are deployed as *sites*. Sites in this sense are not statistical sample units (although later, when we discuss occupancy estimation, they will be); they merely subdivide a larger target area such that the area can be more efficiently searched for evidence of at least one individual. In this case, the number and distribution of sites is a function of (1) the size of the area of interest relative to the home range of the target species, and (2) a tradeoff between the number of sites and the duration over which they are surveyed.

If the area of interest is much *smaller* than the average home range size of the target species, it is important to survey both within and outside of the target area because individuals that use the target area must also use adjacent areas. Failure to detect the target species in the target area during the survey period may result from the fact that this area comprises only a portion of the resident individual's home range (see Zielinski et al. [1995b] for one approach to surveying American martens [*Martes americana*],

fishers [*Martes pennanti*], Canada lynx [*Lynx canadensis*], or wolverines [*Gulo gulo*] when the target area is the center of a much larger area to be surveyed). Thus, if the target species is the ringtail (*Bassariscus astutus*), which has an average home range size of 136 ha (Trapp 1978), and the target area is 100 ha, we suggest that survey sites be established both within and immediately outside of the target area (figure 2.1). If, however, the target area is *larger* than the likely home range of the target species, detection efforts can be largely restricted to within the perimeter of the target area (figure 2.2).

Regardless of the size of the survey area, logic dictates that multiple sites be surveyed because the probability of detecting at least one individual will increase with the number of sites. The most conservative approach to determining the density of sites is to make sure that at least two are surveyed within each hypothetical home range-sized area for the target species. This will reduce the loss of data that occurs when devices are rendered inoperable by wildlife (e.g., black bears [*Ursus americanus*] destroying track plates), technician error, or environmental circumstances.

When the objective of a survey is to detect the presence of at least one individual in a single location, surveying can conclude when the presence of the target species is confirmed. More difficult to judge, however, is how long to continue the survey if the species has *not* been confirmed. This issue relates to **detectability** ($p$: the probability of detecting a species when it is indeed present), and $1 - p$: the probability of committing a **false-negative error** by failing to detect a species that is, indeed, present. Formal occupancy estimation (discussed later in this chapter) provides a method for estimating detectability via repeat **sampling occasions** (sometimes referred to as **visits**) at multiple sites. Single-location surveys afford no such opportunity because the sites are not always independent (i.e., the detection of the target species at one site may affect the detection of the species at an adjacent site) and the surveys usually conclude as soon as the target species is detected. If it is possible to conduct repeat

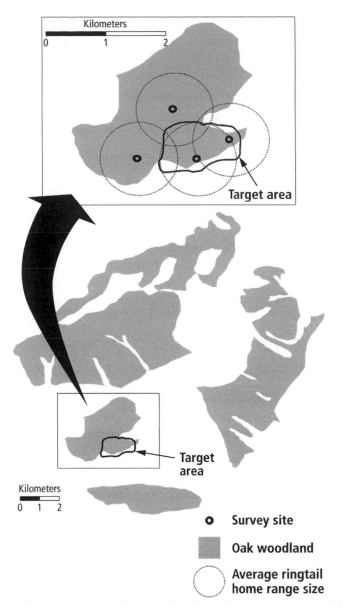

*Figure 2.1.* This hypothetical, single-location survey for ringtails is focused on a target area *smaller* than the species' home range. Thus, survey sites are located both within and immediately outside of the target area.

surveys at a number of independent sites, or to acquire **detection history** data from another more comprehensive survey of the target species using similar methods, a quantitative analysis of detectability (MacKenzie et al. 2002; Tyre et al. 2003; see *Estimating Occupancy* later in this chapter) can be conducted, resulting in device- and **occasion**-specific detectability estimates.

In general we recommend that the **survey duration** be chosen such that the probability of detection for each site exceeds 0.80. Assuming that more than one site will be included in the survey and sites can be considered spatially independent, the total probability of detecting at least one individual within the target area will be $p^n$ (where $p$ is the probability of detecting at least one individual at a given site and $n$

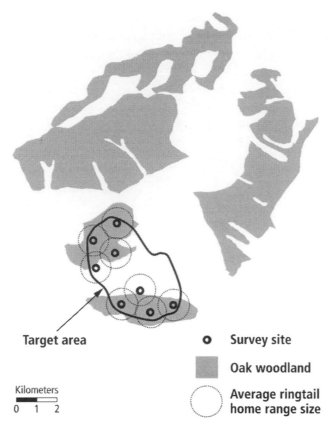

**Target area**                          ○ **Survey site**

                                         ▨ **Oak woodland**

Kilometers                               ⬭ **Average ringtail**
0    1    2                                **home range size**

*Figure 2.2.* In contrast to figure 2.1, this hypothetical single-location survey for ringtails targets an area *larger* than the species' home range. Therefore, detection efforts are primarily concentrated within the perimeter of the target area.

is the number of sites). If detectability is unknown, the frequency distribution of time-to-detection can be used to estimate an appropriate search period. A survey of swift foxes (*Vulpes velox*) in Kansas, for example, found that—after a certain point—as search time elapsed, detection rate declined (G. Sargeant, US Geological Survey, pers. comm.). For species that are readily detected, an abrupt decline in detection rate can provide clear guidance as to the length of search periods. **Latency-to-first-detection** (LTD) has been proposed to have similar value in some circumstances (Gompper et al. 2006).

The goal of a single-location survey is achieved if a detection is recorded *anywhere* within the target area, therefore leaving little reason to deploy more than one device at each site. Nonetheless, if survey devices are expected to sustain relatively high rates of failure or disturbance, or if accessing sites is difficult or expensive, it may be beneficial to deploy more than one device at each site. Because considerations for allocating effort within sites are almost identical to those for allocating effort at true sample units (e.g., when estimating occupancy from data collected across many locations), we refer readers to *Estimating Occupancy* later in this chapter for a more detailed discussion of this topic.

As representative sampling is not relevant to this survey objective, sites should be spaced as evenly as possible but in such a manner that the most likely habitat types are targeted. For the ringtail example mentioned earlier, sites should be disproportionately placed in preferred habitat (i.e., oak [*Quercus*]-

dominated cover types in close proximity to stream-courses [Poglayen-Neuwall and Toweill 1988]), as illustrated in figure 2.2.

## Mapping Distribution from Surveys at Multiple Locations

Many issues relating to carnivore conservation and management require spatial information about where species occur (and do not occur) at regional or continental scales. This type of information is known as a species' distribution or **range**. Gaston (1991, 1994) distinguishes two primary objectives when assessing geographic range. The **extent of occurrence** is the area within the outermost limits of the occurrence of a species, and the **area of occupancy** is the area over which the species is actually found. The methods we discuss here can relate to either of these objectives. Plotting the locations of species occurrence is sometimes referred to as *distributional recording* or simply *mapping* (Macdonald et al. 1998), and is the basis of atlas programs (Robbins et al. 1989; Arnold 1993). Examples of such mapping for carnivore conservation include Zielinski et al. (1995a), Kucera et al. (1995b), Aubry and Lewis (2003), and the USDA Forest Services National Lynx Survey (see chapter 6). For many surveys, distribution mapping is a post hoc or opportunistic exercise, relying on data collected for other survey objectives. But in some situations, mapping is *the* primary survey objective, and the associated choice of survey method and design is vital.

Defining a species distribution based on **detection-nondetection data** requires surveying (1) a large enough area to be relevant (i.e., the extent), (2) evenly enough to ensure few unsampled regions (i.e., evenness), and (3) a large enough number of sufficiently small areas within the extent to maintain resolution (i.e., the grain). Figure 2.3 illustrates a hypothetical survey for ringtails that would generate data suitable for mapping species distribution within the defined survey area. Given limited resources, surveys cannot maximize both extent and grain (Zielinski et al. 2005). Thus, researchers must

typically evaluate the tradeoff between these goals and choose a survey design that best meets their needs. Most important, a sufficient amount of survey effort must be expended at each location to ensure, with some a priori level of confidence, that false-negative errors are minimized. Zielinski et al. (2005) used mapping approaches to qualitatively compare contemporary and historical distributions of California carnivores using detections from track stations and remote camera surveys, and Manley et al. (2004) proposed a quantitative method to detect change in the spatial distributions of multiple species.

Opportunistic data collected over time, while often plentiful, must be evaluated carefully before being used to map distribution (Johnson and Sargeant 2002). Such retrospective datasets often represent the contributions of numerous surveyors working within a variety of settings (e.g., agencies, universities, the public sector), representing a wide range of skill levels, and utilizing many different methods. Quality control checks must be conducted whenever possible to help ensure that species detections are accurate (Aubry and Houston 1992; Aubry and Jagger 2006). In addition, if a standard survey protocol was not used, then very little can be said about locations where the species was *not* found to occur. This is because (1) it is unclear how much effort was expended to detect the species at locations lacking detections (i.e., was the location unoccupied, or was the species indeed present but not detected?), or (2) the resulting occurrence data were not collected using a statistical sampling framework. In some cases information is not recorded at locations where the target species was not detected, resulting in what are known as *presence-only* data.

If there are means for experts to independently verify the identity of the species detected, and when detection data include accurate geographic locations, then the synthesis of opportunistic survey results (much like the retrospective analysis of fur-trapping data [e.g., Gompper and Hackett 2005]) can produce valuable information about the general distribution of a species of interest, the potential

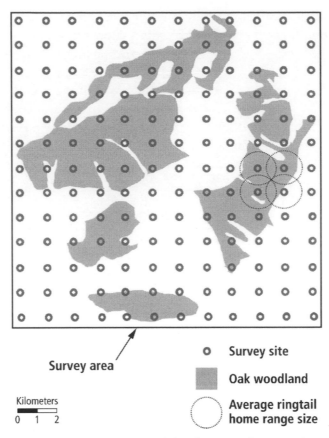

*Figure 2.3.* This hypothetical survey design for mapping ringtail distribution with systematic, nonindependent sites illustrates the intensity and spacing of effort necessary to achieve full coverage with sufficient grain.

vulnerability of isolated populations (e.g., Kucera et al. 1995b; Aubry and Lewis 2003), and hotspots of species richness (e.g., Williams et al. 2002). A number of factors, including the popularity of distribution surveys and access to the World Wide Web, are allowing the development of interactive web-based geographic information system (GIS) interfaces where researchers can enter spatial survey data remotely and view the locations and results of other surveys (Aubry and Jagger 2006). Last, in some cases, it is now possible to predict species distribution from data consisting of only detections (i.e., without information about where the species does not occur). Such presence-only methods (Rotenberry et al. 2002; Zaniewski et al. 2002), however, have important limitations, rely on other site-specific data, and are not technically mapping exercises (see *Estimat-*

*ing Occurrence and Distribution via Spatial Modeling* later in this chapter).

## Estimating Occupancy: A Nonspatial Method for Assessing Distribution

Modern occupancy estimation methods strive to estimate the proportion of the survey area that is occupied (or used) by the species of interest. At the most fundamental level, many carnivore surveys have focused on reporting presence or (observed) absence at individual sampling locations and, thus, a binomial response has been the variable of interest. Indeed, most historical track-based surveys were conducted at multiple locations over a specified survey period and summarized the binomial responses with statistics such as the proportion of surveyed sites with a detection (Kendall et al. 1992; Zielinski and Stauffer 1996) or visitation rate or index (Linhart and Knowlton 1975; Roughton and Sweeny 1982; Sargeant et al. 2003a). These early efforts to index abundance or to describe carnivore distributions employed methods and designs that were precursors to what has developed into modern occupancy estimation (e.g., MacKenzie et al. 2006).

In an attempt to collect more than **binary** data, some early studies assumed that the number of track stations with at least one detection was a faithful representation of the number of individuals in the study area (e.g., Linhart and Knowlton 1975; Conner et al. 1983; Robson and Humphrey 1985), even though this was unlikely to be the case for wide-ranging species when stations were close together and the behavior of individuals could bias the response. Sargeant et al. (1998) found that visits were spatially correlated within survey lines and joined Zielinski and Stauffer (1996) in recommending that the results of surveys be reported as either a detection or nondetection at the level of the independent sample unit. These recommendations helped move the field closer to using a data structure that was compatible with modern occupancy estimation methods.

Another stride toward more accurate occupancy estimation entailed recognizing that the failure to detect a species at a sample unit did not necessarily mean that the species wasn't present. Zielinski and Stauffer (1996) addressed this matter by incorporating the uncertainty of detection success into their proposed monitoring program. Still lacking, however, was a way to explicitly incorporate detectability (discussed in *Assessing Occurrence at a Single Location* earlier in this chapter) into occupancy estimates. Although a number of researchers were concurrently striving to tackle this issue (e.g., Stauffer et al. 2002; Tyre et al. 2003; Wintle et al. 2004), MacKenzie et al. (2002) took the final step on the road to modern occupancy estimation when they recognized that the history of detections and nondetections at repeatedly surveyed sample sites (the **encounter history**) could be analyzed within a framework similar to that employed for **capture-recapture** data. More specifically, by integrating the estimation of detectability and occupancy, it is possible to estimate detectability directly and to use it to adjust occupancy estimates to account for sites where the species was likely present but not detected. Occupancy estimation (MacKenzie et al. 2006) assesses the proportion of sites that are either occupied—or typically in the case of wide-ranging carnivores, used—by a species of interest. Occupancy—essentially the number of sites where the species was estimated to be present divided by the total number of sites surveyed—is also often interpreted as an estimate of the proportion of an area occupied (MacKenzie and Nichols 2004). Modern occupancy estimation can therefore be viewed as a nonspatial assessment of distribution because it represents an unbiased assessment of the area of occupancy.

Modern occupancy estimation (MacKenzie et al. 2002; Tyre et al. 2003) has had an important effect on the field of animal population assessment. We refer readers to the recent volume by MacKenzie et al. (2006), which provides a comprehensive treatment of occupancy estimation that includes history, theoretical and statistical foundations, study design, single- and multiple-season models, single- and multiple-species models, and the use of occupancy in community-level studies. Here we attempt to distill and highlight the essential components of this approach as it relates to carnivore surveys. As occupancy estimation is rapidly changing, however, with new designs and analysis tools being developed on an ongoing basis, readers should also consult the most current literature for new information.

### Incorporating Detectability into Occupancy Estimates

Recent advances in occupancy estimation and modeling have focused largely on methods for explicitly incorporating detectability into occupancy estimates. The key to estimating probability of detection is conducting multiple, independent sampling occasions at all or a subset of sites (MacKenzie et al. 2002), which permits the construction of an encounter or detection history. For example, a three-occasion encounter history of *1, 1, 0* would indicate that the species was detected on the first two sampling occasions but not on the third. A multinomial probability framework and maximum likelihood approach can then be used to estimate occasion-specific probabilities of detecting the species *given that it was present*, and to incorporate these estimates into final occupancy estimates (MacKenzie et al. 2006; also see chapter 11). Further, researchers can compare different models via standard model selection methods (Burnham and Anderson 2002) to explore specific hypotheses about detectability and occupancy and can also include site- or occasion-specific variables in their analyses, thus allowing the comparison of various methods or combinations of methods for detecting the target species (e.g., Campbell 2004; Gompper et al. 2006; O'Connell et al. 2006; Long et al. 2007b). Occupancy analysis assumes the following:

1. Occupancy status at each site does not change over the survey season (i.e., sites are "closed" to changes in occupancy, although see *Allocating Survey Effort at Sites* later in this chapter for examples of situations in which this assumption can be relaxed).
2. The probability of occupancy is constant across sites, or differences in occupancy (i.e., heterogeneity) are modeled as a function of site covariates.
3. The probability of detection is constant across all sites and surveys, or is modeled as a function of site- or sampling-occasion-specific covariates.
4. The detection of species and detection histories at each site are independent.

*Types of Sampling Designs for Occupancy Estimation*

At least three sampling designs have been proposed for implementing multiple sampling occasions: standard "all sites" sampling, double sampling, and removal sampling (MacKenzie et al. 2006). Each of these designs can be applied to any type of survey method, as long as the effort at each site remains constant across sites or can be accounted for with

sampling occasion-specific covariates (see chapter 11).

Standard all-sites sampling is the optimal approach, such that every site is subjected to multiple sampling occasions and detection probability is estimated using data from all sites. Alternately, double sampling entails conducting repeat sampling occasions at only a subset of sites, with the remaining units surveyed only once. Detection probability is then estimated from the repeat occasions. Last, removal sampling describes a scenario where sites are sampled repeatedly until a species is first detected, and then no further sampling is conducted at that site. This is an efficient approach, allowing the most important information (i.e., species presence) to drive the amount of survey effort expended at each site, while still providing data to estimate detectability. Removal designs require that detections be confirmable within a short period of time (e.g., methods requiring genetic confirmation would be precluded), and may be difficult to implement if multiple species are being surveyed concurrently. MacKenzie and Royle (2005) and MacKenzie et al. (2006) discuss the strengths and weaknesses of each of these sampling approaches, and suggest a hybrid approach in which some sites are surveyed with a standard design and the remainder with a removal design.

*Site Size, Location, and Spacing for Occupancy Estimation*

In general, a site is a location or area where data are gathered. For occupancy analysis, a site also represents a statistical sampling unit (as opposed to the non-sampling-based units described in *Assessing Occurrence at a Single Location* earlier in this chapter), and it is assumed that the observed outcome at any given site will be either a species detection or nondetection. One or more detection devices are deployed at each site. Non-station-based methods (e.g., scat detection dogs, snow tracking) typically define effort at sites based on transects or routes. The area comprising a site will depend largely on the target species, and specifically the scale at which an observed detection or nondetection of that species is meaningful (MacKenzie et al. 2006). For example, conducting track surveys for wolverines—a very wide-ranging species—within sites of 1 ha would likely yield mostly nondetections, and would thus permit little to be inferred about wolverine use of the survey area. Occupancy is also a scale-dependent parameter, and occupancy estimates will tend to be higher for larger sites, especially when sites are arbitrarily sized and located in contiguous habitat (MacKenzie et al. 2006; figure 2.4), or when individual home ranges of the target species overlap considerably. This situation is in contrast to that in which the habitat of a species comprises discretely sized entities (e.g., ponds, disjunct forest stands) that are typically either occupied or unoccupied and can directly help to guide decisions about site size. MacKenzie et al. (2006) suggest that sites for occupancy estimation should be large enough to have a reasonable probability of the species being there (i.e., a probability between 0.2–0.8), which, for some wide-ranging carnivores, may require very large sites (e.g., 1-km diameter area [Zielinski et al. 2005]; 2-km transects [Long et al. 2007a]).

If occupancy estimates are to be generalized to the entire survey area, site selection must be based on a statistical probability model. While simple random sampling is always recommended (figure 2.5 illustrates a hypothetical random survey design), this is not necessarily the most efficient approach; other sampling schemes might result in more precise parameter estimates with fewer samples. For example, if qualitative differences exist among groups of sites (e.g., land cover type, expected abundance), a stratified random sampling approach may be a more efficient and precise—and still minimally biased—method of sampling (Krebs 1998; Quinn and Keough 2002). Systematic sampling (figure 2.6), which follows some regular pattern (although unrelated to biological patterns in the landscape) and provides uniform coverage of the statistical population, is also commonly used in wildlife surveys. This approach may be better for long-term monitoring (Morrison et al. 2001) and can provide more precise estimates than simple random sampling (Schaeffer et al. 1990), but the mathematical properties of systematic samples are less straightforward than those of randomly chosen samples (see Krebs 1998 for a thorough discussion of basic sampling designs). Krebs (1998) describes a number of nonstatistical sampling approaches (e.g., accessibility sampling, haphazard sampling, judgmental sampling) but cautions strongly against adopting them.

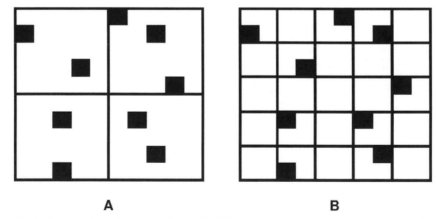

**A**        **B**

*Figure 2.4.* These hypothetical survey areas illustrate the tradeoff between site size and occupancy, with larger sites resulting in a higher occupancy estimate. Both survey areas are 1 ha (100 × 100 m) in size, and show identical detection locations indicated by black cells. A survey of 50 × 50 m sites (A) results in an occupancy estimate of 4 occupied sites/4 total sites = 1.0, whereas a survey of 20 × 20 m sites (B) results in an occupancy estimate of 9 occupied sites/25 total sites = 0.36. Reprinted from MacKenzie et al. (2006), with permission from Elsevier.

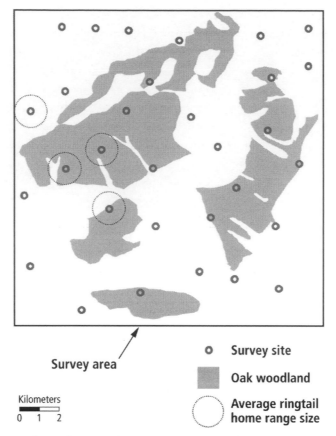

Survey area

Kilometers
0   1   2

○  Survey site

█  Oak woodland

○  Average ringtail
   home range size

*Figure 2.5.* This hypothetical survey design for ringtail occupancy estimation features independent site-spacing (with a minimum site separation distance of 2 km) based on a simple random sampling approach.

As with site selection, site spacing should be based on rules of sampling, and occupancy methods assume that samples are independent (i.e., the detection of the target species at one site is assumed to be independent of detections at all other sites [MacKenzie et al. 2002]). Note the distance between sites in figure 2.5 as compared with those in figures 2.1–2.3. Dependencies among sites may cause data to be autocorrelated, with occurrences at more than one site resulting from detections of the same individual (Swihart and Slade 1985; see *Assessing Relative Abundance* later in this chapter). Although such dependencies may yield inaccurate inferences, possibly affecting bias (i.e., estimates are too high or too low) and precision (i.e., the reported preci-

sion is too high), the actual effects of such dependencies will likely vary depending on the particular scenario and survey objective. In cases where the objective is simply to estimate area use by a wide-ranging species and sites are selected randomly, some dependencies among sites may not be a problem. But in other situations, such as when a non-random design is employed or the objective is predictive occurrence mapping, site independence may be more critical (D. MacKenzie, Proteus Wildlife Research Consultants, pers. comm.). Further, some methods explicitly permit the relaxation of the assumption of site independence (see *Predicting Occurrence Without Site Specific Covariates* later in this chapter). Estimating and addressing the effects of

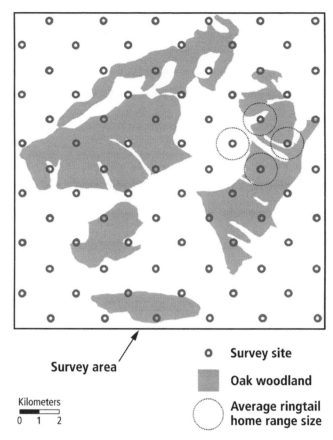

*Figure 2.6.* This hypothetical survey design for ringtail occupancy estimation illustrates the systematic, independent spacing of sites.

spatial dependencies is clearly an area of ongoing research.

### Allocating Survey Effort at Sites

Once the number and spatial distribution of survey sites is decided upon, an efficient surveyor will need to carefully determine how many devices to deploy at each site and for how long (see the final paragraph in this section for a discussion of allocating survey effort when using non-station-based methods). The number of devices deployed per site is determined by evaluating two criteria: (1) the rate at which individual devices are likely to fail or be rendered useless, and (2) knowledge of how the number of devices at a site affects the probability of detecting a target spe-

cies that occurs there. If, for example, the failure rate is negligible and the site is small relative to the area within which an animal is likely to be detected by the device, it is likely that a single device will detect the target species most of the time when the species is present. Although this is an ideal situation, accurately estimating occupancy (as well as relative abundance and abundance, for that matter) lies in the ability to maximize the overall probability of detecting the species during each sampling occasion. Both the number of devices per site and the number of sampling occasions per site affect the overall probability of detecting the target species (figure 2.7). Note that different combinations of temporal and spatial effort can result in the same probability of detection.

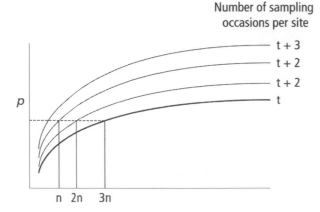

*Figure 2.7.* This schematic illustrates the tradeoff between the number of devices per site and the number of sampling occasions per site in relation to the probability of detection (*p*). Note how different combinations of spatial (n) and temporal (t) effort can result in the same probability of detection (dotted line). The curves are hypothetical; their characteristics will vary by species and context.

In many cases, deploying multiple survey devices at a site can increase detectability while not contributing substantially to effort or cost. Decisions about the optimal number of devices and sampling occasions per site are part of a much larger set of questions that must be addressed when determining the probability of detecting the species (Campbell 2004). For example, one must also consider the total number of sites to be surveyed and whether the objectives require a single survey compared to a longer-term monitoring program. The quantitative evaluation of such information to create an optimal survey design has not yet been undertaken for any carnivore species that we are aware of—and realistically, few surveyors will have access to datasets for which simulations can be executed a priori to determine the optimal combination of factors to increase detectability (although see discussion of GENPRES simulation software [Bailey et al. 2007] later in this chapter under *Allocating Effort: More Sites or More Occasions?*). Thus, most surveyors choose to err on the side of caution, deploying more devices and con-

ducting more sampling occasions per site than may be necessary. After some surveys have been conducted, it may be possible to estimate probability of detection as a function of the number of devices and sampling occasions per site and to reduce effort accordingly. For example, in reassessing fisher survey protocols in California, Zielinski et al. (2005) analyzed their existing data and determined that only five sampling occasions (versus the eight comprising the protocol at the time) were necessary to achieve a >95% probability of detection and to sufficiently minimize the standard errors of the occupancy estimate such that a targeted fisher monitoring program could be conducted (see case study 4.1).

The length of time for which a site should be surveyed—essentially a function of the number and length of sampling occasions—is largely dependent on the need to maximize detectability while also minimizing expense and ensuring closure. More sampling occasions at a site (assuming constant occasion length) typically translate to a greater probability of detecting the target species (figure 2.7). If the occasion-specific probability of detecting the species is known, the number of sampling occasions that will achieve a predefined overall detectability can be calculated (Campbell 2004; Gompper et al. 2006; O'Connell et al. 2006; Long et al. 2007b, Smith et al. 2007). For example, O'Connell et al. (2006) estimated that the probability of detecting a raccoon (*Procyon lotor*) at least once at occupied sites approached 95% after six to seven weeks of sampling with remote cameras and eleven weeks with enclosed track plates. In contrast, Smith et al. (2007) required only twelve days to achieve >95% probability of detecting a marten when whey were abundant. Where marten density was low, however, a detectability of only 33% was achieved with the same survey duration. Long et al. (2007b) estimated that only one sampling occasion with a scat detection dog was required to achieve a 95% probability of detecting fishers when they were present (although note that a survey design specifying a single sampling occasion at all sites would not permit the estimation of detectability nor its incorporation into occupancy estimates). Alternately, if the occasion-specific probability of detection is unknown, LTD can be used to roughly estimate an appropriate number of sampling occasions for detecting species when present.

Tyre et al. (2003) highlight the fact that, for species such as wide-ranging mammals, territories are typically much larger than survey sites. Therefore, even if a survey site lies within an individual's home range, there is no guarantee that the individual will be present while the site is being sampled—especially if the survey consists of few sampling occasions. The larger the home range of the target species, or the slower that individuals move around it, the less likely they are to be present within the site when it is sampled. For such species, researchers must also consider the probability that an individual will be within range of the survey site at the time the site is being visited (Tyre et al. 2003). Regardless of detectability issues, occupancy methods assume that sites are closed to immigration and emigration during the survey period (Karanth and Nichols 2002; MacKenzie et al. 2006). If it can be assumed that movement to and from sites is random, this requirement can be relaxed and the results interpreted as "sites used" (versus "occupied" [MacKenzie et al. 2006]). Shorter survey durations, however, are still preferred for statistical as well as budgetary reasons.

When non-station-based methods are used (i.e., scat detection dogs, snow tracking), the decision about effort per site relates to search time and effort per sampling occasion, rather than to the number of devices or stations per site. Effort can be standardized by following a predefined survey route—or, if roads, trails, or other navigable features such as streams or canyons are present, the survey may proceed along these features for a predefined length of time. If the animal sign of interest is morphologically unambiguous (e.g., most black bear scats in regions where grizzly bears [*Ursus arctos*] do not occur), surveys can be immediately halted after the first detection at a site, thereby maximizing survey efficiency. But because many types of wildlife sign are often difficult or impossible to confirm via morphology alone, it is usually advisable to either complete a predefined route or to collect the minimum number of detections necessary to provide a predefined probability of species confirmation (see Harrison 2006; Long et al. 2007a; chapters 3 and 7, this volume). Site size and shape considerations for non-station-based methods are similar to those discussed earlier for station-based methods.

## Temporal Considerations for Sampling Occasions

Multiple sampling occasions at a site need not be conducted on different days, nor is it a problem for sampling occasions to be conducted concurrently if survey methods allow. For instance, remote cameras and track stations could be deployed simultaneously at a single site, or two snow trackers could survey independently at a single site. Each case would result in two sampling occasions, assuming that detections by each of the methods or surveyors could indeed be considered independent. Further, emerging advances may soon allow data collected with multiple *dependent* devices to be explicitly modeled. These new approaches treat multiple devices similarly to the secondary samples used for the robust sampling model (see *Monitoring* later in this chapter, as well as chapter 11), and build on the idea of temporary emigration to deal with device dependency (L. Bailey, US Geological Survey, pers. comm.; also see chapter 11).

## Allocating Effort: More Sites or More Sampling Occasions?

Although it is clear that the repeat sampling of at least a subset of sites is important for estimating detectability, adding repeat sampling occasions necessarily decreases the number of total units that can be sampled within the study area given limited resources and effort. Much recent work has focused on how to balance the number of sites and sampling occasions to achieve an optimal survey design (Tyre et al. 2003; Field et al. 2004; Field et al. 2005;

MacKenzie and Royle 2005; Rhodes et al. 2006; Bailey et al. 2007), but there still exists no consensus on the relative tradeoffs between spatial and temporal replication. Field et al. (2005) suggested that, in general, when the goal is to detect a **trend**, two sampling occasions at each site usually suffice—except for common species that occupy many sites. Rhodes et al. (2006) suggested that increasing the number of sites increases power and is therefore a better investment than increasing the number of sampling occasions at each site. Sargeant et al. (2003b) came to the same conclusion when estimating the relative abundance of swift foxes. Tyre et al. (2003) agreed, but only when false-negative error rates were <50% (i.e., detectability was relatively high). When these error rates increased above 0.50, the authors found that more information could be gained by increasing the number of sampling occasions at each site. As a general rule, Tyre et al. (2003) recommend using at least three replicate sampling occasions.

Despite these general suggestions, the key to making decisions about more sites versus more surveys lies in the level of detectability that can be expected from the survey method and protocol. Field et al. (2005) suggested that researchers try to limit the probability of a false absence (i.e., $[1 - p]^K$, where $p$ is the probability of detection and $K$ is the number of independent sampling occasions) to 0.05–0.15. By knowing how many total sampling occasions a given budget will permit, and given an estimated value for detectability, researchers can work backward by first determining the number of occasions required at each site and then evaluating how many total sites can be surveyed. For example, if the probability of detecting a species via a search for natural sign is 0.40, the probability of a false absence during a single sampling occasion is $1 - 0.40 = 0.60$, and a total of four independent sampling occasions per site would be required to reduce the probability of a false absence to <0.15 (i.e., $0.60^4 = 0.13$). Thus, if there was sufficient funding to pay for 400 sampling occasions, then it would be possible to survey a total of $400/4 = 100$ units. If the probability of detecting the species at each site could be increased to 0.70, per-

haps by increasing the number of searchers per site, then the number of sampling occasions required at each site would decline to two (i.e., probability of false absence $= [1 - 0.70]^2 = 0.09$), and the total number of sites that could be surveyed would increase to $400/2 = 200$. This example illustrates the importance of maximizing the probability of detecting the species during any single sampling occasion.

MacKenzie and Royle (2005) explored the tradeoffs of more sites versus more sampling occasions in considerable detail, and their results suggest that for a given survey design (i.e., standard, double, or removal), there is an optimal number of sampling occasions that depend primarily on the probabilities of detection and occurrence. For instance, when the landscape is almost fully occupied by a relatively common species (see box 2.1), it is more useful to sample intensively at relatively fewer sites than to sample more sites (Field et al. 2005; MacKenzie and Royle 2005) because a minimal number of sites should provide sufficient information for reducing the variance component associated with imperfect detection to an acceptable level for accurately estimating occurrence (Field et al. 2005; MacKenzie and Royle 2005). Conversely, the best approach for a rare species, whose likely number of occupied sites is low, is to sample broadly across the study area in order to increase the number of occupied sites in the sample (MacKenzie et al. 2006).

Ultimately, it seems that the tradeoff between the number of sites and the number of sampling occasions depends on the biogeographic characteristics of the species and the population under study—thus rendering it difficult to provide general advice on this topic. Clearly, there is a complex relationship between (1) the optimal number of sampling occasions required to maximize precision, (2) detectability, and (3) occupancy. In general, however, survey designs that increase the probability of detection (i.e., decrease the probability of false absences) such that it approaches 0.95 for common species and 0.85 for rare species, should yield the most accurate occupancy estimates (D. MacKenzie, pers. comm.). Researchers must therefore consider their survey

# Box 2.1

## Design issues for rare versus common species

Although optimal survey designs differ for each species, situation, and objective, many key design considerations are largely informed by whether the target species is rare or common. Most carnivores are rare compared to, say, herbivores, but some carnivores exist at much lower densities than others. Densities of wolverines and fishers, for example, can be an order of magnitude lower than those of raccoons or striped skunks (*Mephitis mephitis*; Copeland and Whitman 2003; Gehrt 2003; Powell et al. 2003; Rosatte and Lariviére 2003). It follows that per capita detection rates are much lower, and the effort per detection much higher, for rare species versus common ones. These differences affect survey design in a variety of ways, four of which are briefly considered here.

First, some objectives may be unachievable for rare versus common species because of the increased effort required. Estimating abundance via capture-recapture methods, for instance, may be unachievable for rare species, which is probably why occupancy and distribution estimation have become very attractive goals for estimating the status of the more uncommon forest carnivores (e.g., Zielinski and Stauffer 1996; Campbell 2004; also see case study 4.1). On the other hand, common species may be so ubiquitous that occupancy and distribution may be largely meaningless, and methods with higher resolving power (e.g., count-based indices, true population size estimators such as capture-recapture methods) may be required to assess population status with any precision at all.

Second, visitation rates are partially a function of abundance, thus, rates that are achievable for common species may not be so for rare species. The extent to which this affects the efficiency of designs for rare species has not yet been adequately addressed. A number of studies, however, have found that indices of relative abundance and estimates of occupancy will be imprecise if a species is not detected at ≥20%–30% of sample units (Roughton and Sweeny 1982; Sargeant et al. 2003a). This level of detection may be difficult to achieve for rare

species without considerable sampling effort. Importantly, the rarity of a species can affect the precision of estimates of abundance even if the probability of detecting a given individual is high. This can occur because extremely low population densities can lead to situations where the number of detections and redetections is insufficient to allow sufficiently precise estimates of abundance (Williams et al. 2002). Furthermore, low density—and hence, low abundance per site—would appear to prohibit the application of methods that exploit variation in probability of detection to estimate abundance (i.e., Royle and Nichols 2003).

The third way in which rarity can affect survey design is by influencing the optimal distribution of spatial and temporal survey effort. Rare and common species can require different designs even when attempting to achieve the same objective. For example, occupancy estimation for rare species may demand more survey sites and fewer sampling occasions, whereas surveying fewer sites and visiting them more often is often the best choice for a common species (MacKenzie and Royle 2005). Rarity also can affect decisions about stratification in that stratified random sampling is seldom an option for rare versus common species (Thompson 2004b).

A final consideration is the link between commonness and habitat specificity. Common species are often habitat generalists, and a weak relationship between occurrence and specific habitat types may limit the ability to develop models to predict species occurrence using habitat covariates. In such cases, it may be better to choose approaches that use the spatial pattern of occurrences themselves (Sargeant et al. 2005; and see *Predicting Occurrence Without Site-Specific Covariates* in this chapter). The expected population size, density, extent of occurrence, and area of occupancy of the target species can all affect how a carnivore survey is designed. Many of the implications of rarity on carnivore survey design have not been evaluated or synthesized and thus represent important future research opportunities.

objectives, target species, and available resources when designing occupancy surveys. For example, if abundance is assumed to be low, then detectability may also be lower (e.g., Smith et al. 2007) and survey effort may need to be adjusted accordingly. Bailey et al. (2007) discuss alternative survey designs and their respective effects on estimator precision and also provide software (GENPRES) that enables users to evaluate and rank the precision of different survey designs prior to beginning large survey efforts.

## Estimating Occurrence and Distribution via Spatial Modeling

While basic occupancy methods can be used to estimate the proportion of the survey area that is occupied and the probability of detecting the target species, they are unable to provide spatially explicit maps of predicted occurrence without additional information. Two different methods for predicting occurrence and estimating distribution are currently available, with each having their own requirements for data collection.

### Including Site Covariates to Predict Occurrence

If site-specific information, such as elevation, vegetation, or human disturbance, is available, the relationship between these covariates and detection-nondetection survey data can be used to predict species occurrence at unsurveyed locations (e.g., Fleishman et al. 2001; Boyce et al. 2002; Rotenberry et al. 2002; Johnson et al. 2004). In addition to statistical models, the output may include a map of predicted probability of occurrence (e.g., Carroll et al. 1999; Gibson et al. 2004; Hoving et al. 2005; Long 2006). While detection-nondetection data are the typical input for most such models (see Scott et al. 2002; MacKenzie et al. 2006 for extensive treatments of various models), prediction based on *presence-only* data is also possible (e.g., Rotenberry et al. 2002; Zaniewski et al. 2002). Such presence-only models do not permit estimation or incorporation of detectability, however, and are therefore typically inferior to models that use high-quality detection-nondetection data.

### Predicting Occurrence Without Site-Specific Covariates

In some situations, predictive occurrence models suffer from what are known as errors of commission—that is, they predict areas to be occupied when they are actually not. Commission errors may result when a species is generally associated with a particular site characteristic (e.g., amount of conifer cover) that is correctly identified by the model, but the species is absent at some locations for reasons not captured by the model (e.g., overtrapping, competition). If errors of commission are expected, or the relationship between species occurrence and site covariates is unknown, methods that explicitly evaluate the spatial relationship between species detections may be used to estimate distribution (e.g., Klute et al. 2002). Bayesian approaches, once too analytically demanding and computer-intensive for practical application, are now being exploited to model such relationships (e.g., Sargeant et al. 2005; Magoun et al. 2007).

One such approach is based on methods used to reconstruct degraded digital images. Sargeant et al. (2005) employed Markov chain Monte Carlo (MCMC) image restoration to estimate distribution from (detection-nondetection) swift fox surveys. Although this method produces predicted values between 0 and 1, these estimates are not based on site-specific covariates. Instead, predictions are founded on survey effort, the results of surveys from neighboring sites, a measure of spatial contagion among sites, and the probability of detecting swift foxes when present. MCMC image restoration circumvents many of the negative aspects of other methods for mapping distribution while combining their strengths and accomplishing similar objectives (Sargeant et al. 2005). For instance, sites are not required to be independent, detectability is explicitly incorporated, habitat covariates are not required but may be incorporated, estimates can be calculated for unsampled locations, and outputs are expressed in ecologically useful units (i.e., probability of occupancy). Such spatial approaches are an area of active research, and results from surveys of a few species (e.g., the swift fox example described earlier, and wolverines [Magoun et al. 2007]), suggest that they hold much promise.

## Assessing Relative Abundance

Indices of relative abundance are "statistics assumed to be correlated to the true parameter of interest in some way" (Thompson et al. 1998: 77). Indices include methods that do not typically identify individual animals (but see mention of *Minimum Number Alive* methods later in this section), and are assumed to yield important information about the relative magnitude of abundance using the rate of detection rather than the number of individuals of a particular species. Thus, indices of relative abundance can be based on detections via any of the methods described in this book (i.e., tracks, photographs, hair and scat collection). Historically, indices of relative abundance have been the method of choice for assessing the population status of carnivores (e.g., Wood 1959; Clark 1972; Linhart and Knowlton 1975; Pulliainen 1981; Roughton and Sweeny 1982; Conner et al. 1983, 1984; Knowlton 1984; Stanley and Bart 1991; Kohn et al. 1993; Diefenbach et al. 1994; Beier and Cunningham 1996; Sargeant et al. 1998; Sargeant et al. 2003a, b; Andelt et al. 1985; Robson and Humphrey 1985; Leberg and Kennedy 1987; Nottingham et al. 1989; Best and Whiting 1990; Van Sickle and Lindzey 1992; Smith et al. 1994; Allen et al. 1996; Travaini et al. 1996; Brown and Miller 1998; Mahon et al. 1998; Thompson et al. 1998; Engeman et al. 2000; Warrick and Harris 2001; Schauster et al. 2002; Gehring and Swihart 2003; Sadlier et al. 2004; Engeman et al. 2005).

There is a particularly rich history of attempts to use **counts** of natural sign (e.g., tracks, scats) on survey routes as indices of population size (e.g., Dice 1941; De Vos 1951; Pulliainen 1981; Douglas and Strickland 1987; Thompson et al. 1989; Pellikka et al. 2005). We consider most of these surveys to be "classical" relative abundance efforts, and differentiate them from the precursors to modern occupancy estimation discussed earlier in the chapter (e.g., Kendall et al. 1992; Zielinski and Stauffer 1996; Sargeant et al. 1998). Again, these latter studies—which are largely discussed under *Occupancy Estimating*—were unique in attempting to achieve spatial independence among sample units or to account for the probability of detection when calculating occupancy.

The majority of classical attempts to develop indices of relative abundance deployed scent stations (see chapter 4) throughout a study area, recorded a count of total detections or estimated a *visitation rate* (number of visits/number of station nights), and assumed that visitation rate had some relationship to the number of individuals in the area. Such indices have their advocates (e.g., Engeman 2003; Hutto and Young 2003), who often argue that they serve as a practical substitute for the more demanding estimates of actual abundance. Critics, however, cast doubt on any measure that cannot be linked to the parameter of interest (in this case, abundance) that are not derived from well-established theory (Seber 1982; Pollock et al. 1990), or that may vary with abundance in unpredictable ways (e.g., Anderson et al. 2001; Anderson 2003; Ellingson and Lukacs 2003).

Historically, most classical indices for carnivores were developed using data collected with scent stations methods (Roughton and Sweeny 1982; Diefenbach et al. 1994). As such, the limitations of such indices, as well as considerations for developing and interpreting indices, are discussed at length in chapter 4. One key issue that deserves repeat mention, however, is that in many cases survey locations used with classical scent station methods were so close together that an individual animal could be detected at more than one site (see figure 2.8 for a schematic illustrating such spatial dependence). This situation makes it difficult to evaluate whether visitation rates are a function of population size or repeat visitation by the same individuals. Fur-ther, even when visitation rates and abundance are correlated, the

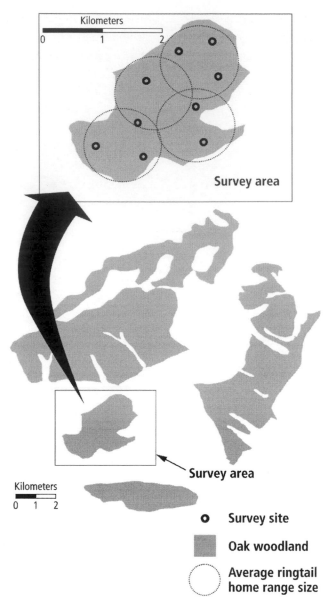

*Figure 2.8.* Dependence among sampling sites can affect parameter estimation. In this hypothetical example, sites are located within range of multiple individual ringtails. Thus, detections at all nine survey sites would result in detections of only four individuals.

relationship may not be linear. For example, Sargeant et al. (2003b) found that, because kit and swift foxes visited survey stations at higher rates when population size was low, visitation rates declined proportionally less than population size.

Uncertainty about the relationship between abundance and an index suggests that classical indices of relative abundance are likely to be eclipsed by modern detection history methods for estimating occupancy and distribution (see *Occupancy Estimating* earlier in this chapter). Occupancy, when adjusted for probability of detection, is a more defensible population metric than classical measures of relative abundance. Regardless, the extent to which an index of relative abundance has value as a population metric may be related to the rarity of the target species, as traditional indices of relative abundance may be more sensitive to changes in the abundance of common species than rare species. Conversely, occupancy estimation methods may better reflect population status for species that are naturally rare (see box 2.1). Indices such as visitation rates have been assumed to reflect abundance for ubiquitous species because probability of detection, as opposed to probability of occupancy, is more likely to be related to abundance for such species (G. Sargeant, pers. comm.). If visitation rates are to be used as indices of relative abundance, Diefenbach et al. (1994) recommend the following guidelines: (1) surveys should be conducted multiple times per year; (2) proportions of sites with detections should be transformed to reduce statistical issues; (3) if strata are used, they should contain as many sites as possible; (4) sites should be located as far apart as possible; and (5) statistical power analysis should be conducted prior to data collection. Last, some have argued that counts of natural sign (as opposed to visitation rates) can faithfully represent population size if such indices can be calibrated to an independent measure of abundance or if multiple indices are correlated (e.g., Allen et al. 1996; also see chapter 4). Counts of carnivore sign per sample unit, however, are typically too variable to serve as a useful index, with a better choice being the proportion of sample units having at least one detection (e.g., Harrison et al. 2004).

One potentially justifiable use of relative abundance indices may be to monitor relative change in abundance over time, as opposed to characterizing population size at a single point in time (G. Sargeant, pers. comm.). For example, despite their poor performance as accurate predictors of population size, Choate et al. (2006) suggest that track-based indices may still be used to detect large changes in population size (this topic is discussed further in *Monitoring* later in this chapter, as well as in chapter 4). In general, if relative abundance indices must be used to monitor population size, it is best to incorporate data from multiple, independent survey methods whenever possible. If all methods are measuring real changes in population size, results should be correlated.

Finally, we recognize a unique population index that results from genotyping individuals via DNA extracted from noninvasively collected samples. *Minimum Number Alive* (MNA), which has a history in the capture-recapture literature (e.g., Krebs 1966; White et al. 1982), can be determined by sampling hair or scat and identifying the unique number of individuals in the sample. MNA is often presumed to be related to the actual number of individuals in the population and, thus, has been considered an index of relative abundance (e.g., Pocock et al. 2004; Schwartz et al. 2004; chapter 6). Similarly, the number of individuals identified from hair snare samples collected per unit of effort (i.e., *catch per unit effort* [CPUE]; Romain-Bondi et al. 2004) has also been considered an index of relative abundance.

## Estimating Abundance

Direct estimates of abundance are typically more defensible than relative abundance indices as measures of population size given that they do not make assumptions about the relationship between the population and the detection statistic (e.g., numbers of tracks or photographs). Capture-mark-recapture methods—the most frequently employed approach for estimating wildlife population size—require that individual animals be identified. Such methods have been effectively adapted for use with noninvasively detected individuals (see *Noninvasive Capture-Recapture*). Other approaches for assessing population size from noninvasive survey data have been developed as well, although none has yet received comparable use or interest. For example, Royle and Nichols (2003) exploited the relationship between detectability and animal abundance to permit population size estimates from detection-nondetection data and occupancy techniques.

Distance sampling (Buckland et al. 1993) is a common method for estimating population size when animals can be observed directly during surveys. Certain characteristics shared by most carnivores (i.e., low population density, elusiveness), however, generally prohibit the use of this approach. And while distance sampling can also be applied to carnivore sign, such sign is usually difficult to detect (although see chapter 7). We thus suspect that efforts to use sign abundance as an index would be better spent acquiring detection-nondetection or individual identification data for use within either an occupancy or capture-recapture framework.

## Noninvasive Capture-Recapture

Traditional capture-mark-recapture methods utilize the total number of animals captured during two or more trapping occasions, along with the associated probability of capturing the individual animals, to estimate population size within the surveyed area. Capture probability is estimated by analyzing the pattern of captures and recaptures during the survey (see chapter 11 for more details of this method). More recently, capture-mark-recapture approaches have been adapted for use with data generated from noninvasive methods (Mills et al. 2000; Boulanger et al. 2002; Mowat and Paetkau 2002; Eggert et al. 2003; Boulanger et al. 2004b; Waits 2004; Bellemain et al. 2005; Lukacs and Burnham 2005a, b; Miller et al. 2005; Karanth et al. 2006; Petit and Valiere 2006; Boulanger et al. 2008). Such noninvasive capture-recapture methods employ the number of individual animals detected, along with the associated probability of detecting the individual animals (calculated from redetections)—the noninvasive equivalent of recaptures—to estimate population size within the surveyed area. Rather than actually marking animals, these methods exploit existing unique "marks" (e.g., DNA from hair, pelage patterns discernable from remote photos). As the ability to individually identify animals is a hallmark of this approach, it is imperative that researchers evaluate whether or not meeting this requirement is realistic (see survey method-specific chapters, as well as chapter 9, for

discussion of individual identification with various methods).

Capture-recapture models are classified based on assumptions about population closure. *Closed* models assume that there are no changes to the size of the survey population (i.e., births, deaths, immigration, or emigration) between sampling occasions, and are used to estimate abundance. In contrast, *open* models can accommodate these types of changes and provide estimates of survival and recruitment in addition to abundance. Open models, however, are more complex than closed models and typically require larger sample sizes and longer study periods. Given such characteristics, and also because the primary objective here is to estimate population size, closed population models will likely be the best choice for most carnivore survey applications. We thus will limit our discussion to these approaches (but see "robust" models in *Monitoring* later in this chapter).

When using closed-population models, the assumption of closure must be carefully evaluated. Because carnivores are typically long-lived, have well-defined birthing periods, and are often territorial, standard violations of closure are probably unlikely over survey durations of even many weeks. Given that some survey methods detect sign, however, the actual survey period may not overlap with the period during which the sign was deposited—resulting in "virtual" emigration or immigration. For example, with surveys relying on DNA from scat, it may be difficult to define the survey period because the timing of scat deposition is unknown (Lukacs and Burnham 2005b; Proctor and Boulanger 2006). In such cases, it is important to develop protocols that help to ensure that collected samples can be assigned to a known time frame (e.g., scats deposited since the previous winter).

The most basic closed-population approach (known as the *Lincoln-Peterson model*) involves only two sampling occasions. In addition to population closure, this approach assumes that all individuals are correctly identified and that all animals are equally likely to be sampled during each of the two occasions (i.e., equal capture probability or detectability; Williams et al. 2002). This last assumption ensures that all segments of the population are available for detection. As it is impossible to detect or account for detection heterogeneity in capture probability with two-occasion methods, multiple sampling occasion (or **K-sample**) methods have also been developed to permit explicit modeling of heterogeneity in detection probability. Note that, although K-sample methods require more sampling occasions, they are still considered closed models and must meet the assumption of closure.

K-sample approaches require that an encounter history be compiled for each detected individual (Williams et al. 2002), with each history composed of a series of detections or nondetections recorded at discrete sampling occasions for a given individual. During any occasion, an individual can be either "captured" (i.e., detected for the first time), "recaptured" (i.e., detected again after being first detected during a previous sampling occasion), or not detected. Although the number of sampling occasions can vary, a minimum of five occasions is suggested, with seven to ten being preferable (Otis et al. 1978). Further discussion of two-occasion and K-sample methods, as well as other analytical approaches (e.g., Pledger 2000), can be found in Williams et al. (2002) and Amstrup et al. (2005).

### Maximizing Detection Probability and Minimizing Detection Heterogeneity

A successful capture-recapture survey is achieved by maximizing detectability while minimizing detection heterogeneity. Whether a two-occasion or K-sample approach is used, detection probabilities must be quite high to provide all animals a reasonable chance of being detected (Lukacs and Burnham 2005b). Further, many individuals must be marked to achieve an adequate level of precision, and, with K-sample methods, to detect and accommodate detection heterogeneity that may be present in the data (Burnham et al. 1987; Williams et al. 2002; chapter

6). Indeed, increasing detectability rates as much as possible has been referred to as *the big law* of capture-recapture (Lukacs and Burnham 2005b). In many cases, especially with low-density and wide-ranging carnivores, the sampling intensity required to produce sufficiently high detection and redetection rates that will in turn produce precise estimates may be prohibitively expensive (Boulanger et al. 2002; Boulanger et al. 2008). It is critical, therefore, to realistically evaluate the potential for detecting and redetecting sufficient numbers of individuals.

The other important rule of capture-recapture approaches (as with most sampling methods) is to minimize detection heterogeneity. Maintaining equal detection probabilities between individuals or subgroups during a given sampling occasion will help to minimize bias, provide more options for analysis, and maximize precision if detectability is lower than ideal. Violation of the equal detectability assumption can occur when individual animals or groups within the population have varying detectibilities (e.g., if male bears are much more likely to use trails, a trail-based scat survey would result in biased estimates; Lukacs and Burnham 2005b), or when animals exhibit a *detection response* such that animals captured at one occasion are more or less likely to be captured during later occasions (Williams et al. 2002). For example, Wegge et al. (2004) found that individual tigers avoided cameras after first being photographed—a response that could complicate data analysis and affect abundance estimates. Sufficient coverage of the survey area with detection devices (or effort, when using non-station-based methods) can help to ensure that no individuals have a zero probability of detection.

One way to minimize detection heterogeneity is to employ different detection methods for initial detections versus redetections (Williams et al. 2002). Since detectability need not be the same for each sampling occasion (only during each sampling occasion are all animals assumed to have the same detection probability; Williams et al. 2002), two independent noninvasive survey methods can be used. With multiple detection methods, there is less chance that an animal detected on the first occasion will be more or less likely than a previously undetected animal to be detected on the second occasion. Boulanger et al. (2008) employed baited hair collection corrals and DNA extraction methods to collect initial detection data, and then unbaited rub trees to collect redetection data (again from hair), for individual grizzly bears in the greater Glacier National Park area of Montana. Although these researchers reported some correlation of capture probabilities between methods, they concluded that the use of both methods together within a two-occasion framework produced more precise estimates of population size than would have been possible had only one method been employed. Similarly, Dreher et al. (2007) used hairs sampled from five baited hair snare sampling occasions, and then a final occasion comprising tissue samples from harvested individuals, to estimate black bear abundance. A final option for minimizing detection heterogeneity is to move survey devices between occasions, decreasing the likelihood of behavioral responses to device locations (Boulanger et al. 2008).

*Designing K-Sample Surveys*

In most cases, it will not be possible to eliminate all forms of detection heterogeneity. Well-designed K-sample surveys provide flexibility for dealing with some forms of heterogeneity (Karanth and Nichols 2002), including variation introduced by time, behavior, and individual responses to detection. The Huggins approach (Huggins 1989, 1991) will allow capture probability to be modeled with population- or site-specific covariates (Manly et al. 2005). For instance, if detectability varies by sex, and the sex of detected individuals can be ascertained via the survey method (e.g., DNA assessment of sex from hair samples), then detection variability can be accounted for during data analysis (see chapter 11). Software packages for capture-recapture analysis (also summarized in chapter 11) provide considerable flexibility for incorporating covariates into esti-

mates of population size. It is therefore important to consider potential causes of detection heterogeneity prior to sampling, and to be sure that related information is carefully recorded. If detection is thought to vary by the density of vegetation at the site, for example, some measure of vegetation density should be recorded. Unfortunately, K-sample approaches, while powerful, take more time to conduct than two-occasion surveys and require larger sample sizes if modeling of detectability is to be successful. These requirements may render K-sample approaches unrealistic for many survey applications. Miller et al. (2005) have developed a small-population capture-recapture approach that is robust to some violations of the assumption of equal detection probability and that provides a possible alternative when more standard approaches are not feasible.

A variety of methods can be used to construct encounter histories, which typically consist of detections, redetections, and nondetections of individuals over time. With cameras, the time and date of each photo is recorded, essentially affording an unlimited number of potential sampling occasions (e.g., hourly, daily, every two days). Sampling occasions should be defined such that encounter histories do not comprise an excessive number of nondetections. This may require that the overall survey duration be increased to provide a sufficient number of occasions from which to model variability in detection probability. For example, if only a single image of a jaguar (*Panthera onca*) is obtained during an unbaited remote camera survey comprising ten sequential, one-day sampling occasions, detection probabilities may have been too low to allow for redetections or for the detection of other individuals. Increasing the duration of the sampling occasion to three days and sampling for a total of thirty days would result in the same number of sampling occasions (ten), but would provide a longer time frame over which to detect or redetect individuals (Henschel and Ray 2003). Last, with baited survey devices, additional care must be taken to adjust the length of sampling occasions to minimize detection heterogeneity and to maintain independence between occasions. For instance, if the occasion length is three days but bait is depleted after day one, detectability may decline on days two and three. Alternately, if an animal is provided a bait reward at a camera station on day one, and is thus more likely to return on day two (i.e., trap-happiness), detection heterogeneity will have been introduced.

Some methods yield detections that effectively occur continuously and thus cannot be assigned to discrete occasions. For example, sampling hairs deposited on rub trees at unknown times could present a scenario in which capture and recapture events appear to occur simultaneously. Relatively little attention has been devoted to such *continuous-time* approaches (Thompson 2004a), and surveys employing a single detection method typically define arbitrary occasions and discard any redetections of individuals that occur during the same occasion. Bellemain et al. (2005), for instance, defined weekly sampling occasions for hunter-collected scat samples and allowed only one detection per individual within each occasion. Combining two or more independent methods can alleviate such problems (e.g., Boulanger et al. 2008). Alternately, both Miller et al. (2005) and Petit and Valiere (2006) have developed methods that allow abundance estimates from a single sampling occasion (i.e., all samples are collected during a prescribed survey period, with captures and recaptures being undifferentiated). While both approaches show promise, neither has been extensively tested.

Another assumption with all capture-recapture methods is that marks are not lost. The marks used with noninvasive capture-recapture represent natural characteristics of the individual animal (e.g., unique DNA code, pelage pattern), thus rendering it very unlikely that marks would truly be lost. For DNA-based surveys, however, misidentifying individuals (i.e., genotyping errors; see chapter 9) can be a common occurrence (Waits and Leberg 2000; McKelvey and Schwartz 2004b). A number of laboratory and analysis approaches have been devised to either minimize genotyping errors (e.g., Taberlet et al. 1996; McKelvey and Schwartz 2004b) or to incorporate such errors into capture-recapture estimates (Lukacs and Burnham 2005a; see chapters 9 and 11).

## Device Placement and Surveying Large Regions

Simple random sampling is not required for collecting capture-recapture data and indeed may be detrimental to an effective capture-recapture survey (Williams et al. 2002). For instance, with grizzly bears, it appears that a stratified random approach—with devices placed in high-quality habitat within randomly selected cells—may be optimal (J. Boulanger, Integrated Ecological Research, pers. comm.). The important point is that each individual in the population must have some chance of being detected. Thus, at least one detection station (and sometimes more) should be located within each home range-sized area (Otis et al. 1978; Karanth and Nichols 2002; Williams et al. 2002; figure 2.9). An individual whose home range overlaps with the boundary of the survey area may have a lower probability of being detected, thus introducing detection heterogeneity into the dataset. The magnitude of such heterogeneity, and potential options for dealing with it, are discussed in other publications (e.g., Otis et al. 1978; Williams et al. 2002; Boulanger et al. 2008).

The ideal scenario for generating encounter histories for capture-recapture analysis is to deploy detection devices across the entire survey area and then to check each one during each sampling occasion for the duration of the survey. But in many cases logistical constraints may prohibit sampling large regions concurrently while still achieving high detection probabilities and meeting the assumption of population closure. Surveys can therefore be designed to allow the survey area to be sampled over time and still provide a single set of encounter histories for analysis. Karanth and Nichols (2002) review three methods for accomplishing this goal with K-sample surveys using unbaited remote cameras. Briefly, the first involves placing devices at randomly chosen locations and moving them each day to a new set of randomly chosen locations. Such an approach is labor-intensive and unlikely to be feasible in most

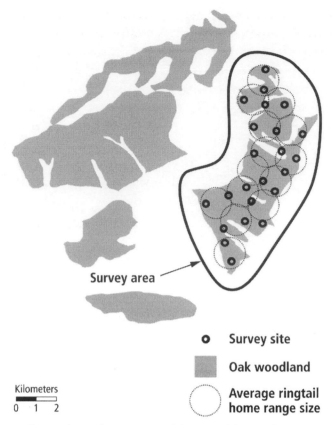

**Survey site**

**Oak woodland**

**Average ringtail home range size**

Kilometers
0    1    2

*Figure 2.9.* This hypothetical survey design shows the spacing and density of devices for capture-recapture estimation of ringtail abundance.

cases. Alternately, the survey area can be divided into several large subareas, with devices deployed throughout one subarea for a number of days (corresponding to the number of desired sampling occasions) and then moved to the next subarea. Last, if stations are difficult to access or set up, devices can be deployed throughout subareas and left for much longer periods before being moved to the next set of subareas. Although all approaches will allow estimation of abundance, the latter two are less desirable as they do not enable certain types of variation in detectability (e.g., over time) to be detected and modeled (Karanth and Nichols 2002).

*Final Thoughts About Noninvasive*
*Capture-Recapture Methods*

The effort required to noninvasively collect high-quality data for individuals within a sampling framework suitable for capture-recapture analysis can be daunting. It is valid, therefore, to wonder whether employing such methods is worthwhile. Despite the hurdles, in many cases abundance estimates from such analyses will be more accurate and defensible than more easily obtained relative abundance indices. Generally, if (1) a target population is not of extremely low density; (2) sampling methods exist that permit high detectability of individuals with minimal sampling bias (i.e., no population segment is unsurveyed); (3) the survey area is sufficiently accessible; (4) the population can be assumed to be closed during the survey duration; and (5) laboratory or analytical methods exist to accurately assign individual identification, then capture-recapture approaches may be successful (see chapter 6 for a discussion of potential ramifications if these criteria are not met). Furthermore, if an estimate of population size is required, noninvasive capture-recapture may be more effective and affordable than trap-based, capture-mark-recapture efforts, especially for low-density and difficult-to-capture carnivores. A large-scale and long-term effort recently conducted by Karanth et al. (2006) illustrates the substantial potential of this approach.

## Abundance-Induced Heterogeneity Models: Estimating Abundance from Detection-Nondetection Data

Given the difficulties inherent to capture-recapture approaches, attempts have been made to develop methods allowing direct estimation of population size from detection-nondetection data. Common sense, and some empirical evidence (e.g., Smith et al. 2007), suggest that detectability and abundance are positively related, and at least one promising method exploits this relationship to permit the estimation of the average number of individuals present across a collection of repeatedly sampled sites (Royle and Nichols 2003). Multiplying this *average abundance* by the number of surveyed sites yields a total population size estimate for the sites sampled. Area expansion techniques can then be used to estimate the total population size in the region of interest (Royle and Nichols 2003). This abundance-induced heterogeneity model is derived directly from standard occupancy estimation (see chapter 11 for model details), but it has not to our knowledge been tested on carnivore datasets.

As many of the limitations and assumptions of this approach are identical to those of occupancy estimation, estimating abundance in this way may not require substantial increases in effort over standard occupancy estimation. But this method does critically depend on accurately characterizing variability in detectability. As discussed earlier, estimating detection probability depends on both the number of sites surveyed and the number of repeat sampling occasions at each site or a subset of sites. Royle and Nichols (2003: 788) speculate that "when heterogeneity exists as a result of variation in abundance,

several hundred sample sites may be necessary to provide reasonable estimates of mean abundance and occupancy rate, although it may be possible to realize success with many fewer." These authors also suggest that five sampling occasions should be suitable for most situations, although (as with standard occupancy estimation) repeat occasions need not be conducted at all sites. Abundance at the site should not be too high, however, as this would result in little variability in detectability. And while abundance at the site can be effectively lowered by reducing the size of the site (see figure 2.4), Royle and Nichols (2003) point out that this is not optimal either as relatively high detectability is helpful for accurately estimating occupancy (MacKenzie et al. 2002).

## Estimating Density

Density is simply defined as the number of individuals per unit area. All methods for estimating population size produce actual abundance estimates (i.e., numbers of animals) that correspond to either the survey area (for capture-recapture) or the site (in the case of abundance-induced heterogeneity models). Technically, therefore, these estimates reflect densities (e.g., individuals/survey area). It is generally more useful, however, if density can be expressed in terms of universal units, such as square kilometers. Calculating density in this way is more difficult, as it must be possible to estimate the actual area surveyed—the **effective survey area**. Except in cases where the survey area is physically bounded, home ranges of some individuals—particularly wide-ranging carnivores—may extend beyond the edge of the survey area as defined by the outermost detection device or survey route. Methods to estimate this additional *boundary width* are typically based on a measure of animal home range size that is itself estimated by examining the average distance between repeat detections of given individuals (i.e., *mean maximum distance moved* [MMDM]). This distance, or one-half MMDM, is then applied as the boundary width (Karanth and Nichols 2002; Williams et al. 2002; see Wilson and Anderson 1985 and Soisalo and Cavalcanti 2006 for evaluations of this approach). Other methods for estimating boundary width and density are discussed in Williams et al. (2002), and briefly mentioned in chapter 5.

## Monitoring

Briefly, *monitoring* is the "repeated assessment of status of some quantity, attribute or task within a defined area over a specified time period" (Thompson et al. 1998: 3). Monitoring is often conducted as part of an adaptive management program (Holling 1978) to decide how a particular management activity should be altered (Goldsmith 1991; Gibbs et al. 1999). In the absence of a management context, surveys designed to collect information about population size over time have alternately been referred to as a trend studies, baseline monitoring, inventory monitoring, or long-term ecological studies (Macdonald et al. 1998; Elzinga et al. 2001).

The term monitoring has been used to describe a wide range of activities. Most carnivore surveys, however, do not actually constitute monitoring. For example, monitoring is *not* the act of setting up a remote camera in a particular place for a set period of time with the goal of detecting a particular species, nor is it deploying a series of track plate or hair collection stations for the ill-defined purpose of evaluating the effects of a proposed forest management activity on a target species. Furthermore, monitoring is *not* repeating a carefully designed survey a second or third time. Monitoring should, instead, be seen as the thoughtful and preplanned execution of a calculated number of surveys over time, with the primary purpose of achieving multiple estimates of occupancy, distribution, or population size over time. It is a prospective exercise which, if conducted properly, is preceded by statistical exercises to help determine the sampling effort necessary to provide results with sufficient precision to detect the magnitude of change of interest (e.g., Kendall et al. 1992; Zielinski and Stauffer 1996;

Gibbs et al. 1998; Elzinga et al. 2001). Monitoring has its own special design considerations that exceed the design considerations for a single estimate of population status (Thompson et al. 1998; Elzinga et al. 2001).

Any of the single-point estimates of population status discussed earlier in this chapter can be monitored over time. Detailed guidance regarding appropriate designs for monitoring animal populations is provided in general texts (e.g., Thompson et al. 1998; Elzinga et al. 2001; Williams et al. 2002), as well as in the primary literature pertaining to carnivores. Here we attempt to highlight the most important considerations.

### Detecting Change in Distribution, Occupancy, Relative Abundance, or Abundance

Change in population status is typically evaluated via one of two methods: (1) comparing estimates of population status at two points in time, or (2) evaluating whether the slope of the relationship between population size and time $\neq 0$ (i.e., suggesting either a positive or negative trend in population size). The first method relies on t-tests, confidence intervals, or nonparametric two-sample tests to evaluate the null hypothesis that there has been no change during the interval between surveys. Although a minimum of two points are theoretically necessary to detect change, these are too few points for most monitoring purposes. For instance, two data points will be insufficient to account for natural temporal variation in carnivore populations, regardless of how far apart in time they may occur. Most monitoring, therefore, is focused on detecting trends, with parametric and nonparametric regression, randomization tests, and Markov models (Thompson et al. 1998; Rhodes et al. 2006) used to evaluate the null hypothesis that the slope of the relationship between the population metric and time $= 0$. In the carnivore literature, examples of testing for significant trends in monitoring data include

Karanth and Nichols (2002) and Boulanger et al. (2004b). A number of additional good examples of trend assessment come from work on marsupials (e.g., Piggott 2004) and, of course, birds (e.g., Sauer et al. 2003).

Although not technically intended for detecting changes in population size, some classes of capture-recapture models are well suited for monitoring. When detectability is high, and individual identification of surveyed animals is possible, capture-recapture methods may provide the most accurate point estimates of population size. Most capture-recapture surveys for carnivores employ closed-population approaches which, by definition, are inappropriate for monitoring change in a population over time. So-called robust methods (Pollock 1982), however, combine periods of short-term sampling (over which the population is assumed to be closed) with longer sampling periods where gains and losses are expected to occur (Williams et al. 2002). The open models estimate survival and emigration rates between years and are referred to as the primary sampling occasions, whereas the closed models estimate abundance by relying on a number of secondary sampling occasions separated by short periods of time (e.g., days; see figure 11.1).

Robust models are able to incorporate and estimate changes in population size during the time when the population is considered open to births, deaths, immigration, and emigration. Such an approach (which has also been extended to occupancy approaches using detection-nondetection data; Karanth and Nichols 2002; Barbraud et al. 2003; MacKenzie et al. 2003) should lend itself well to monitoring population size and vital rates with noninvasive methods (e.g., Karanth et al. 2006). Alternatively, an extension of the Pradel capture-recapture model (Pradel 1996) provides estimates of population growth directly from capture-recapture data, without first requiring abundance estimation. While this and the standard robust approach may prove extremely valuable for long-term monitoring, both models are subject to a rigorous

set of design considerations and require substantial sample sizes (Franklin 2002; Hines and Nichols 2002; Williams et al. 2002).

## Design Considerations for Monitoring

Most efforts to evaluate survey design for carnivore monitoring have focused on comparing two points in time and have used data collected from natural sign surveys (e.g., Kendall et al. 1992; Beier and Cunningham 1996) or track station surveys (e.g., Roughton and Sweeny 1982; Zielinski and Stauffer 1996; Sargeant et al. 2003a). Trend detection entails considerations that exceed those required for comparing two points in time, including specific design considerations (Thompson et al. 1998) that are largely influenced by the expected magnitude of variation in the sample. We, therefore, first discuss the estimation of this variation.

### Variance and the Design of Monitoring Surveys

When count-based metrics are used, variance in counts among sample units is critical to determining the number of sample units that should be sampled; the greater the variation, the greater the number of units. If possible, variance should be estimated in advance by either conducting a pilot study, using estimates from previous similar studies, or, alternately, assuming that the variance of counts is distributed according to a common statistical distribution (Thompson et al. 1998) such as the Poisson (where, by definition the variance of $N$ equals the mean of $N$) or the more flexible negative binomial distribution (Skalski and Robson 1992). Estimating variance among plots for count-based population assessments is an important precursor to estimating sample size.

Understanding variance among sample units is not a relevant concern when occupancy, distribution, or abundance is estimated directly. For these objectives, the allocation of survey effort is influenced more by the variance in the proportion of sites predicted to be occupied or, in the case of estimating abundance, the number of individual animals captured and recaptured.

## Selecting Sites for Monitoring

Due to the diverse nature of natural environments, most ecological metrics will vary more *among* monitoring sites than *within* them over time (Gibbs et al. 1999). This is probably especially true for many carnivores given the largely territorial nature of their social systems (Bekoff et al. 1984). On average, the most precise estimates of population status will be based on a random sample of sites in the first year, with surveys conducted at the same units in subsequent years (Macdonald et al. 1998; Thompson et al. 1998). This approach has been criticized, however, due to the concern that temporal autocorrelation will lead to lack of independence over time (e.g., Usher 1991). Based on the analyses of Cochran (1977), Thompson et al. (1998) argue the contrary position—that is, that a high correlation of results at a site from year to year is a desirable design feature. In addition, permanent plots actually make it possible to use the *difference* between values at the same site over successive survey periods as the test statistic. This is simpler than having to calculate and compare the separate means and variances for each survey period when sites are reselected each year. With a few exceptions (noted in Thompson et al. 1998), permanent plots are thus recommended when estimating trend, and are particularly effective at detecting *lack* of change from year to year (Elzinga et al. 2001). More elaborate reselection methods (e.g., sampling with and without replacement, panel designs [Urquhart and Kincaid 1999], revisit and membership designs [McDonald 2003]) do not appear necessary in most cases. In this regard, Thompson (2004a: 392) advises that "simplicity of design is an important component of an effective monitoring protocol."

Stratification can reduce variation in surveys if strata are relatively homogenous with respect to the probability of detecting the species of interest. Zielinski and Stauffer (1996), for example, assumed that stratification would benefit a statewide monitoring program for fishers and martens in California because distribution and abundance, as well as habitats used, varied substantially across the historical range. But in general, stratification is not recommended for long-term monitoring programs for rare species (Thompson 2004b). Stratification can be an important component of survey design for single time periods, but it may be a liability for long-term monitoring because population centers and habitats used can shift over time (Overton and Stehman 1996).

Some researchers have suggested that statistical inference from monitoring can be maximized by sampling only select portions of the landscape and certain habitats. For instance, simulations conducted by Rhodes et al. (2006) demonstrated that, for changes in population size of 20%–30% over ten years, a substantial improvement in power was achieved by targeting high-quality habitat to detect declines and intermediate-quality habitat to detect increases. In general, optimal statistical power was achieved by targeting the habitats that exhibited the largest changes in the probability of occupancy. In addition, researchers using simulation models have found that power for discriminating change in populations is maximized by sampling sink rather than source habitats (Bowers 1996; Jonzén et al. 2005) because sink habitats tend to be occupied only at high population levels. This extreme form of stratification would call for redefining the target population (and sampling frame). It has not, to our knowledge, been applied to a carnivore monitoring program.

## Using Statistical Power to Inform the Design of Monitoring Projects

A number of software programs are available to help evaluate the effects of various survey components on the power of a particular monitoring design (see box 2.2 for a general discussion of power and precision). Gerrodette (1987) proposed a linear and multiplicative method for detecting trends that is based on an equation with five parameters ($\alpha$, $\beta$, CV [coefficient of variation for the first estimate of population size in the series], number of surveys, and $r$, the specified trend). Program TRENDS (Gerrodette 1993) allows the user to estimate any parameter when the value of the remaining four are specified. Gibbs et al. (1998) have developed a similar tool, called MONITOR, which uses a Monte Carlo simulation approach to evaluate the tradeoffs between sampling effort, statistical power, and logistic constraints. Bailey et al. (2007) discuss precision and the design of occupancy surveys, and provide software (GENPRES) to allow the comparison of competing designs.

Monitoring requires considering sample size at two levels: the number of sites to include in a particular survey, and the number of surveys to conduct over time. These considerations are paramount, as the precision of individual estimates must be sufficient to detect change when it has indeed occurred. Simulations used to explore the effects of sample size on statistical power for detecting change between two points in time have resulted in a wide variety of estimates, ranging from more than 1,000 sample units to detect a 20% decline in fisher or marten populations in California (Zielinski and Stauffer 1996) to two hundred, 1.6-km trail segments to detect a 20% decline in occurrence of grizzly bear sign in Glacier National Park (Kendall et al. 1992). Such examples illustrate the importance of determining a priori the acceptable power and the effect size that must be detected. These decisions will allow the estimation of a sample size sufficient to detect such effects. While sample size requirements are substantial—even for detecting differences between two points in time—they can be mitigated by choosing one-sided alternative hypotheses (e.g., exploring whether a trend is positive as opposed to simply testing for a trend) and tolerating increased Type I error rates (i.e., Type I errors result when a difference in population size or a trend is detected when none actually exists). The shortcoming of this approach is that, if the null hypothesis is not rejected, all that can be concluded is that the population is not in decline; increases in population trend are undetectable.

For detecting trends, the number of independent surveys over time and the time period over which they are conducted are also important considera-tions. The number and frequency of surveys to be conducted depends on (1) the variability in the number of individuals, counts, or proportion of sites with detections; (2) the probabilities of detection; (3) the monitoring objectives; and (4) funding. Gibbs et al. (1998) summarized 512 time-series datasets pertaining to population sizes for various animal taxa. Most groups, including "midsized mammals" (including many mesocarnivores), had coefficients of variation of population size between 50%–100%. Thus, variation in the populations of carnivores is such that credible monitoring programs need to plan for longer versus shorter durations to be able to distinguish real changes in population sizes from noise. Detecting small rates of change in a population requires more surveys conducted more frequently over a longer time period (Thompson et al. 1998). Gibbs et al. (1998) found that, for taxa with highly variable population sizes, three to five surveys per year at a relatively small number of plots (i.e., <30) allowed for the detection of a 50% change in the trend of occurrence over a ten-year period. Many more plots were necessary to detect 25% and 10% changes in the trend, however, even with the same number of surveys per year.

## Best Approaches for Monitoring Carnivores

To detect a trend, successful monitoring requires accurate point estimates for population size, as well as sufficiently high numbers of independent surveys over time. If financial resources are sufficient, populations are at relatively high density, and detectability is high, capture-recapture approaches—especially if robust methods can be employed—will probably provide the best results. For low-density or cryptic species (e.g., most carnivores), occupancy approaches using detection-nondetection data are superior, especially when resources are limited (Joseph et al. 2006). Occupancy approaches also permit the mapping of distribution. To our knowledge, the use of abundance-induced heterogeneity models (Royle and Nichols 2003) for monitoring has

## Box 2.2

### Precision and power in carnivore surveys

All else being equal, studies with larger (versus smaller) numbers of samples representing the statistical population are more precise. Precision is the inverse of uncertainty, such that high precision suggests low uncertainty. The higher the precision, the more confidence there is in the estimate of the parameter of interest (e.g., population size, occupancy). It should be made clear, however, that precision and bias are not the same—bias indicates how far an estimate is from reality, whereas precision indicates confidence in the estimate. For example, a survey with a flawed design (e.g., one that samples only near roads) but a large sample size could result in a very precise (i.e., almost no uncertainty) estimate that is highly biased (i.e., far from reality). Surveys should therefore be designed such that they are minimally biased and maximally precise. Unfortunately, strict adherence to design considerations that lead to low bias (e.g., randomization) often result in more time and effort being expended to collect samples, which necessarily decreases the sample size if a budget or research timeframe is fixed. Researchers must achieve a balance between bias and precision that allows them to meet their research or management needs.

The precision of a parameter estimate, and thus the sample size required to provide this precision, is a function of the variability in the population. Assuming one possesses some knowledge about the population's variability, it is possible to calculate the sample size required to achieve a specified precision in a parameter estimate. For instance, one might wish to estimate the proportion of forest patches occupied by black bears within an area of interest. If the precision necessary to achieve the survey's objectives is ± 0.10, and the expected variability in occupancy can be estimated (even crudely), the sample size required to achieve this precision can be calculated. In the case of occupancy, software has been developed to specifically evaluate the predicted precision of occupancy values estimated from data collected with various survey designs (Bailey et al. 2007). Similarly, if a capture-recapture framework and noninvasively collected genetic data will be used to estimate the abundance of bears in this same area, it is possible to calculate the number of bears that must be "captured" and "recaptured" to achieve a predefined level of precision. If there

are insufficient funds to achieve the required sample size in either of these cases, it may be unwise to attempt the surveys.

The concept of statistical power can also assist researchers in deciding how many samples must be collected to detect differences or changes in populations. Power, a concept related to precision and sample size, is defined as the probability that the null hypothesis is rejected when it is, indeed, false (Morrison et al. 2001). Many carnivore surveys are designed to detect differences between populations (e.g., are there more bobcats on this side of the river, or on that side?), or to detect changes in a population over time (e.g., is more of the area occupied by fishers now, or was more occupied before?). These types of questions are essentially tests of hypotheses, with the null hypothesis being that the population state (e.g., abundance or amount of area occupied) in one area is equal to that in another area (or that states are equal between time periods). Thus, power in these scenarios is the probability of being able to detect a difference in population state when it exists.

Because power, given a predefined probability of committing a Type I error, is a function of sample size, variability, and effect size (i.e., the amount of difference in the state variable that you wish to be able to detect), it is straightforward to calculate one of these components if the other two can be specified. For example, if there were information available about the variability in the percentage of sites occupied by martens in a given region, a researcher could calculate the sample size required to detect a 20% decline (i.e., the effect size) in site occupancy 80% of the time (i.e., the power of the test). Similarly, one could estimate the difference in abundance between two striped skunk populations (i.e., the effect size) that would be detectable at least 80% of the time using capture-recapture methods, assuming a given number of individual skunks could be captured and recaptured during the study (i.e., the sample size).

Entire books have been devoted to power analysis (e.g., Kraemer and Thiemann 1987; Cohen 1988), and software, equations, and examples of the use of power for the design of surveys and experiments are becoming more common (Kraemer and Thiemann 1987; Cohen 1988; Thomas and Krebs 1997; Quinn and Keough

| **Box 2.2** (Continued) |
|---|

2002). Power is a particularly important concept to consider when conducting monitoring efforts, as the ability to detect actual differences of a given magnitude between populations over time is highly dependent on the sample size of the monitoring surveys. Finally, note that while power analysis can be a powerful tool for planning (i.e., a priori), its use for post hoc survey evaluation has been criticized as being biased and misleading (Gerard et al. 1998).

---

not been explored—but this approach appears promising.

An alternative to monitoring population size via the methods discussed thus far involves monitoring the *integrity* of a population on the basis of genetic characteristics. Schwartz et al. (2007) propose the monitoring of population genetic parameters (e.g., allele frequencies, effective population size, gene flow), which will permit changes in abundance to be detected by concomitant changes in genetic diversity. This approach also seems to offer promise, as genetic parameters can be estimated with precision using fewer samples than necessary for estimating population size (Schwartz et al. 2007). We recommend that carnivore researchers explore genetic monitoring as an alternative to the traditional population assessment procedures outlined here.

In summary, monitoring requires careful consideration of variance components, particularly with regard to spatial and temporal variability and sampling variance due to imperfect detectability. Values selected for the number of sites, the number of sampling occasions, the number of surveys, and the magnitude of trend to be detected all affect the efficiency and the cost of a monitoring program. It is sobering, indeed, to consider the implications of these factors for monitoring carnivore population size.

We have primarily discussed monitoring as an independent exercise and have not emphasized its link to management activities or the testing of biological hypotheses. We are cautioned, however, by Krebs (1991: 3) that "monitoring of populations is politi-cally attractive but ecologically banal unless it is coupled with experimental work to understand the mechanisms behind system changes." Monitoring data can be correlated with environmental changes, but these data are rarely satisfactory for establishing cause and effect. This is an important concern in an era when carnivore habitat and populations are under assault from so many threats, and when the most pressing of these threats must be identified and mitigated. Monitoring can be a scientific enterprise, but this will most likely occur when the monitoring data are collected for the purpose of discriminating among competing, a priori hypotheses about how the system under study functions (Yoccoz et al. 2001; Pollock et al. 2002).

## Final Thoughts and Recommendations for Designing Effective Carnivore Surveys

The most appropriate design for a given survey will depend on the survey objectives, the ecology and population status of the target species, the region in which the survey will take place, and logistical constraints (e.g., funding, time, personnel). Nonetheless, here we offer a few rules of thumb for designing a successful survey:

1. Clearly articulate the survey objectives at the outset. MacKenzie and Royle (2005: 1107) note that "the lack of clear objectives will often lead to endless debate about design issues

as there has been no specification for how the collected data will be used in relation to science and/or management; hence judgments about the 'right' data to be collected cannot be made."

2. Don't conduct single-location surveys (also referred to by land management agencies as "project surveys") at the expense of participating in more coordinated regional surveys to estimate occupancy.

3. Maximize the probability of detecting the target species or individual animal, whether using detection-nondetection or individual identification methods.

4. For occupancy surveys, make sure to include repeat sampling occasions for at least a subset of sites to allow the estimation of detectability.

5. When conducting either occupancy or non-invasive capture-recapture surveys, attempt to minimize sampling heterogeneity between subsets of animals (e.g., males and females, subadults and adults) or sampling occasions. In addition, try to identify and record those factors or characteristics that may introduce heterogeneity into the dataset so that they can be modeled during data analysis.

6. Pay careful attention to assumptions about independence among sites, sampling occasions, and devices.

7. Do not use measures of relative abundance if it is feasible to estimate occupancy or abundance. Relative abundance methods require understanding the relationship between the index and true abundance, which is extremely difficult or often impossible.

8. If relative abundance methods must be used to monitor population size, it is best to incorporate multiple, independent methods whenever possible. If all methods are truly detecting changes in population size, results should be correlated.

9. Simple designs, if effective at achieving primary goals, are superior to more complicated ones.

10. Be realistic about the survey's objectives and the intended methods. Most survey objectives require more effort than one might typically expect, and some (e.g., capture-recapture) may clearly be unrealistic given budgetary and time constraints.

Finally, MacKenzie and Royle (2005: 1113) remind us that "designing a study is as much an art as a science. Theoretical and simulation results provide useful guidance about the expected outcome of a study given certain assumptions, analytic techniques and designs. But these results must be tempered with common sense, expert knowledge of the system under study and, occasionally, lateral thinking."

## Acknowledgments

We are indebted to a number of people who graciously assisted us while we were writing this chapter. John Boulanger, Paula MacKay, Kevin McKelvey, Glen Sargeant, and Justina Ray reviewed early drafts and provided invaluable input. Paul Lukacs, Darryl MacKenzie, Andy Royle, and Ted Weller lended their expertise by reviewing various sections of the chapter, or by offering input on particular topics. Each of the other chapters in this book provided material that helped us to consider and frame the topic of survey design, and we therefore thank the respective authors. We offer a tremendous thanks to Paula MacKay for her capable and tireless effort editing this chapter in its various forms—it is a much better resource as a result. Bill Zielinski's time was supported by the Pacific Southwest Research Station (PSRS) of the USDA Forest Service. A portion of Robert Long's time was also supported by the PSRS of the USDA Forest Service and Wildlife Conservation Society Canada.

# Appendix 2.1

*Survey design terms that are used throughout chapter 2. Definitions are tailored specifically to noninvasive survey methods and may not represent the best definitions for other applications. Words in italics are defined elsewhere in the glossary.*

**abundance.** Number of individual animals in a given area. Synonymous with *population size*.

**area of occupancy.** Actual area occupied or used by a species. Often used synonymously with *distribution*. Area of occupancy is never larger than the *extent of occurrence*.

**binary data.** Data compiled from one or more *sampling occasions* at a *site*, representing either a *detection* (commonly recorded as *1*) or a nondetection (commonly recorded as *0*). When multiple detections are recorded at a site, they are nonetheless represented as a single *1*. Multiple detections at a site may also be presented as *counts*.

**capture-recapture.** Method for estimating the number of individuals in a population based on the patterns of detecting and redetecting a sample of individuals over multiple *sampling occasions*.

**counts.** Data representing the number of animal *detections* recorded at a *station*, *site*, or during a *sampling occasion*. Multiple binary results at a site, or during a single *sampling occasion* (e.g., the number of stations or devices registering a detection), can also be used as count data. Likewise, multiple detections at a site can be collapsed and represented as *binary data* (i.e., *1*).

**design.** Overall layout of the *survey*, including the size and location of the survey *sites*, the number and arrangement of *devices* at sites, the *survey duration*, and the *protocol* used to detect species during each *sampling occasion*.

**detectability** (or **probability of detection**). Parameter representing the probability of actually detecting a species or individual animal given that it (or its sign) is present at the survey *site*. Detectability can be estimated given an appropriate survey *design* and is important for accurately estimating both *occupancy* and *abundance* via *capture-recapture* methods.

**detection.** Evidence (i.e., track, hair, scat, or photo) confirming the occurrence of a target species at a *station* or *site* during a *sampling occasion*.

**detection history.** See *encounter history*.

**detection-nondetection data.** *Binary data* representing the results of one or more *surveys* or *sampling occasions* and indicating whether the species was detected (*1*) or not detected (*0*) at each location. Although commonly referred to as "presence-absence data," detection-nondetection data acknowledges that *detectability* is never equal to *1* (i.e., in most cases, it is impossible to assure that a species is actually absent at a location).

**device.** Specific entity (e.g., enclosed track plate, remote camera) that registers a *detection*. By this definition, a scat detection dog is considered a device. Some *methods* (e.g., natural sign surveys) do not use devices but instead entail detections recorded by an observer.

**distribution.** Actual area of species occurrence, typically expressed on a map as either "occupied" or "unoccupied" (or estimated to be in one of these states). Distribution can be displayed as a continuous surface (e.g., as with vector-based map elements) or as a surface divided into subunits (e.g., grid cells)—each indicating *presence* or (inferred) absence. In either case, a given location's state (*1* or *0*) can be inferred based on binary *detection-nondetection* surveys at the actual location, or predicted via occurrence models in concert with a rule-based assignment method (e.g., all sites with predicted *occupancy* of >0.80 are assumed to be occupied).

**effective survey area.** Portion of a *site* or area around a site within which an individual animal is detectable by a given *method*, *device*, or array of devices. Methods that do not use attractants (e.g., remote cameras on trails, scat surveys) have small effective survey areas that are limited to a short distance from the device or surveyor. Methods that employ attractants have a much larger, but often unknown, effective survey area.

**encounter history.** String of *1s* or *0s* (e.g., *01001*) that represents either the pattern of *detections-nondetections* of a species at a *site* or the history of *detections* of an individual at a *station*, over a series of *sampling occasions*. Encounter histories are used in both *occupancy* estimation and *capture-recapture* approaches for estimating *population status*.

**extent of occurrence.** Area within the outermost limits of species *presence* or *occurrence*, often used synonymously with species' *range*.

**false-negative error.** Incorrect assignment of a "nondetection" to a location where the species or its sign is actually present.

**k-sample.** *Capture-recapture* approach employing more than two *sampling occasions* to detect individual animals. Such an approach permits *detection* of and adjustment for heterogeneity of detection probabilities between *occasions*.

**latency-to-first-detection (LTD).** Amount of time between the deployment of a detection *device* and the first *detection* of a particular species.

**method.** Type of survey approach or *device* that is used to detect a target species and, in some cases, to collect information for estimating *population status*. Method is distinct from *design*, which refers to how one or more methods are employed (e.g., survey *sites* located 5 km apart, five track plate stations per site, visited every other day for ten days).

**monitor.** Performing repeat surveys over time, with the goal of quantifying change in *population status* (i.e., *trends*). Monitoring should not be confused with repeat *sampling occasions* (sometimes referred to as checks or *visits*), which are conducted within a single *survey* and either allow the estimation of detection probabilities or provide an increase in overall *detectability* at the site.

**occasion.** See *sampling occasion*.

**occupancy.** Population state variable representing the proportion of *sites* estimated to be occupied (or in the case of wide-ranging species such as carnivores, the proportion used) by a species of interest. If an appropriate survey *design* (e.g., randomly chosen *sites*) is employed, occupancy is also considered an estimate of the proportion of the *survey area* occupied (or used) by the species. Occupancy is not estimated for an individual *site*, but only for *surveys* with multiple sites. Thus, it is differentiated from *presence* in that it is an estimated parameter whose value falls between *0* and *1*.

**occurrence.** Typically a synonym for *occupancy*, occurrence is also synonymous with presence—as in the phrase *extent of occurrence*.

**population status.** Some attribute of a population, such as *area of occupancy*, *distribution*, *relative abundance*, or *abundance*.

**pre-project survey.** *Survey* aimed at *detecting* or characterizing the *population status* of at least one member of a target species and executed prior to the implementation of a natural resource management activity.

**presence.** State of a *sample unit* being occupied by a species (or in some cases, by the species' sign)—regardless of whether the species (or sign) is *detected* by a *survey*. Presence is evaluated at a single *sample unit* (e.g., at a *site*) and differs from *occupancy* or *occurrence*, which are typically parameters that can take on any value between *0* and *1* and are estimated with *detection-nondetection data* from a series of sample units.

**probability of detection.** See *detectability*.

**protocol.** Specific actions taken to carry out a *sampling occasion*. Protocol differs from *design* in that the latter refers to the larger-scale considerations of *survey* layout, while protocol refers to detailed instructions that are repeated at the *sample unit*. Examples of protocols include baiting a track station with a single chicken leg, or testing a remote camera prior to departing a *station*.

**range.** See *extent of occurrence*.

**relative abundance.** Parameter or metric assumed (or proven) to be correlated with population size.

**sample unit.** Statistical unit of analysis. For example, if a *site* comprising five survey *stations* is the *sample unit*, then a *detection* at any number of the stations results in a single detection recorded for the site. Elsewhere, the term has sometimes been used synonymously with site, and also in a nonprobability sense to refer to the subunits of a *survey* aimed at detecting a target species in an area of interest. We do not use it in this manner.

**sampling.** Recording or measuring characteristics of a portion of the sample population in order to infer something about the entire population.

**sampling occasion.** One sampling effort at a *site* or at a *station*. A sampling occasion (sometimes referred to as a *visit* or a "check") occurs when an observer returns to a site or survey station to record whether or not a *detection* has been registered at a detection *device*, or alternately when a non-station-based approach is used in a single attempt to detect a species. Similarly, the recorded sampling history of devices capable of remotely tracking the dates of detections (e.g., remote cameras that imprint dates on photographs) can be divided into any number of sampling occasions between researcher visits. Sampling occasions can be repeated to increase the *probability of detection*, and the results of multiple sampling occasions can be used to estimate overall *detectability* if they are assumed to be independent (i.e., a detection during one occasion does not influence a detection during another *occasion*). The results of multiple sampling occasions are represented as an *encounter history*.

**site.** Statistical unit of analysis when using *binary data* or sign counts to assess *occupancy or relative abundance*. The site is the area within which binary *detection-nondetection data* or *counts* are pooled for analysis. For instance, a single sampling occasion at a site may result in a *detection*, a nondetection, or a count (e.g., number of track stations with a detection, number of scats located). Sites are typically selected based on a statistical probability model and in many cases are considered independent, although this may not be the case for some applications.

**station.** Location within a *site* or *survey area* at which a *detection* attempt is made during a *sampling occasion*. Stations are typically assumed to be dependent (i.e., a detection at one station may affect detection at other stations within the site or survey area), and detections at multiple *stations* are often collapsed into *binary data* or *counts* at the level of the site for *occupancy* or *relative abundance* assessments. Alternately, stations are the locations at which individual animals are detected for *capture-recapture* approaches used to estimate population size.

**survey.** One or more attempts (i.e., *sampling occasions*) to *detect* a species at either a single location or across many *sites* with the intention of making inferences about species *occurrence* or population size. Survey outcomes can include assessments of species *presence*, estimates of *occupancy*, predicted *distributions*, mean count per unit area or per survey time, or estimated population size.

**survey area.** Area within which the *survey* results and resulting inferences are relevant. This is analogous to the statistical population. Survey *sites* should be distributed appropriately within the survey area and based on a statistical probability model if inferences gained from site data are to be unbiased.

**survey duration.** Amount of time or the number of *sampling occasions* comprising the *survey*. The sampling duration affects *detectability* and should be chosen based on both the home range of the target species and the size of the survey *site*.

**trend.** Sustained, directed change in *population status* over time that can be detected by *monitoring* when statistical power is sufficient.

**visit.** Synonym for *sampling occasion* that is often encountered in the carnivore literature. Visit is becoming a less accurate descriptor with the advent of non-invasive methods that permit *sampling* durations to be subjectively parsed after the fact (e.g., when remote cameras provide continuous sampling) and because a visit by the observer can be confused with a "visit" by the target species.

*Chapter 3*

# Natural Sign: Tracks and Scats

*Kimberly S. Heinemeyer, Todd J. Ulizio, and Robert L. Harrison*

Verifying the presence of wildlife by observing natural sign—such as footprints, scats, burrows, and tree rubs—dates back to the earliest human hunters (Liebenberg 1990). When wildlife management emerged as a profession, efforts to document the presence of species in North America relied primarily on the direct observation of animals or the interpretation of their sign (Grinnell 1876; figure 3.1). Given the inherent uncertainty of identifying species via natural sign, such interpretations have often been called art rather than science (Liebenberg 1990; Rezendes 1992). This uncertainty, combined with inconsistent survey designs and quality control procedures, has resulted in criticism of natural sign surveys and the need to improve survey efforts to meet more rigorous standards. In this chapter, we discuss survey approaches and protocols that may decrease the uncertainty associated with species detections and increase the robustness and usefulness of natural sign surveys.

For the purposes of this book, natural sign surveys do not require an individual animal to respond to an attractant or to leave sign at prepared stations (e.g., track stations, as described in chapter 4). In other words, natural sign is deposited over the course of an animal's normal daily activities. While carnivores can leave a diversity of natural sign, we focus exclusively on tracks and scats. Track and scat surveys have a long history, can be applied across large geographic areas, and are applicable to many different carnivore species. We subdivide track surveys into three prominent types: snow track surveys conducted on the ground, aerial snow track surveys, and dust or mud track surveys.

## Background

In their most basic form, natural sign surveys are aimed at determining the presence of a species. Under certain conditions, however, they can also be standardized to delineate species distribution, estimate relative and absolute abundance, and monitor trends in population status. Natural sign surveys have been implemented by government agencies and academic institutions as a relatively low-cost approach to gathering data on regional and local carnivore populations. Additionally, natural sign surveys initiated by citizen and community groups—often in cooperation with government agencies—have popularized such surveys as fun, informative outdoor activities that provide important data on resident wildlife species (e.g., Keeping Track, www.keepingtrack.org; Clyde et al. 2004).

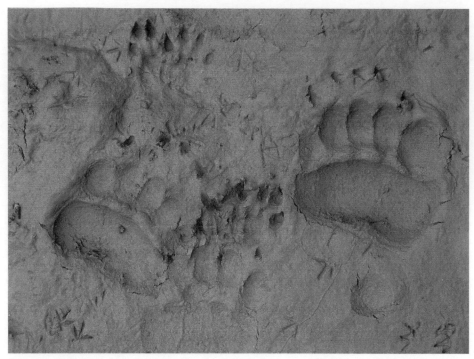

*Figure 3.1.* Tracks, such as these grizzly bear, mustelid, and canid tracks on a river bank in northwestern British Columbia, have long been used to document the presence of species. Photo by K. Heinemeyer.

## Track Surveys

Ground-based snow track surveys have enjoyed many decades of use in the central and northern regions of North America (Seton 1937; Murie 1940) and continue to be widely employed to obtain information about carnivores (see table 3.1; figure 3.2A)—despite concerns about the rigor and utility of some survey approaches. The USDA Forest Service (Zielinski and Kucera 1995a) has developed standard design and data collection protocols for snow tracking efforts to detect American martens (*Martes americana*), fishers (*Martes pennanti*), Canada lynx (*Lynx canadensis*) and wolverines (*Gulo gulo*). Similarly, the British Columbia Ministry of the Environment (1998, 1999) and the Alberta Biodiversity Monitoring Program (Bayne et al. 2005) have recommended snow tracking survey standards for detecting and monitoring the relative abundance of a wide diversity of species, including coyotes (*Canis latrans*), red foxes (*Vulpes vulpes*), lynx, bobcats (*Lynx rufus*), wolverines, fishers, gray wolves (*Canis lupus*), and mountain lions (*Puma concolor*). These protocols and other recent developments in survey design (see chapter 2) and noninvasive genetics (see chapter 9) have helped to strengthen traditional track surveys.

In open landscapes, planes and helicopters have proven effective tools for surveying snow tracks of midsized and large carnivores (table 3.1; figure 3.2B). Aerial snow tracking has been most broadly used for presence and distribution surveys (Heinemeyer and Copeland 1999; Heinemeyer et al. 2000; Gardner 2005; Magoun et al. 2007) and to assess relative abundance (Magoun et al. 2007) within remote and otherwise inaccessible study areas. Recently, biologists in Alaska have explored the ability of digital video cameras to document tracks during aerial surveys, allowing tracks to be evaluated later for species identification (H. Golden, Alaska Department of Fish and Game, pers. comm.). Substantial efforts have also been made in Alaska to develop techniques

*Table 3.1.* References for natural sign surveys of North American carnivore species

| Species | Ground snow track surveys | Aerial snow track surveys | Dust or mud track surveys | Scat surveys |
|---|---|---|---|---|
| Coyote | Murray et al. 1994; Earle and Tuovila 2003; Bayne et al. 2005 | | | Clark 1972; Andelt and Andelt 1984; Knowlton 1984; Kohn et al. 1999; Fedriani et al. 2001; Adams et al. 2003; Prugh et al. 2005; Cunningham et al. 2006 |
| Gray wolf | Wydeven et al. 1995; Bayne et al. 2005 | Becker et al. 1998; Hayes and Harestad 2000; Patterson et al. 2004; Smith et al. 2004[a] | | Lucchini et al. 2002; Creel et al. 2003 |
| Red wolf | Forsey and Baggs 2001 | St-Georges et al. 1995 | | Adams et al. 2003 |
| Gray fox | Earle and Tuovila 2003 | | | Cunningham et al. 2006 |
| Arctic fox | Anthony 1996 | | | |
| Kit fox | | | | |
| Swift fox | | | Roy et al. 1999; Roy 1999; Grenier 2003 | Harrison et al. 2002; Schauster et al. 2002; Harrison et al. 2004 |
| Red fox | Pulliainen 1981; Thompson et al. 1989; Helle et al. 1996; Earle and Tuovila 2003; Bayne et al. 2005 | Sargeant et al. 1975 | | Kolb and Hewson 1980; Cavallini 1994 |
| Canada lynx | Thompson et al. 1989; Murray et al. 1994; Halfpenny et al. 1995; Hoving et al. 2004; Squires et al. 2004; Bayne et al. 2005; McKelvey 2006 | | | |
| Bobcat | Wydeven et al. 1995; Earle and Tuovila 2003 | | | |
| Mountain lion | Desimone and Semmens 2004; Bayne et al. 2005; Sawaya and Ruth 2006 | | Smallwood and Fitzhugh 1993; Smallwood 1994; Smallwood and Fitzhugh 1995; Beier and Cunningham 1996; Grigione et al. 1999; Lewison et al. 2001 | Ernest et al. 2000 |
| Striped skunk | Wydeven et al. 1995 | | Engeman et al. 2003 | |
| Wolverine | Hornocker and Hash 1981; Halfpenny et al. 1995; Flagstad et al. 2004; Bayne et al. 2005; Ulizio 2005; Ulizio et al. 2006 | Heinemeyer et al. 2000; Gardner 2005; Magoun et al. 2007 | | Flagstad et al. 2004 |

*Table 3.1* (Continued)

| Species | Ground snow track surveys | Aerial snow track surveys | Dust or mud track surveys | Scat surveys |
|---|---|---|---|---|
| North American river otter | Reid et al. 1987; Helle et al. 1996; Earle and Tuovila 2003; Sidorovich et al. 2005 | St-Georges et al. 1995 | | Swimley et al. 1998 |
| American marten | Raine 1983; Bateman 1986; Ashbrenner 1994; Halfpenny et al. 1995; Robitaille and Aubry 2000; Forsey and Baggs 2001; Earle and Tuovila 2003; Bayne et al. 2005 | H. Golden, Alaska Fish and Game, pers. comm. | | |
| Fisher | Raine 1983; Halfpenny et al. 1995; Serfass et al. 2001; Bayne et al. 2005 | | | |
| Ermine | Thompson et al. 1989; Forsey and Baggs 2001 | | | |
| Least weasel | Sidorovich et al. 2005 | | | |
| American mink | Earle and Tuovila 2003; Bayne et al. 2005; Sidorovich et al. 2005 | | | Bonesi and Macdonald 2004 |
| American badger | Wydeven et al. 1995 | | | |
| Raccoon | Wydeven et al. 1995 | | Heske et al. 1999 | Gehrt 2003 |
| American black bear | Wydeven et al. 1995 | | | |
| Grizzly bear | Green et al. 1997 | | | Taberlet et al. 1997; Lukins et al. 2004; Sato et al. 2004; Bellemain et al. 2005 |
| Polar bear | | Evans et al. 2003; Stirling et al. 2004 | | |

Note: The exclusion of certain species from this table reflects a lack of published accounts of their detection via these survey methods.

[a] Combined aerial track survey with following tracks to visually observe animal.

for combining aerial snow track surveys with following tracks to observe and count individuals. Such surveys have permitted the estimation of abundance or density for wolves, wolverines, and lynx (Becker 1991; Becker et al. 1998; Becker et al. 2004; Patterson et al. 2004).

Tracks recorded in substrates other than snow (e.g., dust, mud) are used to confirm species presence, but inconsistency in the availability of track-recording substrates across meaningful survey areas has limited the use of dust or mud track surveys for standardized detection and monitoring applications (table 3.1; figure 3.3). Dust or mud track surveys for generalist predators (e.g., raccoon [*Procyon lotor*], red fox, striped skunk [*Mephitis mephitis*]) are relatively common (Bollinger and Peak 1995; Heske et al. 1999; Kruse et al. 2001), and are often associated with the monitoring of predation on nesting birds. Dust tracking is also conducted at baited or unbaited prepared track stations (see chapter 4).

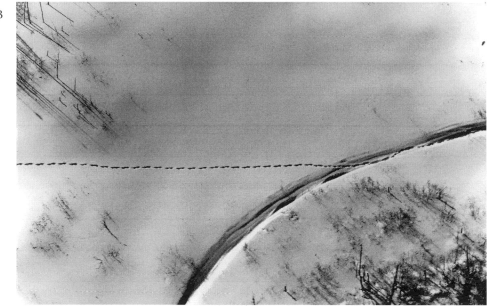

*Figure 3.2.* Snow track surveys can be conducted from (A) the ground to identify species based on track and gait characteristics, as would be possible with these Canada lynx tracks (photo by K. Heinemeyer) or (B) the air, in which case identification is inferred from gait and path attributes. The classic wolverine tracks in (B) exhibit a characteristic, highly linear path (photo by J. Ray).

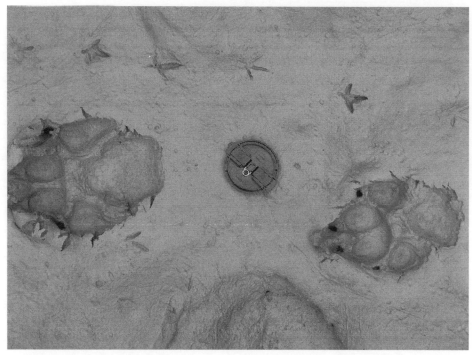

*Figure 3.3.* Tracks in mud or dust can register high detail (as seen here with these wolf tracks along a river in northwestern British Columbia), but a lack of high-quality tracking media often limits systematic survey efforts. Photo by K. Heinemeyer.

Track surveys in dust and mud are more broadly employed in certain regions outside of North America. For example, such surveys have been undertaken for red foxes and European otters (*Lutra lutra*) in Europe (Servin et al. 1987; Chautan et al. 2000; Ruiz-Olmoa et al. 2001); ring-tailed coatis (*Nasua nasua*) in South America (Yanosky and Mercolli 1992); tigers (*Panthera tigris*) in India (Karanth et al. 2004b); dingos (*Canis lupus dingo*) and feral cats (*Felis silvestris catus*) in arid regions of Australia (Edwards et al. 2000); and leopards (*Panthera pardus*), lions (*Panthera leo*) and African wild dogs (*Lycaon pictus*) in arid countries of southern Africa (Stander 1998; Funston et al. 2001).

### Scat Surveys

Scat surveys are commonly conducted for carnivores (table 3.1), primarily in the context of diet studies and often in conjunction with surveys for other sign, such as tracks (figure 3.4). Scat surveys are used less often than track surveys to assess carnivore presence or abundance due to the relative difficulty of finding scats and our inability to consistently distinguish the scats of different species based only on scat size and morphology. Nonetheless, the advent of more accurate means of species identification, such as DNA analysis (chapter 9) and thin layer chromatography of bile acids (Fernández et al. 1997), has significantly increased the utility of scat surveys. Additionally, the relatively recent use of dogs trained to detect scats has enhanced our ability to systematically search for and collect scats (chapter 7).

### Target Species

Natural sign surveys have been applied to a wide variety of species (table 3.1), although climate and environmental conditions limit the applicability of certain survey types. Obviously, snow track surveys are inappropriate for carnivores occurring in regions lacking suitable snow conditions or for species that are inactive in winter (e.g., bears [*Ursus* spp.], Amer-

*Figure 3.4.* Identification of species from scat morphology alone is unreliable, but associated tracks or other natural sign can increase the confidence of the identification, as exemplified by this scat accompanied by wolf tracks. Photo by K. Heinemeyer.

ican badgers [*Taxidea taxus*]). Ground-based snow track surveys are most commonly used to survey or monitor furbearers (i.e., carnivores that are traditionally trapped). Aerial snow track surveys are further limited to species that leave snow tracks and patterns that are identifiable from the air, and to places where dense vegetation does not obscure tracks. To date, aerial surveys have mainly focused on mid- to large-sized carnivores occupying expansive survey areas or remote, inaccessible regions of northern North America.

Track surveys in dust or mud receive limited use on this continent, with most efforts concentrated on mountain lions and swift foxes (*Vulpes velox*). Scat surveys for assessing presence or estimating abundance have been most useful for species whose scat can be easily located, including those that habitually travel along roads, trails, or watercourses (e.g., coyotes). Some species leave highly visible scat deposits that further facilitate scat surveys (e.g., North Amer-

ican river otters [*Lontra canadensis*]; Kruuk et al. 1986).

## Strengths and Weaknesses

The broad application and extensive history of natural sign surveys has firmly established their use as a tool for wildlife detection and monitoring—a tool that is sometimes employed inappropriately and can thus yield unreliable results. Additionally, there is a surprising paucity of research literature reporting on the results of such surveys, and even less on the testing or quantification of error rates (e.g., Prugh and Ritland 2005). The inability to confirm species identification or to identify unique individuals (i.e., independent occurrences) based on field interpretations of natural sign has hampered the advancement of these types of surveys for population monitoring (see box 3.1). The development of DNA techniques to identify species and individuals from noninvasively

## Box 3.1

### Strengths and weaknesses of natural sign (track and scat) surveys

| | Ground snow tracking | Aerial snow tracking | Mud/dust tracking | Scat surveys |
|---|---|---|---|---|
| **Strengths** | | | | |
| No behavioral response required. | X | X | X | X |
| No special equipment requirements. | X | X | X | X |
| Multispecies objectives easily incorporated. | X | X | X | X |
| Applicable to a broad diversity of habitat types. | X | | X | X |
| Additional ecological information (e.g., behavior, habitat use) can be acquired. | X | X | X | X |
| Minimal field expense. | X | | X | X |
| Detection unbiased by use of lure or bait. | X | X | X | X |
| Large geographic areas can be surveyed if access is adequate. | X | X | X | X |
| Highly mobile animals leave many tracks. | X | X | X | |
| DNA samples can be collected. | X | | | X |
| Not limited by ground-based access. | | X | | |
| Systematic linear transects easily incorporated. | | X | | |
| Tracks can be followed to identify individual animals or collect DNA. | X | X | | |
| Potential for high-quality track imprints for casts or detailed measurements. | | | X | |
| Long persistence time in environment. | | | | X |
| **Weaknesses** | | | | |
| Field-based species identification sometimes ambiguous or unfeasible. | X | X | X | X |
| Inability to identify individuals in field. | X | X | X | X |
| Additional effort required to validate identification of species or individual. | X | X | X | X |
| Morphology of sign may vary within a species. | X | X | X | X |
| Limited by availability of appropriate tracking mediums or conditions. | X | X | X | |
| Can be limited by lack of access. | X | | X | X |
| Sign is highly ephemeral and weather dependent. | X | X | X | |
| Highly skilled and experienced trackers required. | X | X | X | |
| Inability to collect DNA samples. | | X | X | |
| Sightability may vary with habitat conditions. | X | X | X | X |

collected hair and scat, however, has enabled increased utility and rigor for some natural sign surveys (Kohn et al. 1999; Mills et al. 2000a; Piggott and Taylor 2003; Riddle et al. 2003; Hedmark et al. 2004; Squires et al. 2004; Prugh et al. 2005; McKelvey 2006; Ulizio et al. 2006).

If concerns about species identification are alleviated, natural sign surveys can be a powerful and efficient means for collecting information about a diversity of carnivores. These surveys do not rely upon special technology or equipment, can be relatively straightforward and quick to conduct, and can easily incorporate multispecies objectives. Additionally, natural sign surveys do not require animals to respond to attractants or other survey equipment. As a result, there are potentially fewer species-specific limitations and biases inherent to natural sign surveys as compared with other noninvasive survey techniques detailed in this volume. New advances in survey design and data analysis provide robust options for dealing with multiple detections of a species within a survey without necessitating the identification of unique individuals (Royle 2004a; Stanley and Royle 2005). Further, the identification of unique individuals via DNA analysis can permit survey data to be used for population abundance estimates. We have summarized specific strengths and weaknesses of different natural sign survey types in box 3.1.

## Treatment of Objectives

Tracks and scat surveys are often incorporated into a formal survey framework to assess the presence, relative abundance, or abundance of target carnivore species. Such surveys can be conducted at a single point in time to address a specific management question, or over longer periods as part of a monitoring program.

### Occurrence and Distribution

Natural sign surveys offer relatively inexpensive options for gathering information about the presence and distribution of carnivores, and have been widely utilized in this manner (Melquist and Dronkert 1987; Bollinger and Peak 1995; Smallwood and Fitzhugh 1995; Zielinski and Kucera 1995a; Sovada and Roy 1996; Massolo and Meriggi 1998; British Columbia Ministry of the Environment 1998, 1999; Heinemeyer et al. 2000; Schmitt 2000; Gaines 2001; Kruse et al. 2001; Grenier 2003; Lukins et al. 2004; Gardner 2005; Sargeant et al. 2005; Magoun et al. 2007). The most valuable surveys standardize survey design and effort, and clearly document results and field-based decisions. When coupled with reliable species identification techniques, natural sign surveys have been used to confirm the presence of rare species such as lynx and wolverines in areas where their status was previously unknown (Squires et al. 2004). Harrison et al. (2002) found that collection of scats provided more swift fox detections than scent stations, trapping, searches for tracks, spotlighting, or calling.

### Relative Abundance

Natural sign counts on predetermined survey routes have been used as indices of relative abundance for numerous carnivore species (Dice 1941; De Vos 1951; Pulliainen 1981; Douglas and Strickland 1987; Thompson et al. 1989; Linden et al. 1996; Bonesi and Macdonald 2004; Gusset and Burgener 2005; Magoun et al. 2007). There have been few tests of the relationship between natural sign frequency and population density, however, nor of the sensitivity of a given index to true changes in the population (Van Sickle and Lindzey 1992; Cavallini 1994; Schauster et al. 2002; Bart et al. 2004; Bonesi and Macdonald 2004; Stanley and Royle 2005). Various factors can influence the quantity of natural sign and associated detections, including adjacent habitat characteristics (Stanley and Bart 1991), environmental conditions influencing local animal movement, and survey conditions. Yet, when surveys have been standardized to control for the most important sources of variation, natural sign indices have been shown to coincide with known population density (Clark 1972; Knowlton 1984; Stander 1998; Gompper et al. 2006) and with other indirect indices of population abundance (Gusset and Burgener 2005). Given the uncertainty about factors that may influence natural sign indices, and whether indices accurately reflect population size or changes in abundance, correlations between sign indices and population metrics should be independently verified (Sadlier et al. 2004).

Sampling designs that focus on assessing presence over time can use formal occupancy methods (chapter 2) and associated estimates of detectability to estimate relative or absolute population abundance (Royle and Nichols 2003; Kery et al. 2005). Emerging analytical techniques may further increase the robustness and utility of sign count indices for estimating relative abundance or multiple relative abundance classes if assumptions about the relationship between the index and the population can be justified (Royle 2004a; Stanley and Royle 2005; see chapters 2 and 11 for further discussion related to assessing relative abundance).

### Abundance and Density

Natural sign surveys have rarely been used for estimating carnivore abundance because efforts to identify individuals from natural sign have had very limited success. Discriminant function analyses of detailed footpad measurements have been conducted to identify individual mountain lions (Smallwood and Fitzhugh 1993; Grigione et al. 1999; Lewison et al. 2001), tigers (Sharma et al. 2005), and black rhinoceroses (*Diceros bicornis*; Jewell et al. 2001)—similar efforts for jaguars (*Panthera onca*) were abandoned due to inconsistent results (C. Miller, Wildlife Conservation Society, pers. comm.). All researchers reported variable success in correctly discriminating individuals under field conditions. This method can also be severely hindered by the requirement that a large number of high-quality track sets (ideally consisting of up to twenty prints from the same paw) be collected from several individuals to identify discriminating variables for any given population (Karanth et al. 2003). Indeed, Karanth et al. (2003) have questioned the utility and validity of efforts to monitor abundance of tigers in India based on individual identification.

The individual identification of tracks from physical characteristics and measurements has not been widely pursued for most North American carnivores, nor has it been adequately tested to provide information leading to estimates of abundance. Despite its limitations, this approach continues to generate interest and is currently being explored for a number of species, from lions in South Africa to polar bears (*Ursus maritimus*) in northern Canada (Unger 2006). Researchers have also identified individual animals using indigestible markers, such as radioactive materials or plastic chips, which permit the estimation of population size via capture-recapture analysis (Sadlier et al. 2004; Silvy et al. 2005).

Genetic methods for assigning individual identification from noninvasively collected scats or hairs provide a higher level of rigor than methods based solely on track measurements. Wolverine snow track surveys that included following tracks and collecting genetic samples resulted in estimates of minimum population size (Ulizio et al. 2006) and—using a mark-recapture framework—population size (Flagstad et al. 2004). Mountain lions in Yellowstone National Park have also been individually identified with genetic samples collected from snow tracks (M. Sawaya, Montana State University/Wildlife Conservation Society, pers. comm.). The genetic identification of scats enables estimates of the number of individuals present over the period of scat accumulation (Taberlet et al. 1997; Kohn et al. 1999; Ernest et al. 2000; Bellemain et al. 2005; Prugh et al. 2005), although accumulation time can only be known with certainty if scats are cleared from survey routes prior to survey initiation. When rates of scat decomposition and defecation are known, the probability of detection of scats can be estimated, thus allowing estimation of population size (Lancia et al. 2005). New techniques for collecting large numbers of scats (e.g., scat detection dogs, chapter 7) may enhance the ability of genetic capture-recapture methods to estimate carnivore abundance in the future.

Certain survey designs (chapter 2) and analytical methods (chapter 11) permit the estimation of population abundance or density without requiring that individual animals be identified. For instance, tracking or other means to measure distances moved by target animals within a twenty-four-hour period can be used to estimate the number of individuals that likely produced tracks encountered along random transects (Stephens et al. 2006). While this approach has not been tested with carnivores in North America, it is widely applied in Russia (Stephens et al. 2006). A similar tactic has been used to estimate tiger density based on photographic rates by simulating individual movements with random walk models (Carbone et al. 2001). Such approaches demand that a wide range of assumptions be carefully evaluated (see Carbone et al. 2002; Jennelle et al. 2002; Stephens et al. 2006). Last, occupancy modeling and other analytical methods that allow the estimation of detectability via repeat surveys at sites (chapter 2) may enable more accurate assessments of abundance based on natural sign (Royle and Nichols 2003; Stanley and Royle 2005).

## Monitoring

Monitoring activities employing natural sign surveys range from single-species efforts (Schauster et al. 2002; Flagstad et al. 2004; Harrison et al. 2004; Stuart 2006) to those aimed at multiple species as part of a regional assessment program, such as the triangle surveys conducted in Finland (Helle et al. 1996) and Alberta (Bayne et al. 2005; see box 3.2 and *Spatial Design* later in this chapter). As with all methods, monitoring considerations for natural sign surveys depend largely on the objective (e.g., distribution, relative abundance), and on ensuring that the survey design has adequate power to identify changes at the scale desired (Gerrodette 1987; Kendall et al. 1992; Beier and Cunningham 1996; Zielinski and Stauffer 1996; also see chapter 2).

## Other Achievable Objectives

Data on habitat use can be acquired by assessing habitat at the natural sign site (Kruuk et al. 1986; Thompson et al. 1989; Kurki et al. 1998; Robitaille and Aubry 2000; Forsey and Baggs 2001; Grenier 2003; Lozano et al. 2003), analyzing track locations observed from the air (St-Georges et al. 1995; Robitaille and Aubry 2000; Gardner 2005), or following tracks to record the relative use of different habitats available along the animal's route (Bateman 1986; Murray et al. 1994; Murray et al. 1995; Matyushkin 2000; Heinemeyer 2002; Fuller 2006). Tracks can also provide data that is difficult to gather in other ways, including information pertaining to movement and behavior associated with foraging (Murray et al. 1994; Matyushkin 2000; Heinemeyer 2002; Hebblewhite et al. 2003; Fuller 2006); responses to modified habitats, human infrastructure, and fine-scale habitat characteristics (Alexander and Waters 2000; Van Wieren and Worm 2001; Heinemeyer 2002; Whittington et al. 2004, 2005); predation on sensitive species (Bollinger and Peak 1995; Yanes and Suarez 1996; Heske et al. 1999; Kruse et al. 2001); and other interspecific interactions (Kauhala and Helle 2000; Matyushkin 2000).

The physical inspection or genetic testing of scat allows for analyses of diet (Kelly and Garton 1997; Gau et al. 2002; Zabala and Zuberogoitia 2003), stress (Creel et al. 2002), viruses (Seal et al. 1995), internal parasites (McCleery et al. 2005), hybridization (Adams et al. 2003; Schwartz et al. 2004), predation (Ernest et al. 2002), community relationships (Dalen et al. 2004a), and genetic relatedness (Lucchini et al. 2002). Also, with field assistance from detection dogs, scat survey information has been incorporated into models to predict species occurrence based on habitat characteristics (Long 2006).

# Practical Considerations

There are several practical issues that help to determine the appropriateness and effectiveness of natural sign surveys. Of primary consideration is the tolerance for error in species identification; additional concerns include the logistical requirements of different survey approaches and designs.

## Species Identification

Given the inherent uncertainty associated with species identification from natural sign, careful thought must be given to survey objectives. Surveys to verify the presence of rare species in areas where their status is unknown require absolute certainty in species identification. Such certainty likely cannot be attained by track and gait measurements, photographs, or expert judgment but can potentially be acquired through additional efforts to collect hair and scat for genetic testing. Alternatively, field-based species identification is sometimes acceptable if experienced trackers are employed and appropriate identification protocols are established. For example, absolute certainty in species identification may not be required for monitoring species that are known to be present and relatively abundant. If misidentification errors are relatively rare and error rates are known, both imperfect detection and false

## Box 3.2

### Conducting snow track surveys with the wildlife triangle scheme

The wildlife triangle scheme (WTS) provides an efficient sampling design for nonmotorized natural sign surveys. The WTS was originally developed in 1988 by the Finnish Game and Fisheries Research Institute for snow track surveys designed to detect species, as well as to estimate relative abundance and population trends for a wide variety of wildlife. A total of 1,600 sampling sites were established across Finland. Fifteen years of data collected for thirty species have provided a powerful foundation for monitoring species population trends (Hellstedt et al. 2006) and species richness (Pellikka et al.

2005; figure 3.5), and the ability to compare game management practices between neighboring countries (Danilov et al. 1996).

The WTS employed a 12-km transect in the form of an equilateral triangle (with each side measuring 4 km) that was surveyed twice each winter to identify species based on tracks and other natural sign. While not as statistically optimal as a linear transect, this design is efficient and practical because the observer completes the survey at its starting point, thereby avoiding back-tracking. The locations of sampling areas in Finland were

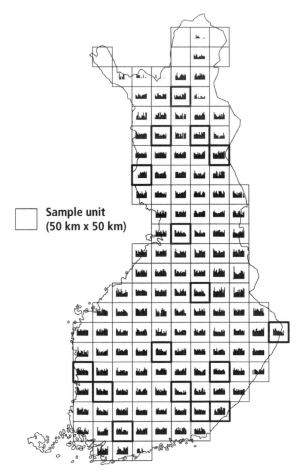

**Sample unit (50 km x 50 km)**

*Figure 3.5.* Pellikka et al. (2005) explored relative trends in species richness across Finland via surveys based on the wildlife triangle scheme. Column charts within sample units show species richness index values for 1989–2003. Sample units framed in bold indicate locations with significant trends (positive or negative) in species richness. Reprinted with permission from *Annales Zoologici Fennici.*

---

## Box 3.2 (Continued)

selected to ensure that the coverage and density of the surveys were spatially and temporally representative (Linden et al. 1996). Approximately 1,000 of the 1,600 triangles were surveyed twice a year by more than 7,000 volunteer naturalists (Pellikka et al. 2005).

Field navigation and repeatability of survey routes were facilitated through the use of GPS and by the flagging of survey routes. During the first survey of each winter, tracks were identified, counted, and then obliterated with a ski pole or conifer branch. Species detections were tallied by the total number of track crossings within 1-km increments. The number of days since the last snowfall were recorded and used to correct for track accumulation time. The second survey was conducted twenty-four hours after the first survey, and the number of new tracks encountered for each species was tallied again.

The practical nature of this survey design has led to its modification and implementation in North America. The Integrated Landscape Management group (ILM) at the University of Alberta performed a field test of the WTS between 2001 and 2004 (Bayne et al. 2005). During this testing period, a total of twenty-six mammal species were recorded. The ILM concluded that, although more expensive than camera surveys or hair snares, the WTS was better suited for detecting species during a single survey effort than were other noninvasive methods. Additionally, preliminary power analyses suggested that, for most species, the tracking protocol was sufficiently precise to detect a 3% annual exponential change in abundance by surveying twenty-five to fifty sampling sites. To improve efficiency and reduce costs, the ILM

has proposed conducting future research to test whether surveys conducted by snowmobile on randomly selected seismic lines or trails could achieve similar results. If the results from the two approaches (i.e., the original protocol versus trail-based surveys) are similar, then a shift towards trail-based monitoring might be desirable. The ILM effort resulted in recommendations for modifying the Finnish design for use by the Alberta Biodiversity Monitoring Program (Bayne et al. 2005):

1. Conduct only a single survey of each sampling site during a given winter, as the second survey provided limited new information on species occurrence but caused a notable increase in cost.
2. Shorten transect lengths to 9 km (3 km per side) due to reduced daylight hours during winter in Alberta and the long distances traveled to reach sampling sites.
3. Conduct surveys between three and ten days after a track-obliterating snowfall.

In areas where natural sign surveys are limited to nonmotorized transportation because of either limited access or study design requirements (e.g., adequate habitat representation), the triangle approach offers an efficient alternative to linear survey designs. Potential biases resulting from increased effort at locations near the corners of the triangle should be evaluated. Such biases should be relatively easy to correct for, however— for example, portions of the transect could be omitted to equalize linear effort in space.

---

positives (Royle and Link 2006) can be addressed during data analysis (chapter 11).

### DNA Analyses of Samples Collected Along Tracks

Following tracks to collect hair and scats can enable the reliable identification of species and individuals via DNA analysis (figure 3.6). Although tracking animals to collect physical evidence of the species in question is not a new concept, the earliest published account of this application appeared only recently— and described the verification of two rare lynx tracks

in Wyoming (Squires et al. 2004). The feasibility of applying similar methods on a broader scale has been evaluated for wolverines (Flagstad et al. 2004; Ulizio et al. 2006), lynx (McKelvey 2006), and mountain lions (Sawaya and Ruth 2006), with positive results.

Following tracks can be time-consuming, and the ability to incorporate this approach into snow track surveys will depend on time and funding availability. While potentially useful samples of hair and scat may be encountered relatively quickly, multiple hair or scat samples should be collected to ensure that at

*Figure 3.6.* Following tracks in the snow can yield hair and scat samples for DNA analysis, as well as a diversity of other ecological information. Photo by K. Heinemeyer.

least one sample contains useful DNA. Reported distances between useful samples include 1,200 m for lynx, (McKelvey et al. 2006), 1,330 m for wolverines (Ulizio et al. 2006; see case study 3.1), and < 2 km for mountain lions (Sawaya and Ruth 2006). In some cases, it is possible to focus on sign-rich locations such as day beds or foraging areas (McKelvey 2006)—or on natural features such as trees and other vegetation (Sawaya and Ruth 2006)—that may contain higher quality samples and help enhance search efficiency.

Incorporating the collection of genetic samples into track surveys relaxes some of the logistical constraints imposed by field-based track identification. For instance, surveys can be conducted under a broader range of conditions, within wider time-frames, or when tracks are somewhat degraded (figure 3.7; McKelvey 2006; Ulizio et al. 2006). Still, the tracker must possess the skill necessary to recognize the putative tracks of the target species and should collect samples from any track likely to have been left by that species. Additionally, field personnel must be cognizant of several factors that can lead to the collection of samples from nontarget species—or samples that are contaminated with DNA from nontarget species. Trails are often used by multiple species (or individuals) that may deposit scats along the path of the target species or urinate or defecate on top of existing scats. Thus, genetic samples should ideally be collected from uncontaminated sections of track. Lastly, DNA samples collected from beds or other high-use areas can be contaminated with prey species brought to the site for consumption, or by animals that visit the site before or after the target individual is present.

*Field-Based Species Identification from Tracks*

In those instances where genetic samples are not collected at each track occurrence, the ability to identify a species based on natural sign requires understanding sign characteristics and how they combine to leave a distinct, species-specific signature (figure 3.8; e.g., Halfpenny 1987; Forrest 1988; Rezendes 1992; Halfpenny et al. 1995; Elbroch 2003). The most reliable field-based species identification depends on highly experienced and careful trackers familiar with the regional species composition and trained in survey protocols (Zuercher et al. 2003). The identifica-

*Figure 3.7.* Following tracks to collect hair and scat samples for DNA analysis permits wider survey windows because species identification can be confirmed from even degraded tracks, such as these wolverine tracks. Photo by T. Ulizio.

tion of species by track characteristics is unreliable where sympatric species have similar or overlapping track measurements (figure 3.9 and table 3.2). Under such circumstances, genetic sampling should be incorporated into the survey design.

Collecting DNA samples or even detailed track measurements is generally not feasible during aerial surveys. Thus, species identification is based primarily on path characteristics, including track pattern (gait, spacing of the gait, and changes in gait), changes in track pattern through different habitats and snow conditions, body prints in deep snow, and the activity and behavior of the animal. Aerial sur-

veyors often have a birds-eye view of the gait and are able to follow a single track for several kilometers, thus permitting them to seek out the best available track section for making a species identification. Minimizing species misidentifications requires sufficiently open habitat to prevent missing tracks and to be able to follow a track for sufficient distance (Magoun et al. 2007). As with ground-based surveys, aerial surveys are not appropriate if there is high overlap in conspecific track characteristics (e.g., table 3.2). Additionally, if absolute species confirmation is needed from aerial surveys, it is necessary to follow tracks on the ground to collect genetic

*Table 3.2.* Overlap in snow track measurements for various pairings of sympatric North American carnivore species

| Species pairs | Footprint[a] (cm) Width | Footprint[a] (cm) Length | Gait (cm) | Stride[b] (cm) | Straddle (cm) | Comments |
|---|---|---|---|---|---|---|
| Fisher | 6.3–7.6[++] | 8.9–12.2[++] | 2–2 | 53.3–127[++] | 7.6–13.3[+] | Tracks of male martens and female fishers may overlap in size (Forrest 1988). |
| American marten | 4.6–6.3[+] | 7.6–10.2[++] | 2–2 | 38.1–83.8[+] | 7.1–10.8[+] | |
| Coyote | 4.1–6.1[+] | 6.1–7.9[+] | Trot | 88.9–132[+] | 6.3–14.0[+] | While footsize is similar, stride has relatively small overlap. |
| Red fox | 4.1–5.3[+] | 5.3–7.4[+] | Trot | 66.0–95.2[+] | 5.1–9.9[+] | |
| Mountain lion | 8.2–10.8[+] | 7.6–12.1[+] | Walk | 86.0–162.5[++] | 20.3–30.5[++] | |
| Canada lynx | 7.6–8.6[+] | 8.2–9.5[+] | Walk | 81.3–162.5[++] | 15.2–22.9[++] | "The [lynx] track pattern and print size are similar to mountain lion; however, the straddle is generally smaller, the foot pads are usually obscured by hair, and the tracks sink no deeper than 8 in." (Forrest 1988: 143). |
| Wolverine | 8.9–12.7[+++] | 10.2–15.2[+++] | Walk | 58.4–63.5[+++] | 15.2–22.9[+++] | "Erratic gaits are diagnostic" (Forrest 1988: 131). |
| Canada lynx | 7.6–8.6[+] | 8.2–9.5[+] | Walk | 81.3–162.5[++] | 15.9–22.9++ | Alternating walk is less common for wolverines than lynx, but gaits are especially similar in soft, deep snow where footpads do not register. |
| Gray wolf | 7.6–10.5[+] | 10.2–13.3[+] | Walk | 102.9–144.8[+] | 7.6–17.8[+] | "Even though their prints are similar in size to wolf, lynx, and cougars, the wolverines' erratic gaits are diagnostic" (Forrest 1988: 131). |
| Wolverine | 8.9–12.7[+++] | 10.2–15.2[+++] | Walk | 58.4–63.5[+++] | 15.2–22.9[+++] | |

[a]Measurements are based on the front foot, which is larger than the hind foot.
[b]All stride measurements are from the front of a given foot to the front of the same foot in the next track (i.e., front left foot to front left foot).
[+]Measurement from Rezendes 1992.
[++]Measurement from Forrest 1998.
[+++]Measurement from J. Squires, USDA Forest Service Rocky Mountain Research Station, pers. comm.

material, or from the air in an attempt to view the individual creating the tracks.

### Error-Checking Species Identification from Tracks

Collecting DNA samples along tracks increases the cost of surveys due to additional observer time and laboratory analyses, and postpones final survey results while samples are analyzed. For surveys of rela-

tively common species, the cost of following each track encountered may be deemed unreasonable and potentially unnecessary. In these instances, we recommend that DNA sampling be integrated into the survey design at a randomly selected subsample of tracks to quantify error rates for species identification. Error-checking increases the robustness of natural sign surveys that do not confirm every species

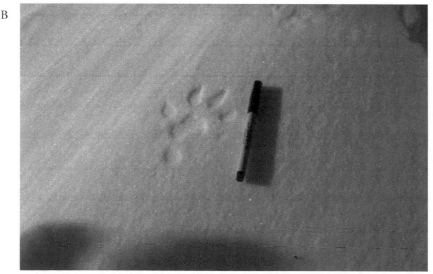

*Figure 3.8.* Species identification based on snow tracks is often limited to gait and behavioral characteristics, as detailed foot-prints are relatively uncommon. These photographs show (A) a classic linear wolverine two-point bound and (B) a rare, highly detailed wolverine print on hard snow. Photos by T. Ulizio.

*Figure 3.9.* Where sympatric species have similar track or gait characteristics, it can be extremely difficult or impossible to identify species based on tracks. These photos illustrate how slightly degraded (A) lynx and (B) wolverine tracks can appear similar. Tracks in soft snow or those that are older and degraded do not usually allow for distinguishing species based on track characteristics (e.g., the number of toes, which would typically distinguish a felid from a mustelid). Photos by T. Ulizio.

occurrence and allows for the incorporation of uncertainty into survey results.

The use of experienced trackers can result in highly accurate field-based species identification, particularly when sympatric species have relatively unique sign characteristics and multiple lines of evidence are used (e.g., tracks, behavior, and scats). In Paraguay, Zuercher et al. (2003) reported 100% agreement between the identification of carnivore species by DNA analysis and by local trackers using natural sign. In Alaska, Prugh and Ritland (2005) tested the ability of observers to correctly distinguish coyote scats from those of sympatric wolves, red foxes, lynx, and dogs (*Canis lupus familiaris*) using scat morphology, associated tracks, and site-specific location. Of those putative coyote scats from which DNA was successfully amplified, 92% were confirmed as coyote scats.

### Species Identification from Scat Surveys

Although scats *combined with* other natural sign can yield accurate species identifications, field-based species identification based solely on scat morphology and size has been found to be consistently unreliable (figure 3.10; Davison et al. 2002; Harrison 2006)—even for identification to the family or genera level (Chame 2003). In a controlled laboratory setting, Bulinski and McArthur (2000) found no difference between experienced and inexperienced field personnel in their ability to correctly identify scat using only scat morphology. Indeed, supplementary natural sign (such as tracks) is often unavailable during scat surveys, and identification based solely on scat morphology should always be confirmed by DNA analysis. Long et al. (2007a) presented a formula to calculate the probability that a scat detection team correctly assigned at least one scat on a given transect to the target species. The formula was based on the number of scats collected, the confidence level of the species identification, and the correct classification rate for scats based on DNA analyses of a subset of all scats collected.

DNA amplification rates for scat may vary by season and species (Piggott 2004; also see chapter 9). Re-

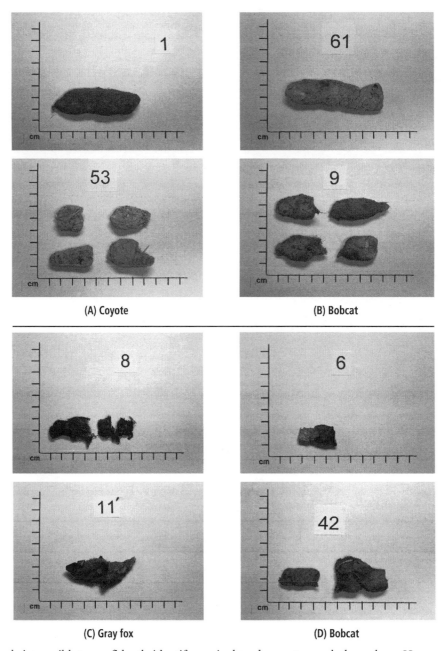

*Figure 3.10.* It is nearly impossible to confidently identify species based on scat morphology alone. Here we show how similar (A) coyote scats can appear to (B) bobcat scats, and the almost identical morphology of (C) gray fox and (D) bobcat scats. Further, these images illustrate the diversity of morphology within a single species. Photos by R. Harrison.

ported success rates for genotyping species identification from scat include 87% for bobcats (Ruell and Crooks 2007), 74% for Eurasian badgers (*Meles meles*; Frantz et al. 2003), 66% for swift fox (Harrison et al. 2004), 63% for mountain lions (Ernest et al. 2002), and 48% for coyotes (Kohn et al. 1999). Ruell

and Crooks (2007) report a substantially higher genotyping success rate for bobcat scat versus hair collected via hair snaring (10%). These researchers suggest that fine felid hair may contain little DNA (see chapter 6).

Scat size may be useful for discriminating between

species with significantly different body sizes (Green and Flinders 1981; Danner and Dodd 1982; Reed et al. 2004; Gompper et al. 2006). For example, among adult animals, fresh bear scats are obviously larger than coyote scats, but fresh coyote, fox, badger, wolverine, bobcat, and raccoon scats can all be very similar in size. In the Adirondacks of New York, a 22-mm size cut-off correctly identified 89% of carnivore scat as coyote (as later confirmed by DNA analysis)—but these results were only possible due to the lack of sympatric large canids and the rarity of bobcats (Gompper et al. 2006). Further, juveniles of larger species may produce scats similar in size and appearance to those of adult animals of smaller species, and characteristics of old or degraded scats may differ from those of fresh samples (see chapter 7 for more scat-related discussion).

Thin layer chromatography of bile acids from scats has also been used to distinguish species (Major et al. 1980; Johnson and Belden 1984; Fernández et al. 1997; Ray and Sunquist 2001). For example, Fernández et al. (1997) reported up to 80% success in distinguishing jaguar and mountain lion scats via bile acid chromatography, and Johnson and Belden (1984) reported 64% and 79% success in identifying known bobcat and mountain lion scats, respectively. Although examination of bile acids is inexpensive (Major et al. 1980) and there are continued research efforts to improve techniques (Cazon-Naravaez and Suhring 1999), this method has been largely supplanted by DNA analysis.

## Mode of Transportation

Track surveys are typically performed via foot, snowshoes, skis, snowmobile, all-terrain vehicle, or aircraft. This flexibility allows them to be tailored to seasonal conditions, available resources, budget, time constraints, and access limitations (e.g., topography, road and trail availability, administrative closures). Tracks in snow offer the best prospects for detection and track following, as they are often continuous and highly visible. Scat surveys, in contrast, are challenging to conduct because scats frequently occur at a single point and require substantial search effort (but see chapter 7). The need to move slowly, especially in areas with poor sightability, limits the modes of transportation that can be used for scat and some track surveys (e.g., those in mud and dust). When appropriate, the use of motorized transportation (e.g., snowmobiles for snow track or winter scat surveys) coupled with adequate access can allow large regions to be surveyed in a relatively short time. In most cases, however, foot travel may also be required to assure adequate coverage of sample units and all available habitat types.

## Safety

Each natural sign survey has unique safety considerations. Winter surveys in remote landscapes (whether conducted from the ground or the air) require training in winter survival and safety, avalanche safety, and backcountry first aid, and each person should carry appropriate personal winter survival gear. First aid, emergency preparedness, and basic survival training are necessary for all personnel carrying out surveys in remote regions during any season. Surveyors using all-terrain vehicles or snowmobiles should be able to execute field repairs and operate machinery in difficult conditions and terrain. Aerial surveys entail their own set of safety concerns and regulations that vary across regions, seasons, and responsible agencies or project partners. Before conducting aerial surveys, researchers should contact the air service to obtain safety and winter preparedness recommendations (and weight limits); ensure that pilots are certified and experienced in conducting winter wildlife surveys and are familiar with the project area; and review all rules, safety issues, and emergency protocols with the pilot.

Field researchers should avoid touching scats directly, as viable eggs of internal parasites may be present—including those of the tapeworm *Echinococcus multilocularis*, which can produce a serious infestation in humans (Reynolds and Aebischer 1991). Gloves or proper instruments such as forceps should be used to handle scats. Also, when collecting scats from dens, researchers should be aware that fleas may be present and can transmit diseases to people (Harrison et al. 2003).

## Materials and Costs

The cost of conducting natural sign surveys is primarily determined by the mode of transportation, the size of the area to be surveyed, the number of survey crew, and the extent to which genetic analyses are necessary. Because natural sign surveys, and particularly snow track surveys, are weather dependent, survey delays and the need to have personnel on call and available when survey conditions are appropriate should be incorporated into the budget. Surveys that employ existing staff and equipment are considerably less expensive than those requiring additional personnel and purchases. State and federal agencies often have skilled biologists on staff who may be able to assist with surveys, as well as vehicles that can be used in collaborative partnerships. A list of major budget items that should be included in a ground-based snow track survey is provided in table 3.3 (also see Halfpenny et al. 1995).

*Table 3.3.* Basic costs associated with a hypothetical, ground-based snow track survey

| Item | Unit | Cost (USD 2007) |
|---|---|---|
| Travel | | |
| 4-wheel drive trucks @ $1,200/mo | day | 40 |
| Snowmobile rentals | day | 185 |
| Snowmobile fuel and oil | day | 28 |
| Personnel | | |
| Technicians–8 hours/day | day | 100 |
| Equipment | | |
| Snowmobile purchase | each | 6,000 |
| Snowmobile trailer purchase | each | 800 |
| Snowshoes | pair | 250 |
| Skis, boots, poles | pair | 300 |
| Avalanche transceivers, probes, shovels | per person | 400 |
| GPS units | each | 100–3,000 |
| Field computer and printer | each | 700 |
| Cameras | each | 200 |

## Survey Design Issues

Natural sign surveys assume that if a species of interest is present, there is some probability of detecting it via its sign. This probability must be sufficiently high to provide confidence in declaring a species absent, or repeat surveys must be conducted to allow the estimation of detectability (MacKenzie et al. 2002; also see chapter 2). Species detection during natural sign surveys is influenced by the spatial design of the survey, the level of survey effort, and the species' density and natural history characteristics.

### Spatial Design

Considerations relevant to the spatial design of natural sign surveys include characteristics of the geographic area (e.g., topography, vegetative cover), access, and the spatial requirements of the target species. Most ground-based surveys for tracks and scats have been conducted along roads or trails due to convenient access, ease of travel, safety considerations, and sightability of sign (Wassmer 1982; Kohn et al. 1999; Harrison et al. 2004). In some regions, it might *only* be feasible to conduct natural sign surveys along transportation routes, as steep terrain and dense vegetation may otherwise limit access and obscure natural sign.

Despite the logistical and safety benefits of using existing trails and roads, there are several concerns regarding their exclusive use for surveys. Foremost is that surveying solely along such features precludes the use of statistically representative sampling designs that help ensure unbiased inferences. For example, in regions with steep topography, surveying only roads and trails—which tend to follow valley bottoms and ridgelines—can yield a pronounced habitat bias. This situation may result in undersampling individuals that avoid roads, and in too much effort being expended in areas of low-value habitat (e.g., logging roads through timber harvest zones; Brody and Pelton 1989; Mladenoff et al. 1995; Lovallo and Anderson 1996). Reciprocally, detectability may increase for species with a propensity for walking along roads and trails. In some cases, representative surveys may be achieved by combining survey routes along roads and trails with randomly selected survey transects located off roads or trails. Closed or little-used routes are recommended, both to reduce potential biases resulting from animal avoidance of active roads and to minimize the probability that tracks or scats will be obscured or destroyed by people using these same routes.

Linear transects placed randomly or systematically across sample units help ensure statistically robust surveys that provide representative sampling of habitat types and areas, and limit the potential biases that can result from surveying roads and trails. Linear transects are relatively easily accommodated in aerial surveys, though visibility limitations in densely forested areas may prescribe alternative sampling patterns. Employing linear transects for ground-based surveys can impose substantial challenges regarding time, access, efficiency, safety, and sightability. The most efficient approach for off-trail, nonmotorized surveys may be the triangle survey design (see box 3.2; Helle et al. 1996; Helle and Nikula 1996; Hogmander and Penttinen 1996; Linden et al. 1996; Kurki et al. 1998). This design consists of three transect lines connected to form an equilateral triangle, eliminating the need for backtracking and therefore increasing efficiency and spatial coverage (box 3.2). Additional considerations for the design of transect surveys are discussed in chapters 2 and 7.

## Selective Sampling and Habitat Stratification

In some cases, it may be appropriate to concentrate surveys on habitats that are preferred by the target species or where sign is most likely to be deposited (Smallwood and Fitzhugh 1995; Sovada and Roy 1996; Heinemeyer and Copeland 1999; Roy 1999; Heinemeyer et al. 2000). Surveys to assess species presence are well served by such selective sampling because a single detection is all that is required (chapter 2). Fishers in the western United States, for example, appear to be tied to closed canopy forest

stands in winter, and to avoid nonforested areas such as recent clearcuts, forest openings, grasslands, and areas above treeline. Concentrating survey effort on the preferred habitat type would maximize the probability of detecting a fisher in the shortest period of time. Similarly, many carnivore species deposit at least some of their scats nonrandomly (Gorman and Trowbridge 1989), and commonly along roads, trails, ridges, streambeds, river banks, lakeshores, and other travel routes—as well as at dens, burrows, saddles on ridges, resting sites, and foraging areas (Elbroch 2003).

Logistical considerations may call for sampling designs that focus on a limited number of habitat types (another form of selective sampling). Mud-based track surveys, for instance, require conditions that are most likely to be found along the banks of waterways, and aerial surveys must target those areas that provide adequate sightability to detect tracks (Patterson et al. 2004; Magoun et al. 2007). Designing surveys based on the ability to detect natural sign severely limits the applicability and value of the survey effort, as this may yield severe biases in the representation of potentially used habitats within the sample unit. The limitations of such surveys should be clearly documented.

Stratification of sampling effort (see chapter 2) may be effective for improving the precision of survey results. For instance, stratifying sampling effort by habitat quality (which may result in surveying intermediate quality habitats) can increase the power of occupancy analyses (Rhodes et al. 2006). As with selective sampling, the assumptions and techniques applied to habitat-stratified survey designs should be well documented.

## Survey Effort

The amount of effort required to ensure an acceptable probability of detecting a target species with natural sign surveys is generally unknown and partially depends upon species density and movement patterns, field conditions and their effect on documenting quality sign, and the objective of the survey.

Within a given sample unit size, transect length translates into survey effort. While the effort required should be based on survey objectives and focal species ecology, few guidelines exist, and a review of published surveys shows an inverse relationship between reported or recommended effort and sample unit area (see table 3.4). Halfpenny et al. (1995) recommended a 10-km snow track transect within each 10.4-km$^2$ survey unit for forest carnivores, while Squires et al. (2004) employed an 8-km transect within each 64 km$^2$ unit for snow tracking lynx.

The British Columbia Ministry of Environment, Lands and Parks (1999) recommends that snow track transect length should be at least equal to the square root of the sample unit size, with the sample unit size defined based on the species' home range size. For example, 10-km transects are recommended for wolverine surveys using a 100 km$^2$ sample unit, and 5-km transects for fisher surveys with a 25 km$^2$ sample unit. If the target species is believed to be rare, it is further suggested that the square of this estimated transect length represents a more appropriate level of effort. Although it is unclear if this specific recommendation has been empirically tested, it is generally advisable to increase effort if detection probabilities are believed to be low.

A survey with the objective of documenting presence can be considered complete as soon as one high-quality track or scat from the target species is recorded and confirmed. For scat surveys, sufficient samples should be collected to allow for some potential failure in DNA testing and for the inadvertent collection of scats from nontarget species. Harrison et al. (2002) found that collecting ten putative swift fox scats per transect provided a 98% probability of confirming at least one swift fox detection. If the species has not been detected after a complete census of survey units, surveys can be repeated at predefined intervals to increase the probability of detection (see chapter 2).

*Table 3.4.* Comparison of survey effort for select surveys of various carnivore species or species groups in North America

| Target species | Survey objective | Survey type | Sample unit size (km$^2$) | Transect length (km) | Effort/area (km/km$^2$) | Reference |
|---|---|---|---|---|---|---|
| Midsized forest carnivores | Occurrence/distribution | Ground snow track | 10.4 | 8 | 0.91 | Zielinski and Kucera 1995 |
| Swift fox | Occurrence | Dust and mud track | 105[a] | 36.8[b] | 0.35 | Roy 1999 |
| Swift fox | Occurrence/distribution | Scat | 808 | 16 | 0.020 | Harrison et al. 2004 |
| Mountain lion | Relative abundance | Dust track | 2,500 | 33.6 | 0.013 | Smallwood and Fitzhugh 1995 |
| Wolverine and Canada lynx | Occurrence/distribution | Ground snow track | 64 | 8 | 0.125 | Squires et al. 2004 |
| Fisher[c] | Occurrence or relative abundance | Ground snow track | 25 | 5 | 0.20 | British Columbia Ministry of Environment Lands and Parks 1999 |
| Wolverine | Occurrence/distribution | Aerial snow track | 1,000 | 34 | 0.034 | Magoun et al. 2007 |
| Wolverine | Occurrence/distribution | Aerial snow track | 1,000 | 32 | 0.032 | Gardner 2005 |

[a] Estimated average size of township (sample units are townships).

[b] Effort based on time: minimum of thirty minutes and maximum of two hours searching for tracks in each township. Value is average resulting length of road surveyed.

[c] Suggested sample unit size and transect lengths are also provided for coyote, red fox, lynx, bobcat, and wolverine.

## Sign Encounter Rates and Reporting Units

Natural sign survey results are frequently reported as tracks or scats found per kilometer of transect, or, less frequently, by hours of search time. To be reliable for monitoring relative population abundance over time, surveys must account for variability that affects sign encounter rates, including accumulation time (i.e., time since last track-clearing event; Cavallini 1994; Kauhala and Helle 2000), distance surveyed (Raine 1983; Theberge and Wedeles 1989), and, potentially, variation in the tracking medium or scat or track visibility (Van Dyke et al. 1986; Beauvais and Buskirk 1999). For example, the kilometer-day calculation (Raine 1983; Theberge and Wedeles 1989) takes the form of (number of detections)/ (kilometers surveyed × days accumulation).

Track survey designs should minimize the repeat counting of individuals (unless the objective is to estimate abundance via capture-recapture analysis). This can be achieved by setting a minimum distance threshold between detections based on the home range size or daily movements of the given species (Hoving et al. 2004), or by delineating appropriately sized sample units and documenting only whether or not the target species is detected in each unit (see chapter 2 for a discussion of sample unit characteristics).

Additional considerations for scat surveys include decomposition rates, which can vary with climate and season (Wallmo et al. 1962; Flinders and Crawford 1977; Cavallini 1994; Bonesi and Macdonald 2004), and variations in defecation rates and decomposition rates caused by diet (Andelt and Andelt 1984; Godbois et al. 2005), individual behavior (Thacker et al. 1995), species-specific factors (Kruuk and Conroy 1987), and habitat (Prugh and Krebs 2004). Additionally, scats may be removed from trails by rodents or other animals (Sanchez et al. 2004; Livingston et al. 2005), or destroyed by vehicle or pedestrian traffic (Kohn et al. 1999). Variability in scat encounter rates may be reduced by conducting surveys during the same season each year and when vegetation height is low, clearing scats from survey routes prior to the designated period of accumulation, and avoiding scat accumulation during periods likely to experience heavy rainfall.

## Sign Accumulation and Survey Timing

The likelihood of detecting sign when a species is present is correlated with the amount of time animals have had to move about the landscape and create sign since the last clearing event (e.g., manual clearing of scats, new snowfall for snow tracking)—as well as with the population density of the target species. Sign accumulation time must be an integral part of survey design. Typically, track surveys should be delayed at least twenty-four to forty-eight hours after a track clearing event (Sovada and Roy 1996; Beauvais and Buskirk 1999; Roy 1999; Kauhala and Helle 2000; Ulizio 2005), and up to three to four days (Bayne et al. 2005). Longer waiting periods increase sign accumulation but also increase the risk of another weather event stalling completion of the survey effort. Additionally, tracks and other sign are ephemeral and at some point degrade or become obscured.

Snow track survey conditions will typically degrade five to seven days after a fresh snow (Kauhala and Helle 2000; Ulizio 2005), or sooner, depending upon wind, freeze-thaw cycles, and snow crust conditions. We recommend that surveys generally be initiated thirty-six to forty-eight hours following a track-obliterating snow (e.g., >2 cm), and completed within seven days—conditions permitting. Bayne et al. (2005) suggest that conducting surveys three to four days after a track-obliterating snow event provides a balance between efficiency (e.g., technician "down time") and allowing tracks to accumulate to increase species detection probabilities. It is important to note, however, that weather conditions and animal traffic differ between geographic areas, so optimal timing of surveys will also differ. Further discussion of snow track conditions and how they affect the timing of snow track surveys is presented by Halfpenny et al. (1995).

Dust-track quality and detection rates may decline three to seven days after tracks are made (Schmitt 2000; M. Grigione, University of South

Florida, pers. comm.), and survey timing should be based on the local tracking conditions. In some regions, suitable dust track conditions may be limited to morning hours if night dew creates the best tracking, thus resulting in a track accumulation time of only a few hours.

For scat surveys, we recommend that survey routes be cleared of scats and allowed sufficient time to accumulate new scats prior to conducting surveys. This ensures that only individuals present during the survey are detected and sampled. To our knowledge, no standards for scat accumulation time have been established. While scats are more robust to weather conditions than tracks, their visibility may vary seasonally. Snow can cover scats, but some species will defecate in snowmobile and other trails (Prugh et al. 2005). High temperatures and strong ultraviolet light in open areas may degrade DNA in scats, as can mold due to excess moisture (chapter 9). In areas lacking snow cover, late winter and spring may be the best time to collect scats, as they will not have been exposed to high temperatures and will not be hidden by seasonal vegetation.

## Sample and Data Collection and Management

Explicit data collection protocols should be formalized and documented. A priori decisions must be made regarding the level of evidence needed for species identification, and appropriate data forms created to record evidence, confidence levels, and field decisions (see appendix 3.1 for a sample tracking data form). Surveys relying upon field-based species identification, for example, must include a rating system for survey and sign quality and conditions (see Halfpenny et al. 1995). It is also critical to incorporate protocols for documenting natural sign through track and gait measurements, photographs (including a scale object), and possibly scaled drawings or casts. For scat surveys, data should include the survey date and time of scat location, sample unit identifier, transect identifier, global positioning

system (GPS) location, putative species, and other pertinent information (e.g., associated sign, habitat).

Detailed methods for the collection and storage of scat and hair intended for subsequent genetic or endocrine analyses are described in chapter 9. We recommend that each survey use a GPS to record not only the location of sign occurrences, but also the actual survey path. This is particularly important if surveys are conducted off roads and trails and cannot otherwise be easily and accurately reconstructed to assess effort and habitats surveyed.

## Future Directions and Concluding Thoughts

Natural sign surveys represent the original noninvasive survey technique, and—with ongoing efforts to increase their rigor and robustness—have a bright future for effectively detecting and monitoring carnivores. Of critical importance will be the continued advancement of survey designs and quality control protocols that ensure a high level of certainty regarding species identity, and provide data that allows for quantitative population monitoring and measurement.

### Track Surveys

The ability to extract DNA from hair and scat collected during track surveys provides an important opportunity to confirm species identification (especially for rare species), or to quantify identification error rates by subsampling track detections when the logistics of genetically verifying all tracks cannot be justified. Sampling protocols utilizing natural sign occurrences and integrating DNA testing may allow population estimation based on capture-recapture analyses (e.g., Flagstad et al. 2004). While such protocols require intensive effort, they may provide new noninvasive alternatives to traditional mark-recapture methods for measuring and monitoring some carnivore populations.

New advancements in analyses based on occupancy modeling and other repeat survey approaches (e.g., Sargeant et al. 2005) provide exciting opportunities to more fully utilize the strengths of natural sign survey data. An approach similar to that of Royle (2004a), who used an extension of site-occupancy modeling with a multistate occupancy distribution to maximize data from frog call surveys, may be useful for deriving other natural sign indices—assuming the abundance of detected sign is believed to be related to underlying animal abundance. Such an approach might be able to incorporate key variables that affect detection rates, including variation in abundance (Royle 2004a). These and other advances may provide the opportunity to revisit and reanalyze existing natural sign survey databases, some of which span many years and provide rare, long-term information about carnivore populations. This opportunity will depend largely on data collected from surveys that are well designed and that maintain acceptable levels of quality control in terms of species identification.

The use of aerial surveys will likely grow, especially given the increasing need to monitor rare or wide-ranging carnivores across large study areas and in relatively remote or inaccessible regions. Digital video cameras, which can be used to record tracks during surveys, hold promise for increasing survey efficiency and allowing analyses of tracks for species identification and confirmation (H. Golden, pers. comm.). As a result, it may become increasingly feasible to conduct effective aerial surveys for smaller carnivores (e.g., martens), and to account for sightability factors such as forest canopy.

Another exciting advancement in track surveys is widely applied in Russia, but not yet in North America. This approach combines track surveys with information about the expected movement distances of individuals to estimate the likelihood of encountering an individual along a random transect (Stephens et al. 2006). Data pertaining to individual movement patterns and distances, acquired from snow tracking (e.g., Heinemeyer 2002; Fuller 2006) or the monitoring of GPS collars, may be increasingly available for some species. The development of movement models based on modified random walk approaches that incorporate changes in movement patterns across different habitat types (e.g., Heinemeyer 2002; Fuller 2006) could further advance such efforts and permit the translation of track encounter rates into estimates of abundance or density.

Mud and dust surveys currently suffer from the limited spatial distribution of appropriate tracking substrates and the inability to follow individuals for sufficient distances to obtain confident species identification via natural sign characteristics or the collection of genetic material. Track surveys in mud and dust have been used successfully outside of North America and can be valuable if appropriate quality control measures are incorporated. The potential for high-detailed footpad imprints in dust or mud may allow for the development of quantitative analyses (e.g., discriminant function analysis) of track characteristics.

## Scat Surveys

The ability to identify individuals from scat has led to important advances in using scat surveys to estimate and monitor population sizes. At present, individuals can most reliably be identified from scats that are fresh, quickly desiccated, or deposited in winter and frozen immediately. Nonetheless, successful DNA-based identification will vary with species, habitat, and environmental variables (M. Schwartz, USDA Forest Service, pers. comm.). As laboratory techniques evolve, it should become possible to identify individuals from older and more degraded scats. Further, laboratory techniques for assessing physiological conditions such as stress (chapter 9) and disease will continue to improve, enabling a much broader understanding of issues important to conservation and management.

The ability to determine the age of scats via laboratory testing would be very valuable indeed. As previously mentioned, in those cases where transects or search areas cannot be cleared of scats prior to surveying, age determination must be made via indirect and imprecise measures. A simple, accurate means of aging scat, linked with individual identification through DNA analyses, would provide information about temporal patterns in habitat use or the duration of residency for individuals. This would also allow investigators to eliminate scats from individuals that are likely no longer present in the study area, thus increasing the performance of abundance estimators.

---

### CASE STUDY 3.1:
### SNOW TRACK SURVEYS FOR WOLVERINES AND CANADA LYNX IN SOUTHWEST MONTANA

**Source:** Squires et al. 2004; Ulizio 2005; Ulizio et al. 2006; T. Ulizio, unpublished data.

**Location:** Four mountain ranges in southwest Montana (figure 3.11).

**Target species:** Primary target species was wolverine; secondary target species was Canada lynx.

**Size of survey area:** ~5,000 km$^2$.

**Purpose of survey:** The primary objectives were to modify a Canada lynx survey protocol published by Squires et al. (2004), and to test this modified protocol for detecting wolverines while also evaluating the feasibility of collecting genetic samples along putative wolverine tracks. Secondary objectives were to collect snow track data for a larger research project on wolverine movement, and to monitor whether the status of lynx, which had been deemed absent in previous surveys, had changed over time.

**Survey units:** A grid of seventy-six units was overlaid on potential wolverine and lynx habitat, which consisted of all forested areas or those above treeline (figure 3.11). Grasslands, shrublands, agricultural lands, and wilderness areas were excluded. Each survey unit was 8 km × 8 km (64 km$^2$).

**Survey method:** Ground-based snow track surveys.

**Survey design and protocol:** A 10-km transect was selected in each survey unit, and existing roads and trails through the highest quality habitat were traversed with snowmobiles. Skis were used to survey units lacking adequate road and trail densities, and to access quality habitats inaccessible by roads and trails. The first unit to be surveyed each day was

randomly selected. If time allowed for more than one unit to be surveyed in a given day, units immediately adjacent to the initial unit were also sampled. Units were selected without replacement until a census of units was completed. This census was then repeated two times each winter for two years. The survey framework was designed for lynx by Squires et al. (2004) and slightly modified (i.e., transect length was increased from 8 km to 10 km per unit) for wolverines (Ulizio 2005; Ulizio et al. 2006).

All tracks encountered during snowmobile-based surveys were identified by experienced trackers using physical track characteristics (i.e., footprint and gait pattern). Project staff included two members with more than three years of experience identifying wolverine snow tracks. To ensure accuracy and provide training feedback, all tracks detected by less experienced personnel were visually verified on-site by one of the two experienced trackers. The majority of verified tracks were then followed by a crew member on snowshoes to collect hair or scat samples for DNA-based species identification.

**Analysis and statistical methods:** Results from genetic analyses provided data on species, sex, and genotype, and allowed the accuracy of trackers as well as tracks reported by the public or other agency staff to be quantified.

**Results and conclusions:**

- Detected sixty-four putative wolverine tracks in three of the four mountain ranges, and no lynx tracks during 1,550 km of surveys.
- Completed fifty-four back-tracks of approximately 1.4 km each and collected 169 hairs and 58 scats.
- Amplification rates of mitochondrial DNA (mtDNA) used for species identification were 62% and 24% for scats and hairs, respectively.
- 64% of tracks were verified as wolverine by genetic analysis. The remainder of tracks either yielded no samples or samples with inadequate DNA for species identification.
- Genetic analysis for all tracks with amplifiable samples verified the initial identification made

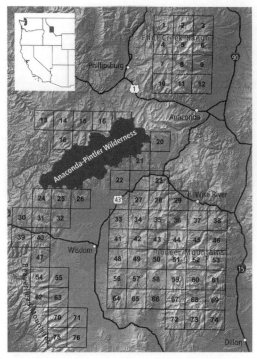

*Figure 3.11.* Study area and sample units for wolverine and lynx surveys conducted in southwest Montana.

by experienced trackers as being correct; some identifications made by the public and staff from other agencies were erroneous.

- An average of 86% of survey units outside of the wilderness were surveyed during both census attempts. Remaining units were not surveyed due to inadequate snowpack, topography, and hazards.
- Wilderness was not surveyed, and would have posed unique issues in that it featured steep topography and excluded the use of motorized vehicles.
- Three visits to each survey unit in a given year were recommended to increase probability of detection.

---

### CASE STUDY 3.2:
### USING SCAT SURVEYS TO ASSESS SWIFT FOX MONITORING PROTOCOLS

**Source:** Harrison et al. 2004.

**Location:** Easternmost thirteen counties of New Mexico (figure 3.12).

**Target species:** Swift fox.

**Size of survey area:** 80,000 km².

**Purpose of survey:** The survey objectives were to determine if an adequate number of swift fox scats could be located to monitor changes in occurrence over time, to establish the percentage of putative swift fox scats confirmed to be from swift foxes, and to evaluate three potential indices of swift fox population size.

**Survey units:** Ninety-nine transects selected to cover the entire known range of swift foxes in New Mexico (figure 3.12).

**Survey method:** Scat surveys.

**Survey design and protocol:** The survey employed 16-km transects along paved and unpaved roads. Searches for scats were conducted on foot at specific locations along roads, including fence corners, cattle guards, culverts, intersections of trails

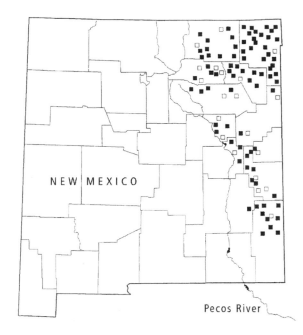

■ Transects with scat

□ Transects with no scat

*Figure 3.12.* Study area and survey locations for a large-scale scat survey conducted for swift foxes in eastern New Mexico.

and roads, and other conspicuous locations. Transects were surveyed in late winter and early spring to reduce damage to scats from summer heat. Surveying from south to north in the state reduced the possibility that snow would cover scats during the surveys. Each transect was searched until a total of seven scats were collected, or a maximum of twelve specific locations along the transect were searched.

**Analysis and statistical methods:** Species depositing scats were identified by mitochondrial DNA analysis conducted at a commercial laboratory.

**Results and conclusions:**

- Scat collection was determined to be an adequate method for monitoring swift fox occurrence.
- Confirmed swift fox scats were found throughout the entire study area.
- DNA analysis confirmed the identity of scats on 59.6% of transects.
- Statistical power was sufficient to detect a 20% decline in the proportion of transects with swift fox scats.
- The proportions of transects with scats and the total number of scats collected had sufficient statistical power to be useful as indices of swift fox population size, whereas the average number of scats found per site was not considered useful.
- A maximum sample size of seven scats per transect was determined to be adequate for verifying the presence of swift foxes.

- Surveys required seventeen field days to complete—far less time than a scent station survey would have required to cover the same area.
- Fewer scats per site were collected in southern New Mexico than in northern New Mexico.

## Acknowledgments

We would like to thank the various researchers who took the time to share their experiences with and insights into natural sign surveys, including Earl Becker, Jeff Copeland, Stan Cunningham, Howard Golden, Melissa Grigione, Kate Kendall, Rebecca Lewison, Marcelo Mazzolli, Kevin McKelvey, Carolyn Miller, Emily Reull, and Shawn Smallwood. Helpful reviews were provided by the editors, as well as by Audrey Magoun and Keith Aubry, and we appreciate their time and effort. Finally, we thank the editors of this book for providing the opportunity to participate in an important effort. In particular, we appreciate Paula MacKay for her tireless coordination and support and Justina Ray for her insights and focused assistance during the final revisions of this chapter.

# Appendix 3.1

*Survey data collection should be standardized to ensure that accurate and complete information is collected. This sample snow track data form includes information on the quality of tracks and the confidence of the technician in making the identification, as well as the track location, snow conditions, and relevant habitat information. Additional data, such as photograph number, would be included in the comments section.*

## Ground-Based Snow Track Survey Sample Data Sheet

Sample Unit ID: _____ Date/Time: _____ Surveyors: _____ Sheet # _____of_____

Last snow ended (Date/Time): _____ New snow depth: _____ Weather since snow: _____

General tracking conditions/comments: _____

| Start Time: _____ Start Odometer: _____ (km or mi?) UTM X: _____ UTM Y: _____ |
| End Time: _____ End Odometer: _____ UTM X: UTM Y: |

Track Detection Log Sheet

| Time | Odometer | UTM X | UTM Y | Track ID | Track Quality[1] | ID Confidence[2] | Habitat Description[3] | Comments |
|------|----------|-------|-------|----------|------------------|------------------|-----------------------|----------|
|      |          |       |       |          |                  |                  |                       |          |
|      |          |       |       |          |                  |                  |                       |          |
|      |          |       |       |          |                  |                  |                       |          |
|      |          |       |       |          |                  |                  |                       |          |
|      |          |       |       |          |                  |                  |                       |          |
|      |          |       |       |          |                  |                  |                       |          |
|      |          |       |       |          |                  |                  |                       |          |
|      |          |       |       |          |                  |                  |                       |          |
|      |          |       |       |          |                  |                  |                       |          |
|      |          |       |       |          |                  |                  |                       |          |
|      |          |       |       |          |                  |                  |                       |          |
|      |          |       |       |          |                  |                  |                       |          |
|      |          |       |       |          |                  |                  |                       |          |
|      |          |       |       |          |                  |                  |                       |          |
|      |          |       |       |          |                  |                  |                       |          |
|      |          |       |       |          |                  |                  |                       |          |
|      |          |       |       |          |                  |                  |                       |          |
|      |          |       |       |          |                  |                  |                       |          |
|      |          |       |       |          |                  |                  |                       |          |
|      |          |       |       |          |                  |                  |                       |          |

[1] Track Quality: 1=poor, 2=fair, 3=good, 4=high, 5=excellent; see associated track survey guidelines.
[2] Track Identification Confidence Level: 1=poor, 2=fair, 3=good, 4=high, 5=excellent; see associated track survey guidelines.
[3] Includes major overstory type, structural stage, and distance to habitat type edge.

# Chapter 4

# Track Stations

*Justina C. Ray and William J. Zielinski*

Since well before the development of formalized survey techniques, people have studied the movement and distribution of wildlife via their footprints. Traditionally, this skill was used to identify and follow animal tracks for great distances—an art form that has given rise to the standardized methodologies described in chapter 3. With the advent of field research, reliance on the chance occurrence of wildlife tracks, which require adequate natural substrate to be followed reliably, evolved into the active manipulation of the environment to enhance prospects for finding such sign. Track stations, the resulting class of noninvasive survey techniques, entail the point-sampling of carnivore tracks to record the presence or activity of a species at a particular location. Importantly, the track station methods described in this chapter are distinguishable from the survey methods discussed in chapter 3 in that the former focus on the creation of track-receptive surfaces at point locations, whereas the latter are aimed at detecting carnivores whose tracks and trails occur on unmanipulated surfaces (e.g., river banks, snow, roads).

Several related techniques fall under the umbrella of track stations, ranging from minimally prepared setups that exploit natural substrates upon which passing animals lay tracks, to manufactured track surfaces placed in the open or within boxes or cubby holes such that animals lured to these devices deposit tracks on prepared media. We have grouped the various track station methods into four general types:

1. A *track plot* (figure 4.1A) is a prepared natural substrate located along an existing travel route or road. The station consists of a smoothed plot (usually circular or square in shape) created out of available soil or some other natural medium. No attractants are employed; rather, detection relies on the passive interception of carnivore movement (e.g., Allen et al. 1996). Tracks are *negatives* in that they comprise the impressions left in the natural substrate.

2. A *scent station* (figure 4.1B) is a prepared natural substrate, usually raked or smoothed into a surface or plot, which includes an attractant to actively draw in carnivores (e.g., Linhart and Knowlton 1975). As with track plots, track impressions are negatives.

3. An *unenclosed track plate* (figure 4.1C) is a thin plate (typically aluminum or wooden) surfaced with an artificial medium (e.g, soot, chalk) and usually deployed with bait or other attractants (e.g., Winter et al. 2000). Unenclosed track

*Figure 4.1.* Various types of track stations, including (A) a track plot on a sand dune in Florida (photo by J. Engemann); (B) a scent station with sifted earth as a substrate (photo by J. Erb); (C) an unenclosed track plate (photo by C. Ogan); and (D) an enclosed track plate, with corrugated plastic as the enclosure material (photo by R. Truex).

plates are unprotected from precipitation or falling debris. Track impressions are typically negatives in that they are created when the foot of an animal removes the medium and thus reveals the foot pattern on the treated surface (but see Pedlar et al. 1997 and Wolf et al. 2003).

4. An *enclosed track plate* (figure 4.1D; figure 6.8A) consists of a wooden or aluminum track plate surfaced with an unnatural tracking medium (usually soot or chalk) and enclosed within a box or tube to protect the track surface from precipitation and falling debris. The devices are baited to entice visitation by carnivores. Unlike the three other types of track stations, enclosed track plates can collect both negative and positive foot impressions because they are designed such that an animal first walks across the black or white tracking medium (thus leaving negative impressions) and then across a white or brown track-receptive surface (thus leaving positive impressions). Track plates sized for mesocarnivores (e.g., Zielinski 1995) and tracking tunnels for small mammals (King and Edgar 1977) are the most common forms of enclosed track plates.

Track plots and scent stations are primarily distinguished from track plates by their general lack of manufactured materials (e.g., aluminum, plastic) other than bait, scent lure, and perhaps some sifted soil or other natural material. All track station methods rely on a behavioral response from target carnivores such that they pass over a track-receptive surface during the survey period, either naturally or via the use of attractants (Harrison et al. 2002). Track plates in particular have diversified tremendously with the increased interest in noninvasive sampling.

## Background

Under natural conditions, the effectiveness of tracking is a function of the quality of the substrate and the skill of the observer (see chapter 3). The emergence of track stations eliminated many chance elements and the effect of tracker experience on survey success. Track stations were first developed to assist in the study of small mammals when Mayer (1957) created a track plate station by placing carbon-sooted kymograph paper into rodent burrow entrances. It was not long before a small plastic enclosure was developed to protect the paper positioned along natural small mammal runways to track movements (Bailey 1969).

Scent stations have a long history of use (beginning with Cook (1949), as cited in Sargeant et al. 2003a); for example, to index the abundance of gray foxes (*Urocyon cinereoargenteus*; Wood 1959) and coyotes (*Canis latrans*; Linhart and Knowlton 1975). The scent station's simple design has remained unchanged since the method was first developed, although the manner in which stations are deployed has evolved significantly. Scent stations have become one of the most widely used methods for estimating the abundance of fur-bearing carnivores.

The modern track plate was first used to survey carnivores by Barrett (1983) for the detection of American martens (*Martes americana*). The impetus for Barrett's using an artificial tracking medium and surface was that his study area in the Sierra Nevada of California was characterized by rocky soils and abundant precipitation. Hence, the author created a sooted surface by smoking a hard and smooth aluminum plate with a kerosene-benzene flame. Protecting this plate from rain called for a shelter, which also provided an attractive cubbyhole entrance. Barrett's method collected negative track impressions that could be photographed or lifted for storage using transparent tape.

Fowler and Golightly (1994) improved the quality of track impressions by adding a tacky, white, track-receptive surface onto which positive track impressions were deposited when an animal walked across the sooted aluminum. This method required an enclosure, however, to guide the animal first over the tracking medium and then over the track-receptive white surface. Track impressions usually have exquisite detail (figure 4.2)—often sufficient to distinguish species of small mammals (e.g., Nams and

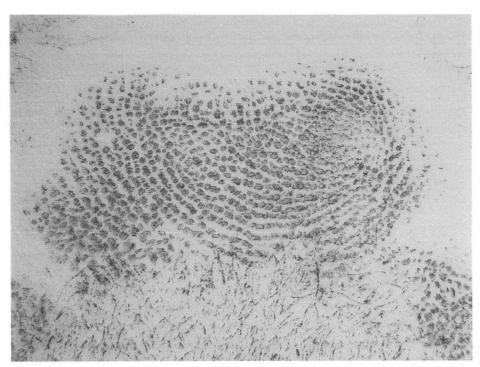

*Figure 4.2.* Close-up of a fisher footprint from an enclosed track station in the Adirondacks. Photo by R. Kays.

Gillis 2003). Such detail has also made it possible to differentiate individuals in some cases (Smallwood and Fitzhugh 1993; Herzog et al. 2007).

On a parallel track, the need to monitor burgeoning stoat (*Mustela erminea*), ferret (*Mustela putorius furo*), and rat (*Rattus norvegicus*) populations introduced to New Zealand brought about the development of a survey device to detect these species. Known as tracking tunnels, the design of these devices follows a similar principle to that of track plate boxes used in North America. Construction material (e.g., corrugated plastic) is folded into a tunnel, inside which is placed an ink-soaked card to record positive impressions of the footprints of animals attracted by a lure (King and Edgar 1977). Tracking tunnels have become the most commonly used technique for surveying and monitoring stoat populations in New Zealand (see references in Choquenot et al. 2001) and have also received some use in North America with rodents and insectivores (Nams and Gillis 2003).

These techniques have been adapted on numerous occasions as a part of ongoing efforts to improve survey methodology. Refinements have occurred where survey efforts have been greatest, often dictated by the urgency of the conservation issue. In the US Pacific Northwest, for example, fishers (*Martes pennanti*) and martens are considered species of critical conservation concern in portions of their range (Ruggiero et al. 1994), and information on their status and geographic range has been collected to assist in their protection (Zielinski et al. 2001; Zielinski et al. 2005). In New Zealand, several exotic carnivores threaten native biodiversity (O'Donnell 1996). Where urgency has required precision (e.g., assessing the status of an endangered or invasive species), there has been more investment in refining the technology itself (e.g., King et al. 2007), combining it with other survey methods such as hair snares (Zielinski et al. 2006), or further developing analytical approaches (Zielinski and Stauffer 1996; Nams and Gillis 2003).

## Target Species

From among the various track station types, a surveyor will find at least one that can function to detect just about any carnivore in North America (table 4.1) for a variety of purposes (table 4.2). Enclosed track plates limit entrance to some species based on size, but larger species are detectable at unenclosed track stations—as long as they do not actively avoid them (see *Behavioral Consideration* later in this chapter). The standard-sized enclosed track plates (Zielinski 1995) can be designed to easily detect most species of mustelids (excluding perhaps American badgers [*Taxidea taxus*] and wolverines [*Gulo gulo*]), procyonids (with the possible exception of white-nosed coatis [*Nasua narica*]), and the smaller species of felids and canids. Forest-dwelling mustelids (martens, fishers, and weasels [*Mustela* spp.]) are frequently surveyed with enclosed track plates (King and Edgar 1977; Zielinski 1995; Gompper et al 2006). Foresman and Pearson (1998) suggested that martens visited unenclosed track plate stations earlier in the survey period than they did enclosed track plates, but Campbell (2004) found that *latency-to-first-detection* (LTD) was similar for enclosed and unenclosed track plates when surveying thirteen species, including martens and fishers. Scent stations are very successful for coyotes and other canids (Linhart and Knowlton 1975), and at least one of the small canids (i.e., gray fox) is also commonly detected in enclosed track plates (Zielinski et al. 2005).

Among the methods considered here, track plots and scent stations placed in potential travel routes are probably the most likely to detect felids and ursids (table 4.1). Felids have a reputation for being relatively indifferent to meat baits, hence, they are not usually targeted by baited track stations. On the other hand, felids can be responsive to scent lures and visual attractants (see chapter 10). For canids and felids, the quality of the natural substrate along the travel route will determine the necessity of preparing track stations versus surveying via continuous tracking (see chapter 3). Ursids are detectable

at track plots, scent stations, and unenclosed track plates—especially those with large tracking surfaces. More typically, a bear visit is evident due to significant disturbance of the detection device (Zielinski et al. 2005).

## Strengths and Weaknesses

Surveyors should consider the relative strengths and weaknesses of track stations, which differ to some degree between types (box 4.1). The major strength of all track stations is that they are inexpensive to construct (with the primary expense for track plots being labor) and can be successfully deployed by individuals with very little training. Further, their operation is quite straightforward and the process of establishing a station involves minimal site preparation. As a result, track stations can be set up in great numbers across very large survey areas, thus having the potential to detect many individuals from a population. Because most types of track stations employ attractants that require regular reprovisioning, however, and given that visits by too many animals can obscure individual tracks (i.e., overtracking), these stations need frequent checking (e.g., every few days) and thus demand more personnel than methods permitting longer sampling occasions (e.g., remote cameras). Last, tracks at track plots and scent stations are not necessarily as well defined as those deposited on prepared track plate surfaces and can therefore require more training and experience for track identification.

When track plates are used, detection evidence is easy to collect, store, and ship. This means that data collection does not rely upon the track identification capabilities of field technicians. As long as technicians follow basic data storage procedures and label records correctly, it is possible for someone with greater expertise to identify the tracks later. This advantage is diminished with track plots and scent stations because methods for preserving the tracks are more problematic. In such cases, evidence of occurrence can be carried away from the field only with plaster casting or photography. Due to impracticality,

*Table 4.1.* References for track station surveys of North American carnivore species

| Species | Track plot | Scent station | Enclosed track plate | Unenclosed track plate |
|---|---|---|---|---|
| Coyote | Engeman et al. 2000; Zoellick et al. 2005 | Linhart and Knowlton 1975; Oehler and Litvaitis 1996; Sargeant et al. 1996; Crooks 2002; Gehring and Swihart 2003; Randa and Yunger 2004; 2006; Zoellick et al. 2005; Choate et al. 2006; Erb 2006a, b; Gompper et al. 2006; Gehrt and Prange 2007 | | Clevenger et al. 2001; Campbell 2004; Green 2006 |
| Gray wolf | Clevenger and Waltho 2000, 2005 | Sargeant et al. 1996; Erb 2006a, b | | |
| Gray fox | | Conner et al. 1983; Smith et al. 1994; Oehler and Litvaitis 1996; Sargeant et al. 1996; Crooks 2002; Gehring and Swihart 2003; Sinclair et al. 2005; Erb 2006a, b | Dark 1997; Hilty and Merenlender 2000; Campbell 2004; Talancy 2005 | Campbell 2004 |
| Kit fox | | Warrick and Harris 2001; Sargeant et al. 2003b | | |
| Swift fox | Schauster et al. 2002; Sargeant et al. 2005 | Schauster et al. 2002; Sargeant et al. 2003b | | Sargeant et al. 2003b |
| Red fox | Zoellick et al. 2005; Ascensão and Mira 2007 | Oehler and Litvaitis 1996; Sargeant et al. 1996; Gehring and Swihart 2003; Sargeant et al. 2003a; Sinclair et al. 2005; Zoellick et al. 2005; Erb 2006a, b; Randa and Yunger 2006 | Talancy 2005; O'Connell et al. 2006 | |
| Domestic cat | Bider 1968; Zoellick et al. 2005; Ascensão and Mira 2007 | Oehler and Litvaitis 1996; Sargeant et al. 1996; Crooks 2002; Gehring and Swihart 2003; Sinclair et al. 2005; Zoellick et al. 2005 | Hilty and Merenlender 2000; Talancy 2005; Gompper et al. 2006 | Campbell 2004 |
| Canada lynx | Bider 1968 | | | |
| Bobcat | Engeman et al. 2000; Zoellick et al. 2005 | Conner et al. 1983; Smith et al. 1994; Diefenbach et al. 1994; Oehler and Litvaitis 1996; Sargeant et al. 1996; Crooks 2002; Zoellick et al. 2005; Erb 2006a, b; Harrison 2006 | | |

*Table 4.1* (Continued)

| Species | Track plot | Scent station | Enclosed track plate | Unenclosed track plate |
|---|---|---|---|---|
| Mountain lion | Bider 1968; Van Dyke et al. 1986; Fjelline and Mansfield 1989; Van Sickle and Lindzey 1992; Smallwood 1994; Cunningham et al. 1995; Smallwood and Fitzhugh 1995; Beier and Cunningham 1996; Clevenger and Waltho 2000; Choate et al. 2006 | Crooks 2002; Zoellick et al. 2005; Choate et al. 2006 | | |
| Striped skunk | Bider 1968; Zoellick et al. 2005 | Sargeant et al. 1996; Crooks 2002; Gehring and Swihart 2003; Sargeant et al. 2003a; Sinclair et al. 2005; Zoellick et al. 2005; Erb 2006a, b | Hilty and Merenlender 2000; Campbell 2004; Talancy 2005; Gompper et al. 2006; O'Connell et al. 2006 | Heske 1995; Dijak and Thompson 2000; Winter et al. 2000; Campbell 2004 |
| Western spotted skunk | | | Campbell 2004; Manley et al. 2004; Zielinski et al. 2005; Green 2006 | Campbell 2004 |
| North American river otter | Zoellick et al. 2005 | | | |
| American marten | | | Bull et al. 1982; Barrett 1983; Fowler 1995; Zielinski and Stauffer 1996; Zielinski et al. 1997; Foresman and Pearson 1998; Mowat et al. 2000; Zielinski et al. 2001; Fecske et al. 2002; Routledge 2000; Truex 2003; Campbell 2004; Manley et al. 2004; Gelok 2005; Zielinski et al. 2005; Gompper et al. 2006; Green 2006; Kirk 2007; Slauson et al. 2007 | Foresman and Pearson 1998; Clevenger et al. 2001; Campbell 2004 |
| Fisher | | | Fowler 1995; Zielinski et al. 1995a; Kucera et al. 1996; Zielinski and Stauffer 1996; Dark 1997; Zielinski et al. 1997; Foresman and Pearson 1998; Carroll et al. 1999; Zielinski et al. 2001; Aubry and Lewis 2003; Truex 2003; Campbell 2004; Manley et al. 2004; Zielinski et al. 2005; Gompper et al. 2006; Green 2007; Davis et al. 2006 | Foresman and Pearson 1998; Campbell 2004 |

*Table 4.1* (Continued)

| Species | Track plot | Scent station | Enclosed track plate | Unenclosed track plate |
| --- | --- | --- | --- | --- |
| Weasels and stoats (*Mustela* spp.) | Bider 1968; Ascensão and Mira 2007 | | Brown et al. 1998; Mowat et al. 2000; Ragg and Moller 2000; Choquenot et al. 2001; Graham 2002; Campbell 2004; Manley et al. 2004; Talancy 2005; Zielinski et al. 2005; Gompper et al. 2006; Green 2006; O'Connell et al. 2006 | Clevenger et al. 2001 |
| American mink | Zoellick et al. 2005 | Zoellick et al. 2005 | Loukmas and Halbrook 2001; Reynolds et al. 2004 | |
| American badger | Zoellick et al. 2005 | Zoellick et al. 2005 | Campbell 2004 | Heske 1995; Dijak and Thompson 2000; Winter et al. 2000 |
| Ringtail | | | Dark 1997; Campbell 2004; Manley et al. 2004; Zielinski et al. 2005; Green 2006 | Campbell 2004 |
| Raccoon | Bider 1968; Engeman et al. 2003; 2005; Zoellick et al. 2005 | Conner et al. 1983; Smith et al. 1994; Oehler and Litvaitis 1996; Sargeant et al. 1996; Crooks 2002; Gehring and Swihart 2003; Sinclair et al. 2005; Zoellick et al. 2005; Erb 2006a, b; Randa and Yunger 2006; Gehrt and Prange 2007 | Hilty and Merenlender 2000; Talancy 2005; Gompper et al. 2006; O'Connell et al. 2006 | Heske 1995; Dijak and Thompson 2000; Winter et al. 2000; Wolf et al. 2003; Campbell 2004; Pedlar et al. 2007 |
| American black bear | Bider 1968; Clevenger and Waltho 2000, 2005 | Sargeant et al. 1996 | Dark 1997; Campbell 2004 | Campbell 2004 |
| Grizzly bear | Clevenger and Waltho 2000, 2005 | | | |

Note: The exclusion of certain species from this table reflects a lack of published accounts of their detection via these survey methods.

Table 4.2. Survey objectives and principal target species by track station type, from select studies

| Type of track station | Survey objective | Principal target species | References |
|---|---|---|---|
| Track plot | Monitor activity of predators relative to sea turtle nesting | Raccoon | Engeman et al. 2003; Engeman et al. 2005 |
| | Assess the effectiveness of wildlife crossing structures | Gray wolf, mountain lion, American black bear, grizzly bear | Clevenger and Waltho 2000, 2005 |
| | Estimate distribution | Swift fox | Sargeant et al. 2006 |
| | Assess relative abundance | Mountain lion | Smallwood 1994; Cunningham et al. 1995; Smallwood and Fitzhugh 1995; Beier and Cunningham 1996; Choate et al. 2006 |
| | Index changes in activity prior to and following predator control program | Coyote, bobcat | Engeman et al. 2000 |
| | Explore spatial and temporal aspects of community ecology | North American weasels, striped skunk, raccoon, domestic cat, American black bear, Canada lynx, mountain lion | Bider 1968 |
| | Compare survey methods and assess relative abundance | Swift fox, kit fox | Warrick and Harris 2001; Schauster et al. 2002; Sargeant et al. 2003b |
| | Assess activity level of carnivores on islands prior to waterfowl nesting season | American badger, bobcat, coyote, feral cat, American mink, raccoon, red fox, striped skunk, North American river otter | Zoellick et al. 2005 |
| Scent station | Index spatial variation in carnivore abundance and activity relative to cover type and habitat fragmentation | Coyote, domestic cat, gray and red foxes, raccoon, striped skunk, long-tailed weasel, bobcat, mountain lion | Oehler and Litvaitis 1996; Crooks 2002; Gehring and Swihart 2003; Gompper et al. 2006; Randa and Yunger 2006 |
| | Assess relative abundance of carnivores or nest predators | Coyote, swift fox, red fox, gray fox, striped skunk, raccoon, domestic cat, kit fox, mountain lion | Linhart and Knowlton 1975; Roughton and Sweeny 1982; Conner et al. 1983; Smith et al. 1994; Sargeant et al. 1998; Warrick and Harris 2001; Schauster et al. 2002, Sargeant et al. 2003a; Sargeant et al. 2003b; Randa and Yunger 2004; Sinclair et al. 2005; Choate et al. 2006 |
| | Document statewide long-term trends | Red fox, striped skunk, raccoon, coyote, bobcat, gray wolf, American black bear | Sargeant et al. 1996; Erb 2006a, b |
| | Compare survey and detection methods | Bobcat, swift fox, kit fox | Sargeant et al. 2003b; Harrison 2006 |
| | Monitor following reintroduction | Bobcat | Diefenbach et al. 1994 |
| | Assess activity level of carnivores on islands prior to waterfowl nesting season | American badger, bobcat, coyote, feral cat, American mink, raccoon, red fox, striped skunk | Zoellick et al. 2005 |

*Table 4.2* (Continued)

| Type of track station | Survey objective | Principal target species | References |
|---|---|---|---|
| Enclosed track plate | Determine if raccoons avoid coyotes at the microspatial scale | Coyote, raccoon | Gerht and Prange 2007 |
| | Assess contemporary distribution and status | American marten, fisher, western spotted skunk, ringtail, North American weasels, swift fox | Zielinski et al. 1995a; Kucera et al. 1995b; Zielinski et al. 1997; Mowat et al. 2000; Zielinski et al. 2001; Aubry and Lewis 2003; Manley et al. 2004; Zielinski et al. 2005; Green 2006 |
| | Build, test, or validate spatial habitat model | Fisher, American marten, American mink | Carroll et al. 1999; Fecske et al. 2002; Gelok 2005; Kirk 2006; Loukmas and Halbrook 2001; Slauson et al. 2007; Davis et al. 2007 |
| | Assess relative abundance | American marten, fisher | Zielinski and Stauffer 1996; Routledge 2000; Truex 2003 |
| | Detect evidence of or potential for colonization | New Zealand stoat | Choquenot et al. 2001 |
| | Assess repopulation after control operations | New Zealand stoat | Brown et al. 1998 |
| | Map spatial distribution of activity and denning | Feral ferret | Ragg and Moller 2000 |
| | Assess abundance relative to vole density | North American weasels | Graham 2002 |
| | Compare survey and detection methods | Raccoon, striped skunk, gray fox, American marten, swift fox, fisher, American black bear, western spotted skunk, striped skunk, North American weasels | Bull et al. 1992; Foresman and Pearson 1998; Hilty and Merenlender 2000; Campbell 2004 |
| | Develop habitat models that account for detection probability | Domestic cat, raccoon, striped skunk, gray fox, red fox, North American weasels | Talancy 2005 |
| | Assess distribution, co-occurrence, and habitat associations | Fisher, gray fox, ringtail, American black bear, western spotted skunk, striped skunk, American marten, fisher, North American weasels | Campbell 2004 |
| Unenclosed track plate | Monitor passage through road drainage culvert or crossing structure | Coyote, American marten, raccoon, North American weasels | Clevenger et al. 2001; Wolf et al. 2003 |
| | Assess mammalian activity to evaluate edge and area effects of nest predation | American badger, striped skunk, raccoon | Heske 1995; Dijak and Thompson 2000; Winter et al. 2000 |
| | Assess distribution, co-occurrence, and habitat associations | Fisher, gray fox, ringtail, American black bear, striped skunk, western spotted skunk, American marten, fisher, North American weasels, raccoon | Pedlar et al. 1997; Campbell 2004 |
| | Compare survey and detection methods | Fisher, gray fox, ringtail, American black bear, striped skunk, western spotted skunk, American marten | Foresman and Pearson 1998; Campbell 2004 |

---

## Box 4.1

### Strengths and weaknesses of track stations

| Strengths | Weaknesses |
| --- | --- |
| • Materials and construction costs are low, particularly relative to remote cameras.<br>• Can be deployed with little training.<br>• Track impressions (from track plates) can easily be removed from the field and stored as permanent records.<br>• Easy to set up.<br>• Can be deployed in great quantities over large areas.<br>• Can target a diverse array of species. | • Currently difficult or impossible to distinguish between tracks of some species.<br>• Even to skilled trackers, prints at track plots and scent stations can appear ambiguous.<br>• In most cases, it is impossible to identify individuals.<br>• Require relatively frequent visits to rebait and check.<br>• Track plots, scent stations, and unenclosed track plates do not function well in wet conditions.<br>• Some animals are wary of approaching artificial surfaces (i.e., track plates). |

---

this is seldom done; rather, field personnel must be skilled enough to identify tracks on the spot, as with more traditional tracking approaches (see chapter 3). There are numerous guides available to assist in track identification (e.g., Orloff et al. 1993; Taylor and Raphael 1988; Zielinski and Truex 1995; Elbroch 2003), and methods have also been developed to distinguish sex within some species (e.g., mountain lions [*Puma concolor*], Fjelline and Mansfield 1989; fishers and martens, Slauson et al. 2008). Tracks subjected to this latter analysis, however, must be complete and legible, and not all track impressions—even those on contact paper—are sufficient for sex identification.

Track stations have a number of other potential limitations that should be considered when planning a survey. They do not function well under wet conditions, for example, although the potential to collect identifiable track impressions in wet weather is greatest with enclosed track plates. Enclosed track plates have even been deployed in the snow, but the fact that they can be covered by deep snow (and that wet paws leave indistinct track records) reduces their utility in the winter.

Like all devices that rely on attractants to improve detection rates, most track stations have the potential to record species at locations where they might not otherwise occur—and attraction distance typically cannot be quantified (see chapter 10 for further discussion). Moreover, track stations are generally of limited value for deriving population density estimates because they do not usually provide sufficient evidence to distinguish individuals (see, however, Smallwood and Fitzhugh [1993] and Herzog et al. [2007] for exceptions). Finally, tracks deposited on a hard surface (e.g., aluminum track plates) differ markedly in size and shape from those left in softer substrates, appearing much smaller and more detailed. Measurements cited in field tracking guides (e.g., Elbroch 2003) will thus not always agree with the dimensions of track impressions recorded on sooted aluminum surfaces. Importantly, the successful identification of species from tracks deposited on the various surfaces described in this chapter requires considerable study and field experience—more than may be necessary for species identification from photographs, for example.

## Treatment of Objectives

Track stations have been deployed for a variety of purposes, ranging from carnivore detection at a single location to a comprehensive assessment of the status of a population and its habitat. The most common survey objectives are to assess occupancy, distribution, and relative abundance, and to evaluate habitat relationships from plot to landscape scales (table 4.2). Track stations are also frequently employed in studies designed to compare their efficacy with other methods (e.g., Bull et al. 1992; Foresman and Pearson 1998; Hilty and Merenlender 2000; Silveira et al. 2003; Campbell 2004; Gompper et al. 2006; Barea-Azcón et al. 2007).

Although all types of track stations have been used to meet these various objectives, some are identified with certain objectives more than others. Scent stations, for example, have been the most frequent choice for addressing population objectives, whereas the majority of habitat assessments have relied on enclosed track plates. The tracking tunnels used in New Zealand have a very specific purpose for which other enclosed devices have not been used: to detect and monitor the presence of stoats as part of a larger effort to eradicate this invasive species (e.g., Brown et al. 1998; Choquenot et al. 2001). Creative applications of track stations have also included monitoring culverts as road crossings (Clevenger et al. 2001; Ascensão and Mira 2007), assessing edge and forest fragmentation effects (e.g., Heske 1995; Dijak and Thompson 2000; Winter et al. 2000), evaluating the magnitude of interference competition (Gehrt and Prage 2007), evaluating the effect of density on probability of detection (Smith et al. 2007), and quantifying predation risk (e.g., Connors et al. 2005; Engeman et al. 2005).

## Occurrence and Distribution

Track stations are most suitable for establishing whether a species occurs at a particular location (i.e., occupancy). When occupancy data are assessed from multiple locations, these data can also be utilized to determine a species' geographic distribution. Enclosed track plates have been widely used in the US Pacific Northwest to determine occupancy and to describe the geographic distribution of fish-

ers and martens (Kucera et al. 1995b; Zielinski et al. 1995a; Zielinski et al. 2001; Aubry and Lewis 2003) as well as other carnivores (Zielinski et al. 2005). Standardized track station surveys throughout the Pacific Northwest, many following a recommended protocol (Zielinski 1995), have identified isolated carnivore populations and described their boundaries, even with different personnel executing surveys at different times. On Cape Breton Island in Nova Scotia, failure to detect martens led to the conclusion that this provincially threatened species occurred at extremely low numbers—if it persisted at all—because similar levels of effort detected martens in other places where they were common (Nova Scotia American Marten Recovery Team 2006).

Some track station methods were precursors to modern occupancy estimation methods (i.e., MacKenzie et al. 2006), as they employed independent sample units to report binomial results at the level of the sample unit (Zielinski and Stauffer 1996; Sargeant et al. 1998; also see chapter 2). In contrast to occupancy estimation methods, however, these studies lacked the recognition that the history of detections and nondetections at sample sites (i.e., the encounter history) could be used to estimate probability of detection, and that this information could in turn be used to adjust occupancy estimates to account for sites where the target species was present but not detected. Zielinski and Stauffer (1996) addressed this issue and incorporated uncertainty of detection success into their proposed monitoring program. But it wasn't until MacKenzie et al. (2002) applied the methods of mark-recapture analysis to detection-nondetection surveys, and adjusted estimates of occupancy ($\psi$) to account for varying levels of detectability ($p$), that occupancy estimation was fully realized (see chapter 2 for a more detailed discussion of estimating occupancy).

Track stations have been used by Campbell (2004), Talancy (2005), Gompper et al. (2006), and Smith et al. (2007) to estimate probability of detection, and by O'Connell et al. (2006) to estimate both probability of detection and occupancy. Only Smith et al. (2007) estimated probability of detection using exclusively track stations (i.e., enclosed track plates). These authors found that the probability of detection for martens, given a twelve-day survey period, was significantly lower ($p = 0.333$) in areas where marten density was low compared to where it was high ($p = 0.952$). Other studies included devices other than track stations, so track station-specific occupancy rates were confounded. Nonetheless, Campbell (2004) estimated track plate–specific, per visit detection probabilities for thirteen species, which ranged from 0.13–0.36 for enclosed track plates and 0.005–0.68 for unenclosed track plates. O'Connell et al. (2006) found that estimates of probability of detection and occupancy varied from 0.15–0.39 and 0.10–0.85, respectively, for the four species of carnivores detected by track stations. These authors concluded that eleven weeks of track station sampling were required to achieve a 95% probability of detecting a raccoon (*Procyon lotor*) at an occupied site. They also noted that occupancy estimates suffered when detectability did not exceed 0.15.

Ivan (2000) took a different approach to estimating probability of detection, monitoring visits to enclosed track plates by individual martens using unique toe pad marks and radio telemetry. The probability of detecting martens when they were present during a twelve-day survey was 0.70 (measured as the ratio of sample units where martens were detected to sample units where martens occurred, with sample unit defined as a spatially independent grouping of track stations), whereas the probability of detecting a particular marten was much lower (i.e., 0.07–0.13). Some martens appeared to avoid track stations. This study concluded that occupancy could more readily be established where marten densities were relatively high than where martens were uncommon, thus supporting the results of Smith et al. (2007). These findings are also in accordance with the variable detection rates

for martens at the southern end of their distribution in the Adirondacks (Gompper et al. 2006) versus high marten detection rates in the more northerly Algonquin Park, Ontario, where the species is presumably more abundant (Gelok 2005).

Sargeant et al. (2003b) achieved low detection rates for kit (*Vulpes macrotis*) and swift foxes (*Vulpes velox*) at scent stations and recommended against using this method to determine geographic distributions for these species. Implementing a novel approach in subsequent work, Sargeant et al. (2006) estimated the actual distribution of swift foxes with a method based on Markov chain Monte Carlo (MCMC) image restoration. With MCMC, prediction of occurrence at a particular site is based on survey effort, the results of detection-nondetection surveys at neighboring sites, a measure of spatial contagion among sites, and the probability of detecting swift foxes when present (see chapter 2 for additional details).

## Relative Abundance

It is tempting to believe that, because many animals leave more tracks than few animals and no animals leave zero tracks, track stations can assess relative abundance. Indeed, relative abundance has been one of the most common objectives professed by researchers who use track stations (table 4.2). Measures of relative abundance have been represented as the proportion of track stations that receive a visit (e.g., Zielinski and Stauffer 1996)—sometimes referred to as *tracking rate* (Brown and Miller 1998) or *visitation rate* (Sargeant et al. 1998). A positive relationship between these indices and actual abundance is expected when individuals are relatively consistent in the number of track stations they encounter, the number of times they visit them, and their propensity to leave tracks. An ideal index should detect a consistent proportion of the target population with little variance (Thompson et al. 1998). These conditions, however, are not expected to occur with regularity—differences in population dispersion, sex-specific and individual behavior, habitat use, and weather and seasonal effects could make a track station index difficult to interpret.

Because individuals typically cannot be identified on the basis of track characteristics, it is customary to ignore the *number* of detections and to simply report whether or not a species was detected at a track station. The resulting dichotomized data (i.e., *0* or *1* results at a sample unit; see chapter 2) are more robust to statistical analysis than the total number of detections recorded at a station or a multiple-station sample unit (Sargeant et al. 1998). Collapsing data in this fashion has characterized most recent track station efforts (e.g., Kucera 1995b; Zielinski et al. 1995a; Beier and Cunningham 1996; Sargeant et al. 1998; Carroll et al. 1999), and has generally replaced earlier methods that used the absolute count of detections as the metric of interest (e.g., Linhart and Knowlton 1975, but see Stander 1998 for an exception). Indices (both raw and dichotomized) have their advocates as well as their critics (see chapter 2 for additional details).

By comparing scent station results with those of independent methods for estimating population size (i.e., usually live-trapping), some authors have found that scent station indices accurately reflected the population size of several midsized carnivores (e.g., Wood 1959; Conner et al. 1983; Leberg and Kennedy 1987; Best and Whiting 1990; Diefenbach et al. 1994; Engeman et al. 2000; Warrick and Harris 2001), while others have not (e.g., Robson and Humphrey 1985; Nottingham et al. 1989; Van Sickle and Lindzey 1992; Smith et al. 1994) or have challenged previously published results (i.e., Conner et al. 1984; Minser 1984). The value of estimating relative abundance from track stations can also differ among species. Mahon et al. (1998), for example, found that tracking on roads versus random sand plots overestimated the abundance of introduced red foxes (*Vulpes vulpes*) but not feral cats (*Felis catus*).

Many early studies of indices were small-scale explorations that—if tested against independent measures of abundance at all—were evaluated across short time periods, small areas, and limited condi-

tions, and were never replicated. Moreover, these early indices were established without exhibiting control over the effects of behavioral heterogeneity, awareness of the effects of pseudoreplication, or knowledge of probability of detection. We now know, for example, that single individuals can visit multiple stations (e.g., Herzog et al. 2007), and we expect that this phenomenon varies across time and space. A measure of relative abundance requires calibration with estimates of actual abundance within a range of habitat conditions, but rigorous calibration of relative and actual abundance has rarely been conducted. Sargeant et al. (1998: 1243) reviewed a number of the experimental attempts to relate a track station index to population size, and concluded that "the general validity of scent-station indices has been neither proven nor called into serious question by objective validation experiments." These authors suggested that logistical constraints will preclude conclusive, experimental validation of track station indices for many carnivores.

In our opinion, the most reliable evaluations of the benefits of indices have emerged from two of the more comprehensive and recent assessments—those of Sargeant et al. (2003b), pertaining to kit and swift foxes, and Choate et al. (2006), pertaining to mountain lions. Both studies compared track indices to concurrent and independent measures of population abundance and found that, without additional research, it was premature to consider track indices a population metric for management purposes. Indices had potential value as assessments of relative abundance, but with important caveats. For example, swift and kit fox indices appeared to reflect increases in population only in spring and summer, and fox visitation rates declined proportionally less than population size such that individuals visited stations at higher rates when population size was low (Sargeant et al. 2003b). Further, the mountain lion track plot index exhibited a nonlinear and unpredictable relationship to population size and did not perform well when densities were expected to vary greatly. Also, scent stations indexed lion densities very poorly compared to track plots.

Both of these studies experienced what appeared to be density-dependent visitation rates (i.e., indices performed better over a much smaller range of abundances than would be expected). This reinforces conclusions from other studies that relative indices from traditional track station designs have low resolution because reliable inferences require independent and much larger samples (Zielinski and Stauffer 1996; Sargeant et al. 2003a, b). Choate et al. (2006: 796) concluded that "despite the allure of lower-cost index measures, the lack of sensitivity to population changes by indices, particularly over time scales involved in management decisions ... warrants considerable caution in their application." Sargeant et al. (2003b: 102) found conditional support for use of indices in some circumstances but suggested that even with the effort expended by these significant studies, "the manipulation of populations on a scale sufficient to validate an index is neither feasible nor acceptable." Thus, it may be impractical to achieve any better understanding of the relationship between an index and actual population size than we have today.

We believe that indices of relative abundance will continue to be viewed with skepticism, especially when similar objectives can be met via formal occupancy estimation (MacKenzie et al. 2006). But there will still be circumstances under which indices of relative abundance are the method of choice. In these situations, Diefenbach et al. (1994) offer the following recommendations: (1) surveys should be conducted multiple times per year; (2) the proportions of stations with detections should be transformed to reduce statistical problems; (3) if strata are used, they should contain as many stations as possible; (4) stations should be placed as far apart as possible; and (5) statistical power analysis should precede the collection of field data.

## Abundance and Density

Very few studies have deployed track stations to estimate abundance because individuals cannot usually be identified from their tracks, and the aforementioned difficulties bedevil measures of even relative abundance. Sargeant et al. (1998) conducted a rigorous evaluation of scent station programs and methods and concluded that scent station indices should not be converted to estimates of abundance. Nevertheless, in a study of source-sink dynamics for culpeo foxes (*Dusicyon culpaeus*) in northwestern Patagonia, Novaro et al. (2005) calculated densities from scent station indices using a conversion factor obtained from simultaneous calibrations with line transect estimates. Indeed, current innovations that permit identifying individuals from track stations using track identification (e.g., Herzog et al. 2007), DNA methods (e.g., Zielinski et al. 2006; also see chapter 9), or chemical bait markers (e.g., Jones et al. 2004) may significantly advance the goal of estimating abundance. In addition, promising new analytical approaches have suggested that abundance can be estimated from repeated detection surveys (Royle and Nichols 2003; also see chapters 2 and 11). These approaches rely on the relationship between variation in probability of detection and variation in abundance (e.g., Smith et al. 2007), making it possible to estimate abundance from track plate data without determining individual identity.

## Monitoring

Monitoring is the assessment of a metric over time and, as such, can apply to any of the objectives discussed earlier (i.e., occupancy, distribution, relative abundance, abundance). With track stations, the most common approach is to monitor the trend (i.e., > 3 time points) in the metric of interest over time. Some track station monitoring proposals have focused on comparing two time points (e.g., Zielinski and Stauffer 1996; Manley et al. 2004; Engeman et al. 2003), such as in evaluating population-level changes before and after an event. But with track station data, the statistical properties of trend estimates (i.e., ≥ 3 time points) via regression appear to have less bias, making trend a more attractive statistical model for monitoring (Sargeant et al. 1998).

Track stations are among the most commonly

proposed methods for monitoring changes in relative abundance, and scent stations have been widely used for this purpose. Sargeant et al. (1998) explored the statistical properties of a Minnesota survey that employed scent stations (case study 4.2), and found that the way in which data were aggregated (i.e., either by station or by lines composed of multiple stations) and the spacing of stations affected the interpretation of statistical trends. These authors recommended using rank-transformed data at the level of the survey line (comprising ten stations) to test for trends, as trends identified in this manner must be large (i.e., larger than annual fluctuations) to be detectable.

Sargeant et al. (1998) concluded that, although scent station indices cannot currently be used to estimate abundance, they can help identify trends and supplement data from other sources. Indeed, counts and visitation rates can vary widely even when there is no population change, in part because counts and rates cannot be distinguished from activity—a potentially serious problem when track stations are to be used for estimating population levels as well as distribution. Warrick and Harris (2001) had access to a particularly unique data set from twelve years of simultaneously collected scent station data and capture-recapture population estimates for kit foxes (*Vulpes macrotis mutica*) in California. Track station visitation had a significant positive relationship with population size in this long-term study, but precision was low because counts could vary widely when there was little change in the population—a finding shared by Diefenbach et al. (1994) for bobcats (*Lynx rufus*). Warrick and Harris (2001) concluded that scent stations were useful for monitoring kit foxes, but were only capable of detecting changes of rather large magnitude.

Zielinski and Stauffer (1996) proposed a monitoring program using enclosed track plates to detect change (in this case, declines) in the occurrence of fishers and martens throughout their range in California. These researchers conducted statistical power analyses to estimate the probability of detecting a 20% change in an index (i.e., the proportion of sample units with at least one detection). Strata were geographic regions and habitat types where occupancy rates were expected to differ. They concluded that 115 sample units for each of ten strata would be necessary to detect the specified declines, should they occur. This work served as the foundation for a long-term fisher population monitoring program in the Sierra Nevada of California (see case study 4.1) and was also an important component of a multiple-species monitoring plan employing enclosed track plates as one of a number of wildlife detection methods (Manley et al. 2004; Manley et al. 2006).

Meanwhile, in New Zealand, where the nonnative status of stoats makes monitoring their colonization a high conservation priority, Choquenot et al. (2001) found that more than two hundred tracking tunnels were needed to detect five or fewer stoats in an area of 10,000 ha. These examples demonstrate the enormous logistical and financial challenges that managers face when they consider using track stations as a tool for monitoring and indexing population status over time, particularly in cases where population levels of target species are low. The challenges are diminished, however, if managers are satisfied with detecting large magnitude changes over multiple-year intervals at a landscape or regional scale of resolution (Traviani et al. 1996; Warrick and Harris 2001).

Support for indices of relative abundance appears to be waning with the increasing interest in occupancy estimation (see *Relative Abundance* earlier, and chapter 2). Currently, the most justifiable use of indices of relative abundance may be to monitor relative change in abundance over time rather than to estimate population size at one point in time (G. Sargeant, US Geological Survey, pers. comm.). Choate et al. (2006) found that track estimators performed poorly as individual indices of population size, but that proportional changes in population size over time were strongly related to proportional changes in indices.

The success of a track station-based monitoring program depends not only on the index's relationship to population size and the number of sample units, but also on the proportion of sample units with detections. The number of sample units expected to receive detections will strongly affect the ability to detect change (Beier and Cunningham 1996; Zielinski and Stauffer 1996; also see chapters 2 and 11). For example, if over a specified period of time (i.e., the survey period), 10% of track stations yield detections and then the population doubles, this index has a chance of detecting the change. If, however, 90% of the stations originally yield tracks, then the index can only increase minimally if the population increases. Ideally, the sampling period should be chosen such that the target species is detected at an intermediate proportion of sample units (Nams and Gillis 2003)—the number of sample units must be high enough to guarantee sufficient samples for analysis (Roughton and Sweeny 1982; Sargeant et al. 2003b).

## Habitat Evaluation

The first use of track stations to understand the habitat of wild animals was probably the work of Bider (1968), who created long narrow plots across ecological gradients and recorded the time and location of tracks of dozens of species, including carnivores. This descriptive research has been supplanted more recently by quantitative studies in which track stations have been used to both create and test habitat models for individual species. Multivariate models are usually developed by contrasting environmental features (e.g., vegetation, topography) in the vicinity of the track stations where target species were detected with those of track stations lacking detections, or with random locations. The most common application is to develop models for landscape or regional habitat suitability, probably because telemetric methods—which provide more resolution for addressing microhabitat or patch-scale selection—are too expensive to apply when study areas become large. In some cases, however, track plates have been used to address small-scale habitat preferences (e.g., Mowat et al. 2000).

Carroll et al. (1999) were the first to employ an extensive array of track stations for the purpose of building and then testing a regional habitat suitability model. Their logistic regression model for fishers in northwest California was developed with preex-

isting survey data gathered over an area of 67,000 km² and related to vegetation, topographic, and precipitation data at various spatial scales. Independent data achieved high (80%) classification success, and the model has since been used for various land management planning and assessment purposes and as a basis for conservation planning. This work demonstrates the multifaceted value of well-planned systematic track station surveys, as the data used to test the model were also used to compare historical and contemporary distributions of carnivores in California (Zielinski et al. 2005), and served as a foundation for a population monitoring program (see case study 4.1).

Davis et al. (2007) employed an approach similar to that of Carroll et al. (1999), relying again on systematically collected track station data from California to generate models to predict fisher habitat suitability. Their work revealed different environmental predictors for California's two fisher populations, and also identified candidate regions for habitat restoration and for potential reintroduction. Similarly, a number of investigators have used track plates as a means for collecting occurrence data to develop or test habitat models for martens (Fecske et al. 2002; Gelok 2005; Green 2006; Kirk 2006; Slauson et al. 2007).

Campbell (2004) utilized a subset of the enclosed track plate data used by Carroll et al. (1999) and Davis et al. (2007) to created habitat models for five carnivores (fishers, striped skunks [*Mephitis mephitis*], gray foxes, American black bears [*Ursus americanus*], and ringtails [*Bassariscus astutus*]) in California's central and southern Sierra Nevada. Campbell's analysis relied on classification tree methods, and considered multiple spatial scales by evaluating vegetation and topographic variables in areas of increasing radius surrounding sample units. Talancy (2005) also used track plates as part of a multiple-device sample unit to evaluate the effect of scale on habitat selection by nine species of medium-sized carnivores (and the Virginia opossum [*Didelphis virginiana*]). More specifically, this researcher used detection-nondetection data to model habitat selection with likelihood-based occupancy models and discovered that landscape factors—in particular, the amount of human disturbance and habitat fragmentation—were important predictors of occurrence.

Several studies have used response variables derived from track stations to assess target species' distribution relative to landscape-level fragmentation. For example, species-level distribution models have been developed at various scales for midsized carnivores using scent station data from agricultural landscapes in Indiana (Gehring and Swihart 2003) and New Hampshire (Oehler and Litvaitis 1996), and from along an urban-rural gradient in northern Illinois (Randa and Yunger 2006). These models assessed species distribution relative to forest edges, landscape composition, anthropogenic disturbance, and other habitat attributes. Other studies have deployed track plates to assess nest predator activity relative to forest edges as a means of evaluating how such edges affect nestling survival (Heske 1995; Dijak and Thompson 2000; Winter et al. 2000).

Track stations are an ideal tool for developing habitat models for large areas because of the relative ease of deploying and checking them in the field and interpreting their results. We caution, however, that detection results collected in this fashion can only address population-level habitat selection (Johnson 1980). Without also collecting genetic samples (e.g., hair, feces) for individual identification, analyses cannot be conducted using the individual as the sample unit—a standard that characterizes more intensive and thorough habitat selection analyses conducted over smaller areas with telemetry. Making inferences at large scales, with low resolution, also protects against the shortcoming that animals are attracted to baited track stations from unknown distances.

## Description and Application of Survey Method

Track plots and scent stations, by virtue of their simplicity, have changed little over time (figure 4.1A, B);

the most sophisticated innovations have related to project objectives and sampling design. On the other hand, track plates (figure 4.1C, D) have evolved considerably, with new materials being introduced and tested on a regular basis and ongoing modifications being made to the general setup. The following section describes the various types of track station devices and offers recommendations with respect to their design and deployment.

### General Setup

For all track stations, site preparation should be concentrated on maximizing the chance that an animal's visit to the station will be recorded. This involves, for example, ensuring that the tracking medium is in the best possible condition, that debris or other obstructions do not fall on the tracking surface, and that the setup itself is stable (i.e., not vulnerable to wind or collapse as soon as an animal enters or begins to explore).

Attractants are an important consideration for track plates and scent stations. Commonly deployed baits include a chicken leg or other carrion, a can of tuna, or jam (see chapter 10 for a more detailed discussion of attractant selection and deployment). Lures are also assumed to enhance the attractiveness of track stations. Typically, researchers employ commercial scent lures that have been deemed appropriate for the target species by the manufacturer, but lures can also be homemade from various scents, or can be visual in nature (e.g., a spinning pie tin [Weaver et al. 2005] or mylar strip [Choate et al. 2006] for felids in particular). Some circumstances require that no attractant be used; for instance, when a track station is deployed to detect visits by predators to bird nests (e.g., Engeman et al. 2005). Lastly, given that some animals are more hesitant to approach devices than others by nature (see *Behavioral Considerations* later in this chapter), prebaiting may occasionally be called for (see chapter 10)—and may in turn reduce LTD once the device is officially deployed.

Track stations are subject to the same superstitions as trapping activities, with many surveyors deploying signature setups that have incurred perceived personal success. Sometimes this results in different station styles within one project. The extent to which camouflage is utilized, for example, is widely variable. Some individuals insist on camouflaging the station if an enclosure is used (e.g., by draping it with branches and leaves), while others do not; there is little evidence that any particular camouflage technique yields the greatest success. On the other hand, camouflage can help to conceal devices from passersby, and branches and boulders can add extra stability.

The first enclosed track plates were actually placed in trees to detect martens (Barrett 1983). Bull et al. (1992) used this deployment tactic as well, and compared track plates with camera traps and snow tracking. Arboreal placement was largely intended to prevent the track plates from being covered by snow in the winter. By and large, however, track stations are easier to set up on the ground and can detect a larger number of species there versus attached to the bole of a tree.

## Enclosures

Enclosures can be made from a variety of materials. Track plate boxes are traditionally manufactured from plywood (Barrett 1983; Zielinski 1995), and these units have been broadly deployed (Fowler and Golightly 1994; Fecske et al. 2002; Aubry and Lewis 2003; Belant 2003b; Hamm et al. 2003; Loukmas et al. 2003). Plastic canopies, which are considerably lighter but less stable, have also been presented as an option (Zielinski 1995), but there is no evidence in the published literature of their extensive use. Plywood has the distinct disadvantage of being heavy and unwieldy to transport by hand, and it has a relatively short lifespan when exposed to the elements for prolonged periods of time—especially in environments characterized by heavy rainfall.

When conducting surveys in areas that are inaccessible to vehicular traffic, and particularly in rugged terrain, minimizing weight is highly desirable. Such conditions led to the recent adoption of corrugated plastic sheets or "fluted polypropylene" (e.g., Coroplast) for creating track plate boxes (Gompper et al. 2006; also see figures 4.1D, 6.8A). As these units are foldable, lightweight, and durable over multiple seasons in wet weather, little should stand in the way of their widespread application (Gelok 2005; Zielinski et al. 2006). It is unknown whether the color of the plastic sheet influences detection rates; black or brown sheets yield dark tunnels that may appear safer from the perspective of the animal and blend in better than brighter colors (e.g., white) in forest environments. The tracking tunnels first developed in New Zealand (King and Edgar 1977) consisted of long and narrow aluminum enclosures, tailor-made for stoats and weasels. These too have evolved in recent years and are now made from folded sheets of corrugated plastic (Corflute; C. M. King, University of Waikato, New Zealand, pers. comm.). There are times, however, when such materials are not needed at all, with existing structures such as drainage culverts (Clevenger et al. 2001; Wolf et al. 2003) or natural enclosures (e.g., crevices, caves, and rock formations) serving as effective enclosures for track plates (figure. 4.3).

The cross-sectional shape of track plate enclosures can be square, domed, or triangular. Triangular enclosures are a recent innovation (M. Schwartz, USDA Forest Service, pers. comm.) and have been found to be quite rigid—thus requiring relatively little vertical support and easily accommodating the addition of hair snares (figure 6.8A). For boxes with square openings, few authors have deviated from the original size recommendations (25.4 × 25.4 × 81.3 cm) suggested by Zielinski (1995). When using the triangular version, it may be necessary to increase the length of the three sides, as the resulting roof angle provides less room than a square entrance with vertical sides. Loukmas et al. (2003) evaluated enclosure size and clarity of view, comparing two sizes of boxes. They concluded that the larger box (23 × 23 × 81 cm) with a clear view was most effective for medium-sized riparian mammals, and that most species (regardless of size) preferred the large boxes

*Figure 4.3.* Track station in a natural enclosure formed by a basal hollow in a red fir (*Abies magnifica*). Photo from Sequoia-Kings Canyon National Park, California, by R. Green.

over the smaller ones (5 × 5 × 81 cm)—particularly feral cats and raccoons. Further, the majority of species did not display a preference for enclosures with a clear view, with the exception of feral cats. Indeed, particularly when targeting terrestrial mustelids, we recommend that the enclosure be blocked off at one end—either with the same material used for the enclosure, hardware cloth attached to the back, or a natural barrier such as a tree or fallen log. Mowat et al. (2000) suggested extending the roof and sides at both ends of the box to provide greater protection from the rain.

## Tracking Surface

A wide array of natural and artificial tracking surface materials have been used with track plates, resulting in track impressions of varying quality (figure 4.4).

*Figure 4.4.* Examples of tracks and track recording surfaces, including (A) a negative coyote print at a scent station, with sifted earth as the tracking medium (photo by J. Erb); and (B) American marten tracks on contact paper (with carbon soot as the tracking medium), collected from an enclosed track plate (photo by J. Ray).

For both track plots and scent stations, the tracking surface is composed of natural material available at the site, sometimes augmented by off-site materials. In certain cases, on-site material is removed to a variable depth and replaced with supplemental material (Randa and Yunger 2004), or tracking material is smoothed over a cardboard disk (Randa and Yunger 2006). Because unenclosed stations do not accommodate the use of a removable track-receptive surface (e.g., contact paper), the choice of substrate hinges on substrate quality and whether precise track measurements or permanent records will be required. One should not underestimate the time required to prepare and test the surface to maximize its ability to record tracks—careful attention to substrate preparation reduces the loss of data (Engeman et al. 2000). The area needs to be graded flat, vegetation and rocks removed, and the surface sifted (Roughton and Sweeny 1982; Harris and Knowlton 2001). The size of the tracking surface varies with the target species; Choate et al. (2006) suggested that the diameter of the plot should be larger than the average stride of the species (in this case, mountain lion) to ensure that the animal is forced to step onto the tracking medium while investigating the scent.

The extent to which existing material (usually soil or sand) in the vicinity of the station can be utilized depends on whether an appropriate consistency can be maintained and prints can be recorded with enough detail for species identification. Often, manipulation of the substrate will enhance print definition, through sifting or the addition of a moistening agent (e.g., mineral oil, water). When existing material is not suitable, fine soil, sand, clay, or another substance (e.g., lime, gypsum powder) can be hauled into the area (e.g., Diefenbach et al. 1994; Crooks 2002; Randa and Yunger 2006; Ascensäo and Mira 2007).

In one of the few studies to compare different types of track stations in one study area, Sargeant et al. (2003b) discovered that swift fox visitation rates to scent stations composed of natural materials were considerably higher than to track plates, or even to those stations where a natural medium was mixed with mineral oil. Randa and Yunger (2004) placed a strip of aluminum flashing around the rim of their track plots at ground level to prevent surrounding vegetation from invading the tracking surface. Other researchers (Allen et al. 1996) have made a discrete

mark in the corner of track stations to determine if rain, livestock, vehicles, or other disturbance agents were responsible for destroying tracks between visits.

Track plates have two or three tracking surface components: the hard plate, the medium covering the plate (which animals pick up on their paws while also leaving negative track impressions), and for enclosed stations, the recording surface onto which positive track impressions are deposited. Aluminum is most commonly used as the track-receptive surface. Indeed, the minimal weight, durability, low price, and availability of aluminum make it difficult to recommend anything else for larger devices. Zielinski (1995) recommended 1.59 mm (14 gauge) flat stock, but others (R. Long, Western Transportation Institute, pers. comm.) have utilized aluminum sheets as thin as 0.59 mm (23 gauge). Wooden hardboard has been used as a substitute in at least one study (Loukmas et al. 2003), while Spehard and Greaves (1984, cited in Raphael 1994) employed vinyl floor tiles, and Wolf et al. (2003) used foam core boards. In the case of New Zealand tracking tunnels, wood is used as the base, onto which a removable aluminum or molded plastic tray is placed (King and Edgar 1977; C. M. King, pers. comm.).

There has been considerable variation in the tracking media used with track plates, although the most common choice in North America has been carbon soot. As long as the tracking medium is deployed evenly across the plate, the surface can be effective for weeks—although tracking media should be tested for their uptake of prints each time they are checked and rebaited. In a comparison of tracking media, Belant (2003b) tested the relative efficacy of carbon soot, photocopy toner, and talcum powder. He found that the ability to identify tracks to species was similar for all three, but toner was most effective overall because it did not require specialized equipment to apply and was easier and safer to use than soot. Belant did not, however, comment on the relative environmental toxicity of toner versus carbon soot (primarily from the standpoint of ingestion by the target species). On this score, we suggest that surveyors obtain a copy of the Material Safety Data

Sheet (MSDS) for the toner in question to explore the possible risks—toxicity will vary, and old toners are generally more toxic. Nevertheless, it is unknown how the amount of toner typically used on a plate relates to those levels that have demonstrated toxicity in the laboratory; every effort should be made to use the least toxic variety. Researchers who have recently used toner for track plate projects have recommended that the dust be applied in a very thin layer. Although this technique does not render the plates black (as does carbon soot), it still produces readily detectible tracks. Moreover, preparation time is reduced with plates prepared with toner on site, and touch-ups are quick and easy (S. Yaeger, US Fish and Wildlife Service, pers. comm.).

Soot has many advantages in terms of track quality and lasting power in the field, but its drawback is that it must be applied under carefully controlled conditions—often off-site—because the need to ignite a flame renders soot application potentially hazardous. In general, soot is applied from an acetylene torch or with kerosene. As it can be considerably more convenient (especially in remote areas) to soot the plates on-site during station preparation rather than to prepare them in advance, J. Malcolm (University of Toronto, pers. comm.) developed a lightweight kerosene device from a paint can and a wick figure 4.5). Although the portability of such devices can make soot application an easy and efficient enterprise, this method is no less hazardous than a torch and should be especially discouraged in fire-prone areas. In fact, its use may be forbidden in some areas.

Moisture absorbed by talcum powder has resulted in limited efficacy in track definition (Belant 2003b), and a similar problem has been noted with chalk (Zielinski 1995). Clapperton et al. (1994a) sprayed plates with a chalk/alcohol suspension. Another unique surfacing approach combined a track plot with a track plate by smearing the plate with grease and covering it with fine sand (Angelstam 1986). Covering aluminum track plates with acetate sheets coated with a graphite/alcohol/oil mixture was shown to enhance water resistance in small mammal

*Figure 4.5.* Device for applying kerosene soot to a track plate, adapted by J. Malcolm (University of Toronto) from a handmade Brazilian lantern. Photo by J. Ray.

surveys (Connors et al. 2005) and holds promise for larger target species as well.

Ink (preferably water-resistant) is the most commonly utilized tracking medium for small mammal tracking tubes (Glennon et al. 2002; Nams and Gillis 2003), while a brown ferric-nitrate solution has been the principal medium used for tracking tunnels (King and Edgar 1977; Murphy et al. 1999; Ragg and Moller 2000; Jones et al. 2004). Inks are generally soaked in a pad rather than applied directly to the plate. More recently, New Zealand researchers have been using diluted blue food coloring as a substitute for the two-component chemical dye, as the former is inexpensive, simple to apply, and raises no environmental concerns (C. M. King, pers. comm.).

When tracks are to be deposited on a recording surface rather than imprinted as negatives on the tracking medium itself, contact or shelf paper is most commonly used (Fowler and Golightly 1994;

*Figure 4.6.* View from the inside of an enclosed track plate, with a fisher taking bait while depositing prints on contact paper. Photo by Hoopa Tribal Forestry.

Zielinski 1995; also see figures 4.4B and 4.6)—although paper is also suitable (Pedlar et al. 1997; Wolf et al. 2003). Attached with duct tape near the end of the plate (figure 4.7), sticky side up, contact paper provides a durable surface on which footprints will be faithfully recorded as the soot-covered paws of the target species walk over it. White contact paper is most sensible when the tracking medium is black (e.g., soot), whereas brown is the color of choice if the medium is white (i.e., chalk; Belant 2003b). In enclosed stations, the tracking medium and recording surface are usually placed in linear sequence such that the animal steps on the former and deposits prints on the latter while proceeding through the tunnel to investigate the bait at the far end. For unenclosed track plates, an alternative arrangement has been used by Pedlar et al. (1997) and Wolf et al. (2003), who covered the outside border of the plate with tracking medium and placed the recording surface and bait in the middle.

In tracking tunnel systems, paper usually serves as the recording surface. For the initial design of this device, King and Edgar (1977: 207) recommended a "coarse grade of Kraft or brown wrapping paper . . .

*Figure 4.7.* Coauthor J. Ray applies contact paper to the back of a track plate. Photo by R. Kays.

rough side up." With some inks, it is necessary to apply a solution (e.g., tannic acid/ethanol) to the paper itself to maximize the deposition of the ink onto the surface (King and Edgar 1977). Jones et al. (2004) noted that, while glossy white paper might yield superior-quality tracks, this paper could potentially deter some animals from entering the tunnel.

A completely different enclosed track plate setup deserves special mention for its innovative use of materials in an aquatic context. Reynolds et al. (2004) recently described a "floating raft" designed to record the tracks of American mink (*Neovison vison*). This device consists of a tracking medium set in a tray, which is placed on a buoyant plywood raft base to enable it to remain above the waterline in shoreline habitats. The entire device is enclosed in a wooden tunnel. The tracking medium is a plasticine-like compound (i.e., a 2:1 mixture of sand and clay) smeared on top of a piece of highly absorbent foam saturated with water to maintain a continuously moist tracking surface.

## Practical Considerations

If track stations are the device of choice, a number of key questions must be addressed prior to beginning the survey: Which type of track station should be used? What time of year should stations be de-ployed? Does the behavior of the target species affect how the station should be set up? What data should be recorded? Such practical issues are discussed in the following section.

### Selecting the Device

The four general types of track stations are not inter-changeable, with decisions on which device to use hinging upon the target species and survey conditions. There are three broad levels to consider during the decision-making process: (1) the target species group, (2) whether the survey is targeting a single or multiple species, and (3) characteristics of the study area and environment (figure 4.8).

Selection of the track station type depends largely on the taxonomic family. Mustelids (e.g., weasels, stoats, ferrets, fishers, minks, and martens) and mephitids (i.e., skunks) generally respond best to devices that are enclosed, which exploit the need of such species to investigate confined spaces for food, shelter, and denning (Loukmas et al. 2003). Success

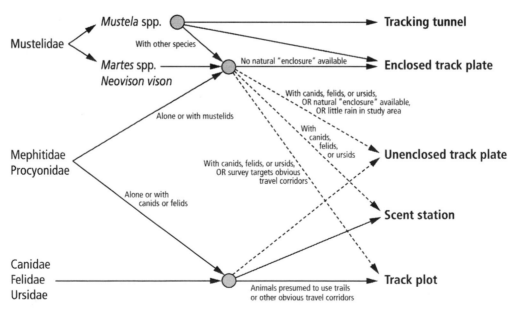

*Figure 4.8.* General decision tree for establishing which type of track station to deploy based on the target species and other circumstances. Solid arrows indicate the preferred choice, while broken arrows are secondarily suitable.

rates for detecting most of these species with unenclosed devices are considerably lower, with mustelids seldom appearing in relevant lists of target species (but see Foresman and Pearson 1998). Skunks (*Spilogale* and *Mephitis* spp.) are an exception in that they appear readily detectable by track plots and scent stations (Sargeant et al. 1996; Crooks 2002; Gehring and Swihart 2003).

Body size also comes into play. For example, weasels are most efficiently surveyed with tracking tunnels (King et al. 1994), which are not appropriate for larger species. Although track plate boxes (Zielinski 1995) were designed for the medium-sized martens and fishers, they are also quite effective at detecting weasels (e.g., Mowat et al. 2000; Zielinski et al. 2005). Other than smaller-bodied domestic cats (Loukmas et al. 2003; Gompper et al. 2006), felids and canids are only occasional visitors to enclosed track plates—hence, unenclosed devices are the preferred option for these taxa. Canids in particular seem hesitant to make contact with the artificial surfaces of track plates (e.g., swift fox; Sargeant et al. 2003b), and are better surveyed with track plots or scent stations. An apparent exception is the gray fox, which was one of the most commonly detected species at enclosed stations in California (Zielinski et al. 2005).

Planning for multiple versus single target species entails different considerations. If weasels are to be surveyed in conjunction with larger mustelids (e.g., fishers, martens), for instance, the track station must accommodate the size of the latter. Tracking tunnels, which were designed specifically for small-bodied mustelids, are most appropriate when weasels are the sole target carnivore, with the main advantage being that small tunnels prevent larger tracks from obliterating smaller ones. For some species, like martens and fishers, enclosed track plates are the only viable track station option; if additional species are to be surveyed (e.g, canids, felids), it will be necessary to deploy unenclosed track plates in addition to devices that better target the other species (e.g., Gompper et al. 2006; also see case study 8.3). Campbell (2004) compared the probabilities of detecting multiple species of carnivores in the Sierra Nevada mountains of California and found that the sample unit design that maximized the probability of detecting all thirteen target species comprised equal numbers of enclosed and unenclosed stations and no remote cameras (Campbell 2004; also see case study 8.2). Additional discussion concerning the use of multiple methods for detecting individual or multiple species is provided in chapter 8.

Environmental and other circumstances also influence track station selection. For example, frequent precipitation or logistical constraints that mandate longer sampling occasions call for the use of enclosures. On the other hand, detection devices that are deployed to record carnivore use of specific sites, rather than to sample more general areas, are most likely to be set passively (e.g., without bait) and to be unenclosed. For example, one study documented nest predator activity on a beach with track plots (Engeman et al. 2003b, 2005).

## Seasonal Timing

Track stations are typically deployed in the summer, when snow is not an issue, although some research suggests that detection rates may be greater in winter than summer (Bull et al. 1992). Carroll (1997) also found detection success to be somewhat lower in summer than other seasons, but the effect of season was eliminated after the effects of vegetation variables were accounted for. Zielinski et al. (1997) discouraged winter surveys for fishers and martens, both because snow interfered with the station-checking schedule and the target species were presumably less active in winter—and would thus encounter fewer stations. Sargeant et al. (2003b) found autumn to yield the highest swift fox visitation rates but noted that this might reflect dispersing juveniles rather than resident adults. Indeed, important seasonal factors abound; for example, surveys should be timed to avoid hunting seasons or human traffic on roads (Roughton and Sweeny 1982), target periods when species' movement is greatest, coincide with low prey availability, and minimize the spoilage of bait.

## Behavioral Considerations

Although track stations are designed to maximize the chance of recording a species, the probability of an individual visiting a station rests on a number of factors—including those relating to environmental characteristics, chance, and behavior. When a station is set up to detect animals on a travel route (e.g., in the middle of a trail), the tracking medium should record them as long as they are not repelled by the device. If an animal skirts or avoids the device, however, there will be no record of its occurrence. This would be characterized as a *neophobic* reaction to either the visual or olfactory aspects of the device. Ironically, bait or other attractants that are deployed to draw in members of a target species may also serve to repel them (Dobbins 2004).

Any number of socioecological factors might negatively or positively influence animal response to track stations, including species density, home range size, daily and seasonal activity patterns, movement patterns, foraging behavior, and prior experience (Hamm et al. 2003; Harris and Knowlton 2001; Smith et al. 2007). For example, Harris and Knowlton (2001) found that coyotes were more likely to visit stations placed along territorial boundaries or home range peripheries than within their home ranges. This might be explained by a greater tendency to avoid a novel stimulus when encountered in familiar as opposed to unfamiliar surroundings, and a consequence may be higher detection rates for transient or dispersing coyotes than for residents. Social status also affects detection probability at camera stations (Sequin et al. 2003) and presumably at track stations as well. Behavior-dependent variation in detections at track stations has been demonstrated for canids, felids, and raccoons (Allen et al. 1996; Winter et al. 2000; Harris and Knowlton 2001; Harrison et al. 2002; Choate et al. 2006; Gompper et al. 2006).

Remote videography has demonstrated that fishers sometimes use their abdominal gland to scent-mark in the immediate vicinity of bait stations (K. Aubry, USDA Forest Service, pers. comm.). Thus, it is very likely that species that use scent-marking to communicate will deposit scent at track stations and, furthermore, that track station locations could become social information centers. Support for this hypothesis has emerged from sites encompassing relatively small areas where track stations have been used for long periods of time to monitor fishers. During a study of fishers on the Hoopa Indian Reservation, California, enclosed track plates that contained no attractants recorded almost as many detections as baited stations (M. Higley, Hoopa Tribal Forestry, pers. comm.), presumably because individuals visited stations to maximize their encounters with social information. Thus, track stations may be attractive due to both food and the social information that can accumulate at the site.

## Materials and Costs

Costs associated with setting up and maintaining track stations are low relative to most of the techniques described in this volume, with the principal expenditure being labor and field transportation rather than materials. The simplest type of track station—track plots—can require little more than a handheld digging and raking implement to prepare stations using the existing substrate on the ground. The addition of other material (e.g., soil, clay, sand, mineral oil) to augment or replace the natural tracking substrate, along with manufactured lure or bait, modestly increases cost. These materials can be either purchased (generally in bulk amounts) or brought in from elsewhere; the per-station cost therefore depends on the number of stations to be deployed.

Track plates require more manufactured materials than track plots or scent stations and thus incur higher costs. Basic purchases include materials to create plates, tracking surfaces, enclosures, and tracking media, as well as baits and lures. Cost estimates per station are elusive because such a wide variety of materials can be used for each of the components but might range from \$2–\$20 USD/station. Additional supplies are necessary for removing and storing tracks, including acetate document protectors and permanent felt markers. A description of materials and methods required for the proper documentation of tracks and other sign can be found in Halfpenny et al. (1995). Meanwhile, commercial track plates have become widely available for homeowners and researchers alike (e.g., Trakka Kits [Connovation 2006]).

Although most track stations can be constructed with basic and readily available materials, the number of stations deployed and the need to replenish bait and tracking media will affect total costs through labor expenses. The remoteness and accessibility of sites will, of course, determine transportation costs. With the introduction of corrugated plastic to replace heavier construction materials for track plate enclosures, the prospects for traveling to track station sites by bicycle or on foot have increased.

## Survey Design Issues

Given that track stations have been extensively deployed for a plethora of purposes, it is not surprising that survey designs have been highly variable. Survey design must be governed by the survey objectives (see chapter 2) and the characteristics of the study area rather than the type of track station. Although protocols do exist that result in the more or less consistent application of track station devices (e.g., Zielinski 1995; Sargeant et al. 1996), there are many exceptions.

As with other detection devices, there is no rule book dictating how many track stations to deploy at a site, how long to deploy them, or the best arrangement for maximizing detections of target species. Track stations can be distributed with one or more stations per sample unit. Deploying multiple (versus single) stations per sample unit usually assures a greater chance of detecting target animals (Roughton

and Sweeny 1982; Diefenbach et al. 1994; Zielinski et al. 1997) by providing added detection opportunities and minimizing the loss of data when a station is rendered inoperable by wildlife (e.g., bears) or severe weather.

Sargeant et al. (2003a) evaluated various sampling designs for scent station surveys, including single-stage sampling (i.e., random or systematic sampling of single stations; reporting visitation rate as a proportion of *stations* with a detection) and cluster sampling (grouping stations within a sample unit, usually a transect or line; reporting visitation rate as a proportion of *sample units* with a detection). These researchers found that (1) single-stage and cluster designs were similarly sensitive indicators of change in visitation rates (for both red foxes and striped skunks); (2) cluster designs inflated variances such that larger samples were required to achieve acceptable precision; and (3) single-stage sampling required fewer samples. Further, Sargeant et al. (2003a) recommended that scent station survey designs can be compared by determining achievable sample sizes for each design and then comparing ratios of these sample sizes to ratios of variance inflation factors. In general, the number of stations to deploy depends on the probability of detection and the survey objective, and represents a tradeoff between economic and logistical constraints, the amount of information gathered, and the precision of the information required to address the survey objective.

The positioning of stations within a sample unit also varies. Different studies have distributed stations at regular intervals along line transects or other linear features (e.g., Bull et al. 1992; Zielinski et al. 2001; Graham 2002; Hamm et al. 2003, Gompper et al. 2006; Randa and Yunger 2006), as clusters (e.g., Zielinski et al. 1995b; Zielinski and Stauffer 1996), inside patches of vegetation to evaluate habitat configuration (Winter et al. 2001), along likely animal travel corridors, or based on ease of access to sites. Randomly placed sample units, without respect to roads or other linear features, achieve the least biased outcomes. Inference from road-based transect surveys, for example, can only be extrapolated to other roaded areas. Regardless of the arrangement of the sample unit and whether it has one or more stations, statistical independence is most likely achieved if the sample units are separated by at least one average home-range diameter in all directions (Harris et al. 2002; but see Hamm et al. 2003).

Survey effort is usually represented as the number of stations per sample unit multiplied by the number of twenty-four-hour periods for which track stations are functional (i.e., survey duration). Within bounds, the spatial and temporal dimensions of a survey can be substituted somewhat for wide-ranging carnivores (see also chapter 2, figure 2.8). If, for example, a unit of effort represents either a twenty-four-hour period or a single site (regardless of how many stations/site), one could achieve similar levels of effort if track stations were deployed at either 50 or 100 sites for either four or two days, respectively. The critical consideration is whether the probability of detection is high enough to expect that most detections will occur during the survey period. Thus, if the probability of detection is low (and LTD is relatively long), the 50-site/four-day scenario will be preferable.

The substitutability of temporal and spatial elements of effort has been questioned by some researchers. Sargeant et al. (2003a, b), for example, explored the design elements of scent station surveys for kit and swift foxes and found that the repeated operation of the same scent stations yielded less information than could be obtained by establishing new stations—a conclusion reached by Rhodes et al. (2006) as well. Regarding total survey duration, Gompper et al. (2006) suggest that a standard two-week survey duration will detect most species of carnivores that are present and that shorter surveys may be adequate, but only for estimating occupancy or for achieving counts of detections for the more common species. The general tradeoff between distributing effort among sites or survey days, for meeting various objectives, is discussed in chapter 2.

Because the success of track station surveys usually depends on the precision and magnitude of visitation rates, there has been considerable discussion on this subject. Some have argued that the precision of track station indices can be improved by designs that increase visitation rates (e.g., Diefenbach et al. 1994), but others suggest that increasing visitation rates also, necessarily, increases sampling variation (Sargeant et al. 2003a). Diefenbach et al. (1994) noted that the variance of a binomial variable is related to the mean, and that the variance of visitation rate increases to a maximum at 0.50. Thus, they conclude that increasing visitation rates such that they approach 0.50 may be detrimental to precisely estimating occupancy. Similarly, Sargeant et al. (2003a) suggested that visitation rates should be >0.10 in order for a scent station to perform well, but they could not defend the benefits of further increases. There is probably no optimal visitation rate for all circumstances and all species, and the effects of visitation rates on surveys designs have not been fully explored. In particular, visitation rates are a function of abundance (e.g., Smith et al. 2007); therefore, rates achievable for common species may not be for rare species. The extent to which this issue affects the efficiency of designs for rare species has not yet been adequately addressed.

## Sample and Data Collection and Management

Track station types differ as to whether track identifications must be made on the spot or imprints can be taken off-site to confirm identification. Permanently recorded tracks allow for careful measurements, comparison with other track samples, and input from multiple individuals—including experts. Distinguishing superficially similar species will often require multiple, careful measurements of track impressions, a process that will be facilitated considerably by being able to remove the track from the field. With track plots and scent stations, removal of negative tracks from the site will obviously not be possible. Not only does this necessitate careful measurements in the field, but photographs of tracks should be considered.

When positive track impressions are permanently recorded on contact or other paper, these recording media must be removed from the field in as intact a condition as possible with absolute certainty of their origin. A common approach for tracks on contact paper is to store them immediately in acetate document protectors (Zielinski and Truex 1995). The process is even simpler with tracking tunnels employing inks, as it takes only seconds to slip the paper from the tray, label it, and insert a new piece. If necessary, track impressions on the sooted portion of a track plate can be lifted by carefully pressing transparent tape (or a piece of contact paper) over the tracks, rubbing the back of the tape with an embossing tool (e.g., a capped pen), and then transferring the tape to a white data sheet (Taylor and Raphael 1988; F. Schlexer, USDA Forest Service, pers. comm.).

The most important data to record from track stations are station identity, presence or absence of tracks, species identity, and date. If the survey includes sample units with multiple track stations, each station should have a unique identifier also directly tied to the sample unit. It is as important to document track absence as presence in order to confirm that each track station has indeed been checked. Using track station detections for indirect estimates of population size requires accurately estimating LTD and probability of detection (see chapter 2). These metrics call for carefully recording the results *every time a station is checked* (i.e., creating an encounter history), not simply a summary of the results after all visits to the station or sample unit have been completed. Further, it is vital to record the approximate number of days that a station has been out of commission for any reason (e.g., collapsed by a bear) so that survey effort can be reported as accurately as possible. Deviations from established protocols (e.g., if a different bait or scent is used) should also be reported. As a general rule, the more data one can record the better, as future unanticipated analyses may be able to incorporate this information. Track records that are not unambiguously related to their time and place of origin will be useless.

## Future Directions and Concluding Thoughts

Thirty years ago, it would have been impossible to predict the form and function of the modern day track station. In New Zealand, for example, the original small mammal tracking tunnel has evolved into the Scentinel (figure 4.9)—an automated monitor equipped with a digital camera, weighbridge, PIT tag reader, remote reporting capability, and bait delivery mechanism, all of which are controlled by a programmable electronic board (King et al. 2007; Technology Transfer, Wellington, New Zealand, www.scentinel.co.nz).

In North America, the most recent advances have involved outfitting track plate stations with devices that can collect genetic material (i.e., hair) from their visitors (see chapters 6 and 8). Zielinski et al. (2006) attached a hair snare made from commercially available, glue-impregnated cardboard sheets designed to trap mice (Catchmaster, Atlantic Paste & Glue, Brooklyn, NY). Sheets were cut into strips and tacked to a wood furring slat inserted through the sides at the front of the enclosure (chapter 6; figure 6.10B, C). The ability to identify individuals via genetic material, track measurements, or other means

*Figure 4.9.* The Scentinel, a device used to target stoats in New Zealand, has evolved from a traditional tracking tunnel (King and Edgar 1977) into an automated monitor. Photo by C. King.

will increase the capacity of track stations to monitor population sizes directly via application of mark-recapture methods rather than indirectly through indices.

The identification of tracks from track stations has become an increasingly quantitative exercise, but there are still species for which professional judgment and experience must be used. Quantitative methods, which rely on multivariate statistics (Manly 1994), will need to be developed to distinguish between pairs of species with similar tracks, such as marten/mink, mink/weasels, ermine/long-tailed weasel (*Mustela frenata*), coyote/domestic dog (*Canis lupus familiaris*), and bobcat/domestic cat—particularly where the members of each pair are sympatric. We envision a future where the identification of all tracks can be accomplished by biologists with limited field experience but who are trained to use peer-reviewed quantitative methods to distinguish the tracks of similar species.

In the short term, future innovations to track stations will likely continue to focus on enhancing our ability to derive true population estimates from this noninvasive survey method. Doing so will involve not only modifying the setup to acquire both genetic material and tracks, as described earlier, but also pursuing new analytical approaches (see chapter 11). We hope that future innovations will preserve the attributes of track stations that make them so easy to use. The simplicity of this family of methods guarantees that it will continue to be used by scientists, as well as informed citizens, in the service of carnivore conservation.

---

### CASE STUDY 4.1:
### FISHER POPULATION MONITORING
### IN THE SIERRA NEVADA

*Richard L. Truex and William J. Zielinski*

**Location:** Sierra Nevada Mountains, California, USA.

**Target species:** Fisher, with incidental detections of other species.

**Size of survey area:** >50,000 km².

**Purpose of survey:** The relatively small size and isolation of the fisher population in the southern Sierra Nevada, one of two extant populations in California, has raised concerns regarding its viability. The USDA Forest Service identified monitoring this population as a high-priority component of an adaptive management strategy developed in conjunction with the Sierra Nevada Forest Plan Amendment (USDA 2001), and a regional population monitoring program was initiated in 2002. The objectives of this ongoing effort are to:

1. monitor the relative abundance of the southern Sierra Nevada fisher population;
2. document fisher range expansion in the Sierra Nevada;
3. monitor the population statuses and trends of other forest carnivores routinely detected by track plates (including American martens, gray foxes, ringtails, and western spotted skunks [*Spilogale gracilis*]);
4. develop and/or refine habitat models for fishers and other forest carnivores.

**Survey units:** Six enclosed track stations arranged in a pentagonal array, with track stations on the perimeter of the array deployed at 72° intervals and approximately 500 m from the center station.

**Survey method:** Enclosed track stations.

**Survey design and protocol:** The survey design included two components—intensive monitoring of the area occupied by fishers, and targeted sampling at "sentinel sites" in the currently unoccupied central and northern Sierra—designed to document population expansion should it occur. The approach relied on detection-nondetection sampling to detect changes in an index of population abundance over a ten-year period. Zielinski and Mori (2001) proposed annually resampling 288 sites to detect 20% declines in the index of abundance with 80% power. The relationship between the index of abundance and population size was unknown but assumed to be positive.

In the intensive monitoring portion of the study

area, sampling was based on the nationwide Forest Inventory and Analysis (FIA) inventory program, with all grid points occurring within the historic range of fisher at suitable elevations (800–3,200 m) in the southern Sierra Nevada (Grinnell et al. 1937) initially considered for sampling. A random selection of the resulting grid points was chosen for the intensive monitoring portion of the study area. Sentinel sites in the central and northern Sierra were selected opportunistically based on appropriate habitat conditions.

Field sampling procedures followed those of Zielinski et al. (2005), with each sample unit separated by approximately 5 km (figure 4.10). All track stations were baited with chicken, and approximately 10 ml of a commercial trapping lure were applied in the immediate vicinity. Technicians visited each sample unit every two days to collect tracks and rebait each station, and each array was surveyed for ten consecutive days. Sampling occurred annually for five months each year (June through October) from 2002 to 2006.

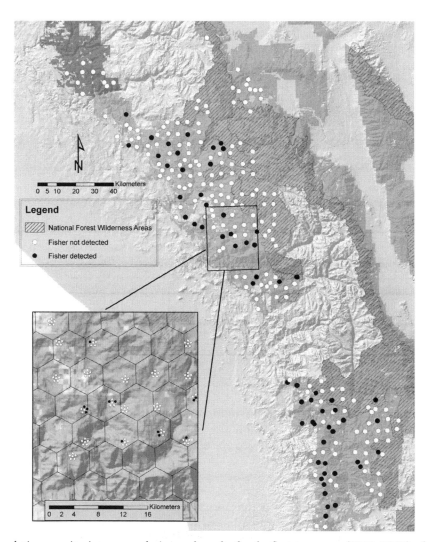

*Figure 4.10.* Fisher population monitoring survey design and results for the first two years (2002–2003) of a proposed ten-year monitoring program in the southern Sierra Nevada, California (Sequoia, Sierra, and southern Stanislaus National Forests; shaded). The inset map highlights the southern portion of the Sierra National Forest, includes the hexagonal grid used to define the FIA sampling frame, and displays sample unit geometry and station-level results for 2002 and 2003.

*Table 4.3.* Detection rates (i.e., proportion of primary sample units with detections) for fishers and other forest carnivores in the Sierra Nevada, California, from 2002 to 2005

| Year | Detection rate | | | | |
|------|--------|-----------------|----------|----------|-----------------------|
| | Fisher | American marten | Gray fox | Ringtail | Western spotted skunk |
| 2002 | 0.268 | 0.176 | 0.205 | 0.087 | 0.199 |
| 2003 | 0.234 | 0.167 | 0.162 | 0.084 | 0.126 |
| 2004 | 0.238 | 0.144 | 0.139 | 0.089 | 0.178 |
| 2005 | 0.248 | 0.084 | 0.175 | 0.143 | 0.162 |

Note: Results for fishers are limited to the southern Sierra Nevada (Sequoia, Sierra, and southern Stanislaus National Forests), while those for other species include sampling conducted throughout the Sierra Nevada. The apparent decrease in marten detections was likely a sampling artifact rather than a true population decline; sampling during 2004 and 2005 emphasized habitats that were less likely to be occupied by martens.

**Analysis and statistical methods:** Preliminary analysis has been limited to calculating the proportion of sites occupied on an annual basis for fishers and select forest carnivores. Fisher population trend (i.e., change over time) will ultimately be analyzed using a logistic model that includes a trend parameter (Zielinski and Mori 2001) and will include a bias adjustment estimated by modeling fisher detectability (sensu Azuma et al. 1990; Mackenzie et al. 2002).

**Results and conclusions:**

- Throughout the study area, 708 sample units were surveyed. In the southern Sierra Nevada intensive monitoring area, 510 sample units were completed.
- The proportion of sites occupied annually by fishers varied from 0.234–0.268, and occupancy rates for other species varied in a similar manner (table 4.3).
- Preliminary results indicated that fishers were well distributed in portions of the Sequoia and Sierra National Forests (figure 4.10), though annual occupancy rates were consistently higher in Sequoia (i.e., 33.3%–41.1%) than in the Sierra National Forest (i.e., 14.5%–22.7%). Fishers were not detected in the central and northern Sierra, where >150 sentinel sites were sampled. There was no evidence to suggest that fisher populations had expanded into this region.
- Achieving the desired 288 sample units per

year in the southern Sierra was logistically and financially prohibitive. Each sample unit was completed every two years rather than every year as originally proposed (Zielinski and Mori 2001). Twenty years will be required to achieve the ten time steps necessary to detect declines in the index of abundance. Detecting change will be possible after ten years, but the required magnitude will be marginally larger and will be detected with slightly lower power.

- Hair snares (Zielinski et al. 2006) were added to the enclosed track stations such that genetic information can complement the occupancy information currently being collected.

---

## CASE STUDY 4.2:
## INTERAGENCY COOPERATIVE SCENT STATION SURVEY

*John Erb and Justina C. Ray*

**Location:** Minnesota, USA.

**Target species:** Red fox, gray fox, striped skunk, raccoon, coyote, bobcat, gray wolf. Domestic cats and dogs, while not explicitly targeted, were often recorded as well.

**Size of survey area:** Approximately 225,000 km$^2$ (area of Minnesota).

**Purpose of survey:** Minnesota's Interagency Cooperative Scent Station Survey (also known as the Predator and Furbearer Scent-Post Survey) was initiated in 1975 and utilized the US Fish and Wildlife Service's (FWS) protocol for monitoring coyote populations in the western United States (Linhart and Knowlton 1975; Roughton and Sweeny 1982). Although the FWS ultimately abandoned their ambitious effort in 1981, Minnesota Department of Natural Resources (DNR) personnel increased the number of survey routes and have continued this terrestrial carnivore survey—with various modifications to the methodology—without interruption for thirty years. This is currently the only long-term survey of its kind in North America that has deployed scent stations at this scale.

Because surveys of this nature are more tenuous for monitoring rarely detected species or wide-ranging carnivores at local scales (Sargeant et al. 1998), the chief focus is limited to monitoring broad temporal trends of the primary target species at intermediate (regional) scales. While the data may offer additional insights (e.g., detections of rarely detected species), uncertainty is appropriately acknowledged.

The objectives of this ongoing survey are to

1. track long-term population trends of terrestrial carnivore species in forest, transition, and farmland habitat zones of Minnesota;
2. determine the distribution and interspecific overlap (e.g., fox versus coyote, wolf versus coyote) of numerous carnivore species at intermediate spatial scales;
3. document annual changes in relative abundance indices of some species;
4. educate the public about carnivore distribution and relative abundance trends in the state of Minnesota.

**Survey units:** Survey routes are approximately 4.3 km in length, with each accommodating ten track stations spaced 500 m apart.

**Survey method:** Scent stations composed of a 0.9-m diameter circle of sifted soil, with a fatty-acid scent tablet placed at the center.

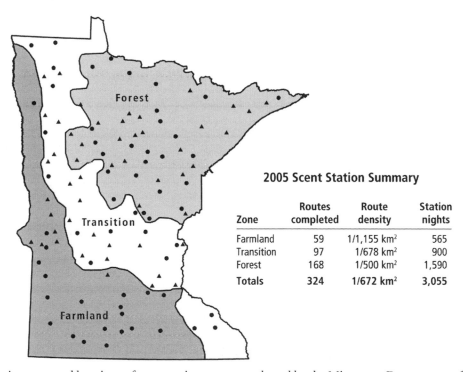

**2005 Scent Station Summary**

| Zone | Routes completed | Route density | Station nights |
|------|------------------|---------------|----------------|
| Farmland | 59 | 1/1,155 km² | 565 |
| Transition | 97 | 1/678 km² | 900 |
| Forest | 168 | 1/500 km² | 1,590 |
| **Totals** | **324** | **1/672 km²** | **3,055** |

*Figure 4.11.* Approximate central locations of scent station routes conducted by the Minnesota Department of Natural Resources (circles) and interagency cooperators (triangles). Each marked location represents one to six actual routes, and the inset table summarizes the 2005 routes.

**Survey design and protocol:** Survey routes are evenly distributed throughout the state of Minnesota (figure 4.11). The routes are nonrandomly placed along existing unpaved roads and trails with a minimum spacing of 5 km. Routes are initially selected by DNR field staff, who are instructed to place them, to the best of their ability, in habitat types representative of their overall work areas. While the number of routes has increased over time (most rapidly from the mid-1970s through the early 1980s), locations of existing routes have been maintained whenever possible. The sampling design is considered a two-stage cluster design. Although the goal is to complete all routes each year (September–October), the number completed varies from year to year (e.g., 381 in 2005 and 324 in 2006; Erb 2006a, b) depending on factors such as weather and personnel availability.

**Analysis and statistical methods:** Track presence and nondetections are recorded at each station. While Minnesota DNR generally analyzes results on the basis of track stations as sample units (e.g., Erb, 2006a, b), Sargeant et al. (1998) conducted analyses using survey routes as the sample units. Track indices were recorded as the percentage of track stations visited by each species (Erb 2006a), and route indices were calculated as the percentage of survey routes visited by each species (Sargeant et al. 1998). Arguments can be made for both the usefulness and limitations of each index (Sargeant et al. 1998; Sargeant et al. 2003b). Spatial dependence testing revealed spatial correlations between stations that extended to approximately 2,000 m (Sargeant et al. 1998), particularly for wide-ranging carnivores.

Over time the relatively even distribution of survey routes throughout the state has afforded the opportunity to stratify analyses by regional unit, primarily on the basis of broad habitat designations (e.g., forest, farmland, and transition). Some analyses have been based on biogeographic sections (see Sargeant et al. 1998; also see figure 4.11). Since 2001, confidence intervals have been computed using bootstrap methods (following Thompson et al.

1998), with an aim towards improving interpretations of annual changes in population indices.

**Results and conclusions:**

- Results of each year's scent station surveys are posted on the Minnesota DNR website during the subsequent year. Generally speaking, results are presented for statewide and habitat-specific route visitation rates (i.e., percentage of routes visited) for red foxes, domestic cats, skunks, raccoons, dogs, and coyotes. Additionally, long-term station visitation indices (i.e., percentage of stations visited) have been graphed for each species in each habitat from 1975 to the present. Other analyses (e.g., for other species) are periodically conducted but are not typically presented in annual reports.

- In the most recent survey (2005), 324 routes and over 3,000 track stations were completed statewide. Among the target species, striped skunks were most commonly detected (on 38% of all routes), followed closely by raccoons (35%), domestic cats (35%), and red foxes (34%), and then by domestic dogs (19%) and coyotes (18%). Red foxes were the only target species among the six reported with highest detection rates in the forest zone; all other species were detected much more frequently in farmland and transition zones.

- Regarding long-term trends, the most compelling results to date have been the recorded changes in red fox and coyote populations. Red fox indices, after increasing for more than ten years, began a steady decline in about 1990 in the farmland and transition zones, at the same time that coyotes began to increase in farmland zones (figure 4.12). Raccoons increased in the farmland zone from 1981–98, then subsequently declined and possibly stabilized coincident with apparent increases in disease prevalence. Generally speaking, trends for other species have exhibited fluctuations over time but without any discernible long-term

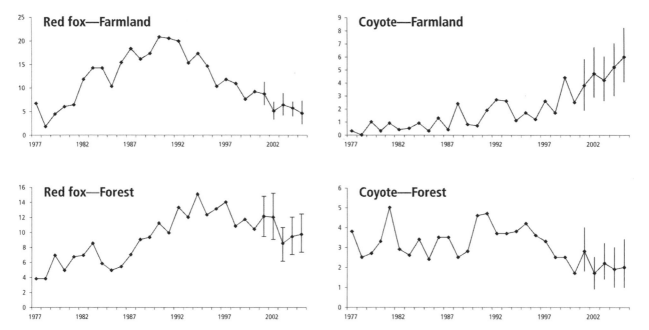

*Figure 4.12.* Percentage of scent stations visited by red foxes and coyotes in farmland and forest zones of Minnesota, 1977–2005. Confidence intervals have been computed using bootstrap methods since 2001.

trends. Wolf indices, however, have increased from 1976–99.

---

## Acknowledgments

We are grateful to the many individuals—too numerous to mention—who have worked with and made continual improvements to the devices detailed in this chapter and have documented lessons learned along the way. Kim King, Jerry Belant, and Kevin Crooks made comprehensive and insightful comments on earlier versions that stimulated marked improvements to our manuscript. We thank John Erb and Rick Truex for providing case study material, Elizabeth Andrews for discussions that improved the chapter, and Nick Buckler and Michelle McKenzie for editorial assistance. We are indebted to Paula MacKay and Robert Long for their editorial prowess and continuous feedback, which greatly improved the chapter.

# Chapter 5

# Remote Cameras

*Roland W. Kays and Keith M. Slauson*

Observing animals firsthand is one of the most rewarding aspects of being a field biologist or a natural resources manager. Although most animals are seldom seen in the wild—especially rare and shy carnivores—such interactions have incredible power to connect humans with nature. Unfortunately, wildlife sightings are too few and far between to address modern scientific and conservation issues. For these purposes, we need large data sets documenting which species live where and when.

Remote cameras allow us to extend our observations through time and space, and to create a permanent record of passing animals. Here we review the use of remote cameras (also known as camera traps or game cameras) as a scientific method for studying carnivores. First, we provide some brief history and background on remote cameras as a wildlife research tool, and we then present numerous objectives that can be met with cameras. Choosing which remote camera to purchase is increasingly complicated, and technology is changing quickly. Thus, rather than recommend particular brands, we discuss the components involved and the tradeoffs associated with different trigger and camera (i.e., film, digital, and video) types. Next we describe various approaches to deploying cameras in the field, as well as some practical tips for maintaining cameras and sensors and

storing camera data. Finally, we offer our ideas on the future of remote cameras for wildlife research.

## Background

Not long after the invention of flash photography, an automatic trip-wire camera was engineered to photograph animals (Guggisberg 1977). The resulting photographs were so exciting and unique that they regularly won awards and appeared in *National Geographic* magazine between 1906–21 (Sanderson and Trolle 2005). A few zoologists recognized the scientific potential of this device and learned how to use bulky camera equipment, wires, and flash powder to document the variety of wildlife in an area (Carey 1926; Chapman 1927; Gregory 1929). Unfortunately, the equipment remained unwieldy and complicated for another half century, despite Abbott and Coombs's (1964: 327) boast that their system was "only 47 pounds." Scientific uses were therefore limited to studying animal activity at a handful of sites (Pearson 1959; Osterberg 1962). Generations of scientists and land managers viewed remote cameras as fun but impractical tools because of the cost, time, and effort they required. By the time cameras became more straightforward and less expensive in the 1980s, biologists were slow to integrate them into their work. Initial attempts to do so combined inexpensive, 110 cameras and other basic materials (e.g., wood stakes, fishing line) to create a line-triggered camera system. Despite the shortcomings of these simple systems (e.g., no automatic film advance, frequent failure in rain or snow), their brief legacy produced detections of American martens (*Martes americana*) and fishers (*Martes pennanti*) in the US Pacific states (Jones and Raphael 1993; Kucera et al. 1995a).

In the late 1980s, deer hunters began to adopt remote camera technique en masse, using them to scout potential hunting grounds for trophy bucks. Manufacturers scrambled to meet this new demand for game cameras by combining 35 mm cameras with active and passive infrared motion sensors. Remote cameras quickly transitioned from being bulky, complicated, and expensive to small, simple, and cheap. Biologists finally rediscovered remote cameras in the 1990s, having realized that statistical tools developed for other survey methods could also be applied to camera data if a sufficient number of cameras were used. Early scientific publications reflect three of the main objectives we accomplish with remote cameras today: detecting the presence of a species (Kucera et al. 1995a; Foresman and Pearson 1998), estimating animal abundance (Mace et al. 1994; Karanth 1995; Jacobson et al. 1997; Karanth and Nichols 1998), and recording animal behavior (Carthew and Slater 1991; Bull et al. 1992; Vanschaik and Griffiths 1996).

## Target Species

Any animal that moves can be studied with a motion-sensitive remote camera. Although most camera-based carnivore studies have focused on medium and large species (see table 5.1 for a comprehensive overview), body size is not a design limitation. Indeed, some of the first quantitative remote camera studies targeted mice (Pearson 1959; Osterberg 1962). There is no single camera setup that will detect all carnivore species at a survey site, however. For any given target species, the motion sensor must be set in an area where the species is likely to visit, and the camera positioned such that it can capture the full animal with appropriate detail. This requires not only an understanding of natural history, but also of the technical function of remote cameras.

## Strengths and Weaknesses

In the context of wildlife research, remote cameras offer many advantages (box 5.1). Reviews of noninvasive survey techniques repeatedly grant remote cameras the highest recommendations for meeting study objectives (Foresman and Pearson 1998; Harrison et al. 2002; Silveira et al. 2003; Gompper et al. 2006; O'Connell et al. 2006). Remote cameras are often favored because they return abundant data on elusive species with relatively minimal labor (depending on the accessibility and size of the study area). Further, because the equipment does not typically require the target animal to step on or rub against the device, cameras are also minimally intrusive—especially if used without a visible flash. Indeed, by not demanding any special behavior from an animal other than its walking in front of the sensor, remote cameras probably have the lowest inherent bias of device-based survey methods and regularly photograph even the shyest species.

Although the position of the camera and sensor need to be customized for the target animal's size, most camera setups are likely to return data on multiple species—including both predators and prey. The resulting photographs provide a permanent record and typically yield unambiguous species identification. In addition, the information in a photograph (or video) often exceeds species identification, and sometimes includes individual identification (Karanth and Nichols 1998; Magoun et al. 2005) or behavioral details (Otani 2001; Hegglin et al. 2004; Rovero et al. 2005; Stevens and Serfass 2005). The capacity for the latter will increase with the evolution of remote video cameras (see *Video Cameras* later in this chapter). Last, photographs and videos produced through remote cameras are often beautiful and

*Table 5.1.* References for remote camera surveys of North American carnivore species

| Species | References |
|---|---|
| Coyote | Lyren 2001; Moruzzi et al. 2002; Sequin et al. 2003; Pina et al. 2004; Ng et al. 2004; Reed and Leslie 2005; Gompper et al. 2006; O'Connell et al. 2006 |
| Gray wolf | Clevenger and Waltho 2005 |
| Gray fox | Kucera and Barrett 1993; Moruzzi et al. 2002; Stake and Cimprich 2003; Kelly 2003; Pina et al. 2004; Zielinski et al. 2005 |
| Island fox | Clifford 2006 |
| Swift fox | Harrison et al. 2002 |
| Red fox | Moruzzi et al. 2002; Hegglin et al. 2004; González-Esteban et al. 2004; Zielinski et al. 2005; Mata et al. 2005; Gompper et al. 2006; O'Connell et al. 2006 |
| Ocelot | Maffei et al. 2002; Trolle and Kery 2003; Pina et al. 2004; Ziegler and Carroll 2005; Srbek-Araujo and Chiarello 2005; Maffei et al. 2005; Dillon and Kelly 2007 |
| Margay | Kelly 2003 |
| Canada lynx | Foresman and Pearson 1998; Poszig et al. 2004 |
| Bobcat | Kucera and Barrett 1993; Lyren 2001; Moruzzi et al. 2002; Cain et al. 2003; Heilbrun et al. 2003; Ng et al. 2004; Reed and Leslie 2005; Heilbrun et al. 2006 |
| Jaguar | Maffei et al. 2002; Wallace et al. 2003; Kelly 2003; Silver et al. 2004; Silver 2004; Pina et al. 2004; Maffei et al. 2004; Rizzo 2005; Maffei et al. 2005; Soisalo and Cavalcanti 2006 |
| Cougar | Chapman 1927; Kucera and Barrett 1993; Pierce et al. 1998; Maffei et al. 2002; Ng et al. 2004; Pina et al. 2004; Srbek-Araujo and Chiarello 2005; Clevenger and Waltho 2005; Maffei et al. 2005; Reed and Leslie 2005 |
| Jaguarundi | Maffei et al. 2002; Srbek-Araujo and Chiarello 2005 |
| American hog-nosed skunk | Kelly 2003 |
| Striped skunk | Moruzzi et al. 2002; Ng et al. 2004; Zielinski et al. 2005; O'Connell et al. 2006 |
| Western spotted skunk | Jones and Raphael 1993; Kucera and Barrett 1993; Ng et al. 2004; Reed and Leslie 2005; Zielinski et al. 2005 |
| Wolverine | Copeland 1993; Foresman and Pearson 1998; Zielinski et al. 2005; Magoun et al. 2005 |
| North American river otter | Stevens and Serfass 2005; O'Connell et al. 2006 |
| American marten | Bull et al. 1992; Kucera and Barrett 1993; Jones and Raphael 1993; Kucera et al. 1995a, b; Brooks 1996; Jones et al. 1997; Foresman and Pearson 1998; Moruzzi et al. 2002; Moruzzi et al. 2003; Zielinski et al. 2005; Gompper et al. 2006; Zielinski et al. 2007 |
| Fisher | Jones and Raphael 1993; Kucera and Barrett 1993; Aubry et al. 1997; Foresman and Pearson 1998; Fuller et al. 2001; Moruzzi et al. 2002; Moruzzi et al. 2003; Zielinski et al. 1995a; Zielinski et al. 2005; Gompper et al. 2006 |
| Ermine | Moruzzi et al. 2002; Zielinski et al. 2005; Gompper et al. 2006 |
| Long-tailed weasel | Moruzzi et al. 2002; Zielinski et al. 2005; Gompper et al. 2006; O'Connell et al. 2006 |
| Least weasel | Osterberg 1962; González-Esteban et al. 2004 |
| American mink | Kucera and Barrett 1993; González-Esteban et al. 2004 |
| American badger | Pina et al. 2004; Reed and Leslie 2005; Zielinski et al. 2005 |
| Ringtail | Kucera and Barrett 1993; Stake and Cimprich 2003; Zielinksi et al. 2005 |
| White-nosed coati | Kelly 2003 |
| Raccoon | Jones and Raphael 1993; Page et al. 2001; Yasuda and Kawakami 2002; Moruzzi et al. 2002; Wolf et al. 2003; LaPoint et al. 2003; Ng et al. 2004; Zielinski et al. 2005; Gompper and Wright 2005; Gompper et al. 2006; O'Connell et al. 2006 |
| American black bear | Kucera and Barrett 1993; Moruzzi et al. 2002; Bridges et al. 2004; Zielinski et al. 2005; Clevenger and Waltho 2005; Gompper et al. 2006 |
| Grizzly bear | Mace et al. 1994; Clevenger and Waltho 2005 |

Note: The exclusion of certain species from this table reflects a lack of published accounts of their detection via this survey method.

---

**Box 5.1**

### Strengths and weaknesses of remote cameras

| Strengths | Weaknesses |
|---|---|
| • Low inherent bias.<br>• Can be used with or without bait.<br>• Species identification usually unambiguous.<br>• Detect multiple species.<br>• Photos serve as permanent record.<br>• Convey multiple types of information (e.g., age, behavior, time of day).<br>• Photos provide outreach opportunities.<br>• Can require minimal labor (depending on checking frequency). | • Costly equipment.<br>• Complicated equipment.<br>• Equipment malfunctions.<br>• Limitations of batteries and film.<br>• Technology is changing rapidly.<br>• High risk of theft.<br>• Individual identity only possible for species with distinctive marks or pelage. |

---

captivating and can thus be valuable for public outreach and fundraising.

The methodological disadvantages of remote cameras stem from their reliance on electronics and, in some cases, film. Compared with other methods, equipment costs to initiate a study are substantial, and equipment may fail in harsh field conditions—resulting in replacement or repair expenses. Some camera models can be complicated to use, requiring careful training of field staff to maintain consistent operation. The frequency with which cameras must be checked can be a drawback if high animal activity consumes a roll of film too quickly or unpredictably, although this is becoming much less of a problem with the evolution of digital models. Further, unless the target species' pelage pattern varies among individuals (e.g., tigers, *Panthera tigris*, Karanth and Nichols [1998]) or animals are previously tagged with visible markers, camera methods do not permit individual identity—unlike genetic methods that sample scats or hair. Thus, estimating abundance and density can be difficult with data collected using cameras (but see *Abundance and Density*). Finally, there is little standardization among most models of remote cameras, and components change over time as camera and sensor technologies advance. The re-

sult is that comparisons among studies using different equipment may be compromised.

## Treatment of Objectives

It is important to consider one's objectives carefully before beginning a remote camera survey. Each survey will have its own set of assumptions and requirements for survey design and camera setup.

### Occurrence and Distribution

At the most basic level, remote cameras can confirm which species are present at a particular site. In some extreme cases, a single image can verify that a species is not extinct (Brink et al. 2002; Moruzzi et al. 2003). More often, an array of cameras is deployed throughout a region to document the diversity of species, or the geographical pattern of occurrence for a target species. These detection-nondetection data are suitable for various analyses, including those pertaining to the determinants of animal distribution, habitat preferences, and response to fragmentation (Lomolino and Perault 2001; Farhadinia 2004; Zielinski et al. 2005; Linkie et al. 2006).

## Relative Abundance

For most species, multiple visits from the same animal are difficult to distinguish via remote cameras; thus, methods to derive meaningful indices of relative abundance from camera data have been elusive. One approach that may prove useful analyzes the temporal pattern of species detections to calculate more accurate estimates of relative abundance (MacKenzie et al. 2006; also see chapter 11). *Latency-to-first-detection* (LTD, i.e., the amount of time required for a camera or other device to first detect a species) has also been put forth as a possible measure of abundance, although it was not valid in one test with coyotes (*Canis latrans*; Gompper et al. 2006).

In some cases, abundance can be estimated even if animals cannot be individually identified from photographs. Although the photograph rate for a species may be related to its local abundance, this has only been calibrated for tigers and remains a controversial metric (Carbone et al. 2001; Carbone et al. 2002; Jennelle et al. 2002). Such an approach is essentially the inverse of traditional visual surveys where an observer walks down a trail and counts the number of animals seen, noting their distance from the trail as an indicator of the area surveyed. With remote cameras, the animals do the walking, and the area covered by the camera's sensor is a measure of the area surveyed. Further, recent work suggests that an adjustment for the typical movement rates of each species can be used to convert raw detection rates into real density estimates, presuming randomized camera placement (J. Rowcliffe, Institute of Zoology, London, pers. comm.).

## Abundance and Density

Data from cameras are increasingly used to estimate animal abundance and density, and potentially produce accurate estimates if proper attention is given to study design, camera deployment, and data analysis. The gold standard for abundance estimation is the *mark-recapture* analysis of the pattern of detections of individual animals (Nichols and Dickman 1996; see chapters 2 and 11 for discussions of capture-recapture methods). This standard can be achieved with remote cameras only if individual animals can be identified via patterns in their pelage (Karanth and Nichols 1998; figure 5.1), or conspicuous external marks (e.g., colored or numbered ear tags; figure 5.2) applied during prior live-trapping (Mace et al. 1994; Fuller et al. 2001). Other than spotted cats (Heilbrun et al. 2003; Heilbrun et al. 2006), skunks (Kays and Wilson 2002), and perhaps, in some cases, coyotes (Larrucea et al. 2007) and cougars (*Puma concolor*; M. Kelly, Virginia Tech University, pers. comm.), most North American carnivores appear to have little individual variation in pelage. Creative camera sets, however, can capture even small, unique marks (e.g., Magoun et al. 2005; figure 5.3).

## Monitoring

As remote cameras become more reliable and easier to use, they are an increasingly attractive option for monitoring the abundance and distribution of a species over time. For instance, Moruzzi et al. (2003) employed remote cameras to evaluate the success of a reintroduction of martens by monitoring the reintroduction site for two years. Data from remote cameras can also be used to monitor more complex population metrics, especially if individuals can be identified from the photos. Karanth et al (2006) analyzed nine years of remote camera data to estimate tiger abundance, density, annual survival rates, number of new recruits, and rate of population change over time.

## Other Achievable Objectives

Still and video photography are indispensable for the study of animal behavior, and motion-sensitive cameras are often the only way to record the natural behavior of shy animals. Although relevant pub-

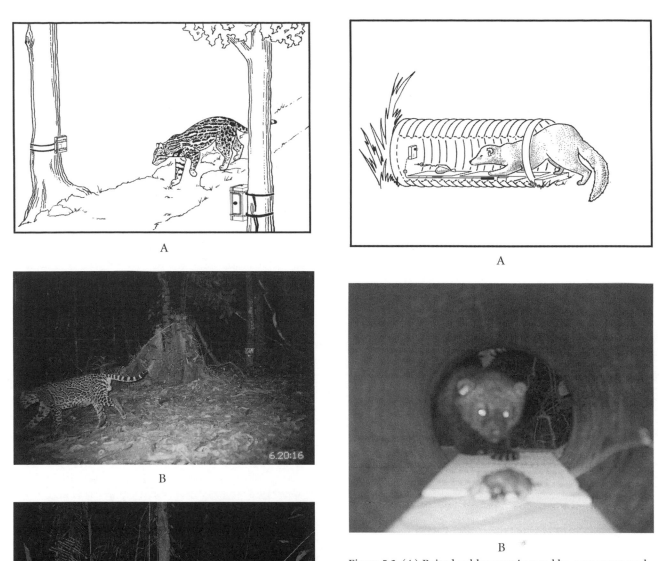

*Figure 5.1.* (A) Unbaited trail set using paired (PIR) cameras to simultaneously photograph (B, C) both sides of an ocelot for individual identification. Illustration by P. Kernan; photos from Belize, courtesy of the Wildlife Conservation Society Great Cats Program.

*Figure 5.2.* (A) Baited cubby set triggered by a pressure pad (illustration by P. Kernan). This particular set is designed to capture photos of the ear tags on (B) a marked fisher. Photo from northern California by J. Thompson; research described in Thompson (2007).

lished studies have typically employed film cameras, the increasing affordability and simplicity of video systems should allow them to be deployed in large enough quantities to open up a new series of questions to this methodology in the future.

To date, the two most common behavioral objectives addressed with remote cameras have been the documentation of wildlife activity patterns (Oster-

B

A

*Figure 5.3.* (A) Camera (PIR) set with hanging bait, intended to capture the unique gular color patch on (B) wolverines when animals look up at bait. Illustration by P. Kernan. Photo from northern Ontario by A. Magoun; research described in Magoun et al. (2005).

B

*Figure 5.4.* (A) Unbaited (AIR) set designed to detect animals, such as this (B) raccoon, moving through a culvert under an interstate. Illustration by P. Kernan. Photo from Adirondack Park, New York, by R. Kays; research described in LaPoint et al. (2003).

berg 1962; Vanschaik and Griffiths 1996; Hicks et al. 1998; Bridges et al. 2004) and animal use of special features such as highway underpasses (LaPoint et al. 2003; Ng et al. 2004; see figure 5.4), den sites (Jones et al. 1997), or other survey devices (Osterberg 1962; Kucera et al. 1995a; Vanschaik and Griffiths 1996; Bridges et al. 2004; see figures 8.1, 8.2). The behavior of animals at specific feeding sites has also been recorded with remote cameras, providing information on a broad array of topics such as prey handling (Otani 2001; Poszig et al. 2004), species interaction

(Pittaway 1983; Macdonald et al. 2004), seed dispersal (Miura et al. 1997), disease transmission (Page et al. 2001), and the effectiveness of vaccine baits (Wolf et al. 2003; Hegglin et al. 2004). An assortment of other interesting observations can be made with remote cameras if photo quality is sufficient. For example, Wong et al. (2005) related the physical condition of photographed bearded pigs (*Sus barbatus*) with fruit famine, and Rovero et al. (2005) recorded the surprising behavior of a herbivorous duiker (*Cephalophus spadix*) eating a frog.

In recent years, video cameras have been used to document bird nest predators (McQuillen and Brewer 2000; Williams and Wood 2002; Stake and Cimprich 2003). Time-lapse video is typically employed for this purpose, as the constant activity of resident birds around the nest precludes motion-sensitive triggering. Remote video and still cameras have also been used to monitor carnivore reproduction, recording data such as litter size, time of weaning, and pup survival (Aubry et al 1997; Clifford 2006)

Finally, remote cameras can produce remarkable pictures that excite scientists and the general public alike. For example, recent photographs of two jaguars (*Panthera onca*) in southeastern Arizona (figure 5.5) have been widely distributed by the media, and have prompted calls for new conservation measures in the area where jaguars have been detected (Rizzo 2005). Remote cameras have also captured striking images for popular articles and books about carnivores (e.g., Nichols and Ward 1998; Ziegler and Carroll 2005). The importance of such pictures for communicating with citizens and policymakers should not be underestimated (Clifford 2006).

## Description and Application of Survey Method

Remote cameras comprise two primary components: a camera and a trigger that is tripped when an animal moves into the camera's view. Unfortunately, there is no consensus regarding the "best" remote

*Figure 5.5.* Remote camera photo of a jaguar in Arizona, documenting this species' return to the area after having been extirpated approximately fifty years prior. Such photos have generated excitement and good will toward the conservation of jaguars along the Mexico-USA border. Photo by E. McCain.

camera model to recommend, and the wealth of new units on the market creates confusion when deciding which one to purchase. Given that the specific type of remote camera used and how it is deployed can make or break a study (Kucera and Barrett 1995; Rice 1995), this uncertainty has slowed the growth of the survey method.

Camera equipment reliability is critical for scientific applications. Failed units are not only costly to repair but require time and labor to replace in the field. Equipment failure also results in lost or compromised data. Personal recommendations and relevant publications are invaluable for identifying reliable remote cameras (Kucera and Barrett 1995; Rice 1995; York et al. 2001; Yahoo Camera Trap Group 2007). The camera and sensor features we highlight in tables 5.2 and 5.3 (and review in the next two sections) should be considered when exploring remote camera models and equipment for purchase. Online equipment reviews and discussion forums are also useful for comparing newer models (e.g., ChasingGame 2006; Jesse's Hunting and Outdoors 2006a; Yahoo Camera Trap Group 2007; professional society listserves). Those with an aptitude for electronics may decide to build units themselves to ensure quality and save money; there are a variety of "home-brew" kits and instructions available through the Internet (e.g., PixController 2006; Jesse's Hunting and Outdoors 2006a).

## Motion Sensors and Triggers

Cameras triggered by pressure pads (i.e., when an animal steps on the pad, figure 5.2) or by string tied to bait (i.e., when an animal pulls at the bait; figure 5.6) were preferred in early remote camera studies and continue to be used for certain designs (York et al. 2001; González-Esteban et al. 2004; Thompson et al. 2004). Pressure pads possess two important advantages. First, they can be designed

*Table 5.2.* Comparisons of key features of active infrared and passive infrared motion sensors

| Feature | Active infrared (AIR) | Passive infrared (PIR) |
|---|---|---|
| Size | Two larger units (separate from camera) | One smaller unit (housed with camera) |
| Models | One company | Many companies |
| Price | Higher | Lower |
| Ease of use | More complicated | Simpler |
| Sensitivity | High (but flexible) | Medium (can be flexible) |
| Detection beam width | Narrow | Narrow or wide |
| False triggers | Usually fewer | Usually more |
| Sensitivity in tropical climates | Not affected by temperature | May be lower |
| Damage by wildlife | Highly susceptible | Lower risk |

*Table 5.3.* Comparison of the features of current film and digital cameras as they relate to remote camera surveys

| Feature | Film | Digital |
|---|---|---|
| Trigger delay | Short | Usually longer |
| Battery use | Low | Usually higher |
| Fragility | Lower | Higher |
| Infrared illumination | Not practical | Possible |
| Photos between checks | 36 | Hundreds or thousands |
| Photo development | Slow, costly | Instantaneous |

*Figure 5.6.* (A) Baited string set to mechanically trigger a film camera; in this case (B), the target species was a fisher. Illustration by P. Kernan. Photo from northern California by W. Zielinski; research described in Kucera et al. (1995a).

such that a minimum weight is needed to trigger the system, therefore excluding smaller-bodied, nontarget species. Second, pressure pads provide precise triggering distances—an important benefit when specific features on an animal's body (e.g., ear tags, gular pelage pattern) must be clearly visible in the photograph. Recently, King et al. (2007) extended this trigger type by integrating a digital scale to weigh each visitor to the camera and an automated bait dispenser that only offers bait when visitors are in the range of the target species (figures 5.7 and 4.9).

Due to ease of deployment, most modern studies use active infrared (AIR; figure 5.8A) or passive infrared sensors (PIR; figure 5.8B). Both AIR and PIR sensors involve infrared energy, but they each function differently and entail their own set of tradeoffs (table 5.2). AIR is essentially a narrow pulsing beam of light energy that, when broken by any object, registers motion. PIR sensors can have a broad or narrow detection area but only detect moving objects that differ in temperature from the environment (e.g., warm-blooded animals). The sensitivity of both sensor types can be adjusted, with AIR

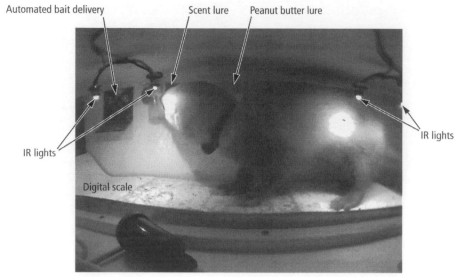

*Figure 5.7.* A ferret (*Mustela furo*) sniffs the scent lure in a Scentinel monitoring device (also see figure 4.9), which records the date, time, weight, and digital image of every animal that enters. Bait is only delivered once per visit, and only for species within the weight range of the target species. Photo from New Zealand by R. McDonald; research described in King et al. (2007).

*Figure 5.8.* Top-down remote camera baited sets using (A) AIR and (B) PIR sensors. Note the corrugated polypropylene snow-shields installed above both units and the flexible tubing to protect camera cables on the AIR set. Photos by K. Slauson; research described in Zielinski et al. (2007).

having a higher maximum sensitivity. AIR sensors for remote cameras are patented and can therefore be purchased solely from one company (Trailmaster, Goodson & Associates, Lenexa, KS). PIR sensors, on the other hand, are manufactured by dozens of companies and in hundreds of varieties, offering a diversity of options, beam widths, and sensitivities. While this diversity yields flexibility, it can also hinder comparisons across studies.

The main advantage of the PIR system is that it requires only one sensor component, which can be housed with the camera. PIR sensors are therefore very easy to set up and are not as easily disturbed as sensors with multiple components. AIR sensors consist of two components that must be aligned and are typically not housed with the camera. Although aligning AIR sensors in the field is not too difficult—especially with two people—small movements can cause misalignment after setup. Such misalignments, plus equipment malfunction and a narrow detection beam, have resulted in AIR sensors generally performing less reliably than PIR sensors in side-by-side comparisons (Kawanishi 2002; M. Kelly, pers. comm.). The fact that the AIR system requires two sensors that are separate from the camera also makes it difficult to lock to trees (i.e., three cables and locks are needed per set), and exposed connecting wires are at risk of damage from chewing rodents or falling branches.

The main advantage of AIR sensors over most PIR designs is that they can be set to a wider range of target areas, with greater specificity and at different angles from the camera. When using AIR sensors, efforts must be made to minimize the possibility of misalignment. For example, AIR sensors can be fixed with metal brackets around a very narrow target area (figure 5.9).

## Still Cameras

Compact, 35 mm film cameras have been standard equipment for remote camera setups over the past two decades, although this is gradually changing. Wildlife research has been slower to transition to digital cameras than have most other photographic applications, in part because of the need for quick triggers. At this time, most off-the-shelf compact digital cameras do not function well as remote cameras because they require too much of a time delay between triggering and capturing an image. A delay of two to four seconds (typical for most models; ChasingGame 2006) would result in missing the majority of animals that pass in front of an unbaited camera—but may be sufficient in cases where animals linger to inspect bait. Additionally, off-the-shelf digital cameras usually require more battery power, have weaker flashes, and are more sensitive to harsh field conditions than film cameras. This situation will likely change as digital cameras improve and engineers design custom cameras to meet the needs of wildlife researchers. Indeed, as of 2007, two high-end (>$800) digital remote cameras have been developed that feature short (1/10 sec) trigger times and long battery life (Buckeyecam, Athens, OH; Reconyx, La Crosse, WI).

The promise of digital cameras for improving scientific research conducted with remote cameras is substantial. A low-cost, reliable, digital remote camera with infrared illumination, long battery life, and a high capacity for picture storage would revolutionize the field by allowing more cameras to be deployed for longer periods of time with less labor. Until such a camera is developed, wildlife researchers will need to consider the tradeoffs of various features (table 5.3) as they relate to project objectives and budget.

Meanwhile, film cameras will continue to be important for remote camera studies for the foreseeable future. Surveyors employing such cameras should use film with the greatest number of exposures possible (typically thirty-six) to maximize the length of time a unit can operate in the field between station visits. Further, employing the most light-sensitive film available will extend the range of the flash at night. ASA/ISO 800 speed film is usually the most cost effective, although ASA/ISO 400 is nearly as fast and results in fewer problems with airport screening machines. Most camera studies use print rather than

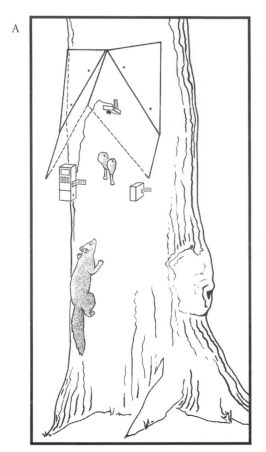

*Figure 5.9.* (A) Single-tree, baited AIR set with metal brackets and a corrugated polypropylene snow-shield and (B) an American marten captured on film. Illustration by P. Kernan. Photo from northern California by K. Slauson; research described in Zielinski et al. (2007).

slide film to minimize developing costs and to make it easier to view images.

## Video Cameras

Video cameras are an increasingly attractive option for wildlife researchers (Aubry et al. 1997; Jones et al. 1997; Macdonald et al. 2002; Macdonald et al. 2004; Gagnon et al. 2005; Jansen and Den Ouden 2005; Clifford 2006; MacCarthy et al. 2006). The behavior recorded by these cameras adds a new dimension to data collection and provides information that is not always available from single still images (figure 5.10). For example, the video monitoring of highway underpasses revealed key behavioral responses by elk (*Cervus elaphus*), which stopped and stared up at ledge features but did not enter the structure (Gag-

non et al. 2005). These responses, which would not have been captured using still images, influenced the design of existing and future underpasses. Last, most video cameras include a zoom lens, so the entire system can be placed at a distance from sensitive areas of interest (e.g., den entrances) to reduce disturbance. Zoom lenses can also be aimed at elevated sites, such as cavities in trees.

Unfortunately, the disadvantages of still cameras (table 5.3) are amplified with video cameras, which are even more expensive and complicated to set up, possess longer trigger times and shorter battery life, and can be very sensitive to harsh field conditions. Again, these drawbacks are likely to diminish as video camera technology improves. In one promising development, many digital still cameras have begun to include pseudovideo options of one to three

*Figure 5.10.* Frames taken from high definition video footage of a female grizzly bear (*Ursus arctos*) and cubs of the year interacting with a gray wolf (*Canis lupus*) at an elk carcass site in Glacier National Park, Montana. Video footage by J. Stetz.

still frames per second, which may yield cheaper, lighter, and more reliable alternatives to video-only systems (figure 8.1).

## Deploying Remote Camera Systems in the Field

The two most basic camera sets respectively aim cameras at a game trail (figure 5.1) or a baited area with vegetation cleared away (figure 5.11). While these approaches are commonly used (Cutler and Swann 1999; Sanderson and Trolle 2005), some creative researchers have customized their designs to target certain species or even to photograph a specific feature of a target animal. By positioning an AIR sensor on a tree just below the bait, for example, Zielinski et al. (2007) were able to set the sensor using just one tree and to reduce the number of photos of nontarget species while still acquiring excellent pictures of climbing target animals (i.e., fishers and martens; figure 5.9). Magoun et al. (2005) conducted trials with zoo animals to perfect a design for photographing wild wolverines (*Gulo gulo*) as they look up at hanging bait, thus capturing the details of a white throat patch that can be used to identify individuals (figure 5.3). Similarly, Thompson (2007) used bait to draw fishers into single-entrance tubes to acquire close-up photographs of their heads, which enabled

the discrimination of ear-tagged individuals and, in some cases, the determination of the sex of untagged individuals by analyzing head width (figure 5.2). Even the smallest carnivores can be surveyed with an appropriate camera set; Osterberg (1962) photographed least weasels (*Mustela nivalis*) using rodent runways (figure 5.12).

Cameras and sensors are typically affixed to large trees with straps, nails, bungee cords, brackets, or tape. If appropriate trees are not available, cameras can be set up on rock piles, hidden in fake rocks, or fastened to posts installed specifically for this purpose. The critically important requirement (especially for AIR sensors) is that the sensors and their anchor points do not move in the wind or shift over time, thus producing false triggers.

The sensitivity of, and area covered by, motion sensors can often be adjusted. Remote cameras usually have a test mode that allows adjustments to be made without actually shooting photographs. Before deploying cameras, researchers need to "tune" sensor settings to fit their objectives and target species by walking, crawling, and waving at sensors from different distances and heights. Bottles of hot water can be used to realistically test PIR sensors for smaller animals (Swann et al. 2004). Digital cameras are typically run at maximum sensitivity, but film cameras may need to be set at lower sensitivity to

*Figure 5.11.* (A) Multi-tree baited PIR set and (B) a resulting photo of a black bear. Illustration by P. Kernan. Photo from Adirondack Park, New York, by R. Kays; research described in Gompper et al. (2006).

*Figure 5.12.* Unbaited AIR set along a mouse runway, with approaching weasel. Illustration by P. Kernan.

avoid wasting film. Sensor settings must be consistent across all equipment deployed in a given study and should be described in the methods section of papers and published reports. Sensors can also break or deteriorate over time. We suggest testing cameras before each deployment by lining them up side-by-side and passing in front of them ten times. Any units that don't trigger reliably (e.g., at least nine of ten times) should be repaired or replaced.

The distance between the camera and the target area varies depending on the size of the target area, the size of the target species, and the sensitivity of the motion sensor. For midsized mammals, this distance is typically 2–5 meters. Closer sets are required for smaller animals, or for capturing fine details. Motion sensors need to be aimed at the height of the target species and should always be tested at the expected distance from the target area. Lower sets record the greatest range of wildlife in terms of size, as they are triggered by both smaller animals and the legs of larger animals. Alternatively, arboreal carnivores can be targeted by mounting cameras higher on trees (figure 5.9A; Zielinski et al. 2007). Last, tape

can be placed across part of the PIR sensor to narrow or fine tune the detection window.

Nearly all remote cameras are positioned to produce horizontal photos. This is the easiest alignment to orchestrate, and side views of animals often maximize information for species or individual identification. A few studies have employed top-down views, which may be useful in some cases (figures 5.8, 5.9). Prior to departing from a camera site, a test photo should always be taken to ensure that the complete system is functional. Further, shooting a test photo immediately upon arrival to a camera site will indicate whether or not the system was functional over the sampling occasion. Such tests also provide date/time references for camera checks to reinforce written records.

Remote cameras usually have two settings that can be used to help limit the number of photos taken: activation time and camera delay. These features are necessary when film or batteries are limiting, or when risk of theft is high. Activation time configures the unit to function only when desired (e.g., daytime, nighttime, dawn and dusk, day and night). Researchers typically leave remote cameras turned on continuously but may restrict their operation according to their target species (e.g., cameras might be operational only at night for nocturnal creatures). Using a night-only mode may also reduce the number of people photographed on trail-sets and minimize the risk of theft since most people do not walk trails after dark. Similarly, flashes can be set to go off only at night, when they are less likely to be noticed by people.

The camera delay option limits how often a photograph is taken by implementing a pause between consecutive pictures. This is most important for sets around bait, den sites, or other areas expected to experience long periods of continuous animal activity. Setting a camera delay can prevent a roll of film from being exhausted by one animal but can also result in missed opportunities for unique photographs. Generally, this delay should be as short as is practical given the set type, target species, frequency of checking the camera site, and type of camera. Two to five minutes is a common interval used in baited sets (Kucera et al. 1995a), while unbaited trail-sets usually impose no delay interval to avoid missing groups of animals walking together.

Researchers using digital systems with sufficient battery and storage space may actually employ the opposite strategy by setting options such that multiple photographs are taken at each triggering event, with no delay time between triggers. The resulting, consecutive images can provide a pseudovideo effect, documenting animal behavior and recording the target individual at a variety of angles (figure 8.1). Some digital systems also include a time-lapse option that triggers photographs at set intervals. These photos can be useful to compare with motion-triggered images lacking obvious animals in the frame to help determine what may have triggered the motion sensor in the latter. Further, the time-lapse option can be used without the motion sensor to regularly sample an area known to experience very frequent animal activity—for example, to count the number of animals at a feeding station or to

record nest predators (Pietz and Granfors 2000; Williams and Wood 2002). Last, if recording detailed behavioral activity is desirable, a remote video camera with a PIR sensor should be considered.

## Practical Considerations

Beyond the fundamental requirements of equipment selection and setup, numerous practical issues must be considered in planning a remote camera survey. Several of these considerations are addressed in detail here.

### Site Selection and Baiting

The exact placement of remote cameras in the field depends on the survey objectives and the target species. Some objectives (e.g., occupancy estimation) require that sample units be selected with a representative or random method. Once a sample unit (e.g., a coordinate or quadrant) is selected, however, cameras are often placed such that they will have the best chance of being visited by the target species—either nearby features thought to be used by this species, or at stations with attractants designed to lure animals to the camera.

*To bait or not to bait?* is an important question that all researchers using remote cameras must address. Baits and lures certainly attract some species to a site, but they may also repel others—or attract only young or dominant individuals (see chapter 10 for further discussion about the pros and cons of attractants). For species that do not predictably use features that researchers can recognize (e.g., game trails, dirt roads), attractants help draw animals to remote cameras and reduce LTD. Baiting can also encourage repeat detections of individuals, which may or may not be desirable. Studies seeking to reduce the number of sampling days per location, or to maximize the likelihood of detecting a species with known bait preferences, are good candidates for using attractants.

The use of attractants is problematic for surveys seeking to estimate animal density; the area sampled is difficult to quantify because the distance from which the scent can be detected is completely unknown and is likely to vary with environmental conditions. Studies of fine-scale habitat selection should also avoid attractants, which can draw individuals to otherwise unused locations. And research aimed at animal activity patterns should be equally cautious of using baited sets because, once the bait is discovered, animals may alter their regular activity and revisit sites to exploit an easy meal. Attractants clearly reveal the location of a set, which may actually cause shy species to avoid the area. Gompper et al. (2006), for example, found no relationship between the local density of coyotes and the probability of detecting them with baited cameras across fifty-four sites in northern New York, probably because some coyotes avoided the strong scent lures used at the sets.

Unbaited sets have their own potential biases. Trails or roads may not be used by a representative sample of the population in a given study area if, for instance, dominant animals or species patrol major roads and force subordinate individuals or species (e.g., prey) to avoid such travel routes (Emmons 1988; Palomares and Caro 1999). Ocelots (*Leopardus pardalis*) and jaguars appear to favor larger, cleaner, older roads over newly cut trails (Maffei et al. 2004; Di Bitetti et al. 2006; Dillon and Kelly 2007). Weckel et al. (2006) found a variety of responses to trail type across rainforest species, with jaguars and Baird's tapirs (*Tapirus bairdi*) preferring large, established trails, and nine-banded armadillos (*Dasypus novemcinctus*) avoiding them. Larrucea et al. (2007) also documented effects of human disturbance, road or trail type, and habitat type on the detection rate of coyotes by remote cameras, with adult coyotes using roads more than game trails, and juveniles avoiding human structures. Another concern with sampling major trails or dirt roads is that they are often frequented by humans, who may exhaust film, vandalize equipment, or steal cameras. Unbaited sets should thus also target other topographic or habitat features (e.g., game trails on ridge tops, saddles, riparian corridors), or be deployed using a randomized protocol.

Given that the respective biases of baited and trail sets are not well documented at this time, every study should be explicit about its methods and test for potential biases if possible. Because the potential for bias seems higher with baited sites (although see Larrucea et al. 2007), we prefer unbaited sets if the target species can be detected at a sufficient rate without attractants.

## Flashes and Camera Avoidance

A visible flash is another source of potential bias for a remote camera survey if animals are disturbed by the flash and avoid the camera thereafter. Wegge et al. (2004) found that individual tigers avoided cameras after first being photographed, and Schipper (2007) presents some evidence that kinkajous (*Potos flavus*) exhibited avoidance behavior after being exposed to flashes. Anecdotal observations of animal responses in successive photographs suggest that aversion to flashes is likely to vary among individuals and species (ChasingGame 2006; Jesse's Hunting and Outdoors 2006a). Some species are obviously not bothered by flashes, as they repeatedly visit camera stations—exhibiting apparent curiosity. Digital cameras with infrared flashes should result in neither aversion nor curiosity, although their flashes may still be visible by people if viewed directly.

One study documented that coyotes discovered unbaited cameras in their territories without being photographed and subsequently avoided them (Sequin et al. 2003). This effect was so pronounced that most photographs featured intruding coyotes (i.e., from neighboring territories) that were not familiar with the locations of the cameras. Such findings emphasize the importance of disguising the scent and appearance of remote cameras—although the results of this particular study might be atypical because the study site had so little human activity that coyotes actually tracked the researchers deploying cameras from nearby hills. Furthermore, coyotes are notorious for being one of the most intelligent and wary animals in North America (Parker 1995); other species are probably less concerned about remote cameras.

## Weather

Weather can present challenges for remote camera function. PIR sensors, for example, can experience difficulties with distinguishing mammalian body temperatures from high ambient temperatures, resulting in failure of detection. This effect varies across sensor models and can be addressed by setting sensors to their maximum sensitivity, or switching to more sensitive PIR sensors or to AIR sensors. Conversely, cold temperatures significantly reduce battery life, requiring more frequent visits to the camera for battery replacement.

### Moisture and Precipitation

Precipitation and high humidity are probably the worst enemies of remote cameras because of their effects on electronics. Moisture-related problems can be minimized by taking preventative measures. First, housings need to be completely waterproof. Seals on doors, windows, and cables need to be checked, maintained, and sometimes supplemented with silicon sealants to ensure their integrity in the face of normal wear and tear. Unfortunately, some humidity can still penetrate the equipment, and rechargeable packets of desiccant should be placed inside camera housings in regions with high humidity (e.g., tropical forests) or when using digital or video systems, which feature more delicate electronics. Umbrellas are necessary to check field cameras in the rain without damaging equipment.

Heavy snow can also cause system failures by triggering AIR systems, hindering station access, or potentially altering the attractiveness of lures and baits. Snow shields (figures 5.8 and 5.9A) may be necessary in areas of high snowfall to prevent snow from triggering AIR sensors and accumulating on camera

components and bait. Humidity, rain, mist, or fog can cause camera lenses to cloud, degrading the clarity of the images and reducing the likelihood of species identification. Taking measures to shield the camera lens from precipitation, and applying optical antifogging solutions or membranes (e.g., LensBrite, Peca Products, USA), helps reduce such problems. Heavy rain can also trigger AIR systems, especially as the distance from the beam's sending unit to the receiver increases.

### Sunlight

Infrared motion sensors can be falsely triggered by sun-related effects, but these can be minimized by carefully positioning the sensors during the initial setup. For example, both PIR and AIR sensors will trigger if they are struck by sunlight at direct angles (i.e., at sunrise or sunset for sensors set horizontally). This can usually be avoided by aiming sensors away from the direction of sunrise or sunset. If sensor angle cannot be adjusted as necessary, light can be blocked without affecting sensitivity by placing a small plastic tube (e.g., a 2-cm diameter plastic pipe) around the receiving window of the sensor unit. A less common phenomenon can trigger PIR systems that are set with branches in view. Although blowing vegetation shouldn't typically trigger PIR systems, this can occur if the vegetation is heated by sunlight penetrating the forest canopy. Such issues can be addressed by placing PIR sensors in shady areas and trimming or tying back suspect branches.

## Minimizing Theft

Humans are omnipresent in North America, and every researcher is likely to have a remote camera stolen eventually. Nonetheless, theft can be minimized by simply placing equipment above or below eye level, setting units slightly off trails or roads, and using on-site camouflage (e.g., bark) to conceal system components. While baited sets can be located or oriented away from trails or roads, unbaited sets may target features that are also used by humans, and thus need to be better hidden. Often the flash betrays the location of a camera— even in daylight if the camera is set in shaded forest. Infrared-sensitive digital cameras with an infrared flash do not have this problem as the flash is invisible to humans, although it can be seen briefly glowing red if one looks directly at it when it triggers. Many remote camera flashes can be programmed to be inactive during the day.

Remote cameras should be secured to trees or other features with cables and locks when possible. It is important to use combination locks, or padlocks set to the same key, to avoid having to carry large collections of keys into the field. Most passersby who discover a camera will not disturb the equipment but may be curious about why it is there. A small sign taped to the side of the camera, briefly describing the purpose of the study and providing relevant contact information, is likely to satisfy curiosity and limit vandalism.

## Minimizing Wildlife Damage

Damage by animals is a frequent cause of equipment failure. Remote cameras can be destroyed by bears, and external cables can be chewed by a variety of rodents (Kawanishi 2002). Any animal can simply bump a camera or sensor enough to alter its angle, rendering target areas out of view. Most damage is caused by mammals, which probably detect the units via smell and may even be attracted to them because of salts or other scents deposited by field personnel. Thus, the best prevention is to minimize human scent on cameras by wiping down units with an alcohol solution or spraying them with an activated carbon spray (e.g., Carbon Blast, Robinson Outdoors, USA). Cotton gloves can also reduce the transmission of scent to cameras during deployment. Care should be taken when transporting remote cameras into the field to minimize the transference of food, bait, or scent lure odors to camera equipment. When working in pairs, one person should handle attractants while another handles the camera equipment.

Remote camera systems completely enclosed in a single protective housing (e.g., most PIR systems) are less susceptible to animal damage because they don't have exposed cables and are less prone to sensor misalignment. External cables can be housed in protective flexible tubing (figure 5.8) or buried to minimize exposure to animals (Henschel and Ray 2003). In areas where bear damage is a serious concern, camera surveys should be conducted during seasons when bears are inactive.

## Equipment Management

The equipment needed to set up, check, and remove remote cameras (appendix 5.1) will vary depending on the type of remote camera system used, the particular set (e.g., baited versus unbaited), the survey design, and data (beyond photographs) to be collected. Each field team should carry and maintain their own camera bag containing all required field accessories, as well as a vehicle box that holds essential replacement and backup supplies.

Inspecting, testing, and preparing equipment prior to deployment and after retrieving it from the field greatly reduces system failures and increases efficiency in the field. All remote cameras should be tested before each deployment (see *Deploying Remote Camera Systems in the Field*, earlier in this chapter, for testing procedures). It is especially important to regularly test motion sensors, as these can deteriorate over time. The charge levels of batteries should be evaluated before they are installed, and, whenever possible, film should be inserted into cameras prior to deployment to minimize opening cameras in the field. Placing an external label on each system element to indicate when new batteries were installed allows a quick gauge of battery life. Manufacturers typically provide estimates of battery life for their camera systems under moderate conditions, but each study must develop its own guidelines specific to the equipment and local temperatures.

Other key items to regularly inspect and service include the rubber gaskets that prevent moisture from entering system components, exposed metal contacts and battery leads (for corrosion and dirt), and external wires. Cameras deployed in very wet or humid conditions may require occasional storage in a dry place (e.g., a closet with a dehumidifier or low-watt lightbulb) to function appropriately. In humid environments, care should be taken when moving cameras from air-conditioned rooms into field conditions because condensation will form on electronic parts and lenses. In such cases, units should be left open in a dry closet or in the sunshine for a few minutes so that cameras are properly sealed and dry when deployed in the field.

Some forethought should be given as to how cameras will be carried into the field; mindlessly throwing a number of remote cameras into the back of a vehicle or a backpack will scratch camera and sensor lenses unless precautionary measures are taken. We suggest that each researcher customize a standard, waterproof camera pack to carry cameras safely to and from their deployment sites.

Finally, it is important to be realistic about the durability of remote cameras and to prepare for equipment malfunction. We advise that researchers not deploy every available remote camera but rather have a few extra units at the ready to replace broken equipment. When working in particularly challenging environments, maintaining a reserve of cameras amounting to 20% of the total number deployed may be necessary to maintain consistent sampling effort.

## Number of Field Personnel Required

A single experienced person can efficiently set up, check, and break down most remote camera systems. Given the logistics and challenges of most surveys, however—including weather, accessing the site (e.g., by vehicle, on foot), safety, the number of stations, and the total amount of weight to be carried into the field—most surveys use two-person field teams. Furthermore, if inexperienced personnel or new equipment are being used, two-person teams

will likely improve the success of the effort and provide more learning opportunities.

## Checking Equipment in the Field

Determining the optimal frequency at which to check each camera station usually entails a tradeoff between maximizing efficiency and ensuring that stations will remain functional during the entire sampling interval. Numerous factors, including the type of camera system (e.g., film versus digital), camera programming (e.g., camera delay), whether sets are baited or unbaited, and the expected site activity level, must be considered. If bait consistency is a concern, more frequent sampling occasions will be required during warm (versus cold) months. Cold temperatures, however, will reduce battery life.

We recommend checking new camera sets within three to five days and then judging when to return next based on the number of exposures shot in that first period. Likewise, deciding when to change a roll of film depends on the rate at which the camera is being visited. Our general rule of thumb is to replace a roll of 36 exposures when fewer exposures remain than were shot between previous checks, or when >25 frames have been shot. The impressive storage capacity of digital cameras makes this a moot issue with such cameras, which can be left unchecked for longer periods of time as long as the other concerns in this section are addressed. One enterprising hobbyist ran a solar-charged digital camera in the Maine woods for over a year without checking it (Jesse's Hunting and Outdoors 2006b).

While one of the strengths of remote cameras compared to other devices is that they can be left unattended in the field for extensive periods, this can also lead to a false sense of sampling security. The longer a remote camera remains in the field, the more likely it is that a malfunction will occur, resulting in lost survey time. The experience and knowledge of field personnel, as well as the costs of lost survey time, should be carefully considered when determining the length of sampling occasions.

## Materials and Costs

Remote camera surveys require camera equipment, consumables, and labor to deploy and maintain units. The costs of all of these items vary considerably—for example, remote cameras can currently be purchased for $100–$1,500 USD, depending on the quality and options required. Since camera malfunction is inevitable for any extended project, managers must also budget for extra cameras to replace broken ones, and for repairs. Consumables for film cameras include film and developing. Although prices differ regionally, $10/roll is a ballpark estimate for purchasing and developing film. The costs of the labor and transportation necessary to maintain film cameras can be substantial, especially in remote areas. As digital camera surveys reuse memory cards, their consumables costs are limited to attractants (if used) and batteries. And because digital cameras can store many more photographs than film cameras, they may offer a labor savings—especially if paired with solar panels to extend battery life (Jesse's Hunting and Outdoors 2006b). Overall, we expect that the costs associated with remote camera surveys will decrease in the future as digital camera equipment becomes less expensive and more reliable.

## Survey Design Issues

The basic principles of survey design pertain to all noninvasive surveys, as discussed in chapter 2. Here we review a few design issues specific to remote camera surveys.

### Distribution and Occupancy Surveys

Surveys documenting the distribution and occurrence of wildlife in a given region should use a standard sample unit and establish these throughout the study area, typically spaced relative to the home range size of target species (see chapter 2 for details). The number of cameras used in each sample unit represents a balance between collecting the best possible data and making the most efficient use of a limited number of cameras. Deploying more cameras within a sample unit can increase the probability of detection, shorten LTD, and allow a diversity of set types (e.g., baited and unbaited) targeting different local features (e.g., game trails, ridge lines, riparian corridors). Multiple remote cameras per sample unit also dilute the effect of data loss due to individual camera malfunctions. Because of the cost of equipment, however, most studies use only two to three remote cameras per sample unit, arranged either along a transect or around a central point (Kucera et al. 1995a; Gompper et al. 2006; Zielinski and Stauffer 1996).

Deciding how long to deploy cameras at each site reflects an important tradeoff between improving the likelihood of detecting a species at a given site (i.e., longer is better); increasing the number of different sample units that can be surveyed during the field season (shorter is better); and, for some objectives, maintaining population closure at a site (i.e., no immigration or emigration; see chapter 2). In addition to survey duration, the set type (e.g., baited or unbaited), the number of remote cameras within a sample unit, the geographic spread of the sample unit, and local animal density will also affect detection probability (see chapter 2).

A brief review of two carnivore studies that estimated the probability of detecting target species based on the temporal pattern of photographs taken at a sample unit illustrates some of these design tradeoffs. One survey in the Adirondacks of New York (Gompper et al. 2006) used three baited cameras for one month along a 4-km linear transect, while a California study (Campbell 2004) used two baited cameras over a smaller area (approximately 1 km apart) for a shorter period of time (sixteen days). These surveys yielded similar detection probabilities, which were well above 50% after fourteen days and 75% after thirty days for most target species.

The above results concur with general recommendations that surveys of approximately two weeks will detect most species present, but that approximately one month is required for exhaustive inventories (Moruzzi et al. 2002). It may be possible to employ shorter survey durations for surveys focusing on common species, as the most abundant species in these two studies had a 50%–80% probability of detection after seven days and 74%–96% after fourteen days. The California survey detected a greater number of carnivore species, reflecting the more diverse carnivore community of the Sierras. Both surveys detected black bears (*Ursus americanus*) at similar rates, while different detectabilities for other species likely highlight differences in regional species abundance (e.g., fishers and raccoons [*Procyon lotor*] were likely more common at the Adirondack sites, and martens were more common at the California sites).

## Abundance Estimation

Although new analytical techniques may allow relative abundance to be estimated with remote camera data collected during distribution surveys (Carbone et al. 2001; Carbone et al. 2002; Jennelle et al. 2002; Royle and Nichols 2003), most published abundance and density estimation studies have discriminated individuals so that animals are not double-counted. This can be accomplished with remote cameras if individuals can be distinguished by unique markings that are unambiguously recorded in photographs (Karanth and Nichols 1998; Heilbrun et al. 2006). Abundance and density estimation studies typically deploy paired cameras at each station so that both sides of an animal are recorded (figure 5.1). As mentioned earlier, wildlife researchers have recently developed creative camera sets to photograph unique marks of target species with single cameras (figures 5.2 and 5.3). These customized remote camera sets are valuable, and we encourage researchers to think about how similar modifications might be made to identify individuals, or at least gender, with their target species.

The statistical models used to estimate the true number of animals in a population from capture-recapture data are well developed (Karanth and Nichols 1998; Karanth et al. 2006), and methods even exist to account for photo misidentification during data analysis (see chapter 11). Methods for estimating the area sampled are less refined, however, and remain controversial. This measure is critical because it has an equal effect on the final density estimate and the estimated population size. Simply drawing a polygon around a camera station underestimates the area sampled, as some animals would be moving outside of that area. A common practice is to surround cameras with a buffer that represents half the width of an average home range (Soisalo and Cavalcanti 2006). If local data on home range size are not available, diameter has been estimated as the maximum distance between recaptures of individuals at camera stations (Karanth and Nichols 1998). Specifically, most surveys have used the *mean maximum distance moved* (MMDM) across all animals detected more than once, since this was found to be an accurate adjustment in models of mice movement (Wilson and Anderson 1985). One recent study, on the other hand, showed that MMDM from cameras underestimated movements of large carnivores and therefore produced an overestimate of animal density (Soisalo and Cavalcanti 2006). Dillon and Kelly (2007) also point out that camera density can have a large effect on the observed MMDM, and therefore on the resulting density estimate.

Camera surveys conducted for capture-recapture density estimation need to be run long enough to obtain multiple photographs of at least some individual animals, but for a short enough period to be considered a closed population. That is, there should be no births, deaths, immigration, or emigration in the population during the survey period. For relatively long-lived animals like most carnivores, a survey duration of even a few months can probably be considered closed (Karanth and Nichols 1998). Repeated sampling of known individuals across multiple years can estimate other critical parameters such as density, survival, recruitment, temporary emigration, and transience (Karanth et al. 2006).

## Sample and Data Collection and Management

Keeping track of which photographs emerged from which camera is a relatively simple concept but can easily become confusing in large studies involving a multitude of people and cameras. A number of organizational measures should be taken to minimize confusion, and the implementation of such measures by field staff should be ensured through a series of quality control checks by managers who randomly conduct inspections to confirm that cameras are deployed and monitored as reported.

### Organizing Film and Memory Cards

It is critical to assure that film or memory cards can be linked to the cameras, locations, and survey periods from which they originated. To do so effectively, each film roll or memory card must be individually labeled (e.g., with stickers) and recorded on datasheets. The survey station and date information should be tracked through the entire film development process; cooperation from the film development laboratory is necessary to prevent errors. Shooting a test photograph at the start of each new station, with the station name and date written on a sign, will also serve to confirm the source of photos. When developing film, we advise having photos converted directly to digital media to facilitate image management and to make it easier to magnify or zoom in on images. A strategy for reducing film development costs is to first print only the contact sheets (i.e., a composite of all photos on the roll), and then to develop the subset of images requiring closer examination. Development options and costs vary among film labs and can often be negotiated if a laboratory knows that a substantial number of rolls will be brought to them over the course of a field season.

Ideally, each photograph from a remote camera has the camera location (i.e., the sample unit and the camera number within the sample unit) printed on the image itself, along with a date and time stamp.

This provides a backup of—but not a substitute for—the record keeping described above. Many new cameras allow a custom title to be imprinted on each frame, so the camera location can be permanently registered with each print. If cameras don't allow customized titles, a sign indicating a location code and date can be placed within the camera's view or photographed in the first and last test shots of a roll. Dry-erase boards are useful for such test shots.

Digital photographs pose a clear advantage for data management, as they can be downloaded and directly labeled with key location and date information and do not require third-party processing. Replacing memory cards in the field is probably the most reliable method of image retrieval, although care must be taken to track the various card-camera combinations. Downloading images from memory cards in the field is increasingly possible with personal digital assistants (PDAs) and field-worthy laptop computers. Images can also be viewed in the field with any digital camera that uses the same media type. Whenever electronics are to be used in the field, however, extra care is necessary to ensure that they stay dry (e.g., via waterproof enclosures and desiccant) and function dependably. A supply of fresh memory cards should always be carried as reinforcements.

### Photo Database Management

We prefer storing and working with digital media over print photographs because they are much easier to organize, view, duplicate, and share. Digital records must be backed up regularly with secure media, however, and each backup version should be clearly labeled to avoid confusion over which is the most recent data. Finally, printed copies of digital images stored in binders can be effective for sharing results and safely archiving data and may be useful for comparing many individuals at once when matching coat color patterns (e.g., Karanth and Nichols 1998).

There are a number of free or inexpensive commercial databases available for organizing and managing digital photographs (e.g., Adobe Photoshop Album, ACDSee, Google Picasa, IDimager). Photo database software will be worth the initial time and investment for any large camera project, allowing images to be marked with multiple tags (e.g., species, location, set type) so that they can be queried via these criteria. Software automatically registers the date and time for images from digital cameras, but other tags must often be entered manually.

Photographs are the raw data of remote camera surveys and must generally be summarized at two levels to be useful for analyses. First, each individual photograph should be registered in a spreadsheet that includes the date, time, and species associated with the photo, as well as metadata for the camera station (see *Metadata*). A group of consecutive photos shot from a film camera without a programmed delay or from a digital camera that produced a burst of images with each trigger should be registered as one visit by an animal (i.e., one event). In order for consecutive images to be considered independent, the time interval is typically set at thirty or sixty minutes—although this setting has rarely been empirically evaluated with a time series analysis. Such an interval may in fact be overly conservative; Yasuda (2004) found that one minute was sufficient to reduce the temporal autocorrelation between photographs of all five species of medium- and large-sized mammals detected frequently at sets baited with 200 g of peanuts. We recommend that more researchers analyze data in this way to validate the independence interval for their particular data set. Raw data would thus be analyzed at a higher level to estimate summary statistics such as the number of species and individuals detected, visitation rates, probability of detection, and LTD (Karanth and Nichols 1998; Carbone et al. 2001; Gompper et al. 2006; also see chapter 11). The grouping of these raw data across time, cameras, or camera stations will vary depending on the design, analyses, and objectives of the study (see chapter 2). Linking the raw data of each individual photo to the proper camera metadata allows for reorganization of data to accomplish different objectives.

## Metadata

Remote cameras are used by a wide array of individuals, including those involved in research, land management, and biological consulting—as well as wildlife and outdoor enthusiasts. To be most useful, each survey effort should adequately document details pertaining to

- survey design;
- survey duration;
- types of remote cameras used;
- camera station locations (i.e., global positioning systems (GPS) coordinates and written descriptions);
- dates of survey for each remote camera;
- programmed time delay and activation time;
- species detection results;
- number of lost survey days for each remote camera due to equipment malfunction.

It is important that any remote camera survey effort—no matter how small—record such details. A given land management agency, for example, may possess a few remote cameras and a small budget to support their deployment at a limited number of locations each year. Over several years, the resulting survey data could become quite valuable, but would be difficult to summarize without adequate documentation. Furthermore, small-scale surveys can collectively enhance our knowledge of species across large regions. For instance, the results of standardized remote camera and track plate surveys conducted in Washington (Lewis and Stinson 1998), Oregon (Aubry and Lewis 2003), and California (Zielinski et al. 1995a) played an important role in evaluating whether the Pacific fisher should be listed under the Endangered Species Act (US Fish and Wildlife Service 2004). In the Pacific states, a centralized, interactive web-based database is being established so that survey results can be permanently archived to serve as a resource for management, research, and future surveys (Aubry and Jagger 2006). Such regional databases will ensure that survey efforts are not lost and can make the greatest contribution to the regional understanding and conservation of carnivores.

## Future Directions and Concluding Thoughts

The popularity of remote cameras as a research tool has grown in recent years, in part due to their ability to detect rare species with relatively minimal effort (table 5.1). For their report on the use of cameras in wildlife research, Cutler and Swan (1999) reviewed seventy-eight studies that had used motion-sensitive cameras in the previous century; our recent literature search uncovered another seventy papers published in the six years following their review (i.e., 2000–2005). This increase in popularity has largely paralleled advances in technology, which have diminished the disadvantages of the method (box 5.1) by enabling cameras and sensors to be cheaper, smaller, and more reliable. The evolution of analytical techniques has also been important—remote cameras can now be purchased and deployed in sufficient numbers to collect data that can be analyzed with modern statistical approaches (Karanth and Nichols 1998; Carbone et al. 2001; MacKenzie et al. 2006; also see chapter 11). We expect that future remote cameras will continue to benefit from technological advances that will allow them to be deployed in greater numbers with less cost—ultimately producing more data that can be rigorously analyzed and compared across sites.

### Moving from Film to Digital Sensor Networks

The primary drawbacks to remote cameras are the initial equipment costs and the limitations associated with using film (box 5.1)—both of which are being addressed by the rapid development of inexpensive digital cameras. In addition to increasing power efficiency to lengthen the battery lifespan, we believe that the two most critical features required in any new digital remote cameras are very short trigger times—which will enable photographs of animals passing briefly in front of the unit—and infrared flashes, to avoid frightening animals. At present, only two manufacturers offer remote cameras that meet these basic needs, although these cameras remain expensive and the resulting images are of relatively low resolution. Other features that would improve the performance of remote cameras include video capability, the ability to imprint GPS coordinates on images, and the potential to transmit images back to a central database through a wireless network.

Remote cameras are essentially a visual sensor for monitoring animals. If wirelessly networked, they could collect long-term data over broad areas with less labor. This ability would eliminate costs associated with regular visits to cameras for image retrieval, as well as reduce the amount of time that malfunctioning cameras were off-line (because their condition would be reported instantly). Further, networking would add to the utility of images by producing them in real time. Although ecologists have begun to deploy sensor networks to increase the spatial and temporal scale at which they can collect environmental data (Porter et al. 2005), such approaches have not yet been widely exploited by zoologists (Arzberger 2004). Connecting remote cameras via satellite communication networks would allow a global coverage, but would also be expensive (Arzberger 2004). Other options, such as cellular or local area networks, offer more economical solutions, and companies serving remote camera hobbyists and hunters have recently started to develop such capabilities (Buckeyecam, Athens, OH; PixController, Export, PA).

If new technological developments can further minimize the shortcomings of remote cameras, wildlife researchers will be better able to meet major research challenges (National Research Council 2001) by developing affordable, long-term, broadscale monitoring programs (Porter et al. 2005). The excitement evoked by remote camera images may also foster opportunities to involve volunteer citizen scientists in research projects, further reducing costs and engaging the public. Finally, the networking of remote cameras would facilitate the transparent co-ordination of data across studies within a central database (e.g., Aubry and Jagger 2006). If such databases are to enable rigorous comparisons across surveys, it is critical that researchers using remote cameras precisely document—and whenever possible, standardize—their methods.

## Standardizing Protocols

Wildlife researchers have only begun to identify the many factors that can affect the ability of remote cameras to detect species and individual animals. Indeed, more such research is warranted if results are to be accurately compared across studies. Ideally, a set of standardized protocols should be developed. Short of that, researchers must at least better understand the factors that impact remote camera performance and document them in any published reports.

Camera equipment variability is probably the most obvious source of bias, as different units are likely to possess unique characteristics that can also change over time as older units fail or manufacturers change components in newer models. The among-unit differences that are most likely to affect data comparability are shutter delay and trigger sensitivity. To capture every animal that passes before a remote camera, shutter delay should be as short as possible—preferably instantaneous. Most film cameras are instantaneous, while shutter delays for digital cameras currently vary from 1/10 sec. to 4 sec. or more (ChasingGame 2006). Motion sensor sensitivity is more difficult to standardize because of the many sensor types, each of which varies in performance according to target species and field conditions. Again, we recommend that surveyors test remote cameras before every deployment, using a model appropriate for the target species and only units that trigger reliably. This precaution should be reported in publications, along with the sensor settings and programmed delay between photographs. Reporting the area (angle and distance) covered by

the motion sensor would also be helpful. Last, we encourage researchers to experiment with the height of the motion sensor and the use of bait to produce empirical comparisons of various set designs.

## Final Thoughts on the Future

From the invention of camera flashes to the development of compact cameras, wildlife research conducted with remote cameras has consistently been tied to advances in technology. Typically, there is a lag time between the advent of new technology and its successful application under harsh field conditions. We believe that remote camera-based research is at the cusp of its own digital age, and that we will see a revolution in the field when inexpensive, sturdy, infrared-sensitive, digital remote cameras with short trigger times are produced. The transition to digital technologies will also enable other advances, such as the ability to connect cameras via wireless networks and the merging of datasets through comprehensive digital databases. To make the most of these upcoming advances, researchers must continue to develop innovative statistical tools for analyzing data and work toward standardizing methods such that studies varying in time or place can be rigorously compared.

---

### CASE STUDY 5.1:
### IDENTIFICATION OF INDIVIDUALS AND
### DENSITY ESTIMATION

**Source:** Soisalo and Cavalcanti 2006.
**Location:** Pantanal region of Brazil.
**Target species:** Jaguar.
**Size of survey area:** 460 km$^2$.
**Purpose of survey:** To provide a rigorous estimate of jaguar abundance and density for the Pantanal region of Brazil. The study also aimed to evaluate published methods for determining the effectively sampled area, and to compare those methods to a novel approach using on-site GPS radio-telemetry data.

**Survey units:** Mean female home range size (40 km²) estimated from GPS radio-telemetry data was used to determine the minimum remote camera density of two to three camera stations per 40 km².

**Survey method:** Unbaited Trailmaster 1550 AIR units and one Camtrakker PIR unit.

**Survey design and protocol:** During 2003 and 2004, fifteen AIR units and one PIR unit were deployed at locations determined to be regular jaguar travel routes based on a combination of radio-telemetry data and natural sign. Each camera station included two cameras that were placed opposite one another—thus positioned to photograph the unique spot pattern on both sides of the animal for positive individual identification. The size of the area to be sampled was driven by accessibility and the number of cameras available (*n* =16). In 2003, three similarly sized blocks (mean = 65 km²) were sampled sequen-

tially; sixteen stations per block were each sampled for twenty days. In 2004, reduced access due to high water levels necessitated placing sixteen remote cameras throughout only the accessible areas and leaving them at those locations for the entire sixty-day survey duration (see figure 5.13).

**Analysis and statistical methods:** To minimize bias, each author examined jaguar photographs independently to identify individual animals. Capture histories were analyzed with program CAPTURE to compute estimates of abundance and density. Four density estimates were calculated, based on estimating the effectively sampled area using (1) half of the MMDM; (2) the full MMDM; (3) the actual MMDM calculated from the collared animals' maximum distances moved; and (4) a buffer width based on male and female home range sizes.

**Results and conclusions:**

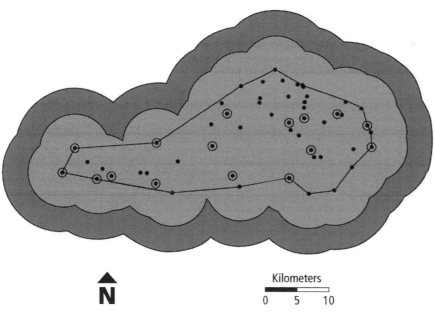

*Figure 5.13.* Survey map showing 2003/2004 remote camera locations for a jaguar survey in Brazil. The outer trap polygon and the effectively sampled area sizes (2003) were calculated using two different buffer-width estimation methods. Solid circles represent 2003 camera locations. Outlined circles represent 2004 camera locations, while --●-- represents the outer camera polygon (area = 165 km²). Light gray shading represents the effectively sampled area estimated with a buffer calculated from the jaguars' mean maximum distance moved (MMDM) or camera information alone (area = 360 km²). Dark gray shading represents the effectively sampled area estimated with a buffer calculated from GPS-telemetry or the jaguars' "actual" MMDM (area = 568 km²). Reprinted from Soisalo and Cavalcanti (2006) with permission from Elsevier.

- A total of thirty-one and twenty-five individual jaguars were identified from 157 and 131 jaguar photos in 2003 and 2004, respectively, from an effort of 960 camera trap nights per year. Of studies to date, this is the greatest number of individual jaguars photographed at a single site.

- Density estimates calculated with MMDM data from GPS-collared animals were 6.5 ± 0.107 and 6.7 ± 1.13 jaguars/100 km$^2$ in 2003 and 2004, respectively.

- Traditional density estimates based on half of the MMDM estimated from camera survey grids significantly underestimated the maximum distances moved by jaguars, ultimately resulting in considerably inflated jaguar density estimates.

- Distances moved and minimum home ranges obtained from camera survey grids collected over short periods were highly unrepresentative of true movement distances and home range sizes. When possible, telemetry data rather than MMDM data from grids alone should be used to inform decisions on buffer widths.

---

### CASE STUDY 5.2:
### CHARACTERIZING THE BEHAVIOR OF FOXES AND BADGERS VIA REMOTE VIDEOGRAPHY

**Source:** Macdonald et al. 2004.

**Location:** Oxfordshire and Gloucestershire, England.

**Target species:** Red foxes (*Vulpes vulpes*) and European badgers (*Meles meles*).

**Size of survey area:** Cameras covered 80 m$^2$ at artificial feeding areas and 35 m$^2$ at badger setts (i.e., dens).

**Purpose of survey:** To understand the behavioral interactions between two carnivores known to share feeding sites. The level and type of aggression between species is relevant to the spread of diseases such as rabies and bovine tuberculosis. Additionally, patterns of bait consumption affect strategies for administering vaccines via bait.

**Survey units:** Six badger setts at two sites.

**Survey method:** Time-lapse and motion-sensitive video cameras were mounted on trees, directed downward toward the target areas, and illuminated with infrared lights.

**Survey design and protocol:** Recording took place during two years at the badger setts and one summer at the artificial feeding site.

**Analysis and statistical methods:** Filmed interactions ($n = 135$) were analyzed frame by frame, characterized as aggression, vigilance, or feeding, and coded using focal sampling at five-second intervals. Behaviors were categorized according to explicit definitions and analyzed with a variety of methods including log-survivorship plots of encounters and a *0/1 method* for interspecific encounters.

**Results and conclusions:**

- As predicted given their larger size, badgers were dominant over foxes and regularly displaced them from food sources with aggressive charges, although no cases of physical contact between animals were observed.

- Badgers fed in longer bouts and consumed most of the bait.

- Many encounters were not aggressive, with both species ignoring each other and remaining at the site. Badgers were more aggressive at feeding sites, especially after food was depleted. At badger setts, once it was clear that an encounter was not going to escalate to aggression, neither species was affected by the presence, proximity, or orientation of the other.

- Badgers and foxes feeding from the same feeder increased the risk of disease spread, especially as both species were seen licking the feeders after the food was depleted. Scattering vaccine baits would reduce this problem.

- There are indications that foxes sometimes seek the company of badgers, loitering around their setts even in the absence of food. The authors speculate that foxes may be attempting to benefit from the interspecific transfer of information (e.g., by following badgers to potentially food-rich sites).

---

### CASE STUDY 5.3:
### ASSESSING TIGER POPULATION DYNAMICS AND POPULATION VIABILITY

**Source:** Karanth et al. 2006.
**Location:** Nagarahole Reserve, southern India.
**Target species:** Tiger.
**Size of survey area:** 644 km².

**Purpose of study:** To build photographic capture histories of individual tigers for estimating tiger abundance, density, annual survival rates, number of new recruits, and rates of population change over time.

**Survey units:** Cameras were placed along forest roads or trails (each 10–20 km long) traversing the study area.

**Survey method:** Tiger images were captured at stations consisting of a pair of AIR remote cameras positioned to simultaneously photograph both flanks.

**Survey design and protocol:** On each road, twelve to fifteen remote cameras were established at 0.8- to 2.0-km spacing intervals. Cameras were placed at optimal locations, based on the presence of tiger sign, to simultaneously maximize capture probability while sampling the entire area of interest. Traplines were run for three to sixteen occasions, which were defined as the four to six successive camera nights during which all traplines were sampled at once. This was repeated once per year for nine years.

**Analysis and statistical methods:** Capture histories of individual tigers were analyzed using programs CAPTURE (Otis et al. 1978; Rexstad and Burnham 1991) and MARK (White and Burnham 1999) to derive the primary population estimates described above via "robust" capture-recapture models. These robust models also permitted the esti-

mation of transience, temporary emigration, and variation in probabilities of initial capture and recapture among individual animals.

**Results and conclusions:**

- Capture histories were collected for seventy-four individual tigers during the nine-year study.
- Tiger annual survival rates were estimated to be $0.77 \pm 0.051$.
- From 1996–2000 the tiger population varied from $17 \pm 1.7$ to $31 \pm 2.1$ tigers, with a mean rate of annual population change estimated as $\lambda = 1.03 \pm 0.02$—representing a 3% population annual increase.
- Estimated recruitment of new animals varied from $0 \pm 3.0$ to $14 \pm 2.10$ tigers; population density estimates ranged from $7.33 \pm 0.8$ to $21.73 \pm 1.7$ tigers/100 km².
- Despite substantial annual losses and temporal variation in recruitment, tiger densities remained at relatively high levels. These results are consistent with the researchers' hypothesis that protected wild tiger populations can remain healthy despite heavy mortality because of the species' inherently high reproductive potential.

---

## Acknowledgments

We would like to thank our colleagues who shared their knowledge of remote cameras by teaching us how to work with them in the field and through correspondence. We are grateful to the reviewers who substantially improved the manuscript: Tom Kucera, Ric Schlexer, and Marcella Kelly. Most of all, we thank the editors, who had the foresight to organize this outstanding volume and the patience and attention to detail necessary to help our contribution meet their high standards.

# Appendix 5.1

*Basic equipment list for deploying and checking remote cameras in the field*

| Application | Item | Application | Item |
|---|---|---|---|
| Multiple use | Hammer | Baits and lures | Bait |
| | Nails | | Rubber gloves |
| | Duct tape | | Lure and applicator |
| | Electrical tape | | Bait fastening material |
| | Zip ties | | Antibacterial soap |
| | Pruning saw or machete | | Climbing gear (for suspended bait) |
| | "Do Not Disturb" signs | | Suspension lines, wire (for suspended |
| | Screwdriver and Allen wrench | | bait) |
| | Spare screws | | |
| | Waterproof indelible pens | Data management | Data forms |
| | Station identification signs or dry-erase | | Mechanical pencils |
| | boards | | Downloading cables |
| | Scent masking spray solution | | Field notebook |
| | Battery voltage tester | | Pocket PC |
| | Plastic baggies (for old bait, scat | | |
| | collection) | Navigation | Maps and aerial photos |
| | | | GPS |
| Sensor system | Extra batteries | | Station location information and |
| | User manual | | coordinates |
| | Brackets and straps | | Compass |
| | Desiccant | | |
| | | Safety | Headlamp or flashlight |
| Camera | User manual | | First-aid kit |
| | Spare camera cables | | Radio |
| | Extra film | | |
| | Antifogging solution | Protocol checklists | Sensor and camera system functions |
| | Extra batteries | | Station setup and checking steps |
| | Locks, cables, and keys | | Data collection steps |
| | Spare digital memory cards | | Computer download commands |
| | Flexible tubing (to protect external | | |
| | cables) | | |
| | Metal contact cleaner (solution and | | |
| | brush) | | |

*Chapter 6*

# Hair Collection

*Katherine C. Kendall and Kevin S. McKelvey*

The identification of species from hair samples is probably as old as humanity, but did not receive much scientific attention until efficient and relatively inexpensive methods for amplifying DNA became available. Prior to this time, keys were used to identify species through the microscopic analysis of hair shaft morphology (Moore et al. 1974; also see Raphael 1994 for a review of pre-DNA approaches to species identification). For North American carnivores, such analyses are reliable primarily at the family level. Canid hairs, for example, can consistently be differentiated from felid hairs (McDaniel et al. 2000), but hairs of closely related species are often difficult to distinguish. Indeed, for most species, DNA analysis is required to confirm species identification from hair samples, as well as to determine individual identification and population characteristics such as abundance (Woods et al. 1999), substructure (Proctor et al. 2002, 2005), movement (Proctor 2003; Proctor et al. 2004), relatedness (Ritland 1996), and population bottlenecks (Luikart and Cornuet 1998).

In this chapter, we describe methods for collecting hair with hair snagging devices that are positioned so that target animals either make contact with them naturally or via the use of attractants. Hair can also be collected opportunistically from den sites, snow track routes, or other areas frequented by the species of interest (McKelvey et al. 2006; also see chapter 3), but here we focus only on sampling devices specifically designed and deployed to collect hair. We assume that sampled hair will undergo genetic analysis.

A hair sampling method is inherently multitiered. At the most basic level, it typically comprises a hair collection device or series of devices—such as barbed wire, glue or adhesives, or brushes—forming or strategically situated within a collection structure, such as a corral or cubby. These structures are in turn sited within a sampling framework, thus permitting the acquisition of meaningful data. For the method to be effective, the collection structure must permit or promote use by the animal, the collection device must snag the animal's hair, and the resulting hair samples must contain useful DNA.

Hair collection methods can be broadly subdivided into *baited* and *passive* (unbaited) approaches. Although baited methods are most frequently used, passive approaches tend to be more effective for sampling certain species and for addressing fine-scale habitat use and a number of other survey objectives because behavior is not influenced by the draw of bait. Passive methods also have the advantage of requiring no induced response from the target animal; samples are collected during normal behavior, and there is little risk of individuals becoming averse or habituated to a hair collection structure. We have further subdivided baited methods into four distinct types:

1. *Hair corrals* are structures that use at least one strand of barbed wire to encircle an attractant and are predominantly used to sample ursids.
2. *Rub stations* are structures saturated with scent lures to induce rubbing, and they typically use one of two types of hair collection devices:
   a. *Barbed rub pads* usually consist of a carpet pad with protruding nails (or, in some cases, stiff natural fibers) and are used primarily for felids.
   b. *Adhesive rub stations* typically consist of blocks of wood covered with adhesives and are used mainly for canids.
3. *Tree and post hair snares* are wrapped with barbed wire or fitted with alternative hair snagging devices and have generally been used to sample wolverines (*Gulo gulo*).
4. *Cubbies* are boxes or tubes containing attractants and fitted with snaring devices at the entries or along the inside walls and are used mostly for mustelids, but can also be effective for other small- to medium-sized species.

Finally, we have grouped passive methods into two categories:

1. *Natural rub objects* are objects found in nature (e.g., bear rub trees) that are fitted with hair snagging devices.
2. *Travel route snares* are hair snagging structures that target animal travel routes or other areas of concentration such as dens, burrows, beds, and latrines. Travel route snares employ one of three types of hair snagging devices:
   a. *Barbed wire* strands strung across travel routes are primarily used to sample ursids, but they have also been used for badgers.

b. *Adhesives* (such as double-sided sticky tape) hung across travel routes are used to sample hairy-nosed wombats (*Lasiorhinus krefftii*) in Australia, and have been employed for some North American carnivores.
c. *Modified snares and traps* are leg and body snares or traps that have been adapted to allow animals to escape but deposit hair samples in the process. These are used for a variety of species.

## Background

Although the DNA analysis of animal hair dates from the early 1990s (Morin and Woodruff 1992), the monitoring of rare North American carnivores via noninvasively collected samples began more recently with the analysis of mitochondrial DNA to identify different species (Foran et al. 1997a, b; Paxinos et al. 1997; also see chapter 9). Foran et al. (1997a, b), for example, discussed reliable and inexpensive methods for identifying many species based on universal DNA primers (Kocher et al. 1989) from scats (Foran et al. 1997a ) and hair (Foran et al. 1997b).

The use of noninvasive hair collection methods to survey wildlife has expanded rapidly since the mid-1990s. Studies of high-profile, rare, and elusive species such as grizzly bears (*Ursus arctos*; Woods et al. 1999; Poole et al. 2001; Boulanger et al. 2002; Paetkau 2003), American black bears (*Ursus americanus*; Boersen et al. 2003), Canada lynx (*Lynx canadensis*; McDaniel et al. 2000; J. Weaver, Wildlife Conservation Society, pers. comm.), and American martens (*Martes americana*; Foran et al. 1997b; Mowat and Paetkau 2002) were among the first to exploit DNA-based hair snaring techniques in North America, and have generated the bulk of literature available in this field. Hair sampling is now common and has expanded to include numerous other carnivore species (table 6.1).

## Target Species

The following section describes the primary carnivore species studied with noninvasive hair sampling

**Table 6.1.** References for hair collection surveys of North American carnivore species

| Species | Barbed wire corral | Barbed wire-wrapped tree or post | Barbed or adhesive rub pad | Cubby, enclosure, box, tube, or bucket | Natural rub object | Barbed wire or adhesive tape on travel route | Modified leg or body snares and traps on travel route |
|---|---|---|---|---|---|---|---|
| Coyote | | | Harrison 2006[a]; Shinn 2002[a]; NLS[b] | | | | |
| Gray wolf | Poole et al. 2001[a] | Fisher 2004[a]; Mulders at al. 2005[a]; Dumond 2005[a] | NLS[b] | | | Clevenger et al. 2005 | |
| Gray fox | | | Harrison 2006[a]; Shinn 2002[a]; Downey 2005[a] | Bremner-Harrison et al. 2006 | | | |
| Arctic fox | | Fisher 2004[a]; Mulders at al. 2005[a]; Dumond 2005[a] | | | | | |
| Kit fox | | | | Bremner-Harrison et al. 2006 | | | |
| Red fox | | Fisher 2004[a]; Mulders at al. 2005[a]; Dumond 2005[a] | | | | | |
| Ocelot | | | Shinn 2002; Weaver et al. 2005 | | | | |
| Canada lynx | | | McDaniel et al. 2000; NLS[b] | | | | |
| Bobcat | | | NLS[b] | | | | |
| Cougar | | | NLS[b]; Sawaya et al. 2005[c] | | | | |
| Striped skunk | | | | Belant 2003a | | | |
| Western spotted skunk | | | | Zielinski et al. 2006 | | | |
| Wolverine | | Fisher 2004; Mulders at al. 2005; Dumond 2005 | | | | | |

**Table 6.1** (Continued)

| Species | Barbed wire corral | Barbed wire-wrapped tree or post | Barbed or adhesive rub pad | Cubby, enclosure, box, tube, or bucket | Natural rub object | Barbed wire or adhesive tape on travel route | Modified leg or body snares and traps on travel route |
|---|---|---|---|---|---|---|---|
| North American river otter | | | | | | | DePue and Ben-David, 2007 |
| American marten | | | | Foran et al. 1997b; Mowat and Paetkau 2002; Cushman et al., case study 6.1, 2006; Zielinski et al. 2006 | | | |
| Fisher | | | | Mowat and Paetkau 2002; Zielinski et al. 2006; Cushman et al.; case study 6.1; Belant 2003a | | | |
| Ermine | | | | Mowat and Paetkau 2002 | | | |
| Long-tailed weasel | | | | Mowat and Paetkau 2002 | | | |
| American mink | | | | | | | DePue and Ben-David 2007 |
| American badger | Frantz et al. 2004 (Eurasian badger, *Meles meles*) | | | | | | |

*Table 6.1* (Continued)

| Species | | | | | |
|---|---|---|---|---|---|
| Ringtail | Zielinski et al. 2006 | | | | |
| Raccoon | Belant 2003a | | | | DePue and Ben-David 2007 |
| American black bear | NLS[a,b]; Long et al. 2007b[a] | Proctor 2003; Proctor et al. 2004; Kendall and Stetz, case study 6.2; Boulanger et al. 2006; Boulanger et al. 2008 | Boulanger et al. 2008; Kendall and Stetz, case study 6.2 | Clevenger et al. 2005; Beier et al. 2005; Haroldson et al. 2005; Mowat et al. 2005 | |
| Brown bear; grizzly bear | | Proctor 2003; Proctor et al. 2004; Kendall and Stetz, case study 6.2; Boulanger et al. 2006; Boulanger et al. 2008 | Boulanger et al. 2008; Kendall and Stetz, case study 6.2 | Beier et al. 2005; Haroldson et al. 2005; Mowat et al. 2005 | DePue and Ben-David 2007 |

Note: The exclusion of certain species from this table reflects a lack of published accounts of their detection via this survey method.

[a] This survey did not target the given species, but was somewhat or quite effective at collecting hair from this species.

[b] National Lynx Survey (NLS; K. McKelvey, unpubl. data).

[c] Produced very low detection rates.

methods and the types of hair collection devices and structures that have most often been used for these species.

*Ursids*

A variety of hair collection techniques are effective for sampling American black bears and grizzly bears (also referred to as brown bears). Hair corrals are used extensively to sample bears (Proctor 2003; Proctor et al. 2004; Boulanger et al. 2006; Boulanger et al. 2008), and bears are also readily detected by hair collected from rub trees and other natural rub objects (Boulanger et al. 2008; case study 6.2). Numerous black bears and a few grizzly bears have been sampled as nontarget species at barbed rub pads deployed to detect Canada lynx and other species (Long et al. 2007b; K. McKelvey, USDA Forest Service, unpubl. data). Hair collection devices erected across travel and feeding routes, such as salmon spawning streams, have been employed to sample bears (Beier et al. 2005; Haroldson et al. 2005; Mowat et al. 2005) or have sampled them incidentally as nontarget species (DePue and Ben-David 2007). Barbed wire strung across highway underpasses and overpasses has been successful for sampling both black bears and grizzly bears (Clevenger et al. 2005). Polar bears (*Ursus maritimus*) have not been surveyed with hair collection methods. In some studies, clearly damaged and disturbed cubbies targeting mustelids were a sure sign that black bears were present (Zielinski et al. 2005).

*Felids*

A landmark effort to detect lynx in the United States, dubbed the *National Lynx Survey* (NLS; K. McKelvey, unpubl. data), employed barbed rub pads as hair snares. Barbed rub pads were originally designed to sample lynx and are fairly effective at detecting their presence (McDaniel et al. 2000). Along with 42 lynx identified during the first three years of the NLS, 166 bobcats (*Lynx rufus*) were also detected in the northern United States—even though the rub pads were deployed in preferred lynx habitat at elevations higher than those generally frequented by

bobcats (K. McKelvey, unpubl. data). The NLS also detected numerous cougars (*Puma concolor*) and a few domestic cats (*Felis catus*). A rub pad survey of ocelots conducted in southern Texas detected twenty-nine bobcats, as well as eight ocelots (*Leopardus pardalis*) and a single cougar (Shinn 2002). Apparent detection rates were also high in another rub pad study of ocelots in Texas, with three of four radio-collared ocelots having been detected (Weaver et al. 2005). Finally, Ruell and Crooks (2007) successfully sampled hair from bobcats with ground-mounted natural fiber pads that did not contain barbs.

Some attempts to use barbed rub pad methods for detecting felids have been less successful. For example, even though the NLS obtained many samples from bobcats (K. McKelvey, unpubl. data), results from rub pad-based bobcat studies have been largely unsatisfactory (Harrison 2006; Long et al. 2007b). Although the NLS collected almost as many hair samples from cougars as from lynx, other studies using this method have either failed to detect cougars known to be present (P. Beier, Northern Arizona University, pers. comm.) or experienced lower than expected detection rates (Sawaya et al. 2005). A study targeting margays (*Leopardus wiedii*) in an area where they reportedly occurred also did not succeed in collecting margay hair on rub pads (Downey 2005). In contrast, barbed wire strung across highway underpasses obtained hair from three of five cougars documented with remote cameras (Clevenger et al. 2005).

*Canids*

There are relatively few published hair sampling surveys that include canids as one of the primary target species. Hair from gray foxes (*Urocyon cinereoargenteus*) and San Joaquin kit foxes (*Vulpes macrotis mutica*) has been collected with cubbies (Bremner-Harrison et al. 2006), however, and adhesive rub stations (see *Rub Stations*) have been used to sample dingoes (*Canis lupus dingo*) in Australia (N. Baker, University of Queensland, pers. comm.). Canids have more routinely been detected during surveys for other target species. In New Mexico, for example, a rub pad study of bobcats collected fifty gray fox, eighteen coyote (*Canis latrans*), and sixteen dog (*Canis lupus familiaris*) hair samples compared with only a single bobcat sample (Harrison 2006). Similarly, rub pads made of natural fiber carpeting affixed to wooden boards and placed on the ground to sample bobcats were highly successful at collecting hair from coyotes and gray foxes (Ruell and Crooks 2007). Of hair samples collected during a rub pad study targeting margays, 44% were genotyped as gray fox and some samples were from dogs (Downey 2005). A survey of ocelots conducted in southern Texas collected ten coyote, three dog, and two gray fox hair samples (Shinn 2002). Nontarget species sampled by the NLS included numerous coyotes, and wolves (*Canis lupus*) or dogs (K. McKelvey, unpubl. data).

In British Columbia, a bear inventory employing hair corrals detected wolves at fourteen sites (Poole et al. 2001). At Banff National Park, three of five wolves that were observed via remote cameras using a highway crossing structure deposited hair on barbed wire strung across the underpass (Clevenger et al. 2005). During three wolverine studies that employed barbed wire-wrapped posts, nontarget species sampled included arctic foxes (*Vulpes lagopus*), red (silver) foxes (*Vulpes vulpes*), and wolves (Fisher 2004; Mulders at al. 2005; Dumond 2005). The foxes were able to climb the post and reach the bait perched on top (figure 6.1), while wolves were sampled when they stood on their hind legs and braced against the post with their front legs to explore the bait (which they could not reach).

*Figure 6.1.* By-catch resulting from a hair collection survey. A red fox climbs a post wrapped with barbed wire and baited for wolverines. Photo by R. Mulders.

## Wolverines

Trees or posts wrapped with barbed wire are currently the most effective hair collection method for sampling wolverines (but see *Cubbies* under *Baited Hair Collection Methods* later in this chapter; Mulders et al. 2005; Dumond 2005). Hair corrals are not effective with this species (Fisher 2004). The NLS detected only a single wolverine with rub pads (K. McKelvey, unpubl. data), and Mowat et al. (2003)

found that rub pads were ineffective with wolverines (although a few wolverine hair samples were collected in box traps fitted with barbed wire across the entrance).

## Smaller Mustelids and Other Mesocarnivores

Cubbies have been used for many years to trap small and mesocarnivores, particularly mustelids. In recent years, these structures have been modified and employed for noninvasive sampling of martens and fishers (*Martes pennanti*) using both track plates (Zielinski 1995; Zielinski and Truex 1995) and hair snares (Mowat and Paetkau 2002; Zielinski et al. 2006). Mesocarnivore studies using cubby-type hair snares have detected long-tailed (*Mustela frenata*) and short-tailed weasels (*Mustela erminea*; Mowat and Paetkau 2002), ringtails (*Bassariscus astutus*), gray foxes, and western spotted skunks (*Spilogale gracilis*; Zielinski et al. 2006).

Other hair collection methods have also been successful with these species. For example, modified body snares and foot-hold traps have collected hair from North American river otters (*Lontra canadensis*), American mink (*Neovison vison*), and raccoons (DePue and Ben-David 2007). Short-tailed weasels have been sampled at barbed wire-wrapped posts (Fisher 2004), and hair from striped skunks (Shinn 2002) and long-tailed weasels (Downey 2005) has been found on barbed rub pads. Barbed wire strung across travel routes has collected hair from mink and raccoons (DePue and Ben-David 2007). Finally, work by Frantz et al. (2004) suggests that hair corrals erected around burrow entrances may be effective for sampling American badgers (*Taxidea taxus*).

### Nontarget Species as Bycatch

Hair collection methods are, to one extent or another, "omnibus" sampling methods that frequently sample nontarget species along with target species (table 6.2). Sometimes this bycatch can provide useful information. Simply knowing that a given species is in the area, for instance, is often of interest. Additionally, genetic monitoring is an expanding field (Schwartz et al. 2007), and samples that vary in quality or degree of population representation can yield a variety of insights into population status. A sample might be of insufficient quality to permit population enumeration, for example, but may allow estimation of effective population size (Schwartz et al. 2007). Further, if a method consistently collects hair from a particular nontarget species, it could potentially be used in more formal surveys of this species.

### Strengths and Weaknesses

Noninvasive hair collection provides a means to obtain genetic samples from animals at known locations and has been especially transformative for the conservation and management of species that are reclusive, potentially dangerous, or that inhabit thick vegetation. For capture-mark-recapture (hereafter capture-recapture) studies, DNA marks offer the advantage that they cannot be lost. Because the rate at which hair sheds (and therefore capture probability) differs with age class, species, and season, however, these factors must be considered when designing studies to estimate population size. Despite the noninvasive nature of hair collection devices, some animals may avoid hair collection structures simply because human odors are present, resulting in detection heterogeneity, although any avoidance effect is likely to be much smaller than with animals that have been live-captured.

While the genetic analysis of hair samples can render hair collection more expensive than other survey methods, it is also generally more reliable. Further, if the enumeration of individual animals is required to meet survey objectives, costs are comparable to or lower than those of other methods. Genetic analyses associated with hair collection are expensive not only due to the high price of labor, materials (e.g., DNA polymerase), and equipment, but also because noninvasively collected hair samples are inherently uneconomical. A hair sample can fail to produce useable information if it is too small or degraded or contains hair from more than one animal (for projects seeking individual identification) or species (for projects seeking species identification). On the other hand, DNA from hair is often less degraded than DNA extracted from scat, and generally provides more consistent results at far lower cost (see chapter 9).

Hair collection methods can yield information about a large number of individual animals representing a significant proportion of the population from vast study areas. Furthermore, hair sampling can lead to reliable detections of rare animals where live-capture and other methods fail and can provide population-level metrics such as abundance, isolation, dispersal rate, and origin that are often only accessible through DNA-based methods (Proctor et al. 2004, 2005; Schwartz et al. 2007). The genetic analysis

*Table 6.2.* North American carnivore species sampled by various hair collection methods

| | Hair sampling method | | | | | | |
| | Baited | | | | Unbaited | | |
| Target species | Barbed wire corral | Barbed wire-wrapped tree or post[a] | Barbed or adhesive rub pad[b] | Cubby, enclosure, box, tube, or bucket | Natural rub object | Barbed wire or adhesive tape on travel route[c] | Modified leg or body snares and traps on travel route[d] |
|---|---|---|---|---|---|---|---|
| Canids | | | | | | | |
| Coyote | | | B | B | | | |
| Gray wolf | B | B | | | | T | |
| Gray fox | | | B | B | | | |
| Arctic fox | | B | | | | | |
| Kit fox | | | | T | | | |
| Red fox | | B | | | | | |
| Felids | | | | | | | |
| Ocelot | | | T | | | | |
| Margay | | | N | | | | |
| Canada lynx | | B | T | | | | |
| Bobcat | | | T[e] | | | | |
| Cougar | | B | N | | | T | |
| Mephitids | | | | | | | |
| Striped skunk | | | B | | | | |
| Western spotted skunk | | | | B | | | |
| Mustelids | | | | | | | |
| Wolverine | N | T | B | N | | | |
| North American river otter | | | | | | T | T |
| American marten | | B | B | T | | | |
| Fisher | | B | | T | | | |
| Ermine | | B | | B | | | |
| Long-tailed weasel | | | B | B | | | |
| American mink | | | | | | B | B |
| Procyonids | | | | | | | |
| Ringtail | | | | B | | | |
| Raccoon | | | | | | | B |
| Ursids | | | | | | | |
| American black bear | T | T | B | | T | T | |
| Grizzly bear | T | B | B | | T | T | T |

T = Method used to target this species.
B = Bycatch species detected with this method.
N = Method tried on this species but not effective.
[a]Requires animals to climb.
[b]Includes barbed rub pads and adhesive hair snare devices baited to elicit rubbing behavior.
[c]Includes barbed wire strung across animal trails (for bears, Eurasian badgers) and double-sided sticky tape hung across travel routes of hairy-nosed wombats.
[d]Leg/body snares and foothold traps modified to collect hair and allow animal to escape easily.
[e]Produced very low detection rates (Harrison 2006; Long et al. 2007b).

---

## Box 6.1

### Strengths and weaknesses of hair collection methods

| Strengths | Weaknesses |
|---|---|
| • Representative sampling can often be achieved.<br>• Can survey large, remote areas and locate rare, cryptic animals.<br>• Allow discrimination between closely related species, individuals, gender.<br>• Genetic analysis of samples enables many population metrics to be calculated.<br>• Applicable in a broad diversity of habitat types.<br>• Often capable of collecting hair from more than one species.<br>• Snagging devices are generally lightweight and inexpensive.<br>• Baited and passive methods can be mixed, improving sample quality and minimizing bias. | • DNA analysis typically required for species and individual identification.<br>• Amount of DNA in hair samples varies widely between species.<br>• Baited methods require a response from the target animal.<br>• Effective hair snagging methods have not been developed for all species.<br>• DNA degradation may be rapid in warm, wet environments.<br>• Hair snares may become snow covered in the winter.<br>• Most designs are largely effective only for the target species and others of similar size and behavior. |

---

of hair cannot, however, furnish information about an animal's age, reproductive status, body condition, or daily movement rates or patterns and is a relatively weak tool for investigating habitat use.

Passive hair sampling methods generally lack spatial representativeness. By definition, hair can only be collected in those places where animals leave it during the course of their normal activities (in areas of high concentration where the chances of obtaining a sample are high), or where it is feasible for people to find it—such as along trails. In most applications, passive sampling is better suited to detecting presence or assessing *minimum number alive* (MNA) than to estimating population size. But sometimes passive methods can be used in tandem with baited methods in a capture-recapture framework (chapters 2 and 11) to estimate population abundance (Boulanger et al. 2008).

Baited methods have different limitations and strengths. Baits can be set out systematically (as on grids) or randomly. This design flexibility allows the application of a variety of approaches to estimate population size, and spatial analyses are enhanced by the regular distribution of sampling locations (see chapter 2 for further discussion). Baited methods, however, must elicit a behavior from an animal to obtain a hair sample; at rub stations, lynx must rub against a baited, barbed pad (McDaniel et al 2000), and hair corrals require that bears cross a strand of barbed wire (Woods et al. 1999). Individual animals that do not engage in these behaviors (e.g., subordinate animals that are less likely to scent mark on rub pads or adhesive blocks) will not be sampled.

Most carnivores are highly mobile, and individuals can be drawn to bait from relatively long—but typically unknown—distances. In studies using attractants, wolverines have been live-trapped 20 km from the boundaries of putative home ranges defined through subsequent relocations (J. Copeland, USDA Forest Service, pers. comm.). Thus, the area surveyed via methods employing bait or lures can be problematic to define, and habitat associations inferred from sample locations are suspect, at least at the local scale. Such complexities must be carefully investigated or considered when attractants are incorporated into hair collection surveys. A summary of strengths and weaknesses associated with hair collection methods is provided in box 6.1.

## Treatment of Objectives

Study objectives that can be successfully addressed by noninvasive hair sampling vary among species (table 6.3). Many factors affecting hair collection methods are species- or survey-specific, such as the thickness of hair, whether hair is readily pulled by snagging devices or only shed hair can be collected, the temperature and moisture of the environment, and the type of snagging device used. These factors will all affect study design and limit potential analyses.

### Occurrence and Distribution

Species presence and broad-scale distribution are the most general and least demanding objectives of hair collection studies. Most species identification is based on mitochondrial DNA (mtDNA; Foran et al. 1997a, b; Mills et al. 2000; Riddle et al. 2003; see chapter 9 for further background on DNA types and approaches), which occurs in many copies per cell

and is more durable than nuclear DNA because it is protected from enzymatic action within the cell (Foran 2006). The majority of hair samples thus contain sufficient DNA for species identification, even when samples are small or weathered. Because most hair samples from most species can be reliably identified to the species level using mtDNA, standard repeat-visit protocols (MacKenzie et al. 2002; see chapter 2) can be used to estimate detection probability and occupancy. The development of universal mammal primers (Kocher et al. 1989), and the fact that all published DNA sequences must be stored in GenBank (see details in chapter 9) allows virtually any species to be identified rapidly with minimal initial development costs (see Mills et al. 2000 and Riddle et al. 2003 for examples).

### Relative Abundance

In hair collection-based studies of relative abundance, microsatellite DNA is usually examined to

*Table 6.3.* Study objectives addressed by noninvasive hair sampling methods for carnivore families

| Study objectives | Canids | Felids | Mephitids | Mustelids | Procyonids | Ursids |
|---|---|---|---|---|---|---|
| Population status | | | | | | |
| Occurrence and distribution | S | S | S | S | F | S |
| Relative abundance | | S | F | S | F | S |
| Abundance and density | | | F | F | F | S |
| Monitoring | | | | | | F |
| Population genetics/structure | | | | | | |
| Effective population size, evolutionary significant unit, genetic variation | S | S | F | S | F | S |
| Connectivity between populations: barriers to movement, interbreeding, recolonization | S | S | F | S | F | S |
| Detection of hybridization | S | S | | | | |
| Ecology | | | | | | |
| Niche or diet via chemical/stable isotope analysis | | | | | | S |
| Identify individuals for management/forensics | | | | | | |
| Livestock predation | S | | | | | S |
| Incidents with human injury or property damage | | S | | | | S |
| Harvest rate and illegal take | S | S | | S | | S |

Note: This is a current list that likely will change with advances in sampling and DNA technology.
S = Successfully applied.
F = Appears feasible but to our knowledge has not been attempted.

identify individuals (see chapter 9). Relative abundance is estimated using a systematic, or rarely a random, distribution of hair sampling stations. Methodologically, relative abundance estimates fall between well-developed occupancy statistics (MacKenzie et al. 2002) and abundance estimation using capture-recapture approaches (Otis et al. 1978) in terms of certainty. In general, we discourage the use of relative abundance to monitor population trends (chapter 2), but we believe that MNA assessments are often useful for management when detection effort is well documented. Particularly for rare and cryptic species whose presence in an area is subject to speculation, the ability to state with high reliability that multiple individuals are present can be powerful and useful. In northern Minnesota, for example, genetic analyses of hair and scat samples yielded an MNA estimate of twenty lynx (Schwartz et al. 2004). Coupled with anecdotal information that breeding was indeed occurring there, this MNA was sufficient to infer that a breeding population of lynx inhabited the area.

## Abundance and Density

Abundance and density estimation from data acquired via hair collection surveys require reliable individual identification based on nuclear DNA (see chapter 9), and—if capture-recapture methods are used—that a substantial proportion of the total population be both captured and recaptured (see chapter 2). In many cases, hair collection methods may not be efficient enough to provide a capture-recapture sample size sufficient for meeting this objective, and extracted DNA may be of too low quality to reliably identify individuals (see chapter 2 for a more detailed discussion of capture-recapture considerations). Thus, for most species, hair sampling is currently less effective for estimating population size than for estimating occupancy.

## Monitoring

Population trends can be obtained by periodically repeating capture-recapture population size esti-mates using hair snare grids, and changes in distribution or relative abundance can be assessed by repeatedly monitoring occupancy if detection probabilities can be estimated (MacKenzie et al. 2006). Alternatively, purely genetic indices of population status can be derived from much smaller and erratically collected groups of samples (Schwartz et al. 2007). Genetic indices of population status may be a desirable objective when the quality of DNA acquired through hair sampling is relatively high but representatively sampling a large proportion of the population is not feasible. Trends in population health can be tracked with statistics such as effective population size ($N_e$), expected heterozygosity ($H_e$), and allelic diversity (A). In a retrospective study of brown trout (*Salmo trutta*) in Denmark, for example, changes in $H_e$ and A were examined between 1944 and 1997; older samples were acquired from museum-scale collections (Østergaard et al. 2003). Similarly, $N_e$ was estimated for grizzly bears in Yellowstone National Park using samples from the 1910s, 1960s, and 1990s (Miller and Waits 2003). The ability to use museum specimens to accomplish such analyses demonstrates the utility of irregularly collected samples in producing these types of statistics and points to the tremendous potential of using noninvasive hair samples to achieve similar objectives.

## Population Genetics

When DNA quality is high, hair can be used to answer questions about population genetics and structure, thereby providing guidance to conservation measures. For instance, hair sampled from both sides of transportation corridors has been analyzed to determine if highways and rail lines pose barriers to grizzly bear movement and breeding (Proctor 2003; Proctor et al. 2004) and to document wildlife use of highway crossing structures (Clevenger et al. 2005). Further, hair samples can be used to identify the source population of individuals recolonizing historic species ranges (e.g., grizzly bears in Montana), and to define distinct population segments that help identify and prioritize populations for conservation efforts. Vinkey et al. (2006), for example, found that Montana fishers contained unique mtDNA haplotypes (see chapter 9), indicating that native fishers—formerly thought to have been extirpated—had survived and formed a population in west-central Montana. Basic research on the genetic characteristics of populations (e.g., effective population size, evolutionary significant units, amount of genetic variation) can also be addressed with hair collection-based sampling (Schwartz et al. 2007; see chapter 9).

### Habitat Assessment

Some hair collection studies have successfully assessed habitat relationships with data collected via attractant-based methods. Apps et al. (2004) applied grid-based hair sampling and the identification of individual animals through genotyping to evaluate relationships of grizzly bear detections with habitat and human activity variables. The resulting predictive model of the spatial distribution and abundance of grizzly bears was used as a strategic planning tool for large (11,000 km²) regions of British Columbia and Alberta. Mowat (2006) examined coarse-scale habitat selection by martens using detection-nondetection data collected via hair snares.

### Diet

While most survey objectives utilizing hair samples rely on DNA analyses, questions regarding ecological niche and differences in diet between and within species can be addressed through stable isotope and elemental analysis of hair. In a study of brown bears, stable isotope analysis documented one population segment that fed upon salmon (*Onchorhynchus* spp.) and another that fed on berries at higher elevations and did not frequent spawning streams (Mowat and Heard 2006). In Yellowstone National Park, the presence of naturally occurring mercury in fish was used to estimate the amount of cutthroat trout (*Salmo clarkii*) ingested by bears through the mercury concentration in bear hair (Felicetti et al. 2004).

## Description and Application of Survey Method

As discussed earlier, a variety of noninvasive hair sampling methods have been used to study carnivores. To be effective, most methods need to be designed or adapted for a particular species or group of animals with similar body size, hair characteristics, and behavior (table 6.4).

### Overview of Hair Collection Devices and Structures

Hair snagging devices vary in effectiveness among species due to differences in hair length and texture. In general, barbed wire is most useful for collecting hair from bears, canids, and wolverines because the hair of these animals is long enough to get pinched between the twisted wires of the barbs. Aggressive barbed wire—four prongs per set of barbs and 6–12 cm spacing between barbs—is the wire of choice for hair snagging and is available in a range of gauges. Ideal between-barb spacing varies with the size of the target species, with smaller animals requiring tighter spacing. Typically, all hair collected on one barb is considered one sample, regardless of the number of hairs present (figure 6.2).

When sampling kit fox hair in cubbies, Bremner-Harrison et al. (2006) found that dog brushes snagged more hair during molting, but lint roller tape was better at sampling hair from winter coats. Short-bristled wire brushes, such as gun-cleaning brushes (case study 6.1), curry combs (Belant 2003a), and glue pads designed to capture mice (Zielinski et al. 2006), are more efficient at snagging the shorter hair of small- and medium-sized mustelids such as martens and fishers. Adhesives and glue work well for both short and long hair, but because they are messy to deal with and time is required to remove the hair from them prior to analysis, alternative collection

*Table 6.4.* Devices determined effective (Y) and ineffective (N) for collecting hair from North American carnivore species

| Species | Barbed wire | Barbed nails | Sticky tape[a] | Adhesives[b] | Tree bark | Combs, brushes[c] | Modified[d] snare cable |
|---|---|---|---|---|---|---|---|
| **Canids** | | | | | | | |
| Coyote | Y | Y | | | | | |
| Gray wolf | Y | | | | | | |
| Arctic or red fox | Y | Y | Y | | | Y | |
| **Felids** | | | | | | | |
| Margay | | N | | | | | |
| Canada lynx | | Y | | | | | |
| Bobcat | | Y | | | | | |
| Cougar | Y | | | | | | |
| **Mephitids** | | | | | | | |
| Skunk | | | | Y | | | |
| **Mustelids** | | | | | | | |
| Wolverine | Y | N | | | | Y | |
| North American river otter | | | | | | Y | Y |
| American marten | N | N | Y | Y | | Y | |
| Fisher | Y | | Y | Y | | Y | |
| Short- or long-tailed weasel | | | | Y | | | |
| American mink | | | | | | Y | Y |
| American badger | Y | | Y | Y | | | |
| **Procyonids** | | | | | | | |
| Ringtail | | | Y | | | | |
| Raccoon | Y | | | | | | Y |
| **Ursids** | | | | | | | |
| American black bear | Y | Y | Y | Y | Y | | Y |
| Grizzly/brown bear | Y | Y | Y | Y | Y | N | |

[a]Types: duct tape, gaffer's tape (similar to duct tape but leaves no residue), commercial lint roller, double-sided carpet tape.
[b]Types: commercial plastic or cardboard-backed glue traps for entangling mice and rats.
[c]Includes gun brushes, curry combs, and dog brushes.
[d]Snares modified by inserting short pieces of wire perpendicular to the cable.

devices are generally preferred if they have been proven effective for the target species. Zielinski et al. (2006) report that hair removal from glue pads and subsequent cleaning with xylene requires twice as much handling time in the lab as removing hair from wire. Less toxic, citrus-based solvents, such as Goo Gone (www.magicamerican.com), are as effective as xylene at cleaning the glue from hair, hands, and equipment (D. Paetkau, Wildlife Genetics International, pers. comm.) and do not require a ventilation hood. Adhesives can be rendered ineffective in wet weather because wet animal fur fails to adhere, and glue can lose its ability to stick to hair when wet (Fowler and Golightly 1994; Mowat and Paetkau 2002). Glue pads, however, remain effective hair collectors at temperatures as low as –28°C (Mowat 2006).

Hair collection structures can be open, such as barbed wire corrals, or can be enclosed, as with cubbies. In addition, collection structures can be designed to become inaccessible after a sampling encounter (referred to as a *single-catch* configuration) or to remain accessible, thus allowing multiple animals to deposit hair. Choices between open versus enclosed and single- versus multiple-catch structures depend on many factors, including the social dynamics of the species, environmental conditions, and study goals (see *Practical Considerations*).

*Figure 6.2.* A bear hair sample is collected from barbed wire. Photo by Northern Divide Grizzly Bear Project, US Geological Survey.

## Baited Hair Collection Methods

The following section describes hair collection methods that employ baits or lures to attract animals to detection devices and to elicit the response necessary for sampling hair.

### Hair Corrals

Hair corrals typically consist of a perimeter of barbed wire supported by trees or posts and centered around a lure or bait (figure 6.3; box 6.2). Wire height is adjusted to the size of the target species with the goal of snagging hair when animals cross under or over the wire (figure 6.4A). To prevent the target species from crossing the wire without touching it, the optimal wire height is maintained throughout the corral by filling in low spots on the ground surface, and by using brush to block high terrain. Wire position is further ensured by securing

it as tightly as possible—one person stretches the wire while another hammers the staples. Placing staples just in front of the barbs prevents the wire from loosening through slippage. If corrals are erected in areas without trees to support the wire, metal or wood fence posts can be used instead; steel T-posts work well, especially if the corners are braced with guy-wires. As bears often step on the wire when entering and exiting hair corrals, two or more staples should be used to attach wire to each tree, and staples should be long enough to penetrate the outer bark.

Wire and attractants should be positioned so that animals are compelled to cross the wire—rather than lean across it—to investigate. For grizzly bears, the attractant should be at least 2 m from the wire. Typically, one strand of wire is used per corral, but some studies have found that using two parallel strands for bears (positioned at 25 and 50 cm above

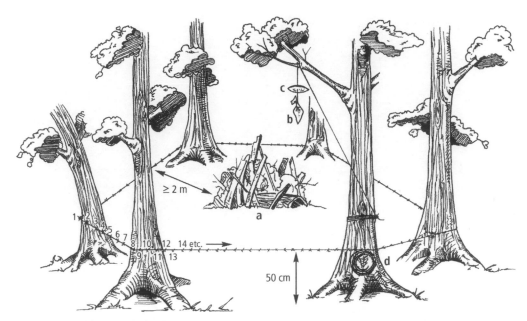

*Figure 6.3.* Components and layout of a barbed wire bear hair corral, showing the (a) debris pile treated with scent lure; (b) scent lure-soaked rag; (c) paper plate or aluminum pie pan hung to protect the rag from rain; and (d) coil of excess barbed wire. Note that barbs are numbered sequentially beginning at one of the trees. Illustration by S. Harrison.

A                                                                          B

*Figure 6.4.* A grizzly bear (A) passing over barbed wire (photo by S. Himmer, Arctos Wildlife Services and Photography), and (B) depositing hair on debris by rubbing its neck on a lure pile (note the hair on the barbed wire in the foreground; photo by M. Maples).

the ground) yields larger hair samples, presumably by forcing more contact between the bears and the wire (T. Eason, Florida Fish and Wildlife Conservation Commission, pers. comm.). Tredick (2005), found no benefit from using a second wire, however.

Although the use of two wires should theoretically increase the sampling of young bears and other smaller species, Boulanger et al. (2006) determined that a single wire placed 60 cm from the ground successfully captured grizzly bear family groups, and

---

## Box 6.2

### Hair corrals for sampling bears

Hair corrals can be formed with a perimeter of barbed wire wrapped around trees or posts and encircling a central bait or lure (see figure 6.3). You'll need

- 20–30 m of barbed wire, placed at a height of 50 cm for grizzly bears, 45 cm for black bears (be sure to maintain optimal wire height throughout the corral). Use four-pronged barbed wire with a 7–12 cm barb interval.
- Fencing staples (3 cm in length for most trees; ≥4

cm for thick bark) to securely attach wire to trees so that it can support the weight of bears when they step on it.
- Nonconsumable, liquid scent lure to apply to the debris pile on the ground, or bait to suspend out of reach from above (see chapter 10 and figure 10.3). For lure placed on debris, the corral should be large enough that the lure is at least 2 m from the closest wire.

---

concluded that single-wire sampling suitably targeted all bears in the population. These researchers also found that adding a second wire increased field and lab costs substantially but did not change population abundance estimates or improve estimate precision. Note that bears often rub or roll on or near debris treated with lure (figure 6.4B). The lure pile can therefore be a productive source of additional hair samples, and hair can often be found on the ground beneath the wire.

Hair corral microsite selection is usually based on habitat quality and species activity patterns. In locations with grizzly bears, we recommend that baited sites be situated ≥100 m from roads and trails and ≥500 m from developed areas for human and bear safety. If hair corrals are deployed in areas frequented by cattle, they must be surrounded by a livestock exclusion fence to prevent trampling. In several Montana studies, most unprotected hair corrals exposed to cattle produced no bear hair because cattle trampled the wire or knocked the bear hair off, and further masked the presence of bears by filling the barbs with their own hair (K. Kendall, unpubl. data; R. Mace, Montana Fish, Wildlife, and Parks, pers. comm.).

Hair corrals have been used with variable success to detect carnivores other than bears. Corrals with three wire strands, intended to survey wolverines in

Alberta, failed to sampled wolverines but collected some hairs from martens and lynx (Fisher 2003). Eurasian badgers (*Meles meles*) were sampled with 20 cm-high barbed wire corrals baited with peanuts and deployed less than 10 m from communal burrow systems (Frantz et al. 2004). Because previous studies suggested that Eurasian badgers would be difficult to attract, bait was placed near burrows up to four months prior to the construction of the corrals. Even though 33% of the hair samples contained only a single guard hair, 93% produced reliable individual genotypes after a single round of amplification. This approach may be useful for sampling American badgers, but their solitary habits suggest that each corral would only have the potential to sample a single individual or a female with young.

### Rub Stations

Rub stations exploit the natural cheek-rubbing behavior of many small felid species (Weaver et al. 2005; figure 6.5A) and the neck-rubbing behavior of canids. McDaniel et al. (2000) provided the first published description and test of this method, the prototype for which was developed by J. Weaver (unpubl. data). Rub pads basically consist of small carpet squares embedded with nails and treated with a scent lure (figure 6.5B; see box 6.3 for details).

The NLS, described earlier, represents the most

## Box 6.3

### Barbed rub pads for sampling felids

To create a barbed rub pad station for collecting hair from felids:

- Cut 10 x 10 cm pads from short, closed-loop carpet of a uniform color that contrasts with the hair of the target species.
- Stud the rub pads with eight to ten nail gun nails, 3.2–3.8 cm in length—depending on the thickness of the carpet. Barbs consist of short lengths of copper connector wire.
- Nail the pad to the tree trunk at a height of 0.5 m above the forest floor for lynx, 0.3 m for ocelots. Trees should be selected for long sight distance.

- Apply 2 tbsp. of liquid lure to the pad, then sprinkle it with 3 tbsp. of crumbled dried catnip. The recipe for the lure is a 1:1:6 ratio of propylene glycol, glycerin, and beaver castoreum. Add six drops of catnip oil per oz. of castoreum/preservative mixture.
- Hang a second small carpet pad baited with liquid lure 0.5 m above the pad on the tree.
- For a visual attractant, mold an S-shaped undulation into an aluminum pie plate (see figure 10.1A) and attach it to a nearby tree limb with a fishing swivel (see figure 10.1D) at a height of 1 m from the ground.

A

B

*Figure 6.5.* (A) Canada lynx rubbing on a barbed pad (photo by P. Nyland). (B) Close-up of a rub pad with barbed nails protruding through the carpet (photo by K. McKelvey).

extensive use of rub pads to date and yielded a number of valuable lessons. For example, carpet pads of a uniform color (such as red or green) that contrasts with animal hair make it easier for field researchers to determine if hair has been deposited on the pad. Closed-loop carpet with tight, short loops eases hair collection and holds liquid lure best, and nails (designed for use in nail guns) should be long enough that barbs are fully exposed and not buried in the carpet. Last, nails snag hair most efficiently when the connecting wire is cut and the nails are pushed through the carpet by hand. This retains the wire barbs and bends them to approximately a 45° angle (figure 6.5B). Otherwise, if nails are fired through the pad with a nail gun, the barbs (as the primary hair collection device) flatten against the nail shaft or are broken off.

The NLS used pie plates as visual attractants and found that twisting them into an undulated form (figure 10.1A) increased their movement in light breezes and reduced entanglement in vegetation. Further, reinforcing pie plates with grommets and fastening them to fishing swivels (figure 10.1D) increased the length of time the plates remained hanging above the rub pad.

At least one published study has extensively tested the effectiveness of various scent lures for use with rub pads. McDaniel et al. (2000) tested five lures in Kluane National Park, Yukon, during a period of high lynx abundance. These tests generated high capture rates, with lynx hair collected on 45% of transects. Although all lures attracted lynx, a simple mix of beaver castoreum and catnip oil was most effective. This lure yielded 39% of lynx detections and was used at only 20% of the stations.

Adhesive rub stations—blocks of wood covered with sticky-side-out tape and treated with commercial canine lures to induce rubbing—were very effective at snagging large quantities of hair from dingoes in Australia (N. Baker, pers. comm.) and could potentially be effective for sampling North American canids. This method worked particularly well in the breeding season but continued to collect hair all year if lures were refreshed frequently. To ensure that nondominant animals were not missed, Baker also sampled DNA from epithelial tongue cells deposited when animals licked blocks of wood wrapped with sand paper and baited with rotting meat.

### Tree and Post Hair Snares

Barbed wire can be wrapped spirally around a tree or wooden post, with bait attached above the wire, to entice the target animal to climb (box 6.4; figure 6.6; Fisher 2004; Mulders et al. 2005). These types of hair snares are potentially useful for any species that climbs trees but seem to work best on medium to large species—probably because barbed wire is more

*Figure 6.6.* (A) Barbed wire-wrapped post showing wolverine hair samples covered with rime ice. (B) Wolverine climbing a baited, barbed wire post. Photos by R. Mulders.

effective at snagging samples from animals with long, coarse hair (case study 6.1).

Wolverine sets, typically baited with meat and sometimes treated with secondary scent lures, have been shown to be highly effective during winter for both live-trapping (Copeland et al. 1995) and hair collection (Mulders et al. 2005). Scent lure alone is not effective for drawing wolverines to posts (Fisher 2004). Various baits have thus far proven ineffective for attracting wolverines in summer (J. Copeland, pers. comm.), and summer sampling can result in bear damage to the hair collection structure (Dumond 2005). For tree setups, if hair snagging devices do not encircle the bole, sheet metal can be mounted on the tree to prevent climbing on surfaces that are not fitted with devices.

The results of the first substantive trial using hair sampling for a capture-recapture-based wolverine population estimate are encouraging. Mulders et al. (2005) report that 284 baited rub posts were de-

ployed in a 3 x 3 km grid for four sampling occasions in the tundra habitat of the Northwest Territories. Capture probabilities were above 0.5 for both sexes, suggesting a high degree of attraction to posts baited with caribou meat and scent lures. The sampling density they used (i.e., one rub post per 9 km² cell) was extremely high considering the large daily movements and home range sizes documented for these vagile animals. Given the high capture rate, it is likely that this population could have been adequately estimated with fewer sampling occasions or a lower snare post density (Mulders et al. 2005).

### Cubbies

Cubbies (referred to as *enclosures* in chapter 4) were one of the earliest structures used for noninvasive hair sampling (Foran et al. 1997b). Cubbies designed for hair sampling are long, thin boxes or tubes containing hair snagging devices and an attractant (figures 6.7, 6.8, 6.9; see chapter 4 for enclosure design

*Figure 6.7.* Marten cubby that can be accessed from both ends. (A) Cubby is vertically mounted on a tree with a roof installed above the top end (photo by J. Stetz). (B) Glue traps are fitted inside to collect hair on either end of the bait attached at the center of the cubby (photo by K. Kendall).

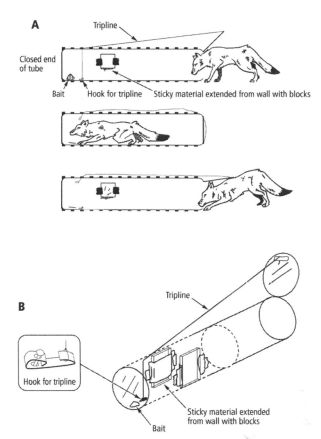

*Figure 6.9.* Single-capture cubby trap for kit foxes, illustrating (A) a side view of a fox entering and exiting and (B) the location of the sticky material and (inset) details of a tripline hook. Figure 6.9A is reprinted from Bremner-Harrison et al. (2006) with permission from The Wildlife Society. Illustrations by S. Harrison.

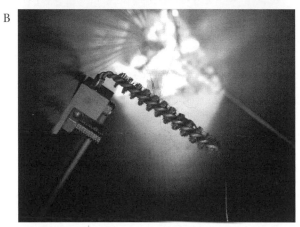

*Figure 6.8.* Triangular cubby (with track plate) installed on the ground, showing (A) the placement of gun brushes serving as hair collection devices, and (B) a close-up of a gun brush mounted on a mechanical lug (with fisher hair). Photos by P. MacKay.

details) and offer two primary advantages for hair collection. First, they improve the reliability of hair capture by orienting the target animal. Further, certain species (e.g., martens and fishers) are detected at highest frequencies when bait is enclosed within a structure (Foresman and Pearson 1998). In North America, hair snaring cubbies have mainly been

used to increase the attractiveness of hair snare devices for martens (figure 6.7; Foran et al. 1997b; Mowat and Paetkau 2002), kit foxes (figure 6.9; Bremner-Harrison et al. 2006) and fishers (figure 6.8; Zielinski et al. 2006; R. Long/P. MacKay, pers. comm.).

Until recently, published methods for the detection of fishers focused on track identification at track plates (Zielinski and Truex 1995; also see chapter 4) versus hair collection. Zielinski et al. (2006), however, tested the effectiveness of modified cubbies containing both hair snagging devices and sooted track plates for detecting fishers and martens. Some cubbies were modified by placing three strands of barbed wire across the opening in a Z formation (figure 6.10A). Barbed wire with four-prong barbs every 7.6 cm was used to prevent target animals from slipping between the barbs unsampled. Other cubbies were modified with glue-impregnated cardboard sheets (originally designed to catch mice) attached to wooden slats and placed in front of the bait near the rear of the cubby at a height of 6 cm above the floor (figure 6.10B, C). The authors concluded that glue was preferable to the wire configuration for snagging hair—particularly for martens. Mowat and Paetkau (2002) also found glue to be highly effective for collecting marten hair. Likewise, in tests with captive wild martens and in a field trial in Michigan,

61 cm-long pieces of 10 cm-diameter plastic French drain tile, with glue pads attached to the top half of both ends, successfully detected martens (J. Belant, National Park Service, pers. comm.). Given that cardboard-backed glue pads can fall apart when wet (Mowat and Paetkau 2002), plastic-backed glue traps (Foran et al. 1997b) should be considered in wet environments.

It is important to note that, because box-type fisher cubbies were originally designed to obtain tracks using sooted plates and sticky paper (Zielinski and Truex 1995) and have only recently been modified to collect hair (Zielinski et al. 2006; case study 6.1), some published design features have been con-

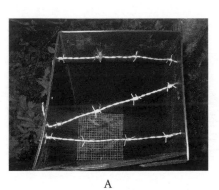  

A                                    B                                    C

*Figure 6.10.* Rectangular cubbies showing (A) barbed wire mounted in a Z pattern to collect hair from fishers; (B) wooden slats fitted with glue traps to snag marten hair; and (C) a close-up of a glue strip on a slat, with marten hair. Photos by F. Schlexer.

---

## Box 6.4

### Wire-wrapped trees or posts for sampling wolverines

Hair from wolverines can be collected from wire-wrapped trees or posts (Fisher 2004; Mulders et al. 2005). To implement this method:

- If possible, conduct sampling for mustelids in winter when baiting is most effective and bears are inactive.
- Remove the lower branches from a tree with a diameter of ≥ 10 cm that is at least 2 m from other trees.

- Where there are no trees available, use 10 x 10 cm wooden posts, 2–2.5 m in length. If sampling in winter, posts should be mounted in snow or on crossed boards weighted with rock to provide a stable base. For summer surveys, mount posts securely in the ground to ensure that they can withstand bear activity.
- Wrap the tree or post with 20 m of barbed wire, starting from the base up to 2 m above the ground/snow pack. Attach the bait above wire.

strained by the requirements for collecting tracks. Since the target animal needs to walk across soot and then paper before reaching the bait, for example, fisher cubbies are very long relative to the size of the animal and are constructed to allow easy removal of both soot plates and paper. In addition, the rectangular cubbies used by Zielinski et al. (2006; figure 6.10A) required hardware cloth and stakes for stability. Such features are likely unnecessary—and may be counterproductive—if the sole goal is hair collection.

Triangular cubbies (figure 6.8A) have also been used to capture hair from fishers and martens (see case study 6.1). The primary advantages of the triangular design are that it does not collapse or need stakes for stability, and it requires less material than box-type cubbies. In the Idaho survey described in case study 6.1, single entry cubbies were fitted with barbed wire at the entrance, behind which three 30-caliber (7.62mm) gun brushes were attached to mechanical lugs (threaded metal connectors that provided support for the brushes) projecting from the walls of the cubby approximately 30 cm from the other entrance (figure 6.8B). Like Zielinski et al. (2006), this survey had no success sampling martens with barbed wire; of the forty-eight marten samples collected at 158 cubbies, all were on gun brushes. Triangular cubbies with gun brushes, based on the same design but with a single entrance, were combined with track plates (as in the Zielinski et al. [2006] surveys) to collect both hair and tracks from fishers during summer surveys in the Adirondacks of New York (figure 6.8A; R. Long/P. MacKay, pers. comm.).

Gun brushes have proven effective for sampling captive wolverines (J. Copeland, pers. comm.) as well. In general, gun brushes offer benefits in terms of ease of use. Lugs can easily be attached to a variety of surfaces, and because brushes are secured by a set-screw (figure 6.8B), they can be removed and replaced with very little handling. And brushes can be deposited directly into desiccant-filled vials, thus eliminating the need to handle hair in the field.

As opposed to setting cubbies on the ground (e.g., Zielinski et al. 2006; case study 6.1), Foran et al. (1997b) and Mowat and Paetkau (2002) mounted cubbies vertically on trees (figure 6.7A). The primary advantages of tree-mounted cubbies are that they are less likely to be covered with snow than are cubbies on the ground, and they may reduce unwanted bycatch. The main disadvantage of vertical mounting is that water can enter the cubby and expose both bait and hair to moisture. This can be addressed by placing a roof above the cubby (figure 6.7A; Mowat and Paetkau 2002) and is likely less of an issue if hair snagging devices other than glue are used.

It is important to size the cubby opening and the distance between snagging devices appropriately for the target species. Structures should also be tested to ensure that certain segments of the population—such as large males—are not excluded, and that smaller animals cannot slip through undetected. Devices must be placed in the cubby such that they make physical contact with the target animal, and there should be no space between them large enough to allow the target species to enter the cubby without contacting at least one device. Mounting devices to the sides of the cubby, as is common with gun brushes, will help to control the maximum width of the entry. Blocks can be used to extend adhesives away from the cubby walls (figure 6.9B; Bremner-Harrison et al. 2006). Pointing gun brushes away from the entrance minimizes resistance to entry while providing a more aggressive snag when the animal exits.

To maximize effectiveness, a combination of hair snagging devices should be considered. The length of the cubby allows for several different devices between the entrance and the bait. Zielinski et al. (2006) found that some animals entered cubbies but failed to leave hair samples. Not only should multiple snagging devices decrease the number of undetected visits, but this approach provides an experimental context for testing the relative efficacy of various devices.

We suspect that as interest in and experience with collecting hair via cubbies increases, the diversity of cubby structures and the types and arrangements of hair snagging devices will expand. For example, J. Belant (pers. comm.) sampled raccoons by attaching barbed wire with 6 cm spacing between barbs in an inverted V at the entrance to a five-gallon bucket lying on its side. He baited the structure with a chicken wing or strip of bacon in a small mesh bag attached to the top rear of the bucket and braced it against a tree or with logs to prevent it from rolling or being moved. The optimal size for cubbies will also continue to be refined. Smaller enclosures can increase hair snagging efficiency and decrease unwanted bycatch, but may discourage entry (see chapter 4). Although most cubbies to date have either exclusively allowed entry from one end or are set up such that entry is primarily limited to one end, there is no intrinsic reason for this design—bait can be located in the center of the structure with hair collection devices at each opening to allow access from either end.

## Passive Hair Collection Methods

This section describes hair collection methods that do not use baits or lures. Again, these methods rely on hair deposited by animals engaged in natural behavior (e.g., rubbing on trees), or snagged as animals pass by devices deployed on travel routes.

### Natural Rub Objects

Although many species rub or roll on natural objects, opportunistic hair collection associated with rubbing behavior has largely been limited to rub trees for bears. Rubbing by bears has not been studied rigorously, but it is thought to represent a form of chemical marking for social communication (Green and Mattson 2003). Surveys in Montana and Wyoming found that grizzly bears and black bears commonly rubbed on trees (figure 6.11A), as well as power poles, sign and fence posts along forest trails and roads, and other structures (K. Kendall, unpubl. data; Green and Mattson 2003). Wolverines and wolves also rub on trees and various other natural and manmade objects and could potentially be sampled with this approach.

The height of the hair deposited on bear rub trees and limited photographic evidence suggest that, although bears sometimes rub the sides of their bodies while positioned on all four feet, they typically stand on their hind feet and rub their back, neck, and head. The most heavily used bear rub trees can be easily spotted by smooth or discolored patches of bark (figure 6.11B), bear trails (track-like depressions worn into the ground by bears repeatedly scuffing or grinding their feet in the same locations) leading to them (figure 6.11C), bare ground at the base, or the presence of large amounts of bear hair. But most rub trees are more subtle, and careful inspection is required to find them. Rub trees that do not occur along human trails or roads are often found on short game trails that become more distinct near the rub trees (Burst and Pelton 1983).

Bear hair naturally accumulates on rub objects, but samples from barbed wire attached to the rub area (figure 6.11D) tend to be of higher quality, require less time to collect, and define discrete samples that help prevent mixed samples containing hair from more than one individual (K. Kendall, unpubl. data). All hair should be cleared from rub trees before sampling to ensure that the period of hair accumulation is known and that genotyping success rates are optimized. Barbed wire should be mounted low enough to sample young bears and bears that stand on four feet when they rub.

Rub tree surveys are problematic in areas heavily used by cattle or horses. Cattle and bears tend to rub on the same trees, making it very difficult to find deposited bear hair. When rub trees are located on trails used by horses, they are often bumped by pack stock. As barbed wire can damage packs, alternative hair snagging devices may be required. In the Bob Marshall Wilderness, Montana, where 15% of rub trees surveyed along trails were bumped by pack stock, the most effective alternative hair device tested was barbless fencing wire mounted vertically on trees (figure 6.12; K. Kendall, unpubl. data). Hair was snared between the split ends of the twisted wire

*Figure 6.11.* Bear rub trees. (A) Grizzly bear rubbing on a tree in Glacier National Park (photo by J. Stetz). (B) Bear rub tree illustrating the discolored bark and damage that usually results from bear rubbing behavior (photo by A. Macleod/J. Stetz). (C) Bear trail leading to a rub tree (photo by A. Macleod/J. Stetz). (D) Barbed wire (with bear hair) mounted on a rub tree to enhance hair collection and the ease with which researchers can remove samples (photo by W. Blomstedt).

*Figure 6.12.* Smooth fencing wire can be mounted vertically on a tree to collect bear hair when the use of barbed wire conflicts with horse use on trails. Photo by Northern Divide Grizzly Bear Project, US Geological Survey.

strands and where staples attached the wire to the tree.

It may be possible to capitalize on the curious nature of bears by installing posts to serve as rub objects. A post-based hair collection survey for wolverines, conducted in a treeless area, found that grizzly bears rubbed and deposited hair on posts that had not been baited for several months (Dumond 2005; see case study 6.2). Abandoned apple orchards and feral crab apple trees that dot rural landscapes in northeastern North America attract black bears in late summer and early fall, providing additional opportunities for the passive collection of bear hair (Hirth et al. 2002). When black bears climb these trees to feed, their claws leave identifiable damage and hair accumulates on the dense, prickly branches

and rough bark. Barbed wire can be wrapped around tree boles to increase the amount of hair snagged and decrease the amount of time required for hair collection.

### Travel Route Snares

Unbaited hair collection devices including body snares, foot-hold traps, and lengths of barbed wire can be positioned on travel routes and runways to sample hair from select species. Such travel route snares are most effective in areas that feature dense concentrations of animals, and therefore single-catch methods (see *Multiple- Versus Single-Catch Structures* later in the chapter) are often employed to circumvent genetic lab costs associated with analyzing mixed samples (Paetkau 2003).

Beier et al. (2005) developed an efficient, single-catch method for collecting hair from brown bears using a modified wolf neck snare (figure 6.13). Snares were hung across bear trails along salmon spawning streams. Short pieces of barbed wire were attached to the snare cable, and a piece of inner tube was inserted to complete the loop and provide the breakaway component. To protect the target animal and enable the recovery of samples, the snare was firmly anchored so that the bear immediately broke the inner tube and the snare dropped where it could be later found. Snares effectively collected hair from bears representing a variety of coat conditions, both sexes, and a wide range of sizes, and they were inexpensive and easy to deploy. Substituting wire brushes for barbed wire increased hair sample sizes and may be worth the investment when maximum capture rates are required (Beier et al. 2005). This method should be used judiciously in areas where other species frequent travel routes and is obviously inappropriate for trails used by people (see *Safety* later in this chapter).

Bears congregating to feed on fish can also be sampled by suspending barbed wire between trees (or posts in treeless areas) across bear trails. Hair snares were set up near spawning streams in Wyoming's Yellowstone ecosystem to assess the importance of cutthroat trout for grizzly bears (Haroldson et al.

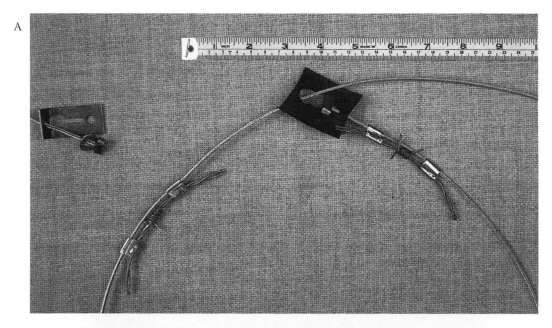

*Figure 6.13.* Modified, self-releasing body snares used to collect hair. (A) Close-up of a bear body snare with a break-away component (photo by S. Lewis). (B) A body snare hung across a trail to snag brown bear hair (photo by L. Beier).

2005). Approximately half of the sampling effort consisted of unbaited barbed wire stretched diagonally across bear trails or fishing sites; the other half comprised baited hair corrals near spawning streams. During four years of hair sampling, seventy-four grizzly bears were identified, and many black bears were detected but not genotyped to distinguish individuals. Although unbaited wire sets were less effi-cient at obtaining hair samples than baited sites, they were useful in areas where it was not appropriate to use bear attractants (see *Safety* for measures to prevent human injuries from snares).

Similarly, a pilot study monitoring highway crossing structures designed for wildlife employed two strands of barbed wire spanning the width of underpasses to collect hair for identifying animals by species, gender, and individual genotype (Clevenger et al. 2005). The strands were respectively suspended 35 cm and 75 cm above the ground to target large carnivores (figure 6.14). Initially, a sticky string or webbing (Atlantic Paste & Glue Company, Quebec, Canada) was intertwined with the barbed wire. Although this method captured hair, many of the samples did not contain DNA—leading the researchers to conclude that hairs collected on the string were largely shed hairs (A. Clevenger, Western Transportation Institute, pers. comm.; see *DNA Quality and Hair Storage* in this chapter for a discussion of shed versus plucked hairs).

Two types of single-catch, unbaited hair snares have been used to sample river otters at river- and ocean-side activity sites (DePue and Ben-David 2007). The first involved setting modified body-snares on otter trails. Two to four microstrands of cable were inserted perpendicularly through the snare cable, with 4 mm lengths protruding at various angles from either side (figure 6.15A), and the snare locking mechanism was replaced with a paper clip to allow the snare to cinch around the target animal and then break free. In the second application, modified foot-hold traps (figure 6.15B) were set at otter

latrine (i.e., scent-marking) sites. Hair capture success for otters was three times higher with body snares than foot-hold traps, but foot-holds may be useful should animals develop an aversion to entering body snares. As otter feces are difficult to locate in the field, hair sampling may be superior to scat collection for the DNA identification of individual river otters (DePue and Ben-David 2007; but see chapter 7). Modified foot-hold traps and body snares could potentially be effective for hair sampling coyotes and other canids (DePue and Ben-David 2007).

Eurasian badgers that did not respond to baiting were sampled by barbed wire strung 20 cm above the ground between stakes set on both sides of a clearly visible badger run (Frantz et al 2004). With a similar design, hair from hairy-nosed wombats was collected by suspending strong double-sided sticky tape between metal posts placed on both sides of burrow runways and entrances (Sloane et al. 2000). Because this method was likely to produce mixed samples, single rather than pooled hairs were genotyped to identify individuals. Given that the distal end of the hairs usually stuck to the tape and the follicles held clear, the samples air-dried rapidly and were easily

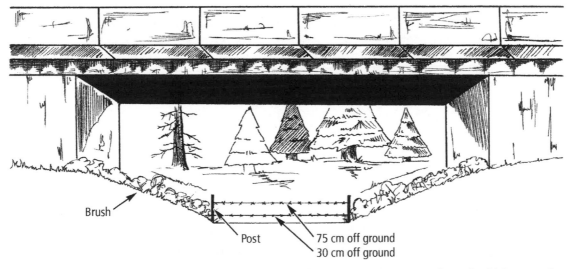

*Figure 6.14.* Ground-level view of a hair sampling method for detecting carnivore movement through a highway underpass (adapted with permission from Clevenger et al. [2006]). Note the brush placed over page-wire material and used to funnel animals toward the wire structure. Illustration by S. Harrison.

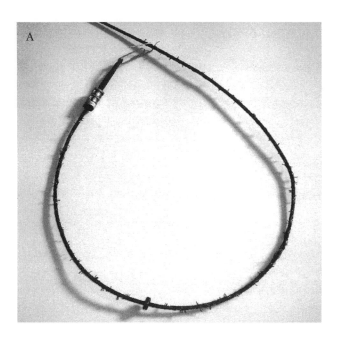

*Figure 6.15.* Single-catch structures modified from live-capture traps designed to collect hair from a single individual and then instantly release the animal. (A) A body snare with a break-away component (i.e, paper clip) used to sample river otter hair; wires that serve as the hair snare device are inserted into the cable. (B) A foot/leg trap fitted with hair collection brushes modified to press brushes against the animal's leg and then allow it to escape. Photos by J. DePue; reprinted from DePue and Ben-David (2007) with permission from the The Wildlife Society.

clipped for extraction (Banks et al. 2003). Unbaited sticky tape corrals and runway barbed wire or adhesive hair snares appear to function best when sampling small species whose travel routes, dens, or nest sites are well defined.

## Practical Considerations

When planning and implementing a hair collection survey, practical considerations abound. Here, we discuss those pertaining to safety, DNA quality and hair storage, single- versus multiple-capture devices, target species-specific behavior, and basic materials and cost estimates for a number of methods.

### Safety

There are a number of safety concerns related to stringing barbed wire across game trails. To protect nontarget species—especially ungulates—a thin, strong pole can be nailed above the wire; ungulates tend to step over the pole while bears duck under the pole and wire (G. Mowat, British Columbia Ministry of Environment, pers. comm.). The pole prevents fast-moving elk (*Cervus elaphus*) and moose (*Alces alces*) from breaking the wire and reduces the collection of hair from nontarget species. In areas where people use game trails, such as along fishing streams or mountain biking and jogging trails, the use of poles—or alternatively, flagging and signing the wire—can prevent human injury and may promote human tolerance of the wire's presence. Careful consideration should be given to employing barbed wire snares in places frequented by people.

### DNA Quality and Hair Storage

Hair must be collected in a manner that allows for subsequent analysis and stored such that DNA degradation is minimized. Successful surveys require detailed planning that takes into account seasonal pelage changes, animal behavior, and the climate and circumstances under which hair will likely be collected (also see chapter 9 for details on obtaining DNA from hair).

As a source of DNA, hair varies in quality and quantity depending on the particular species, the environment in which samples are collected, and whether the method plucks hair or collects shed hair. Follicles are the best source of DNA, and plucked hairs have follicles attached more frequently than shed hairs (Goosens et al. 1998). Hair that is still growing is more difficult to pull out than shedding hair, however, so its collection should be timed to optimize competing trends in hair sample size and quality. For example, bear projects typically collect larger samples in spring and summer because the hair is preparing to shed and is looser than in fall and winter, but samples snagged during cold weather contain more DNA per hair (D. Paetkau, pers. comm.; T. Eason, pers. comm.).

There is no nuclear DNA in hair shafts, but shafts with no follicle can provide useful DNA contributed by dander, saliva, or DNA-containing tissue that adheres to hair as it grows (Williams et al. 2003). Using mitochondrial DNA, Mills et al. (2000) successfully obtained species identification from 84% (91/108) of hair samples without follicles. Occasionally, problems can arise when genotyping hair with no roots from family groups or social animals such as wolves, because it is difficult to discern if the DNA originated from the animal that deposited the hair or from saliva or dander contributed by conspecifics.

Coarse guard hairs yield more DNA than fine underfur because the follicles are larger. Compared with scats, DNA from hair is "cleaner" (i.e., contains few polymerase chain reaction [PCR] inhibitors) and less degraded, but hair produces a relatively tiny DNA sample. Thus, in many cases, only a single extraction can be made from any given sample, and it is possible to exhaust the sample before analysis is complete (Paetkau 2003).

The number of hairs available and the amount of DNA obtained from them varies by species due to differences in density of the coat and fineness of the hair (Goosens et al. 1998). The smaller follicle of fine hair contains less DNA than that of coarse hair and provides less surface area for the surrounding DNA-carrying tissue to adhere to. For example, because felid hair is finer than bear hair, noninvasively collected felid hair samples only sporadically contain DNA sufficient in quality to reliably identify individuals. Genotyping success rates from bear guard hairs are much higher, while success rates for finer bear

underfur are similar to felid hair (D. Paetkau, pers. comm.)

The climate at hair collection study sites influences the amount of useful DNA obtained. Ultraviolet light and moisture degrade DNA, with the degree of deterioration increasing with length of exposure. When sampling with cubbies or in forests with dense canopies, however, ultraviolet exposure is limited and moisture is the chief concern. For best genotyping results, hair should not be left in the field for longer than three to four weeks in dry, sunny climates, and should be collected more frequently in wet climates (D. Paetkau, pers. comm.). It is possible to obtain useable data from some older, weathered hairs, but genotyping success is lower and lab costs are higher than for fresh hair.

Hair is uniformly stored dry, and two approaches have been widely used. For bears, which often provide many samples, hair is pulled from barbs and generally stored in small paper envelopes—with silica gel desiccant if the climate is damp. A second approach is to place each hair sample in an air- and water-tight plastic vial containing desiccant. The advantages of vial storage are that it is more secure and typically minimizes sample handling in the field. For gun brushes and barbed wire used in cubbies, the brush or barb (cut from of the wire with pliers) can be dropped directly into a vial, obviating the need to handle hair. The disadvantages of vials include bulk and expense (of materials and because many labs will charge extra to remove hair from the sampling device)—both of which are important considerations for bear surveys. When glue is used to collect hair, glue pads can be covered with clean plastic (the manufacturer's cover works best), and placed in a paper envelope or bag to protect hair samples until they reach the DNA lab (see chapter 9 for hair removal methods).

## Multiple- Versus Single-Catch Structures

Multiple individuals or multiple species depositing hair on a single hair sampling device can result in a failure to identify individuals or species, respectively.

These problems can be eliminated by analyzing single hairs, but single hairs often provide too little DNA for analysis. In a study of shed hair from chimpanzees (*Pan troglodytes*), Gagneux et al. (1997) showed that 31% of all single-hair amplifications produced allelic dropout (see chapter 9 for a discussion of genotyping errors). Goosens et al. (1998) found that error rates fell from 14% to 4.9% to 0.3% as numbers of alpine marmot (*Marmota marmota*) hairs increased from one to three to ten hairs, respectively. It is therefore advantageous to design hair collection structures and sampling approaches to minimize mixed samples.

Open hair collection methods—such as barbed wire hair corrals—are capable of detecting multiple individuals at a single set, but because there are many barbs available and animal movement is not concentrated, there is a low probability of two animals leaving hair on the same barb. Mixed hair samples are more likely to result when animal movement is either concentrated, such as with hair snares stretched across travel routes, or channeled to relatively few snagging devices, as in cubbies. Thus, in these situations, collection intervals should be shorter than with hair corrals if individual identification is needed. With canids and other social species that overmark, even open methods can yield many mixed samples. If the device is intrinsically likely to collect mixed samples, the only way to reduce their proportion is to reduce the total number of samples collected at each device. This can be accomplished by employing shorter intervals between checks, or by providing an easily removable, single-serving size of bait that is consumed during the first visit by an animal (Foran et al. 1997b; Mowat and Paetkau 2002).

Single-catch structures terminate sampling after one animal visit. Kit fox cubbies can be modified so that a trip wire attached to the bait frees a door that closes the trap after the animal backs out (Bremner-Harrison et al. 2006; figure 6.9). A similar idea was successfully implemented for fishers and martens using a modified box trap in which the door was prevented from locking (Belant 2003a). This allowed captured animals to push the door open to escape, but prevented any other animals from entering. In another application, a mechanism for sampling black bears consisted of an arm that pressed tie plates (used in wood construction) against the target animal when the animal pulled on hanging bait (Immell et al. 2004), and was unlikely to be activated once the bait was gone. Modified body snares, described above in this chapter, have also been employed as successful single-catch structures for bears (Beier et al. 2005) and river otters (DePue and Ben-David 2007).

## Behavioral Considerations

Given that all baited hair collection methods seek to induce a behavioral response, understanding the biology underlying this behavior is important for successful sampling. For instance, if the induced behavior is related to territorial marking, the sex or age of samples may be biased. Alternately, food baits are more effective during seasons when the target animals are hungry and might fail at other times of the year. If the response to bait is linked to its novelty, then initial response rates can provide misleading information when used to design a multisession capture-recapture study. Additionally, interspecies interactions may affect sampling success. Downey (2005), for example, postulated that gray fox visits to rub pads might have interfered with felid marking.

While there is no doubt that induced behavior lies at the core of baited methods, behavioral studies designed to elucidate how animals respond to bait are exceedingly difficult to execute. Most frequently, these studies use captive facilities to study the reactions of a few animals. Captive animal responses, however, may differ from responses of animals in the wild. Captive animals are generally bored, well fed, and situated only a short distance from the bait. Also, captive facilities are saturated with animal scent, and territorial behavior is often absent. For these reasons, results from captive studies are probably most valuable for addressing physical versus behavioral issues. For example, the use of captive animals to determine whether a given hair collection device will reliably produce DNA if an animal enters a cubby will probably be more reliable than testing a variety of baits to determine which bait elicits cubby entry. Due to the logistics and costs of field testing, most baited hair collection methods rely on modified trapper sets whose baits and capture methods have been refined through centuries of trial and error. In fact, the evolution of hair snagging methods and protocols has benefited as much from the rich tradition of trapper lore as it has from controlled experimentation.

## Materials and Costs

The bulk of expenses for hair collection-based population studies are associated with genetic analyses and field technician salaries. Nonetheless, the cost of materials and equipment necessary for hair snaring devices can, in some cases, determine the collection method, the sampling intensity, and the size of area that can be studied. Although space precludes an exhaustive list, we have provided a summary of the materials and costs associated with three commonly used hair collection methods.

### Bear Hair Corrals

MATERIALS. Barbed wire (30 m), fencing staples (0.23 kg), lure, Rite in Rain paper for warning signs, twine to hang lure-soaked cloth, cloth, paper envelopes for hair. Cost: $4.50 per corral.

EQUIPMENT. Fencing pliers, leather gloves, global positioning system unit (GPS).

### National Lynx Survey—Detection Protocol

MATERIALS. Carpet pads, pie pans, dried catnip, liquid lure, forceps, surgical gloves, nails, stove pipe wire for hanging pie pans, swivel hooks, desiccant vials, plastic bags, flagging. Cost: $2.50 per set.

EQUIPMENT. Hammer, GPS, magnifying glass (to look for hair).

### Cubbies for Marten/Fisher

MATERIALS. Corrugated plastic sheeting (0.5–10 m² per cubby); barbed wire, gun brushes, or glue pads; hardware cloth; meat or chicken wing; commercial scent lure; duct tape; gloves; pliers with wire cutters (if barbed wire is used); desiccant vials; plastic bags; flagging. If the cubby is to be mounted on a tree, add wood screws and nails. Cost: $4 per cubby with barbed wire, $7 per cubby with gun brushes.

EQUIPMENT. Hammer, pliers, GPS.

## Survey Design Issues

The efficacy of various hair collection methods is governed by the biology of the target species, the physical characteristics of the hair, and the ability of the devices to collect hair. Method effectiveness in turn determines the types of analyses that can be conducted. Capture-recapture methods, for example, require that a significant proportion of the total population be captured more than once. For sparsely distributed carnivores, achieving this level of capture typically requires a very desirable attractant capable of "pulling" animals from long distances. Species that have an acute sense of smell, like bears and wolverines, can presumably be drawn from great distances to visit bait or scent stations. Thus, high capture rates can be achieved with bears (Boulanger et al. 2002, 2005a, b) and winter-surveyed wolverines (B. Mulders, Northwest Territories Department of Resources, Wildlife, and Economic Development, pers. comm.) using widely spaced detection stations. Felids, in contrast, are thought to be difficult to attract from long distances because they respond primarily to visual stimuli. Therefore, lynx detection stations include visual attractants (see *Rub Stations* and box 6.3) and are set in closely spaced transects (McDaniel et al. 2000). Audio attractants can also be used to enhance detection rates (see chapter 10); Chamberlain et al. (1999) reported that bobcat (*Lynx rufus*) detection was higher at track stations equipped with a mechanical cottontail rabbit distress call than at stations containing a fatty acid scent, bobcat urine, or a visual lure.

Ultimately, the questions that can be addressed using hair collection data are related to the overall detection rate, the recapture rate, and the quality of the DNA extracted. Therefore, a clear understanding of detection rates and the expected quality of the DNA to be collected must be developed prior to designing the study. This will also help ensure that anticipated analyses are consistent with the data.

## Bias

When designing noninvasive surveys based on hair collection, there will almost always be the strong potential for sampling bias in captures, recaptures, or both. Because many carnivores are difficult to sample, however, detection rates may not be high enough to quantitatively assess these biases. Potential sources of bias should be carefully considered, and survey designs should attempt to minimize these biases even if bias has not been demonstrated in previous studies.

As discussed in chapter 10, scent-based attractants can consist of either consumable food or a scent lure that provides no food reward. When animals receive food at sampling structures, they may develop a trap-happy response—thus exhibiting higher recapture rates. Attractants lacking a food reward may have the opposite effect: once the animal determines that there is no reward, it might not be interested in revisiting the site—a situation referred to as trap-shyness. For capture-recapture estimates of population size, models that accommodate a trap-happy or trap-shy response (i.e., behavioral variation) are less precise than simpler models. Hair corrals for bears are commonly moved between sessions to increase novelty and thereby discourage trap-shy behavior (Boulanger et al. 2006).

Unbaited methods are also prone to biased captures. Collecting hair from natural rub objects, for instance, may bias capture toward those sex and age classes that are engaged in territorial marking. For grizzly bear population estimates using combined rub tree/hair corral data (e.g., Boulanger et al. 2008), we advise modeling males and females separately if sample size allows; rub tree samples are biased toward males when collected prior to midsummer (K. Kendall, unpubl. data). Similarly, because ursid so-

cial hierarchy dictates that adult male bears exclude other sex and age classes from the most favorable fishing sites (e.g., in Yellowstone National Park, male grizzly bears consume five times more trout than females [Felicetti et al. 2005]), less productive sampling locations must be sought to ensure that subadults and females with cubs are adequately sampled. Furthermore, barbed wire heights that are best for sampling adults often miss juvenile animals. And where grizzly bears and black bears are sympatric, rub trees detect more grizzlies than black bears—even if black bears substantially outnumber grizzlies (K. Kendall, unpubl. data). This phenomenon must be kept in mind when estimating survey effort in studies targeting both black and grizzly bears.

## Power and Precision Considerations for Capture-Recapture Sampling

When estimating population size, power analysis based on the expected population size and desired precision of the estimate can be used to determine the density of hair snares (Boulanger et al. 2002, 2004b). As most of the published, hair-based capture-recapture studies have been directed at bears, we primarily use bear studies as examples in this section.

For traditional capture-recapture studies, traps are typically placed in a grid—a model that has been followed in most hair corral surveys of bears (Woods et al. 1999; Triant et al. 2004 used hexagonal cells). In capture-recapture analysis, the precision of population estimates increases with the probability of capture, the number of sampling occasions, and the degree to which the capture rates follow "null model" expectations (i.e., equal capture probability for all individuals and across sessions, no behavioral response, geographic and demographic closure; Otis et al. 1978; White et al. 1982). Capture-recapture-based bear studies typically conduct four to five sampling occasions (Boulanger et al. 2004b). To minimize individual capture heterogeneity, the ideal cell size is no larger than the smallest individual home range during sampling (White et al. 1982). A meta-analysis of seven DNA-based hair collection studies of interior grizzly bear populations examined tradeoffs between increasing the precision of the estimate and ensuring geographic closure (Boulanger et al. 2002). As cell size decreased, the recapture rate and precision of the estimate increased, but cost constraints mandated decreasing the size of the study area resulting in an increased likelihood of closure violation. Because monetary constraints often preclude sampling at the optimal intensity, we advise careful consideration of precision requirements before embarking on capture-recapture studies using hair sampling.

Another important design question is whether to move hair collection structures between sampling occasions. If attractants are used, moving structures to new locations between occasions may inhibit habituation to the attractant and decrease individual capture heterogeneity. Relocating structures is also thought to reduce conditioned behavioral responses to sites baited with food. When hair snare density is high (e.g., ≥4 snares per home range), and scent lures (as opposed to reward-type baits; see chapter 10) are used as attractants, moving snare sites between occasions is generally thought to be unnecessary (Mowat and Strobeck 2000; Boersen et al. 2003). In an empirical test of sampling strategies, however, Boulanger et al. (2006) compared moved and fixed site designs using the same sampling density and found that moving sites between sample sessions resulted in more captures and reduced capture heterogeneity.

Hair collection intervals will also be determined by balancing competing goals. Shorter sampling intervals (e.g., one- to seven-day sessions) minimize violations of demographic and geographic closure for closed population models, as well as exposure of hair to DNA-degrading UV radiation and moisture. But the number of individuals visiting a site, and therefore the probability that any given individual will be captured or recaptured during an interval, increases with interval length as long as the attractant remains effective. To complicate matters further, the effectiveness of scent lures and baits fade with time unless they are refreshed. For many hair collection

studies, a fourteen-day sampling interval has been chosen as a reasonable compromise (Boulanger et al. 2005a, b; Proctor et al. 2007; K. Kendall, unpubl. data). These and other considerations for designing capture-recapture surveys are discussed further in chapters 2 and 11.

## Assessing Occurrence and Distribution

For detection-nondetection sampling, the goal is to survey with sufficient effort to reliably detect at least one individual if the area is occupied, or to estimate the probability of detecting an individual, which enables occupancy to be accurately estimated when detectability is low (see chapter 2 for more details). Repeated visits (i.e., multiple independent sampling occasions) are the key to meeting either goal, as the resulting pattern of detections and nondetections furnishes the information necessary to compute detection probabilities (MacKenzie et al. 2002).

The NLS, designed to provide reliable presence-absence information for lynx across large administrative units such as national forests and national parks (McKelvey et al. 2006), presents a good example of design issues that should be considered when using hair sampling to document occurrence. The overall goal of the NLS was to define current lynx range at a relatively coarse scale, and to locate populations. Thus, the initial survey was the first step in a multistep process. The first step of the survey was to collect hair from at least one lynx in each occupied area and to do so with high reliability. To accomplish this objective, rub pad transects were designed to saturate a given area, with twenty-five transects placed on a 3.2 km grid. Each transect consisted of five collection stations 100 m apart, running perpendicular to the slope contour. To satisfy survey requirements, pads were left in the field for at least one month (two sampling occasions) during the summer, and for three consecutive years. Additional grids were located in areas where lynx were known to be present, providing tests of survey effectiveness that ran concurrently with the general sampling. If lynx were detected in an area where they were previously undetected, intensive snow tracking surveys were initiated the following winter (Squires et al. 2004). This follow-up effort was designed to determine whether resident lynx were in the area and to look for evidence of reproduction (i.e., family groups). Further research, in turn, could entail live-capture/radio telemetry to evaluate survival, reproduction, and habitat use patterns.

## Sampling Without a Representative Design

In contrast to surveys such as the NLS, where a grid-based design was used to equalize sampling effort, the objective of some surveys may be to simply confirm the presence of a single individual in an area where a species has been sighted or where putative tracks have been identified (see chapter 2 for pitfalls of such single-location efforts). In fact, much useful information can be gleaned from hair collection efforts even if fully representative sampling is precluded by logistical constraints or the use of passive methods. Establishing that a rare carnivore is present in an area, and particularly that both sexes are present, can be of tremendous importance for conservation. Such goals can often be achieved using nonrepresentative sampling. Further, in certain cases (case study 6.2), passive methods (e.g., rub trees that can be sampled opportunistically while traveling to and from baited survey structures) can increase the total number of individuals counted and contribute to MNA estimates.

Additionally, the genetic monitoring of effective population size and habitat connectivity, or the detection of hybridization (Schwartz et al. 2004) can often be based on small samples, and rules for representativeness may be relaxed when compared with occupancy or capture-recapture sampling (Schwartz et al. 2007). Perhaps the most important point to keep in mind when contemplating nonrepresentative sampling is that meaningful results are produced only if samples are obtained, and results generally become more meaningful as sample size increases. Negative results are not interpretable (i.e., nondetections do not mean that the species is not present).

The variable detection effort associated with passive hair collection methods is generally consistent with the above types of goals but not with the requirements of abundance or occupancy estimation. It may, however, be possible to use nonrepresentative samples in combination with representative samples to produce abundance estimates. The Lincoln-Petersen model, for example, requires only that capture *or* recapture achieve equal capture effort for all animals (Seber 1982). Thus, a rub tree survey can provide the recapture samples for captures made with a hair corral grid if the grid-based captures are uniformly distributed across the sampled population. Recently, more complex capture-recapture models that allow mixing of representative and nonrepresentative samples have been developed that yield estimates of similar magnitude but higher precision than those made with grid-based data alone (Boulanger et al. 2008).

# Sample and Data Collection and Management

Extensive hair collection surveys using the methods presented in this chapter may result in hundreds to thousands of hair samples. With many individuals in the population, and possibly thousands of samples collected, there are countless ways in which errors can creep into a survey—potentially invalidating its results. Errors can be minimized, however, with careful sample and data management.

## Subsampling Hair

Most noninvasive hair sampling methods provide redundant samples for many of the animals sampled. For example, a bear visiting a corral may leave multiple samples upon entry and exit. While genotyping multiple samples from the same individual is one way to check for genotyping errors (through replication of multilocus genotypes; see chapter 9), analyzing all samples collected is seldom desirable. Not only does the analysis of redundant samples increase cost, but genotyping errors can lead to "inventing" spurious animals if stringent measures are not taken to guard against them (chapter 9). Thus, even when single-catch methods are employed, a subsampling scheme is often necessary to minimize analytical costs while maximizing the number of individuals detected. Although not all hair samples will be analyzed, it is important to collect and retain all samples as reserves in case problems arise with genotyping the initial sample.

There are two basic approaches to subsampling, with one rooted in design and the other in analysis. These approaches are not mutually exclusive. Design-based approaches seek to limit redundancy by taking advantage of known characteristics of the target species and hair snagging structure. With bear corrals, for example, two hairs found on adjacent barbs during a single sampling period are more likely to be from the same bear than hairs found on barbs 5 m apart. Most bear studies using barbed wire corrals thus subsample hair based on adjacency. Typically, the largest sample among adjacent barbs is analyzed (M. Proctor, Birchdale Ecological, pers. comm.; R. Mace, pers. comm.; G. Stenhouse, Alberta Fish and Wildlife Service, pers. comm.). This strategy is commonly thought to detect the maximum number of individuals at the least cost, but that assumption has not been tested. An alternative design-based approach employs systematic subsampling. Mowat et al. (2005) usually analyzed every third sample in a group of adjacent samples, and at least one sample from each group of adjacent samples. These researchers did not extract adjacent samples or samples separated by a single barb. Such corral-specific approaches can be adapted for most of the other hair collection methods described in this chapter.

Analysis-based approaches involve subsampling—either randomly or with design considerations—and analyzing samples until the desired output metric is stabilized. For instance, if the goal is to determine the number of individuals represented in a sample, one strategy would be to randomly analyze samples until the total number of individuals as-

ymptotes. The advantage here is that the effects of subsampling are directly related to the desired output metrics. That is, one can estimate the likely change in the output had all samples been analyzed. The disadvantages associated with analysis-based approaches lie in the need for very close collaboration with the DNA lab, and lab-related inefficiencies and resultant higher costs due to running samples in multiple, smaller batches.

## Tracking Hair Samples

For most studies, properly associating a particular sample with a specific time and place is critical. Mistakes in recorded time or location can be made in the field or in the lab. To avoid labeling errors, a recent, large-scale grizzly bear survey in Montana utilized bar-coded labels on hair sample envelopes, with duplicate peel-off sample number labels (i.e., piggyback labels) for field data forms (figure 6.16; K. Kendall, unpubl. data). This system allowed data entry via scanning of the bar codes and ensured that forms and data remained linked to the proper samples in the field and lab. Sample labels should include complete information on the date and location associated with the collection of each sample, so that even if the field data form is lost, the sample can be properly documented. In cases where survey results may be controversial or affect the management of rare or high profile species, hair samples need to be closely tracked and secured in limited-access, locked files.

As discussed, many hair sampling techniques produce multiple samples from one animal visit. For instance, with bear hair corrals, it is common to obtain multiple samples from adjacent barbs associated with a single bear crossing the wire, and when wolverines climb posts wrapped with barbed wire, they can deposit hair on adjacent rows of wire. To identify samples that are likely to be redundant, it is almost always useful to record the position of the hair sample on the device or within the collection structure.

For barbed wire hair corrals, barbs can be numbered sequentially beginning at any of the trees or

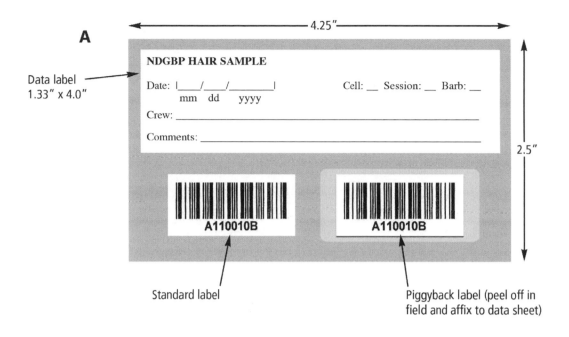

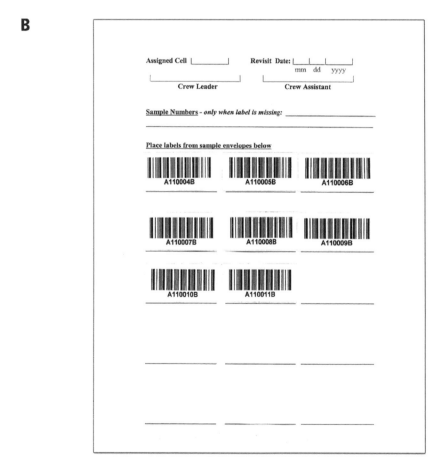

*Figure 6.16.* (A) Bear hair sample envelope with a removable piggyback barcode label and (B) field form with attached bar code labels (K. Kendall, unpubl. data).

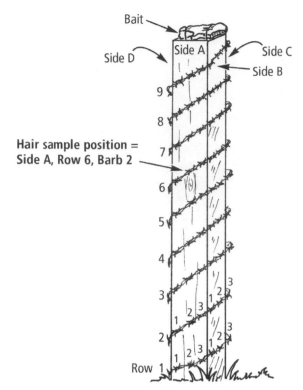

*Figure 6.17.* Method for numbering barbs on a wire-wrapped post to record the position of hair samples, for use in selecting a subset of samples for genetic analysis. Illustration by S. Harrison.

posts supporting the wire (figure 6.3). If two wires are used, the barb number for a hair sample found on the lower wire should correspond with the number assigned to a sample found directly above it on the upper wire. To record sample position on wire-wrapped trees and round posts, it is helpful to divide the wire into four vertical sectors with permanent, waterproof paint (paint pens work well) after the wire is spirally wrapped. Then the row and barb number can be recorded for each sector. If square posts are used, each side can be labeled and considered a sector (figure 6.17).

## Future Directions and Concluding Thoughts

Noninvasive hair sampling is increasingly being used worldwide to enhance our scientific understanding of an ever-widening array of taxa. With hair collec-

tion methods, questions can be addressed that have defied other sampling strategies, or that were not possible to tackle before the advent of methods to analyze small DNA samples (e.g., PCR; chapter 9). In this relatively young field, existing techniques are being continuously refined and new sampling approaches developed. For example, break-away body hair snares that have been used for brown bears and river otters impart unexplored potential as single-catch hair sampling methods for a variety of other species.

Hair will likely persist as a primary source of DNA for mammal studies, particularly where baits and scent lures are used to attract animals. But emerging avenues for the noninvasive acquisition of genetic material also offer promise. For example, saliva samples containing DNA have been collected from tree cambium fed upon by bears and from baited sampling discs (D. Paetkau, pers. comm.); snake, whale, and bird populations have been studied using sloughed skin or shed feathers; and dingoes have been identified from epithelial tongue cells (N. Baker, pers. comm.).

Improvements in DNA extraction and the development of better primers will undoubtedly enhance our ability to identify animals from hair and reduce the costs associated with genetic analyses. Nonetheless, hair collection will remain a multistage endeavor. The collection and subsequent DNA analyses of samples are only the last steps in a lengthy survey process; if no animals visit a collection structure, the effectiveness of the snagging device and the quality of the DNA lab will be of little importance. The utility of hair samples is tightly linked to the overall efficacy of the survey design, which in turn is linked directly to the behavior and biology of the animals. Analysis methods such as capture-recapture are critically dependent on rates of detection. For these reasons, we believe that the greatest advances in noninvasive sampling will likely be associated with better understanding of target species biology. Studies of an animal's behavior when presented with a bait stimulus are a vital and often undervalued component of noninvasive sampling design.

The field of noninvasive hair collection has developed rapidly in the last ten years and will continue to do so with innovations by field biologists. The rate of growth in the future will depend in part on how well experimental studies of new methods are designed, and on how widely the results are disseminated. We encourage experimentation with and adaptation of the methods described in this chapter to create new hair sampling approaches. We also recommend using domestic and captive animals in initial trials of hair snagging devices and structures, as well as testing the efficacy of novel techniques with pilot studies before launching larger projects. The notes sections of journals, and methods-oriented periodicals in general, should be fully utilized to make sure that the details of newly emerging hair collection methods are made available to other scientists and managers.

---

### CASE STUDY 6.1:
### DNA SURVEY FOR FISHERS IN NORTHERN IDAHO

*Samuel Cushman, Kevin McKelvey, and Michael Schwartz*

**Location:** Northern Selkirk Mountains in northern Idaho.

**Target species:** Fisher.

**Size of survey area:** ~1,500 km$^2$.

**Purpose of survey:** Unique haplotypes indicating the presence of a residual native population of fisher were found in central Idaho (Vinkey et al. 2006). Fishers had been detected previously using camera sets in the Selkirk Mountains just south of the Canadian border, but their population status and genetic composition were unknown. The purpose of the study was to provide a comprehensive survey of the northern Selkirk Mountains and to determine the genetic makeup (and therefore population source) of detected fishers.

**Survey units:** Creek drainages ≥30 km$^2$ in area.

**Survey method:** This study used cubbies constructed from folded plastic sheeting. In 2003–4, the cubby design followed Zielinski et al. (2006). The cubbies used in 2005–6 were triangular by cross section, with sides 41 cm in length, and each contained three 7.62 mm gun brushes in addition to the Z of barbed wire described in Zielinski et al. (2006). Both

years, the cubbies were baited with a carpet pad soaked in beaver castoreum and an approximately 125 cm$^3$ cube of deer meat. These items were attached to hardware cloth (i.e., wire mesh) on the inside of the cubby. A sponge splashed with skunk essence was hung above the cubbies as a lure.

**Survey design and protocol:** The Selkirk Mountains are a granite batholith cut by deep canyons. As fisher habitat was located in the densely timbered valleys, surveys were concentrated in the valley bottoms, while the higher elevation areas were not surveyed. Surveys were conducted during the winters of 2003–4 and 2004–5. Cubbies were placed at approximately 1-km intervals along roads and trails in major creek drainages (figure 6.18), and were checked and rebaited once after a period of sixteen to thirty-six days. Total sampling periods varied from thirty to seventy-three days. Snowmobiles were used to set and check hair snare cubbies, with the exception of a single roadless area that was surveyed using snowshoes. Efforts were made to survey all drainages larger than 30 km$^2$, although there were some holes in the survey due to lack of access.

**Analysis and statistical methods:** Hair samples from mustelids were analyzed to the species level using restriction enzymes for all samples (Riddle et al. 2003). A small group of nonmustelid samples were sequenced and compared to published sequences in GenBank (www.ncbi.nlm.nih.gov/BLAST/).

**Results and conclusions:**

During both years combined, 344 cubbies (186 in year one, 158 in year two) were placed along roads and trails in twenty major creek drainages (figure 6.18).

2003–4 Field Season

- Of 300 hair samples, most were collected from the floor of the cubby versus from barbs.
- Only 55% of samples produced DNA of sufficient quality for analysis.
- Eighteen samples collected from eight cubbies were identified as fisher; twenty-two samples collected from fourteen cubbies were identified as marten.

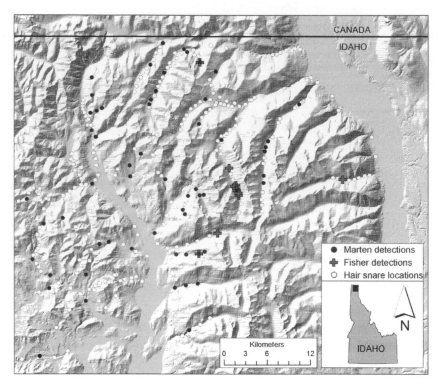

*Figure 6.18.* Hair sampling locations for fishers in the Selkirk Mountains of northern Idaho.

- Of the eighteen fisher samples, one had a haplotype associated with native fishers. The other haplotypes were associated with fishers from Wisconsin and Minnesota (Vinkey et al 2006; Drew et al. 2003).

2004–5 Field Season

- In all, 337 samples were collected; 6 of the samples were taken from barbed wire, 183 from gun brushes, and 148 from the bottom or sides of the cubbies.
- Of the 260 samples tested, 83% yielded sufficient DNA for species identification. The 77 untested samples were deer hair from the bait.
- Eight fishers were detected at three cubbies; all fisher haplotypes indicated Midwestern origin.
- Eighty-three marten samples and one wolverine sample were also collected.
- Other species detected included red squirrel (*Tamiasciurus hudsonicus*), striped skunk,

short-tailed weasel, coyote, wolf or dog, and bobcat.

Synthesis

- At the time of the survey, a relatively small population of fishers occurred in the northern Selkirk Mountains.
- Most of the samples collected were likely associated with an introduction of Midwestern fishers into the Cabinet Mountains in 1989–91 (Vinkey et al. 2006), but at least one fisher was maternally descended from native fishers.

---

**CASE STUDY 6.2:**
**BEAR RUB TREE SURVEY**

*Katherine C. Kendall and Jeffery B. Stetz*

**Location:** Glacier National Park, Montana.
**Target species:** Black bear, grizzly bear.
**Size of survey area:** 4,100 km².

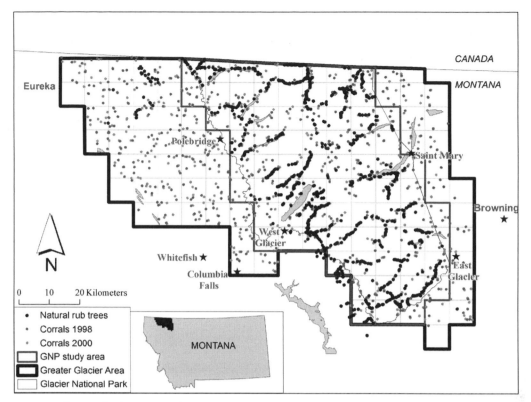

*Figure 6.19.* Locations of natural rub trees and baited hair corrals used to sample grizzly bear and black bear hair in and around Glacier National Park.

**Purpose of survey:** To test rub tree survey methodology, compare detection bias between bear rub tree and barbed wire corral grid sampling methods, and compare the bias and precision of capture-recapture population estimates made using joint rub tree/hair corral data with hair corral-only detections.

**Survey units:** Hair was collected from rub trees identified on maintained trails in the Glacier National Park area. Hair corrals were distributed systematically on an 8 x 8 km grid with one corral per cell.

**Survey method:** Hair snagging devices comprising three to four short (~30 cm) pieces of barbed wire, totaling nine to twelve barbs, were stapled to each selected tree in a zigzag pattern on the rub surface.

**Survey design and protocol:** As part of a study to estimate density and distribution of grizzly bear and black bear populations in the greater Glacier area,

1,185 km of maintained trails were surveyed to identify bear rub trees bears. Based on the level of bear use and geographic distribution, 884 trees were selected for monitoring (figure 6.19). Rub trees were surveyed concurrently with hair corral surveys, which consisted of five, two-week sampling occasions on a grid of 126 baited hair corrals. Rub tree surveys were conducted on foot at approximately four-week intervals in 1998 and two-week intervals in 1999 and 2000. All hair from each barb was placed in its own sample envelope and sent for genetic analysis.

**Analysis and statistical methods:** Genetic analysis was initially attempted on all hair samples with at least five follicles. For those sites where no grizzly bears were identified during the initial analysis, all hair corral samples with at least one follicle, and the two largest hair samples per rub tree survey, were analyzed. The bear species associated with a given sample was determined via analysis of mitochondrial DNA and confirmed with microsatellite analysis. The individual identity of grizzly bears was established using six highly variable microsatellite loci, and gender was determined using the Amelogenin system (see chapter 9). Population estimates using hair corral data alone and joint rub tree/hair corral data were compared using Huggins closed mixture models and the Lincoln-Petersen estimator in program MARK (Boulanger et al. 2008).

**Results and conclusions:**

- The mean number of surveys per tree ranged from 2.46 in 1998 to 6.10 in 2000.
- Two hundred thirty-eight grizzly bears were identified through rub tree sampling during three summers.
- Rub trees were more heavily used by grizzly bears than black bears; the grizzly to black bear ratio was 57:43 at rub trees and 30:70 at hair corrals.
- Male grizzly bears used rub trees more than females during the mid-May through September survey period, however, detection of females increased from virtually no samples in May to 50% or more of the samples from September and October. The male to female ratio of unique grizzly bears sampled was 70:30 at rub trees and 41:59 at hair corrals.

- Of the 231 individual grizzly bears identified in 1998 and 2000, when both hair corrals and rub trees were sampled, 28% were found only at rub trees and another 29% were found at both corrals and rub trees. Thus, including rub trees in the survey significantly increased the number of detected bears.
- The joint rub tree/hair corral data set produced population estimates of similar magnitude but greater precision than hair corral grid data alone (Boulanger et al. 2008).

---

## Acknowledgments

Amy Macleod provided invaluable help by ferreting out information on the latest developments in the noninvasive hair snaring field and creating an electronic bibliography and library. We thank the many scientists around the world that presented us with information on new, unpublished hair snaring techniques in response to our listserve postings. The US Geological Survey and USDA Forest Service provided support for Kate Kendall and Kevin McKelvey, respectively.

# Chapter 7

# Scat Detection Dogs

*Paula MacKay, Deborah A. Smith, Robert A. Long, and Megan Parker*

Domestic dogs (*Canis lupus familiaris*) possess a remarkable sense of smell, allowing them to perceive a vast amount of olfactory information in their environment (Syrotuck 1972). For millennia, humans have exploited this profound scenting ability for tracking game and safeguarding livestock from predators (Coppinger and Coppinger 2001), and the ancient Greeks even employed dogs for forensic work (Schoon and Haak 2002). Today, canine olfaction is widely applied to the detection of drugs, explosives, avalanche victims, agricultural contraband, and myriad other odors of human interest. Adapting training techniques from these applications, biologists are increasingly using dogs in wildlife field research (e.g., Smith et al. 2001; Wasser et al. 2004).

This chapter provides background and guidance for researchers who are interested in employing detection dogs for wildlife surveys. We primarily focus on the use of specially trained dogs for the location of scats (feces) from free-ranging carnivores, although additional research applications are mentioned (e.g., discrimination of species or individuals from scat, detection of live animals). Given the integral role of the dog's nose in this work, we begin with a brief overview of canine olfaction and then provide a historical review of canine detection in wildlife research.

Throughout the main body of the chapter, we discuss specific considerations for planning, designing, and executing a detection dog survey.

While dog and handler performance issues are explored in some detail, this chapter is *not* a proxy for professional training. We believe that the involvement of experienced dog trainers is pivotal to the success of detection dogs as a survey tool and strongly urge researchers to contact a qualified detection dog trainer or organization at the outset of the project planning process (see appendix 7.1). Meanwhile, we hope to provide readers with a solid understanding of how detection dogs can be used to meet numerous survey objectives and to present some of the key issues that may arise in the process.

## Background

In his comprehensive review of dogs in wildlife biology, Zwickel (1980: 535) suggested that "most uses of dogs in the wildlife field involve their sense of smell. If dogs are used properly, this is equivalent to adding a new sense to the observer." While modern humans have pursued animal sign since the dawn of hunting (see chapter 3), canine assistance has made this enterprise much more efficient.

## Canine Olfaction: A Brief Overview

The dog is well designed for olfaction. Most breeds have elongated noses, which house a complex structure of folded turbinate bones covered with epithelial membrane (see figure 7.1). In large breeds, the inner nose may contain as many as 250 million olfactory receptor cells, compared to roughly 5 million receptors in humans (Lindsay 1999). When a dog sniffs a sample of air, odorant molecules are directed deep into the nasal cavities, where they come into contact with the moist mucus of the olfactory epithelium and interact with the embedded scent receptors. The resulting sensory data is ultimately conducted to the olfactory bulbs of the dog's brain, which are considerably larger than those of the human brain (Syrotuck 1972).

The dog's highly sensitive olfactory system enables the detection of a broad range of odors and the discrimination of a target odor in the presence of distracting odors. An estimated 1,000 different types of odor receptors in the mammalian nose collectively respond to an astounding diversity of smells (by comparison, the eye has only three types of photo receptors that enable us to see color as we do; Axel 1995). Humans can presumably distinguish between at least 10,000 different odors, and dogs, a far greater number (Lindsay 1999). Remarkably, it doesn't take much odor to stimulate canine olfaction—Johnston (1999) reports that dogs are capable of detecting certain compounds at concentrations as minute as 500 parts per trillion.

Dogs also have an excellent memory for odor, and have been shown in controlled studies to remember odors on which they were trained at least four months prior (Johnston 1999). D. Smith (unpubl. data) reports that one of her detection dogs alerted her to scats from San Joaquin kit foxes (*Vulpes macrotis mutica*) 671 days after last exposure. Inter-

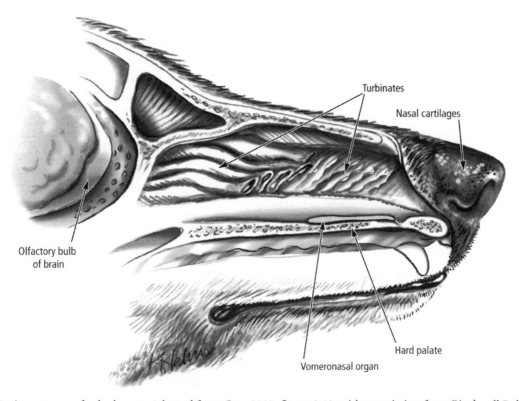

*Figure 7.1*. Basic anatomy of a dog's nose. Adapted from Case 2005, figure 3.10, with permission from Blackwell Publishing and L. Case. Illustration by K. Helms.

estingly, dogs that are systematically trained on certain odors may actually develop more receptors for these odors, thus increasing their sensitivity (Schoon and Haak 2002).

Dogs, like many other mammals, posess an auxiliary scent organ called the vomeronasal organ (see figure 7.1). This pair of fluid-filled sacs, located above the roof of the dog's mouth and behind the upper incisors, is also lined with olfactory receptor cells. Although the function of the vomeronasal organ is not fully understood, it likely serves a role in the perception of pheromones, or body scents (Lindsay 1999; Schoon and Haak 2002). Some dogs exhibit a flehmenlike tonguing behavior—during which the dog's tongue is pushed rapidly against the roof of the mouth—that may help transfer odor to this organ (Lindsay 1999).

### Canine Detection in Conservation

In the 1890s, New Zealand's first conservation officer, Richard Henry, employed dogs to locate flightless kakapo (*Strigops habroptilus*) and kiwi (*Apteryx australis australis, Apteryx owenii*) on the mainland with the intent of moving them to predator-free islands—thus launching this nation's legacy of using dogs for conservation (Hill and Hill 1987). Indeed, dogs have assisted biologists in detecting wildlife and wildlife sign for decades (see Zwickel 1980 for a historical review). For example, dogs have been trained to search for bird carcasses (Finley 1965; Homan et al. 2001) and nests (Jenkins et al. 1963; Evans and Burn 1996), bat carcasses (Arnett 2006), brown tree snakes (*Boiga irregularis*; Engeman et al. 1998, 2002), box turtles (*Terrapene carolina triunguis*; Schwartz and Schwartz 1991), ringed seal (*Phoca hispida*) lairs (Smith and Stirling 1975; Furgal et al. 1996), cougars (*Puma concolor*; Hornocker et al. 1965), and black bears (*Ursus americanus*; Akenson et al. 2001). Most dogs employed in the earlier days of wildlife biology were sporting breeds, which were often trained to track, flush, or retrieve target species (see Zwickel 1980).

In the 1970s and 1980s, pioneer researchers began to use dogs to help them find scats from various target species, including black-footed ferrets (*Mustela nigripes*; Dean 1979; Winter 1981); gray wolves (*Canis lupus*), coyotes (*Canis latrans*), and black bears (P. Paquet, University of Calgary, pers. comm.); and Eurasian lynx (*Lynx lynx*; U. Breitenmoser and C. Breitenmoser-Wursten, The World Conservation Union/Species Survival Commission Cat Specialist Group, pers. comm.). But in the 1990s, advances in genetic and endocrine techniques catalyzed a quest for a more systematic and efficient approach to scat detection—a quest that inspired an innovative team based out of the University of Washington to develop methods for using professionally trained "scat detection dogs" to search for wildlife scats (Meadows 2002). Scat is an easily accessible animal by-product that can convey a wealth of biological information about the individual that deposited it (Putnam 1984; Kohn and Wayne 1997). As an increasingly effective source of DNA and hormones (see chapter 9), scat can also potentially yield reliable data relating to animal distribution and movement, abundance, population density, habitat use, home range size, and other biological and ecological measures.

In the last few years, detection dogs have been used to systematically search for scats from roughly two dozen carnivore species within the ursid, canid, felid, and mustelid families (table 7.1; figure 7.2). In addition to locating scats in the field, such dogs have been trained to detect a variety of carnivore-related target odors, including carcasses (e.g., gray wolves; M. Parker, unpubl. data) and the scent of burrowing animals (e.g., black-footed ferrets; Reindl-Thompson et al. 2006). Last, scat detection dogs have been used for fine-scale discrimination in a number of applications. For example, dogs have been trained to distinguish between scat from closely related species (Hurt et al. 2000; Smith et al. 2003; Harrison 2006) and to identify individual animals by scat (Krutova 2001, as cited in Kerley and Salkina 2007; Kerley and Salkina 2007; S. Wasser, University of Washington, pers. comm.; see figure 7.3 and case study 7.2).

*Table 7.1.* References for scat detection dog surveys of carnivores

| Species | References |
|---|---|
| North American carnivore species | |
| Coyote | L. J. Gormezano and R. F. Rockwell, American Museum of Natural History, pers. comm.; A. Bidlack, University of California, Berkeley, pers. comm.; P. C. Paquet, University of Calgary, pers. comm. |
| Gray wolf | Beckmann 2006; M. Parker, unpubl. data (scat and carcass); P. C. Paquet, pers. comm.; S. Wasser, University of Washington, pers. comm. |
| Gray fox | Boydston 2005; Smith et al. 2006b; A. Bidlack, pers. comm. |
| Kit fox | Smith et al. 2003, 2005, 2006a |
| Swift fox | A. Whitelaw, Working Dogs for Conservation Foundation, pers. comm. |
| Red fox | Boydston 2005; Smith et al. 2006b; A. Bidlack, pers. comm. |
| Canada lynx | M. Parker, unpubl. data |
| Bobcat | Harrison 2006; Long et al., 2007a; G. Fowles, New Jersey Division of Fish and Wildlife, pers. comm.; S. Reed, University of California, Berkeley, pers. comm. |
| Jaguar | C. Vynne, University of Washington/Conservation International, pers. comm. (Brazil) |
| Cougar | Beckmann 2006; F. Bonier, Queen's University/Virginia Tech, pers. comm.; S. Reed, pers. comm.; C. Vynne, pers. comm. (Brazil) |
| North American river otter | J. Packard/B. Alexander, Texas A&M University, pers. comm. (pilot test only) |
| Fisher | Long et al. 2007a; K. Purcell, USDA Forest Service, pers. comm.; S. Wasser, pers. comm. |
| Black-footed ferret | Dean 1979; Winter 1981; Reindl-Thompson et al. 2006; M. R. Matchett, US Fish and Wildlife Service, pers. comm., and D. Smith, unpubl. data |
| American black bear | Hurt et al. 2000 (scent-matching only); Wasser et al. 2004; Beckmann 2006; Long et al. 2007a; P. C. Paquet, pers. comm. |
| Grizzly bear | Wasser et. al. 2004; Beckmann 2006; M. Gibeau, Mountain District National Parks (Alberta), pers. comm. |
| Polar bear | L. J. Gormezano and R. F. Rockwell, pers. comm. |
| Domestic carnivores | |
| Domestic cat | S. Reed, pers. comm. |
| Non-North American carnivore species | |
| Maned wolf (*Chrysocyon brachyurus*, Brazil) | C. Vynne, pers. comm. |
| African wild dog (*Lycaon pictus*, Kenya) | M. Parker, unpubl. data |
| Cheetah (*Acinonyx jubatus*, Kenya) | M. Parker, unpubl. data |
| Eurasian lynx (*Lynx lynx*, Switzerland) | C. Breitenmoser-Wursten and U. Breitenmoser, IUCN/SSC Cat Specialist Group, pers. comm. (scats, dens, kill sites) |
| Andean cat (*Oreailurus jacobita*, Argentina) | Claudio Sillero, University of Oxford, pers. comm. |
| Amur tiger (*Panthera tigris altaica*, Russia) | Kerley and Salkina 2007 (scent-matching only); L. Kerley, Lazovsky State Nature Zapovednik, Russia, pers. comm. (field searches) |
| Amur leopard (*Panthera pardus orientalis*, Russia) | L. Kerley, pers. comm. |

Note: The exclusion of certain species from this table reflects a lack of documented accounts of their detection via this survey method.

*Figure 7.2.* Detection dog "Briar" searches for black bear, fisher, and bobcat scats in Vermont. Photo by R. Long.

*Figure 7.3.* In this scent-matching trial, scat detection dog "Panda" matches the odor of a test sample (i.e., an individual Amur tiger's scat) to a scat from the same tiger in a lineup of seven tiger scats. Photo by J. Dauffy, courtesy of L. Kerley.

## Target Species

In theory, any terrestrial animal that deposits persistent scats can be surveyed with this method—detection dogs have even been used to detect floating feces from marine species such as right whales (*Eu-*

*balaena glacialis*; Rolland et al. 2006) and killer whales (*Orcinus orca*; S. Wasser, pers. comm.). While scat detection dogs have proven effective in temperate (Long et al. 2007a), Arctic boreal (S. Wasser, pers. comm.), and montane (Wasser 2004) forests, tropical savannas (C. Vynne, University of Washington/Conservation International, pers. comm.), desert scrub and grasslands (Smith et al. 2005), prairies (Reindl-Thompson et al. 2006), and other habitat types, environmental and site-specific factors may limit their applicability in some cases. A species inhabiting precariously steep or impenetrable terrain, for example, may not be a feasible target for this method (see *Practical Considerations* in this chapter for other potential limitations).

## Strengths and Weaknesses

Scat detection dogs offer several potential benefits as a survey method (see box 7.1). First, dogs can survey multiple species over extensive areas (e.g., Wasser et al. 2004; Beckmann 2006; Long et al. 2007a), and are notably proficient at recovering small and cryptic scats (e.g., Smith et al. 2001; Long et al. 2007a). These attributes are critical to the study of carnivores, given their wide-ranging and characteristically elusive nature. Scat detection dogs have also demonstrated greater success at locating scats than researchers relying on visual detection alone (Smith et al. 2001) and presumably result in less sampling bias; in some species, scats from dominant individuals are commonly deposited in prominent places as territorial marks (Gorman and Trowbridge 1989) and are thus more likely to be seen. Furthermore, scat detection dogs may provide less spatially biased data than detection devices that use attractants to draw target animals to a site (see chapters 2 and 10) and have been shown to confirm occupancy more quickly and accurately than other noninvasive survey methods (Long et al. 2007b). Finally, the charismatic nature of this method should not be underemphasized—dogs have broad public appeal. Thus, in addition to being highly effective, scat detection dogs can serve as

---

### Box 7.1

## Strengths and weaknesses of scat detection dogs

| Strengths | Weaknesses |
|---|---|
| • Minimal sampling bias compared to human searchers.<br>• Highly effective at locating scats.<br>• Allow quick confirmation of occupancy.<br>• Efficiently survey large, remote areas.<br>• Locate small and cryptic scats.<br>• Discriminate between species (and individuals).<br>• Detect animals after they've left the area.<br>• Applicable in broad diversity of habitat types.<br>• Scat samples can be collected for various analyses.<br>• Effective for multiple target species.<br>• Charismatic "tool" with broad public appeal. | • Substantial time and financial investment for dog and handler selection and training.<br>• Require numerous confirmed scats from target species for initial dog training.<br>• Significant dog care and maintenance responsibilities.<br>• Real or perceived potential conflict with wildlife.<br>• Must work within physical limitations of dog (and handler).<br>• Detection rates can vary between dog/handler teams.<br>• Scat detection can be affected by environmental conditions.<br>• Difficult to assess when target species were present since scats persist over time.<br>• DNA analysis typically required for species confirmation and individual identification.<br>• Rates of species and individual identification from scat are extremely variable.<br>• Very rare species may present methodological challenges.<br>• Confusion can result if target and nontarget species have morphologically similar scats. |

---

captivating ambassadors for carnivore research and conservation.

Scat detection dogs are not without their limitations, however (box 7.1). As with any scat-based method, the persistence of scat in the environment is of obvious importance. Scat decomposition, or the possibility of its being consumed or manipulated by wildlife, must be considered potential sources of sampling bias (Sanchez et al. 2004; Livingston et al. 2005), although the detection dog's remarkable scenting ability maximizes the chance that even small scat remnants will be detected. Reciprocally, the potential long-term persistence of scat can violate closure assumptions of some analytical methods (see chapters 2 and 11) unless scat age can be controlled for (e.g., by attempting to clear all target scats prior to conducting the survey). Further, given that DNA confirmation of species is generally recommended for this and other scat-based methods, laboratory expenses may add significantly to the cost of dog surveys. And despite advances in genetic techniques, scat can still be an unreliable source of DNA (see chapter 9)—sometimes rendering species confirmation difficult or impossible. In addition, detection rates can vary between dog/handler teams (Smith et al. 2003; Wasser et al. 2004; Long et al. 2007a), highlighting the importance of both proper training and a study design that allows detectability to be both estimated and corrected for (see chapter 2 for further discussion on detectability). Handlers as well as dogs can introduce a potential source of bias, as successful detection work relies heavily on the handler's ability to respond appropriately to the dog's behavior.

## Treatment of Objectives

Any objective that can be addressed via scat surveys (see chapter 3) can potentially be met with detection dogs. Also, limitations of scat-based studies that employ only human searchers (e.g., trail bias, insufficient sample size) may be partially or completely circumvented with scat detection dogs—thus expanding the number of objectives that can be met.

### Occurrence and Distribution

Scat detection dogs can be an excellent tool for rapidly and accurately surveying sites with a high probability of detecting the target species, if present (Harrison 2006; Long et al. 2007a)—an important requirement for estimating site occupancy (Long 2006) and distribution (Wasser et al. 2004; Smith et al. 2005). Additionally, detection dogs can efficiently survey for multiple species simultaneously (table 7.2 and case study 7.3). In Vermont, Long et al. (2007a) estimated an 86% probability of detecting black bears and a 95% probability of detecting fishers (*Martes pennanti*), given presence, with off-trail detection dog surveys requiring an average of only 4.1 hours each. Similarly, scat detection dogs had an 86% success rate in determining ferret presence at test prairie dog town sites in South Dakota. Dogs surveyed at an average rate of 26 ha/hr, and the mean time to first detection was 21 minutes (compared to a search rate of 1.6 ha/hr and a mean time to first detection of 208 minutes for spotlighting surveys; Reindl-Thompson et al. 2006).

Several studies have shown scat detection dogs to be more efficient (i.e., detection ability per unit effort) than other methods in establishing carnivore presence. In New Mexico, for example, detection dogs yielded ten times more bobcat (*Lynx rufus*) detections than remote cameras and hair snares combined (Harrison 2006). Further, the probability of detecting black bears, fishers, and bobcats with scat detection dogs during a single visit to forested sites in Vermont was two to three times higher than that achieved with fourteen-day remote camera surveys (Long et al. 2007b). And Smith et al. (2001) found that a detection dog searching through vegetation found approximately four times as many kit fox scats as an experienced human searcher.

### Relative Abundance

Only one study that we are aware of (Smith et al. 2006a) has used data from detection dog surveys to specifically assess relative abundance based on scat abundance. This study used scat abundance to infer location-specific differences in kit fox abundance across eight counties in California's San Joaquin Valley. We suspect that the dearth of detection dog projects with similar objectives reflects the difficulty of relating scat abundance to relative abundance, as well as the availability of better methods for inferring population size (see chapter 2).

*Table 7.2.* Examples of studies using scat detection dogs to survey multiple carnivore species

| Species | Location | Reference |
|---|---|---|
| Black bear and grizzly bear | Yellowhead Ecosystem, Alberta, Canada | Wasser et al. 2004 |
| Fisher, bobcat, and black bear | Vermont, USA | Long et al. 2007a |
| Kit fox, red fox, and gray fox | San Joaquin Valley, California, USA | Smith et al. 2006b |
| Black bear, grizzly bear, cougar, and gray wolf | Centennial Mountains, Idaho/Montana, USA | Beckmann 2006 |
| Maned wolf , puma, jaguar | Cerrado biome, Brazil | C. Vynne, University of Washington/ Conservation International, pers. comm. |

## Abundance and Density

Capture-recapture approaches, although typically applied to the physical capture of animals, can also be used effectively with individuals "captured" via DNA confirmation (e.g., Bellemain et al. 2005; see chapters 2, 6, 9, and 11 for further discussion). High capture probabilities increase precision and decrease bias in abundance estimates from capture-recapture techniques (Williams et al. 2002; chapter 2). This requirement may be less limiting in studies employing detection dogs because of their ability to efficiently survey large areas and to potentially locate large numbers of scats. Smith et al. (2006b) reported that 79% of scat samples detected primarily by dogs and identified as kit fox were successfully amplified to the individual level. Further, D. Smith (unpubl. data; case study 7.1) found that a trained dog recovered more scats representing unique genotypes of kit foxes (i.e., more individuals) than did human searchers. Wasser et al. (2004) achieved a 40%–73% success rate for extracting and amplifying microsatellite DNA for identifying individual grizzly bears (*Ursus arctos*). These grizzly bear data were then used within a capture-recapture framework to estimate a total of twenty-eight individuals (± 7 bears, 95% CI) in the study area. The same study found that scats collected with detection dogs represented more individuals than DNA samples collected via hair snares and thus that detection dogs surveyed a portion of the population that would have otherwise been missed.

Applying methods from suspect discrimination, whereby dogs are trained to match odor at a crime scene to that of the suspected perpetrator (Schoon 1996; Schoon and Haak 2002), some wildlife studies have employed scat detection dogs for *matching to sample* or *scent-matching* (e.g., Krutova 2001, as cited in Kerley and Salkina 2007). In such controlled discrimination tests, the detection dog is presented with a target scat scent and asked to find its match in a lineup of other odor sources (figure 7.3). Scent-matching has been effective in identifying individual animals (Krutova 2001, as cited in Kerley and

Salkina 2007 [see case study 7.2]; S. Wasser, pers. comm.) and may be particularly useful for providing capture-recapture data when the target species has low genetic variability (thus making individual discrimination by DNA unfeasible), or for studies with limited funds for laboratory work.

## Monitoring

At this time, we know of no published studies that have used scat detection dogs to formally monitor a carnivore population (see discussion of monitoring in chapter 2), although data collected from recent studies will ultimately be used to evaluate population characteristics over time (e.g., Beckmann 2006; D. Smith, unpubl. data; C. Vynne, pers. comm.; S. Wasser, pers. comm.). The lack of publications addressing this application reflects the relative infancy of the survey method. Detection dogs are a promising monitoring tool, however, especially when a monitoring program is based on detection-nondetection. Projects using this method for monitoring should arrange for consistent access to reliable and field-tested dogs.

## Other Achievable Objectives

Here we briefly discuss several additional objectives that have been met with assistance from scat detection dogs.

### Stress

Fecal cortisol levels measured from scats located by detection dogs have been used as an index to stress level for bears (Wasser et al. 2004; R. Long, unpubl. data) and right whales (Hunt et al. 2006), and a number of additional detection dog-based stress hormone analyses are underway (e.g., C. Vynne, pers. comm.; S. Wasser, pers. comm.). Similar analyses have been used to assess the effects of various environmental stimuli (e.g., human disturbance) on other species (e.g., northern spotted owl [*Strix occidentalis caurina*], Wasser et al. 1997; African wild dogs [*Lycaon pictus*], Monfort et al. 1998; gray wolves and elk [*Cervus elaphus*], Creel et al. 2002).

## Habitat Use

Scat detection dogs do not require the use of attractants and are therefore particularly well suited for surveys designed to assess habitat use. Wasser et al. (2004) employed scat counts to infer land use by black and grizzly bears and found that results based on scat abundance were similar to indices of use based on radio-telemetry data, counts of hairs collected at hair snare stations, and individual identification of bears from fecal DNA. Scat abundance has also been used to compare kit fox use of paved versus unpaved roads and saltbush scrub versus nonnative grassland (Smith et al. 2005). Finally, occupancy modeling using data collected by detection dogs was used to develop predictive occurrence models based on landscape variables for black bears, fishers, and bobcats in Vermont (Long 2006).

## Home Range Delineation and Animal Movement

Smith et al. (2006b) analyzed the locations of scats found primarily by detection dogs, along with fecal genotype information from those scats, to estimate the diameter of kit fox home ranges and to assess animal movement within the survey area. Carnivore data acquired from scat detection dogs are currently being used to evaluate a potential habitat linkage zone between central Idaho and the Greater Yellowstone Ecosystem (Beckmann 2006; case study 7.3).

## Sex and Sex Ratio

Genetic analysis of scats located by detection dogs has allowed researchers to determine animal sex (Wasser et al. 2004; Smith et al. 2006b; R. Long, unpubl. data), as well as sex ratios within populations (Wasser et al. 2004; Smith et al. 2006b) and differences in the rate of defecation between sexes (Smith et al. 2006b).

## Latrine Use

Scat detection dogs have been effective in locating kit fox latrines (Ralls and Smith 2004), and show promise for locating latrines of North American river otters (*Lontra canadensis*; J. Packard/B. Alexander, Texas A&M University, pers. comm.).

# Description and Application of Survey Method

A scat detection dog is trained to find scats from target species, and to alert the dog handler to the specific location of each scat (figure 7.4). In contrast to tracking dogs, which follow scent on the ground, a scat detection dog primarily samples odor in the air—similar to Search and Rescue dogs and other air-scenting dogs. Further, scat detection dogs are trained to detect scats *only* from target species and to ignore scats from nontarget species. A detection dog team typically consists of the dog, the handler, and an orienteer—all of whom require extensive training to function successfully as a team. It is generally preferable for dogs to work off-leash to maximize searching ability, but they may be worked on-lead for training, safety, or other project-specific reasons (see box 7.2).

## What Characterizes a Scat Detection Dog?

Selecting the correct dog for scat detection work is key to the success of the method. While breed should

*Figure 7.4.* Scat detection dog "Rick" alerts his handler to the location of a scat with a passive sit-alert. Note black bear scat in bottom right corner. Photo by P. MacKay.

---

## Box 7.2

### Ten commonly asked questions about scat detection dogs

1. *What breeds work best?* The type of breed used for scat detection work is not as important as the dog's health, agility, scenting ability, trainability, stamina, and play drive. The dog also needs to have appropriate physical and personality attributes for the particular project. Large working breeds (e.g., German shepherd, Labrador retriever) have proven quite effective. Short-nosed dogs (e.g., pugs) typically don't possess the scenting ability necessary, and some tracking or sporting breeds may be too easily distracted by game scent to focus on their detection work.

2. *My pet dog loves tennis balls—could she be a scat detection dog?* Although many dogs love playing with tennis balls and other toy objects, successful scat detection dogs are usually obsessed with their toy reward to the point of frantically wanting to retrieve it regardless of distractions. And object focus is only one of a number of import attributes. At Working Dogs for Conservation Foundation, over 60% of dogs initially selected for detection work based on the initial criteria of "loving tennis balls" ultimately fail out of the training program (D. Smith, unpubl. data). In general, dogs that have been trained as pets may lack the drive necessary to work in scat detection.

3. *At what age can a dog begin working?* Scat detection dogs must demonstrate a sufficient level of maturity and be able to maintain consistent concentration before working in the field. In most cases, candidates will be at least one year of age. Early socialization and training, however, can begin in puppyhood.

4. *How long does it take to train a scat detection dog?* Although this depends on the given dog's aptitude and ability, initial training generally requires a four- to six-week course with training sessions held four days a week. Ultimately, training is an ongoing process that will continue through the dog's working life.

5. *At what distance can scat be detected?* The distance at which a scat can be detected by a dog depends on many variables, such as age and size of scat, temperature, wind conditions, humidity, vegetation, and terrain. We feel that a reasonable estimate of the potential detection range on land is 1 m to 100 m, although this is *highly* dependent on the variables described above.

6. *Do dogs eat or roll in samples?* Dogs that have successfully completed a training program should not eat, roll in, or otherwise make contact with samples. These dogs are conditioned to associate scat from target species with their reward object and thus should immediately alert their handler to its presence upon location. If a dog does show some interest in licking or consuming scat from a particular species, this can potentially be addressed through training.

7. *How many species can a dog be trained to detect?* Currently, scat detection dogs used on multiple projects have odor repertoires of up to ten different species (D. Smith, unpubl. data). On single surveys, dogs have been successfully trained to detect scats from up to four carnivore species (e.g., Beckmann 2006). In controlled settings, detection dogs have been shown to learn at least ten chemical odors with ease (Williams and Johnston 2002), and explosives detection dogs are typically trained on ten to fourteen odors (B. Davenport, pers. comm.).

8. *Is it better to work a dog off-lead or on-lead?* The decision to work a dog off- or on-lead is project-specific. Dogs obviously have greater freedom of movement working off-lead, potentially maximizing their ability to search. Effectively working a dog on-lead in thick vegetation, for example, can be extremely difficult or impossible. All scat detection dogs should be highly responsive to recall. Working a dog on-lead may be necessary to guarantee safety under certain field conditions (e.g., surveying at the edge of a highway), or to best accommodate particular survey designs (e.g., closely spaced transects in a grid search).

9. *How much distance can a scat detection dog cover in a day?* Again, this is highly dependent upon numerous factors, including weather, topography, the ease with which the dog (and the handler) can move through the terrain, the fre-

## Box 7.2 (Continued)

quency of scat encounters, and the intensity of the search. It is also important to keep in mind that dogs typically walk a much greater distance than their human counterparts given the nature of their search patterns (e.g., based on GPS data, dogs covered 2.1 times the distance of handlers during transects in the Centennial Mountains [Beckmann 2006]). Conducting bobcat surveys in hilly coniferous woodlands in New Mexico, Harrison (2006) estimated that scat detection dogs could effectively search up to 3–5 km per day. In Vermont, Long et al. (2007a) surveyed topographically demanding, diamond-shaped transects totaling 2 km in length—although accessing survey sites often required additional walk-ins of several kilometers. Lastly, detection dog teams surveyed an average of 6 km per day

in the Centennial Mountains of Idaho-Montana (J. Beckmann, Wildlife Conservation Society, pers. comm.; see case study 7.3).

10. *Will dogs detect locations where the target species has scent marked or traveled?* While scat detection dogs can no doubt detect residual scent from an animal's presence, they're rewarded only when they've located scat. Thus, persistent ground tracking of animals is not encouraged or reinforced. In training dogs to detect live desert tortoises in burrows, Cablk and Heaton (2006) found that dogs learned to distinguish between the residual scent of vacant burrows and the scent of burrows containing live tortoises. Similarly, scat detection dogs quickly learn that scat—not residual scent—is the vehicle for their reward.

be taken into account (see box 7.2), a number of other characteristics are of fundamental importance. Similar to dogs in other detection disciplines (e.g., American Rescue Dog Association 2002), scat detection dogs must be of good temperament and exhibit a strong, object-oriented play drive (Smith et al. 2003; Wasser et al. 2004). Most commonly, these dogs are trained with a toy—for example, a tennis ball or tug toy—that is used to reward the dog for performing the desired behavior (figure 7.5). Of course, many playful dogs are passionate about their toys, but scat detection dogs are typically *obsessed* with their reward object. A candidate dog is not only able to focus on this object to the exclusion of other distracting stimuli but must also possess bountiful energy, the ability to focus on work, and the endurance and stamina to *stay* focused while searching for a target odor under various—and changeable—environmental conditions (e.g., hot weather, rain, biting insects). In addition to being in excellent overall health, the dog must be athletic and agile in order to survey sites that might include steep slopes, slippery surfaces, downed trees, river crossings, or other such physically challenging terrain.

*Figure 7.5.* Scat detection dog "Briar" enjoys her reward for a scat detection—a game of tug with a tennis ball on a tether. Photo by R. Long.

It should be noted that scat detection dogs are not average house pets—high drive corresponds with high maintenance. Careful consideration must be given to using one's own dog for detection work (e.g., dogs initially trained as pets may well lack the drive necessary for a work-a-day life), and any potential candidate should be screened by a professional trainer. Once dogs are trained for scat detection, they should be cared for in a manner that maximizes their performance as working dogs (see *Detection Dog Planning and Care* and box 7.2).

## The Handler

In many cases, a detection dog's success or failure can be traced in part to the handler. A common expression among trainers is that "everything travels down the leash," meaning that a handler's bad mood, lack of focus, or frustration will likely result in a dog's not working to full capacity. In general, a successful handler must be able to read canine behavior and respond to it in a way that enhances—rather than detracts from—the dog's performance. The handler must have good timing in presenting rewards and should reward the dog with a level of enthusiasm appropriate to the dog's energy level. Although bonding between the handler and the dog is integral, the handler must simultaneously nurture the dog's ability to work independently. Finally, handlers must be acutely aware of their dog's physical condition and should determine when the animal needs water, rest, and time off (see *Practical Considerations*). The handler is also responsible for the dog's daily feeding and health checks and for monitoring weight, foot pad condition, ears, coat, and other potential health concerns.

In addition to meeting the dog-related requirements above, a handler must have an exceptional constitution for field research. The handler should be physically able to keep pace with an athletic dog for several hours a day—dogs can make comparatively short work of navigating downed trees or ascending steep hills. Handlers must also maintain a positive attitude under even the most challenging conditions, including bad weather or biting insects. And a handler must maintain good communication with the orienteer while constantly attending to the dog.

## The Orienteer

Under some circumstances (e.g., limited personnel resources, open landscapes, nonremote survey areas), a handler may choose to work alone with a scat detection dog (figure 7.6). In complex terrain, however, it can be very difficult for a handler to stay on-transect while remaining fully focused on the dog. Conducting surveys alone can also pose safety risks if either the handler or the dog becomes injured. Thus, we generally recommend that scat detection dog teams include an orienteer (figure 7.7). Like the dog and the handler, the orienteer must be in exceptional physical condition. Further, given that this individual's primary responsibility is to keep the team on course, she or he must be adept at reading maps and using a compass and global positioning system (GPS). The orienteer walks behind or beside the handler, providing verbal guidance as necessary. When the dog alerts the handler to a scat, the orienteer marks the transect line so that the team can resume travel from the same point once the sample and relevant data have been collected. The orienteer's responsibilities might also include collecting and processing samples while the handler rewards and rests the dog.

It is important that the orienteer maintains a calm presence—if he or she is disruptive or overly attentive to the dog, the dog may be distracted from searching. The orienteer should also be able to read the dog's searching behavior in order to assist the handler when the dog is out of the handler's view. On occasion, when scat detections are infrequent and the dog needs a little "boost," the handler may ask the orienteer to hide a training sample nearby.

## Searching for Scat

Every scat detection dog team will establish its own unique working style and protocols. Nonetheless, a typical field search scenario involves the dog ranging and quartering ahead of the handler, with the orienteer guiding the handler from behind (see figure 7.8). Both team members observe the dog's behavior, making sure that the search area is well covered and that the dog is on-task (as opposed to ground-tracking a deer, for instance). The handler keenly watches for a sudden change in direction or body language that indicates the dog's having

*Figure 7.6.* Dog handler D. Smith works scat detection dog "Rio" in California. Photo by A. Whitelaw.

*Figure 7.7.* Orienteer R. Long provides direction to dog handler P. MacKay while surveying carnivores in Vermont. Photo by K. Cross.

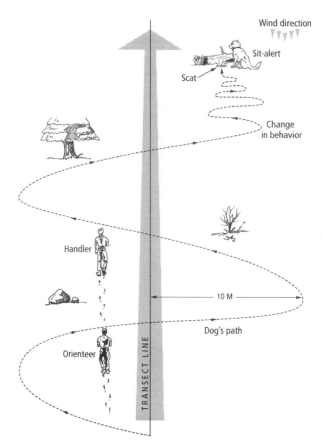

*Figure 7.8.* One typical field search scenario involves the dog ranging and quartering ahead of the handler, working all the while to encounter target odor. The handler guides and carefully studies the dog from behind, noting any behavioral indications that target scat is in the area. Meanwhile, the orienteer helps the handler stay on course by providing direction as necessary. When the target odor is encountered, the dog shows a change of behavior and follows the scent cone to its source—at which point a passive sit-alert is provided. Illustration by S. Harrison.

*Figure 7.9.* Scat detection dog "Pasha" pinpoints a Vermont black bear scat with her nose. Photo by R. Long.

encountered the target odor. When such behaviors are observed, the handler slows down and provides the dog ample space to work the scent to its source. If the dog has difficulty locating the scat site due to erratic wind or other environmental factors, the handler may need to assist the dog in conducting a methodical search (see *Practical Considerations*). Upon pinpointing the scat (figure 7.9), the dog provides a trained alert in expectation of the reward. If the scat

is found at a distant location, the dog either stays with the scat until the handler approaches, or returns to alert the handler—who directs the dog to do a refind (i.e., to lead the team back to the scat).

Generally, a scat detection dog should provide a *passive* alert or indication (see Rebmann et al. 2000; figure 7.4) upon locating a target scat. For example, the dog may be trained to hover and stare, sit, or lay down next to the scat, but should not make physical contact with it. An *active* alert, such as digging at or biting the scat, risks contaminating or destroying the sample. The dog can also be trained to bark when it locates a scat (e.g., M. Gibeau, Mountain District National Parks, Canada, pers. comm.), which may be advantageous in thick vegetation or other settings where the handler cannot see the dog. Regardless of the type of indication used, the dog is immediately rewarded at the scat site once the target scat is confirmed by the handler (figure 7.10).

## Practical Considerations

Many fecal DNA-based study objectives, such as population size estimation and habitat assessment, require large numbers of scats, while others call for

*Figure 7.10.* Detection dog "Camas" receives a reward from handler A. Whitelaw for finding a grizzly bear scat in Montana. Researcher J. Beckmann is serving as orienteer. Photo by Julie Larsen Maher ©WCS, with permission.

the recovery of only a single scat to confirm species presence. In some cases, human searchers alone can acquire sufficient scat samples for meeting study objectives—for example, scats from certain species might be readily collected from trails (e.g., Gompper et al. 2006). But if scats are difficult to locate visually, or human searchers are otherwise unable to meet the needs of the study in an efficient manner, the use of scat detection dogs should be considered.

## Target Species-Related Considerations

Scat detection dogs can be used to survey both common species (e.g., fishers in Vermont; Long et al. 2007a) and rare species (cougars in Montana/Idaho; Beckmann 2006) and can be a particularly important tool for the latter, given the difficulty of surveying low-density populations with other techniques. Extreme scat rarity, however, may present method-

ological challenges, as detection dogs can become bored or unmotivated if worked for many hours or days without locating scats and receiving positive reinforcement. This issue can potentially be addressed through dog training and management (e.g., a more common species can be added to the dog's detection repertoire, or target scats can be subtly planted in the field to catalyze reinforcement).

Some dogs may respond differently to scat from different species. For instance, dominant dogs have been known to urinate on or near wolf scats (M. Parker, unpubl. data), while submissive dogs may avoid them (Beckmann 2006). Other dogs seem to have an affinity for consuming or rolling in felid scat (B. Davenport, PackLeader Detector Dogs, pers. comm.). These examples further illustrate the importance of selecting the correct dog for a given project. Ideally, a pilot study should be conducted when using scat detection dogs on a novel species or in a new study area to help ensure that the method and the dogs selected are appropriate for the target species and the environment. During the pilot study, species identification should be confirmed via DNA analysis for at least a subset of scats collected if there is the potential for confusion with sympatric species.

When exploring detection dogs for surveying multiple species, it may be helpful to consider the relative rarity or commonness of each species. A dog could conceivably become focused on the more abundant scats of a common species and overlook the occasional scat of a rare species—although the authors have observed dogs detecting uncommon species among more common ones in numerous surveys. In addition, if a scat detection dog is to be used in multiple studies across time or space, it is recommended that the dog either be used to detect the same target species for all studies, or to detect only species unique to each study area. Once a dog is trained on a given species, it will be part of its detection repertoire for some time to come and thus may be detected by the dog in subsequent studies even if it is not a target species. In such cases, the dog could be distracted by the presence of nontarget scats on which it was trained in a prior study. This is not necessarily an insurmountable problem, especially if (1) the nontarget species is absent or rare in the new study area; (2) nontarget scats are easily identified by the handler and are not rewarded for or collected; or (3) there is adequate funding to verify all scats collected via DNA.

One common concern is the real or perceived possibility of detection dogs posing risks to wildlife—especially threatened or endangered species (e.g., Cablk and Heaton 2006). While care should be taken to minimize such risks, and dogs must be trained not to chase wildlife, at least two studies have shown that the presence of professionally trained detection dogs did not threaten or disturb target species. In their study of desert tortoises (*Gopherus agassizii*), Cablk and Heaton (2006) reported that no tortoises were harmed during the study, and that tortoises exhibited no visible signs of stress when located by dogs. Similarly, D. Smith (unpubl. data) found that surveys with scat detection dogs did not cause kit foxes to abandon the study site. Data from this latter study also demonstrated that resident foxes continued to scent mark despite the presence of dogs, and handlers observed foxes resting at their dens and exhibiting normal foraging behavior when the dog/handler team passed by.

Researchers should be aware that domestic dogs in the vicinity of wild carnivores can potentially cause the spread of diseases common to both species (Funk et al. 2001). It is essential that all dogs used for wildlife work are fully vaccinated, undergo annual or biannual full veterinary exams, and are continuously monitored for any changes in health (see *Dog Health and Safety*).

## Climate

Climate plays an important role in a dog's ability to detect scent. In discussing the potential affects of climate on game bird research, Gutzwiller (1990) pointed out that direct sunlight, heat, or freezing or near-freezing temperatures can hinder or terminate bacterial action and thus impede scent production. Since the bacteria in feces produce compounds and gases resulting in odor, climate-induced effects on bacterial action no doubt affect the process of scat detection by dogs. Indeed, Wasser et al. (2004) postulated that rising temperatures throughout the day might increase bacterial action in scat and thus the amount and diffusion of odor.

In addition to influencing scent, climate affects biological function in the dogs themselves. Using scat detection dogs in sunny, hot climates, for example, calls for careful planning, as high panting rates can reduce olfactory performance—and overheating is a risk (Case 2005). Working in central California's Carrizo Plain National Monument, Smith et al. (2003) attributed one scat detection dog's relatively low detection rate for kit fox scats to elevated panting levels (average temperature during search hours was 23°C). Nasal tissue dryness can also occur in extremely dry environments, potentially impeding the dog's ability to smell. Dogs should always be kept well hydrated for health and performance reasons (see *Dog Health and Safety*).

Survey areas characterized by extreme cold also require special consideration. Rebmann et al. (2000) noted that well-trained cadaver dogs can most successfully locate human remains at temperatures above freezing (and below ~32°C). Snow obviously affects the mobility of handlers and dogs, as well as the bacterial activity of scat and the movement of scent to the surface (Syrotuck 1972). Nonetheless, snow is not necessarily a prohibitive factor for detection work. Trained detection dogs have been successful at finding scats in deep snow (B. Davenport, pers. comm.; M. Parker, unpubl. data; S. Wasser, pers. comm.), and one dog located a gray wolf carcass buried under a meter of snow (M. Parker, unpubl. data). Search and Rescue dogs have been known to locate avalanche victims buried beneath several meters of snow (Syrotuck 1972; Bryson 1984).

## Terrain

While dogs selected for detection work are capable of traversing a wide variety of landscapes, certain terrains may warrant special consideration. Wetland habitats with large areas of open water will impede movement, for example, and dense forest can hamper the handler's ability to see the dog (although a trained bark alert might help). In addition, maneuvering over extensive downfall can exhaust both dogs and handlers quickly, and areas that feature cliffs or steep slopes may require that dogs be kept close for safety reasons—thus potentially biasing detection. Urban settings or busy parks present their own unique challenges, as dogs can be distracted by people, pets, bicycles, and other human-related activities (S. Reed, University of California, Berkeley, pers. comm.). Scat detection dogs can potentially be effective in all such scenarios, but the proper dog must be selected and protocols should be adapted to the situation.

## Species Confirmation

Scats from sympatric carnivores can be morphologically indistinguishable (figure 3.10). Lucchini et al. (2002) found that gray wolf scats could be confused with dog and red fox (*Vulpes vulpes*) scats; Prugh et al. (2005) suggested that coyote scats could be difficult to distinguish from those of gray wolves, red foxes, Canada lynx (*Lynx canadensis*), and dogs; and Ernest et al. (2000) found that cougar scats could be confused with bobcat scats. Similarly, expert naturalists in Scotland failed to reliably distinguish scats from pine martens (*Martes martes*) and red foxes (Davison et al. 2002). In contrast, detection dogs accurately distinguished kit fox scats ($n = 1,298$) from those of sympatric carnivores such as coyotes, striped skunks (*Mephitis mephitis*), and American badgers (*Taxidea taxus*) at two field sites in central California (Smith et al. 2001, 2003). Further, of seventy-eight potential bobcat scats located by a detection dog in New Mexico, DNA analysis identified 72% as bobcat, 6% as coyote, 3% as fox, and failed for 19% of scats (Harrison 2006). Long et al. (2007a) assigned confidence ratings to scats that were collected based on dog response and scat morphology. DNA analyses conducted on a subset of high confidence scats determined that 100% ($n = 12$) were correctly identified as black bear, 90% ($n = 48$) as fisher, and 90% ($n = 10$) as bobcat.

Accuracy rates can vary between dogs, visits, and sites (Wasser et al. 2004; Smith et al. 2005; Reindl-Thompson 2006; Long et al. 2007a). Given that detection dogs are a relatively new tool for locating scats, we recommend that genetic testing be used to confirm species identification on at least a subset of scat samples sufficient to meet study objectives (e.g., if documenting presence is the objective, confirm at least one scat from each site surveyed). Success rates for species identification from fecal DNA ranged from 60%–98% in four published studies that employed detection dogs for scat location (Smith et al. 2003; Wasser et al. 2004; Harrison 2006; Long et al. 2007a).

## Training the Team: A Brief Overview

The success of the scat detection team relies heavily upon the skills of the dog *and* the handler, and a respectful working relationship between the two. Above all else, the handler and the dog must be able to communicate with each other effectively, and the handler must monitor and support the dog's health, attitude, and working behavior in order to maintain performance. Both dog and handler should be able to adapt to new challenges and ever-changing conditions, thus maximizing the probability of locating target scat samples.

Dogs that are able to perform this level of work possess very high energy—an essential trait that can also result in behavioral challenges. The initial selection of a dog and the intensive training regime it must follow require leadership from an expert in detection dog training. Ideally, the trainer should remain accessible to the team throughout the course of the project such that professional consultation is available when necessary.

For scat and other detection work, a reward (figures 7.5 and 7.10) is used to reinforce the detection of target odor, while nontarget odors are ignored. Scat detection dogs must be trained to stay focused in the presence of wildlife and other distractions likely to be encountered in the field. Obedience training is critical to both the dog's safety and performance. Beyond the basics, there is no single correct way to train a dog for this type of work—a classic joke in the trade is that the only thing two dog trainers can agree on is what a third trainer is doing wrong. Consistency in training is important, however, and the skills and adaptability of the trainer are crucial to the success of the detection dog training program.

One common training method for scat detection—also used in narcotic detection and other canine detection disciplines—introduces the dog to the target odor using a scent box (figure 7.11). The scent box features several compartments, each with a hole in the surface large enough to enable the dog to sniff its contents. A scat sample is placed in one of the compartments. The dog is led along the box by the handler, who taps each hole in order to encourage the dog to smell it. When the dog comes across the compartment containing scat and inhales the scent, it is *immediately* rewarded. The timing and presentation of the reward are integral, and require

experience and skill. Once the dog has learned to associate the target scent with the reward, it is trained to give an alert by the proper compartment.

For most applications, scat detection dogs must learn to generalize their search to include scat from any member of the target species, regardless of the particular animal's diet or other characteristics. Indeed, dogs sometimes show initial hesitancy indicating on scats from a target species when they're first exposed to novel dietary contents (e.g., Smith et al. 2003; P. MacKay, unpubl. data). Thus, during training, dogs should be exposed to *at least* a dozen scats (ideally more) of varying dietary composition and ages and representing as many individuals of both sexes as possible (B. Davenport, pers. comm.). Preferably, training scats should be acquired from numerous free-ranging animals inhabiting the study region—scats comprising the target species' natural diet versus a zoo diet hold the greatest promise for success. If a sufficient number of wild scats cannot be acquired from the survey region, training scats can also be collected from captive animals or free-ranging animals from another region. Regardless of the source, it is essential that any scats used for training are absolutely known to be from the target species.

Upon transitioning from structured scent box training to searches in more natural environments (figure 7.12), the trainer presents the dog/handler team with increasingly difficult sample placement and terrain. This process allows the handler to become more adept at reading the dog's behavior, and at helping to "place" the dog for better success at capturing odor. Reciprocally, the dog learns to work airborne scent to its source under various conditions, and to look for cues from the handler when necessary.

An effective handler must learn the principles and protocols of working a dog on airborne scent. As part of this process, the handler should understand how scent behaves in the natural environment. Scent molecules generally become less concentrated and more dispersed as they diffuse from the source, forming a theoretical "scent cone" (Rebmann et al.

*Figure 7.11.* Many detection dogs first learn to associate target odor with a reward via training on a scent box, shown above. Photo by P. MacKay.

*Figure 7.12.* After mastering the scent box, detection dog "Cedar" transitions to training in the natural environment (i.e., a Vermont woodland). Here, having already provided a trained sit-alert, he responds to the command "show me" by pinpointing the *exact* location of a bobcat scat sample. This command is useful for scats that are hidden or too small for the handler to locate visually without further assistance from the dog. Photo by R. Long.

2000; figure 7.13A). Under ideal conditions, experienced scat detection dogs can follow a scent cone's increasing level of odor intensity to its source from great distances. D. Smith reports one kit fox scat detection at 229 m (unpubl. data), and dogs working at sea detected right whale scat up to 1.93 km from where scat samples were eventually collected (Rolland et al. 2006). But innumerable environmental factors, such as wind, temperature, humidity, and terrain, can affect the behavior of airborne scent molecules and ultimately distort the scent cone. Trees or other barriers to air flow, for example, can cause scent to pool (Rebmann et al. 2000; see figure 7.13B) or swirl (Syrotuck 1972; see figure 7.13C),

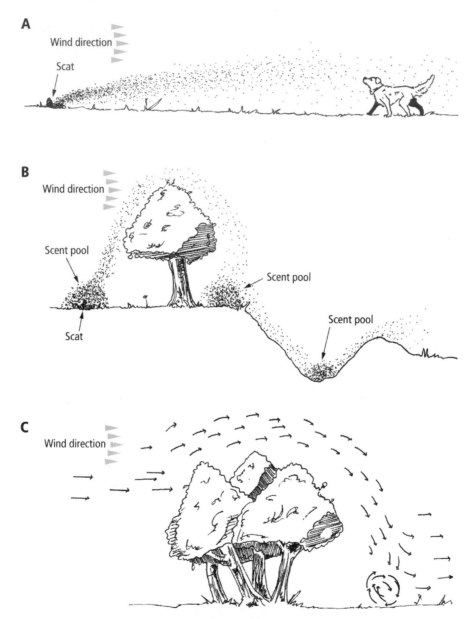

*Figure 7.13.* Basic scent cone theory (see Syrotuck 1972; Rebmann et al. 2000; American Rescue Dog Association 2002). Scent molecules become increasingly dispersed as they move away from their source, forming a theoretical "scent cone." Under proper conditions, a scat detection dog can trace the concentration gradient of the scent cone to its apex (A), but barriers to air flow can cause wind-borne scent molecules to pool (B) or swirl (C)—potentially confusing the dog's efforts to locate the source. Illustration by S. Harrison.

while drainages can funnel scent like water (American Rescue Dog Association 2002). Time of day can also affect the movement of scent molecules (e.g., as the sun warms mountain ridges in the morning, warm air rises and the cooler air from the valley flows uphill to replace it [Syrotuck 1972]). The handler must be able to read and anticipate the dog's response to such environmentally induced anomalies and to guide the dog accordingly.

The orienteer is typically brought into the training process after the dog/handler team has established a solid foundation. A positive working relationship between the handler and the orienteer is paramount to the success of the team. Just as the dog responds to the mood of the handler, tensions between handler and orienteer can be a detriment to the dog's performance. The respective roles of the handler and the orienteer should be clearly identified—and associated protocols developed—prior to the initiation of the survey. In addition to acquiring training and skills in using a map and compass, the orienteer, like the handler, should become familiar with the principles of working a dog on airborne scent. A new handler/orienteer team may wish to practice navigation skills together in the field before adding the dog to the mix. Prior to conducting actual surveys, the entire team should work practice problems in the area of the field site—every site has its own unique challenges. Any issues that do emerge between team members should be discussed and resolved immediately.

## Detection Dog Planning and Care

Detection dogs are a unique wildlife survey method as they are living beings with associated ethical considerations and maintenance responsibilities. Careful thought must be given to how dogs are acquired (see *Materials and Costs*). Further, incorporating detection dogs into a research project comes with a host of logistical concerns, including short- and long-term housing needs, transportation, and proper health and veterinary care. Researchers should address such matters early on when planning a dog-based project. For example, detection dogs

can become sick or injured in the field, and despite sufficient training and testing, there is always the chance that a given detection dog may not perform well on a particular study. As a precautionary measure, it may be warranted to have one or more backup dogs trained and on call, whether they are housed at the field site or the training facility.

### Housing

Detection dogs require adequate housing. At some field sites, temporary kennels have been erected to shelter dogs in a safe environment during nonworking hours (e.g., Wasser et al. 2004). Other projects have housed dogs in local homes or motel rooms with their handlers (e.g., Smith et al. 2003, 2005), or crated them in field vehicles or at campsites (R. Long, unpubl. data). Regardless of form, the housing arrangement must provide dogs with a quiet place to rest and recover from the day's work. Dogs (like people) require ample down time, and too much stimulation may ultimately affect the dog's resting patterns and performance. This issue should be considered when sheltering multiple dogs in a single area.

### Transportation

The logistics of transporting dogs to and within the study area must also be considered. In most circumstances, dogs are easily transportable and can adapt to traveling in cars, trucks, or other vehicles (e.g., planes, helicopters, boats). Additional training sessions may be necessary to condition dogs to new types of transportation, however, and several days of rest should be scheduled between arduous travel sessions. Field vehicles must be of sufficient size (e.g., a minivan or larger) to accommodate a kennel or crate for safe travel (figure 7.14) and must also offer adequate ventilation for the dog.

### University and Agency Requirements

Universities or government agencies may require documentation of proper animal care and use during a detection dog project. Many institutions have specific protocols (e.g., see *Institutional Animal Care and Use Committee* website at www.iacuc.org) for

*Figure 7.14.* Scat detection dog "Briar" rests in her crate in preparation for travel to a survey site. Photo by P. MacKay.

such documentation, which might include a description of the purpose of the research, how dogs will be utilized, steps that will be taken to provide sufficient care and safety for the dogs, and the name of the supervising veterinarian for the project.

### Dog Health and Safety

High-quality veterinary care is essential for any working dog. As discussed earlier, detection dogs used in the field should be kept current on all required vaccinations, receive one or more thorough veterinary exams per year, and be continually monitored for good health. If a dog is suspected of being ill at any time over the course of a survey, it should not be worked in the field. Numerous health conditions—for example, heavy parasite loads, poor nutrition, fatigue (Gutzwiller 1990; Gillette 2004), canine distemper (Myers et al. 1988a), and canine parainfluenza (Myers et al. 1988b)—have been shown to decrease the dog's scenting ability. In addition, because olfaction can be seriously altered by some medications (e.g., steroids; Ezeh et al. 1992),

handlers should consult their veterinarians about drugs prescribed to a dog prior to or during a survey.

Dog handlers and orienteers should be well educated about dog safety in the field, and canine first-aid kits should be carried by all detection dog teams (see appendix 7.2). The American Red Cross offers a course (see www.redcross.org) and a reference book (Mammato 1997) on pet first aid, and there are numerous other resources available that specifically address first aid for working dogs (e.g., Acker and Fergus 1994; American Rescue Dog Association 2002). Of course, such resources are no substitute for professional consultation with or treatment by a veterinarian.

In some study areas, detection dogs may be at risk of injury from wildlife (e.g., grizzly bears, rattlesnakes [*Crotalus* spp.], porcupines [*Erethizon dorsatum*]). It is critical that the handler have verbal control over the dog in the face of potential harm—many wildlife conflicts should be preventable with proper obedience training. Nonetheless, surprise encounters may be unavoidable. Handlers and other team members should establish detailed protocols in case of an emergency and should carry the gear necessary to administer medical care in the field or facilitate the quick evacuation of the dog.

Heatstroke and heat exhaustion are additional concerns, and must be prevented. Dogs transported from another region must be given ample time to acclimatize. During searches, handlers must ensure that dogs are well hydrated, provided with essential rest stops, and physically capable of covering the given terrain and distance. In study areas where heat is a concern, search times should be restricted to the cooler periods of the day—usually early morning and evening hours—and dogs should be encouraged to soak or swim in streams or other water bodies during rest stops. Dogs can also be wetted down or fitted with a cooling vest if necessary. While the appropriate number of hours to work a detection dog varies widely with topography, climate, and other site-specific conditions (see box 7.2), we feel that limiting actual search time to approximately four hours per day is a good general rule (e.g., Smith et al. 2003; Harrison 2006; Reindl-Thompson et al. 2006; Long et al. 2007a).

Finally, dog handlers should carry contact information for veterinary hospitals in the vicinity of the survey area, and, if feasible, notify the nearest veterinarian(s) and emergency clinics that detection dogs are being used in the field. In this way, local veterinarians will be prepared to assist the team in case of emergency.

### Rest Periods

Detection dogs require regular days off to provide them with crucial rest time between work periods. The optimal schedule will vary between surveys, but should be based on the physical needs of the dog (e.g., dog handlers with Working Dogs for Conservation Foundation often employ a varied schedule of three days on–one day off–two days on–one day off to maximize dog performance). Projects involving intensive mountain searches may benefit from dogs working the above schedule for two weeks and then taking a full week off before beginning the cycle

again (D. Smith, unpubl. data). Regardless of the work schedule, dogs should not engage in strenuous exercise during rest days, and handlers should ensure that a dog's activity and interactions during time off are consistent with maintaining its role as a working dog (e.g., no playing with reward objects, nor recreational hiking in areas where scat from a target species might be detected). A dog's reward object should be carried by the handler any time the dog might come across a scat from a target species— even if the dog is not technically supposed to be working—so that the dog can be reinforced on detections as appropriate.

During extended time off, it is important that scat detection dogs are kept in excellent physical condition. Altom et al. (2003) found that athletic dogs enrolled in a regular conditioning program maintained higher levels of olfactory acuity during periods of strenuous exercise, and Gillette (2004) cautions that underconditioning can lead to reduced scenting abilities. Dogs that are "out of shape" breathe more through their mouths (i.e., pant), resulting in decreased airflow through the nasal passages and decreased hydration of the mucosal layer (Altom et al. 2003). Reciprocally, overtraining can result in immunosuppression, chronic inflammation, and reduced scenting ability (Gillette 2004).

### Potential Trouble Areas

Even highly effective scat detection dogs may exhibit behavioral or performance problems. Such problems can be complex, and may develop early in the training process or later in the dog's career. Handlers must be prepared to address these issues at any time during the working life of the dog and should thus maintain an ongoing relationship with a professional trainer. The following are a few examples of problems that may be encountered during a dog survey:

- Dog responds "aggressively" to scat, attempting to roll in, dig at, urinate on, or eat it.

- Dog obviously misses or avoids scat from a target species.
- Dog locates scats from a target species but fails to provide a trained alert.
- Dog alerts handler to scat from a nontarget species.

Many such performance issues may be rooted partially or entirely in handler error. If the handler erroneously reinforces the dog on scats from a nontarget species, for example, the dog is effectively being trained to search for the nontarget species. This can be especially problematic in survey areas where scats from target species are morphologically similar to those from sympatric, nontarget species, as handlers may be uncertain about whether or not to reward the dog for an indication if they are unsure about the identity of the scat. The expeditious confirmation of species identification via genetic analysis of scats is critical in cases where nontarget species are a concern, as false detections must be resolved as quickly as possible via additional dog and handler training or replacement of the dog, if necessary. Another potential option is to reward the dog for all carnivore scat detections and to conduct DNA analysis on the entire collection of scats (A. Bidlack, University of California, Berkeley, pers. comm.). This approach is not generally recommended, however, because (1) it is less cost-effective to pay a genetics laboratory to sort out target and nontarget scat samples than to train the dog to detect only target species; (2) some scats will inevitably fail to yield species identification via DNA analysis; (3) the dog will become quickly exhausted if there are numerous carnivore species depositing a large number of scats in the survey area; and (4) the dog may be "ruined" for other, more species-specific projects if it is effectively trained to detect every carnivore species in the survey area.

Dogs also respond to nonverbal cues from their handlers (Schoon and Haak 2002; Szetei et al. 2003); a handler pointing at or closely investigating scat from a nontarget species may be inadvertently prompting the dog to indicate on this species. Fur-

ther, a team may miss a target scat because the handler failed to read the dog's body language or behavior properly. Dogs encountering a diet-induced variation in target odor, for instance, may initially provide a weak or ambiguous indication due to uncertainty. Handlers must observe the dog carefully to pick up on such subtle behavior.

Certain problems may lie within the scat itself. In moist areas, scats rapidly develop mold, which can cause avoidance behavior in dogs (M. Parker, unpubl. data) or interfere with their ability to detect scent (Krutova 2001, as cited in Kerley and Salkina 2007). Dogs are capable of detecting both fresh and older scats, but degraded scats may provide less scent and cause dogs to miss them during searches. Research from other detection disciplines suggests that, if an environment is saturated with an odor, the dog may be unable to locate the odor's exact source (Schoon and Haak 2002). This may hold with scats as well—for example, a dog might have trouble pinpointing individual bear scats when there are numerous scats clustered in one area (P. MacKay, unpubl. data). Similarly, differences in *amount* of odor may be perceived as *qualitative* differences in odor (Schoon and Haak 2002); a dog may perceive a scat remnant differently from a large, intact scat. Each of these potential scenarios points to the importance of training dogs with scats that vary in quality and quantity of odor.

In some cases, the problems discussed above can be addressed through training. It is beyond the scope of this chapter to present detailed training methods and protocols, but if a detection dog is alerting to scat from a nontarget species, it must be trained off of the nontarget species and consistently reinforced on target species. To the extent possible, the handler should be familiar with the characteristics of target and nontarget scats to minimize potential confusion. Again, it is also critical to budget for DNA confirmation of scats—especially in the early phases of the project when a dog's training is still being solidified. If a dog exhibits frequent or persistent performance issues on a particular project, even if it has a proven track record, it may not be the right dog for the particular species *or* the particular handler (B. Davenport, pers. comm.).

Last, scat detection dogs may occasionally exhibit more general behavioral problems while working in the field, such as chasing wildlife, ground-tracking wildlife scent (see box 7.2), failing to respond to recall, or working too far from or too close to the dog handler. While such problems can typically be avoided with proper screening, training, and testing, they must be corrected immediately if they do emerge. A dog should be replaced if behavioral problems persist or infringe upon dog, handler, or wildlife safety.

## Materials and Costs

Much of the cost associated with this survey method relates to dog selection, training, maintenance, and handling requirements—all of which require substantial effort and expertise. We recommend that an individual or agency interested in using scat detection dogs for carnivore surveys pursue one of the following options: (1) hire a professional dog/handler team to conduct the survey; (2) hire a professional detection dog trainer or organization to help select and train a dog for purchase, and train a handler; or (3) hire a professional detection dog trainer to prepare a dog for lease and train a handler. The best option will largely depend on the nature of the project. For example, a one-week survey to assess whether rare Pacific fishers occur in a particular area may warrant a different approach than a long-term grizzly bear monitoring project in the Canadian Rockies. All approaches have been used with success, and each comes with its own advantages (table 7.3). Ultimately, the decision of whether to hire an experienced dog/handler team or to purchase or lease a dog must be evaluated in the context of available funding, personnel resources, dog care infrastructure, and the longer term vision of the project.

Beyond acquiring dogs, there are a number of additional costs to consider when planning a scat detection dog-based project. DNA analyses and other laboratory testing of scats can be a major expense, especially given that dogs are capable of finding so many samples (but there are potential alternatives to exhaustive DNA analyses; see *Future Directions and Concluding Thoughts*). Funds for veterinary care—as well as for dog food and other dog-related equipment (see appendix 7.3)—must also be included in the budget.

As the cost of conducting a scat detection dog survey varies profoundly with the location, objectives, logistics, and duration of the project, it is not practical to provide general cost estimates for this method. We will, however, briefly summarize cost-related analyses from two recently published studies. Harrison (2006) estimated that the cost of hiring a trained scat detection dog and handler to locate (seventy-eight potential) bobcat scats over two, five-day work periods in New Mexico—and to confirm via DNA analysis that scats were indeed deposited by bobcats ($n = 56$)—was $4,900. The author noted that, while hair snares and scent stations were much less expensive, they were also much less effective. Long et al. (2007b) estimated that the expenses associated with two hypothetical, sixty-site surveys with (1) a leased detection dog, and (2) a purchased detection dog, to be $316 and $257 per site, respectively—including DNA species confirmation of two scats per site. Again, while these estimates were higher than those for comparable remote camera ($254 per site) and hair snare ($153 per site) surveys, the authors suggested that detection dogs would be the more cost-effective method if a high probability of detecting the target species were required. Last, both Harrison (2006) and Long et al. (2007b) pointed out that detection dog surveys, versus the other survey methods mentioned above, required only a single visit to each survey site—resulting in a substantial savings of time as well as travel and personnel costs.

## Survey Design Issues

The design of a scat detection dog survey will be highly dependent on the survey objectives, as well as the attributes of the target species (e.g., rare versus

*Table 7.3.* Three recommended options for acquiring scat detection dogs

| Option | Potential advantages | Special considerations |
|---|---|---|
| Hire a professional dog/handler team | • Skilled team with proven performance.<br>• No dog care required of principal investigator.<br>• Team consistency for monitoring or ongoing projects. | • High cost for "package deal."<br>• Subject to availability.<br>• Lack of history with project. |
| Purchase a professionally selected and trained dog | • Good long-term return on investment.<br>• Dog available for ongoing work.<br>• Dog acclimated to environment.<br>• Team consistency for monitoring or ongoing projects. | • High up-front cost.<br>• Potentially inexperienced dog/handler.<br>• Long-term commitment to dog ownership and maintenance training.<br>• No replacement available if dog is injured. |
| Lease a professionally trained dog and contract associated handler training | • Flexible timing.<br>• Cost-effective for short-term or periodic surveys.<br>• Project can employ handler who is familiar with the project, species, and environment.<br>• Ongoing access to trainer and back-up dogs if necessary. | • Transportation costs and logistics for dog and handler.<br>• Principle investigator incurs full responsibility for dog care during survey period.<br>• Potentially inexperienced handler.<br>• Dog may have inconsistent handlers over time. |

common) and the study area (e.g., climate, topography, vegetation). In general, surveys should be designed to maximize the probability of detecting the target species (or member individuals, depending on the objective) when present, as well as efficiency. See chapter 2 for a thorough review of general survey design considerations. Here we discuss survey timing and route configuration specifically in the context of detection dog surveys, and design issues to bear in mind when integrating other survey methods with detection dogs.

## Seasonal Timing of Survey

As with other scat-based methods, seasonal patterns of wildlife (e.g., hibernation) are important to consider when designing detection dog surveys. Seasonal climate effects may also influence study viability or design. Abundant rainfall, for instance, can quickly wash away scats. Working with grizzly and black bear scats collected via detection dogs, Wasser et al. (2004) found that the success rate for DNA amplification declined with moisture and mold. Reciprocally, dry heat and direct sunshine can desiccate scats and reduce scent transmission—although

dried scats have certainly been detected by dogs in the field (e.g., D. Smith, unpubl. data; R. Long, unpubl. data).

While the extent to which environmental factors ultimately affect scat detection by dogs still remains largely untested, several researchers have quantitatively looked at effects of weather on dog performance. In field tests conducted in northern California, for instance, higher air temperatures and relative humidity correlated with decreased rates of carnivore scat detection (S. Reed, pers. comm.). In contrast, Long et al. (2007a) found little or no effect of temperature (range = 9.4°C–33.9°C), humidity (range = 21.8%–98.8%), rain (i.e., none, light, moderate, high), or wind (i.e., calm, light, moderate, high) on the probability of detecting black bear and fisher scats at forested sites during two Vermont summers. It should be noted, however, that dogs in this study typically worked under forest cover and that survey protocols called for avoiding heavy rain when possible, beginning work very early in the morning (i.e., at the coolest part of the day), and resting and watering dogs more often on hot days. Last, during spring trials in the Mojave Desert, detection dogs located desert tortoises at the same rate across temperatures of 12°C–27°C, relative humidities of 16%–88%, and wind speeds up to 8 m/s (Cablk and Heaton 2006).

## Route Configuration

The configuration and length of the route used to conduct a scat search with a dog depends largely on the objective of the survey. For example, a survey designed to detect a very rare species with high probability (e.g., to confirm the presence of an endangered species) would require a more intensive search than a survey designed to detect a common species. If the objective of a survey is to locate scats with an extremely high probability of detection, or to detect all or most scats in an area (e.g., clearance surveys), the survey route should probably consist of closely spaced transects in a grid pattern. Grid searches are often used in Search and Rescue work and involve coursing back and forth through the search area at

regular intervals. If necessary, the dog can be worked on lead. The search pattern should be oriented perpendicular to the wind to maximize the dog's probability of detecting scent (see Rebmann et al. 2000). Such an approach will maximize coverage by the detection dog.

Carnivore applications often require that a detection team survey substantially more area than can be covered with a systematic grid search in a timely manner. Many scat detection dog projects employ transect configurations designed to search more space less intensively, using landscape features such as roads (e.g., Smith et al. 2005; Harrison 2006) or creek beds (Harrison 2006), or a standardized shape superimposed on the survey area (e.g., Smith et al. 2003; Wasser et al. 2004; Long et al. 2007a; also see figure 7.15, and figure 7.16 in case study 7.3). Such transects should provide a relatively high probability of detecting at least some target scats, even if they occur at relatively low density across a large survey

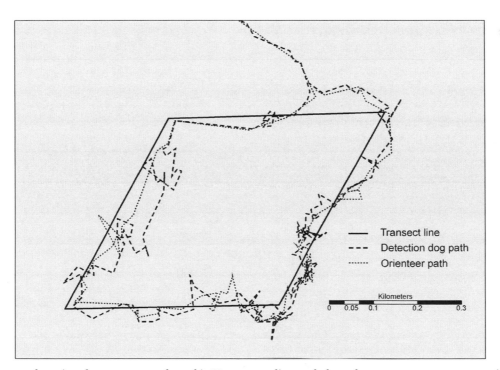

*Figure 7.15.* For scat detection dog surveys conducted in Vermont, a diamond-shaped transect pattern was superimposed on the survey area (Long et al. 2007a). Each side of the diamond measured 0.5 km in length, and the diamond circumscribed an area of approximately 22 ha. Both the scat detection dog and the orienteer carried a GPS unit. The resulting track logs show the respective paths of dog and handler as they walked the transect (R. Long, unpubl. data).

area. To the extent possible, transects that do not cover the entire area of interest should be located based on some probability sampling framework (see chapter 2) to minimize systematic biases (e.g., surveys conducted on transects defined by roads may not effectively sample a target species that is road-sensitive). Regardless of their size or shape, well-defined survey routes allow researchers to control or adjust for survey effort by providing some measure of the area covered by the detection dog team.

As with other design elements, the actual length and configuration of an optimal survey route will vary with the survey objective and the target species. The survey route should provide ample coverage to provide at least a minimum probability of (1) detecting target species (e.g., occupancy studies) or particular individuals (e.g., capture-recapture-based abundance studies), when they are present at a site; or (2) detecting sufficient numbers of scats for effective inference (e.g., relative abundance estimates, studies that require minimum sample sizes). Survey routes for occupancy studies may be relatively low density, with the rationale being that although some scats may be missed, a sufficient number of samples will likely be located to document species presence (see chapter 2). In contrast, survey routes for capture-recapture or relative abundance studies generally require greater transect coverage to help ensure that as many individuals as possible are detected within the survey area.

Route length and configuration are also a function of the rarity of the target species and its natural history. For instance, to achieve the same probability of detection, surveys for rare species or those that may bury their feces (e.g, bobcats) will presumably require greater transect coverage than surveys for common species that deposit scats conspicuously. Before selecting a survey route configuration, researchers are encouraged to consider their target species carefully and to consult the existing literature for guidance from other studies. Once again, we suggest conducting a pilot field season—or at least a number of test surveys—to fine-tune route configuration and other details.

## Using Detection Dogs with Other Detection Devices

Researchers considering scat detection dogs in concert with other detection devices (e.g., remote cameras and hair snares; Long et al. 2007b) must evaluate whether or not one survey method could potentially affect the results of the other(s). In occupancy studies where repeat surveys are assumed to be independent, for example, a baited detection device deployed prior to a dog survey may draw the target species to the area and thereby increase the probability of the dog detecting the species—thus violating the assumption of independence between devices (see chapter 2). On the other hand, a dog survey conducted immediately prior to or concurrent with the deployment of a second detection device could potentially discourage visitation to the device by individuals sensitive to canid scent. Indeed, Sequin et al. (2003) found that when (pet) dogs were present at a camera site during deployment, alpha coyotes were more likely to approach the site than beta coyotes or transients. Although these scenarios may be uncommon, and for some objectives may not present problems, it is important to be aware of potential conflicts early in the design process.

# Sample and Data Collection and Management

The success of a scat detection dog survey relies not only on the team's ability to locate scats, but also on the proper collection and preservation of samples for subsequent analyses. Further, recording accurate and thorough data regarding scats, scat sites, and dog performance and behavior is pivotal to this method.

## Scat Sampling

Due to the valuable biological information contained in scat, most projects will likely choose to collect part or all of each scat detected. For surveys designed to estimate population size via capture-recapture methods, removal of the entire scat (or implementing an effective tagging system that allows surveyors to disregard previously detected scats) will be helpful. Diet studies also warrant removal of the entire scat. In some cases, however, collecting or removing the whole sample may not be desirable. For example, if repeat surveys are to be conducted with the intent of estimating the probability that a given team will detect scats when present (MacKenzie et al. 2002; also see chapter 2), it is advisable to leave at least a portion of each scat at the site for subsequent sampling occasions. In such situations, completely removing detected scats could potentially result in underestimates of scat detectability since these scats would not be available during all occasions. Ultimately, researchers should ensure that their primary study objectives are consistent with scat collection protocols.

Detailed techniques and recommendations for scat sample collection and preservation are presented in chapter 9. Very briefly, scats should always be handled with unused latex gloves—both for surveyor safety and to prevent contamination of the sample. A variety of methods exist for preserving samples until genetic or endocrine analyses are conducted (e.g., air drying, freezing, or immediate placement in a container with ethanol or silica desiccant; see chapter 9). Each preservation method has its own advantages, and researchers should discuss their options with the laboratory that will be conducting the analyses. Obviously, field storage methods must be compatible with field conditions—air drying scats in a very moist environment isn't realistic. Last, it is recommended that DNA be extracted from samples as soon as possible to minimize deterioration.

## Data Collection

In addition to the standard field notes required for virtually any wildlife survey, it is critical to document information relevant to the working condition and performance of the scat detection dog. Such data might include weather and other environmental factors, a description of the terrain, the number of target "finds" and nontarget detections, noteworthy dog behaviors, and the actual duration of the survey. If collected consistently, this data can later be used to help assess the effectiveness of the team and as covariate information in statistical analyses. For example, Long (2006) was able to adjust for variability in detection team performance when estimating occupancy and predicting occurrence for a suite of carnivore species. During carnivore surveys conducted in Vermont, teams recorded scat site-specific information directly onto a scat data form (see appendix 7.4) and also completed a standardized dog data form (appendix 7.5) upon finishing each survey (R. Long, unpubl. data).

For each scat located during a survey, details about the scat and the scat site should be thoroughly documented (appendix 7.4). The particular nature of this data will vary depending on the objective of the survey (e.g., habitat-use studies may require extensive vegetation data at each scat location), but a minimum amount of information should be collected regardless of the survey objective (e.g., a unique identifying number for the scat, the date, the site number or name, the team conducting the survey, a putative species identification, beginning and end times for the sampling process, GPS coordinates taken at the scat location). The scat identification number, the date, and the putative species identification should also be immediately recorded on the sample collection container for sample processing and storage purposes.

Some projects may require a spatial record of the actual route surveyed. Small, waterproof GPS units can easily be attached to the handler and to the dog's harness, enabling recorded movement data to be later downloaded and analyzed (e.g., Beckmann 2006; Cablk and Heaton 2006; see figure 7.15, and figure 7.16 in case study 7.3). This technique yields distances walked by both the dog and the handler and can be used to estimate the sampling area of the survey. Researchers must be aware of the limitations of their GPS units, including field accuracy, performance under high tree-canopy closure and issues related to the loss of satellite-lock (e.g., some GPS units "fill in" gaps in data by connecting the last known location point to the next known location point with a straight line; R. Long, unpubl. data).

## Future Directions and Concluding Thoughts

We have only just begun to realize the potential that scat detection dogs hold for wildlife research. The key to maximizing this potential relies heavily on testing and innovation. In the years to come, researchers must continue to conduct trials with scat detection dogs on new target species across a broad diversity of habitat types and under a variety of environmental conditions. In addition, we should continue to explore the effectiveness of detection dogs in locating other sources of carnivore scent, including dens, urine, hair, and—when necessary—live animals or carcasses. Those involved in such efforts must thoughtfully design, document, and analyze their studies such that the strengths and limitations of detection dogs as a noninvasive survey method are increasingly understood. Dog handlers, trainers, and associated researchers should work to identify and minimize possible sources of bias and error, and to adapt dog- and handler-training protocols to meet desired goals and objectives. Given the relatively small (but growing) number of individuals using this method, information-sharing via publications and more direct forms of communication are especially critical to its advancement. There is also much to be learned from more established canine detection disciplines and from controlled field trials.

The performance of detection dogs and handlers should be carefully quantified whenever possible. Simple summary statistics such as detection rate, false hit rate, and accuracy provide baseline information for others planning dog surveys. More rigorous approaches for quantification should be pursued as well. For example, occupancy estimation (MacKenzie et al. 2002; see chapter 2) allows researchers to estimate the probability of detecting a scat from a target species presuming the scat is present at the survey site (MacKenzie et al. 2006) and also provides a model-based method for evaluating the effects of various site- and survey-specific variables (e.g., temperature, topography; Long et al. 2007a) on detection probability. Further, distance sampling (Buckland et al. 1993) enables researchers to estimate how detectability decreases with distance from the actual survey route.

Knowing how and why scat detectability varies— as a function of site-specific variables, survey conditions, or dog/handler performance, for example— can be helpful both for interpreting survey data and for planning future surveys. Quantitative approaches require establishing an appropriate study design prior to data collection, and extensive time and effort may be required to comply with these designs. For instance, the estimation of detectability at a site requires multiple, independent surveys of at least a subset of sites (MacKenzie et al. 2006; see chapter 2), and distance sampling methods call for accurate measurement of the distance between the survey route and each detected object (Buckand et al. 1993). The ability to quantitatively evaluate dog performance is critical if detection dogs are to be accepted as an accurate and reliable survey method among carnivore researchers.

Despite proven effectiveness, there are many questions relevant to scat detection dogs that warrant further attention and research. For example:

1. To what extent do environmental variables and scat-related factors (e.g., species of origin, dietary contents, sample age) affect dog performance?
2. What (if any) are the performance tradeoffs of training dogs on one versus multiple species?
3. Under what conditions might dogs generalize to nontarget species?
4. How can between-team differences in detection rates best be minimized?

5. Are breed and sex important considerations in dog selection?
6. Which training protocols are most effective and efficient in preparing dogs for this work?

Until the answers to these and other questions pertinent to scat detectability are reasonably well understood, it will make sense to design surveys such that the effect of between-team variability in performance is minimized. Gutzwiller (1990) proposed the following basic guidelines for improving accuracy and precision of dogs in game bird research: (1) use the same dog throughout the study, or balance the use of multiple dogs in time and space; (2) ensure that dogs are physically fit; (3) conduct surveys under similar temperature, wind, precipitation, and barometric conditions when possible; (4) restrict surveys to certain periods of the day; and (5) maintain similar amounts of survey effort at all sites. We suggest that these recommendations also apply to scat detection dogs and that (1) and (2) pertain to handlers as well.

This survey method will also benefit from scientific advances in scat-related methodologies. The ability to accurately age scats, for example, would permit researchers to precisely determine when a species was present in a given survey area. Field-based techniques for confirming species identification via DNA would be extremely helpful, as this would allow handlers to more consistently and accurately reward dogs for appropriate detections. More generally, advances in genetic and endocrine techniques will increase the success rates (and lower the costs) associated with acquiring genetic and hormonal information from scats.

The development and implementation of widely held standards and certification for scat detection dogs and their handlers would further help to advance the scientific effectiveness of this method. While existing wildlife detection dog trainers and organizations typically implement their own strict guidelines for selecting, training, and testing dogs, these have yet to become universal. Again, models can be gleaned from other canine detection disci-plines. The American Rescue Dog Association, for example, has created standards and recommended training methods for air-scenting dogs used in Search and Rescue operations (see American Rescue Dog Association 2002). The Scientific Working Group on Dog and Orthogonal detector Guidelines (SWGDOG; www.swgdog.org)—a partnership of local, state, federal and international law enforcement and homeland security agencies—is developing consensus-based best practice guidelines for detection dogs across various disciplines. As part of this process, SWGDOG has put forth general guidelines for initial training, canine/handler team certification, maintenance training, periodic proficiency assessment, and documentation.

New Zealand has established strict certification requirements for dogs working in conservation at the national level. The Department of Conservation's (www.doc.govt.nz) National Conservation Dog Programme certifies dogs and their handlers in two categories: *Threatened Species Dogs* are used to passively indicate the presence of nationally threatened species so that populations can be adequately monitored, while *Predator Dogs* search for introduced predators (J. Cheyne, New Zealand Department of Conservation, pers. comm.). In order to participate in either program, all conservation dogs and handlers must undergo a two-stage certification process. Interim certification focuses primarily on obedience and control work, while full certification (six to twelve months later) assesses the team on its field search skills. Handlers must also demonstrate adequate knowledge about dogs and their target species.

As scat detection dogs continue to gain visibility and credibility, carnivore researchers will no doubt explore new frontiers for their application. Dogs trained to discriminate between species, for example, can potentially be used to prescreen scats before they are sent to a DNA laboratory for species identification, thus reducing the costs associated with genetic testing. Based on results from one discrimination trial, canine prescreening of red and kit fox scats could have removed approximately 120 nontarget red fox scats from the pool of scats destined for DNA testing (Smith et al. 2003). Scent-matching dogs can also be used to match scat samples from the same individual—including when samples are too degraded for genetic testing, or low genetic diversity may prohibit DNA discrimination at the individual level (S. Wasser, pers. comm.). Wasser suggests that, once dogs sort individuals into groups, a single sample from each group can be selected for DNA amplification to genetically identify the individual—yielding a significant savings in laboratory expenses (see http://depts.washington.edu/conserv/Conservation_Canines.html). Scent-matching dogs have potential field applications as well; dogs trained in discrimination could be used at research stations to test collected scat samples for accuracy, or could even be deployed in the field to search for scats from given individuals. Additional controlled discrimination trials are warranted to further explore the promise of such applications.

In closing, future opportunities for scat detection dogs will likely be limited less by the canine nose than by the human imagination. To the extent that they enhance our ability to survey large, remote areas in an efficient and highly effective manner, detection dogs can be an invaluable tool for those engaged in carnivore surveys.

---

## CASE STUDY 7.1:
### DETECTION AND INDIVIDUAL
### IDENTIFICATION OF KIT FOXES

**Source:** Part of a larger research effort published in Smith et al. 2001, 2003, and Smith 2006b.

**Location:** Carrizo Plain National Monument, California.

**Target species:** San Joaquin kit fox.

**Size of survey area:** ~16 km².

**Purpose of study:** The main objective of the overall study was to determine the feasibility of surveying a kit fox population via noninvasive genetic sampling. Here, results are described for the specific goals of (1) identifying factors that affected recovery

of scats and unique genotypes, and (2) comparing data from fecal genotypes with those gathered using conventional field techniques (i.e., trapping and radio collaring). Further, the effectiveness and efficiency of the detection dog protocol was assessed by comparing detection rates between dog teams and human searchers working alone.

**Survey units:** Eight 2 × 1 km rectangular transects were established, each containing a smaller rectangular transect of 1 × 0.5 km. One side of each (larger and smaller) rectangle was located on the main unpaved road in the area, and all other sides passed through vegetation. Additional road transects were established on unpaved roads branching from the main unpaved road. Transect length varied from 0.5 km to 2 km. Transects included a total of 23.2 km of unpaved roads and 48 km of vegetated area.

**Survey method:** Each transect was searched by both a detection dog team and a human searcher alone.

**Dog details:**

- *Which breeds were used?* One Labrador retriever, one German shepherd.
- *How were dogs acquired for the study?* Owned by the professional handlers.
- *On average, how long did dogs survey per day?* Three to four hours.
- *Were dogs usually worked off-leash or on-leash?* Off-leash.
- *How were the dogs housed during the study?* With handlers at a rented home adjacent to the study area.
- *What type of indication were dogs trained to provide?* Sit-alert.
- *How were dogs rewarded for a detection?* Period of praise and play with a tennis ball and a tug toy, respectively.

**Survey design and protocol:** Prior to surveys, detection dog teams attempted to clear all transects of presumed kit fox scats. Subsequent searches for "fresh" scats were conducted on all transects within an eight-day period and repeated via switching dog teams and human searchers such that the search effort by each dog and person was equal.

**Analysis and statistical methods:** DNA tests were performed to verify species and to establish individual identification and sex, and multiple genetic and geographic information system (GIS) analyses were used to assess the reliability of DNA results. Chisquare tests quantified differences in the sex ratio of detected foxes by type of searcher (i.e., dog team versus human).

**Results and conclusions:**

- A total of 592 presumed kit fox scats were recovered, and DNA extraction and verification of species were successful for 448/592 scats (76%). Only one nontarget scat was identified, and this scat was detected by a human.
- After meeting strict genetic and GIS scoring requirements, 337/447 kit fox scats (75%) were considered reliable scat genotypes. These 337 scats corresponded to seventy-eight unique genotypes (i.e., individuals), of which seventy (90%) were successfully assigned a sex (i.e., thirty-eight males, thirty-two females).
- Information pertaining to estimates of a minimum number of foxes in the area, sex ratio, relatedness, movements, scent-marking behaviors, and home range size inferred from scat genotypes was similar to that obtained in other studies using conventional field techniques.
- One dog experienced a reduction in scenting efficiency presumably due to high summer temperatures, although this dog's scat detection rate was still similar to those of two experienced human searchers. The other dog detected approximately four times as many individually identifiable scats and approximately three times as many unique genotypes as did human searchers. This dog detected 90% of the foxes identified at the study site.

- There was no difference in detection rates of male and female foxes by dog or human searcher.

---

### CASE STUDY 7.2:
#### SCENT-MATCHING TO IDENTIFY INDIVIDUAL AMUR TIGERS FROM SCAT

**Source:** Kerley and Salkina 2007.

**Location:** Lazovsky State Nature Zapovednik, Russian Far East.

**Target species:** Amur tiger (*Panthera tigris altaica*).

**Size of study area:** Dogs were housed, trained, and tested in a 0.35 ha area. Scent trials were conducted in a 5 × 5 m building.

**Purpose of study:** The Amur tiger is extremely difficult to monitor given its secretive nature, low densities, and large home range. Further, genetic analyses have been ineffective for identifying individual tigers due to low genetic variability in the population. Thus, the purpose of this study was to assess the accuracy of dogs at identifying individual tigers from scats and to evaluate the potential application of this method for capture-recapture studies.

**Dog details:**

- *Which breeds were used?* One German shepherd, one German shepherd mix, one German wire-haired pointer, two Labrador retriever mixes.
- *How were dogs acquired for the study?* Dogs were opportunistically selected and rescued as pups.
- *On average, how long did dogs survey per day?* One to two hours.
- *Were dogs usually worked off-leash or on-leash?* On-leash.
- *How were the dogs housed during the study?* In separate outdoor pens (5 × 3 m) with covered shelters.
- *What type of indication were dogs trained to provide?* Sit-alert.

- *How were dogs rewarded for a detection?* Small pieces of hot dog.

**Test design and protocol:** A total of five trained scent-matching dogs and fifty-eight scats from twenty-five individuals tigers were used in trials. Each randomized, independent accuracy trial was conducted by a dog and a handler. An assistant set out scats for each trial. Dogs made identifications by matching a test sample to a scat from the same tiger in a "scent lineup," which consisted of seven scat jars (in scent boxes) arranged in a semicircle, spaced $\geq$ 50 cm from one another. After smelling the test sample, the dog moved around the semicircle with its handler and indicated a matching jar by sitting next to it (see figure 7.3). Dogs worked three to five days per week and performed three to seven trials per day.

**Analysis and statistical methods:** Repeat identification trials were conducted to determine accuracy rates of dogs. Chi-square analyses were used to test the hypothesis that dogs would make correct choices more often than expected by chance. Accuracy rates among dogs were compared using ANOVA and protected t-tests.

**Results and conclusions:**

- A total of 521 randomized and independent trials were conducted to assess the accuracy of dogs in identifying individual tigers from scat.
- Dogs correctly selected one of seven scats to match the test scat at an average rate of 87%. Each of the five dogs made correct choices significantly more often than expected by chance.
- The average accuracy rates for four dogs increased to 98% using repeated-trial tests (i.e., for three trials conducted on different days using the same dog and target scat, the dog was required to correctly match the sample in at least two of three trials; $n$ = eighty-six sets of repeated trials).
- Four dogs were able to match eleven scats deposited from one tiger over a four-year period with an accuracy rate of 100% ($n$ = 40 trials).

- Data derived from scent-matching, used in a capture-recapture framework, may be a useful technique for estimating Amur tiger abundance in the wild.
- Scent-matching may be a valuable alternative to genetic analysis in scat-based studies for which DNA genotyping is impractical or ineffective.

---

CASE STUDY 7.3:
## MODELING LINKAGE ZONE FUNCTIONALITY

**Source:** Beckmann 2006 (ongoing study by the Wildlife Conservation Society).

**Location:** Centennial Mountains along the Idaho-Montana border.

**Target species:** Grizzly bear, gray wolf, cougar, black bear.

**Size of survey area:** Area of inference was ~2,500 km².

**Purpose of study:** In the Greater Yellowstone Ecosystem (GYE), isolation is of particular concern for large carnivore populations occurring within the Yellowstone National Park and Grand Teton National Park core areas. The Centennial Mountains represent a potentially important connection between central Idaho and the GYE, and anchor the southern Yellowstone to Yukon (Y2Y) system. The Centennials have also been identified as a peripheral sink area within the GYE (Noss et al. 2002). It is critical to determine if the Centennials are functioning as a linkage zone for large carnivores. Linkage zones differ from corridors in that they can support carnivores at low densities over time and are not areas strictly used as travel lanes (Servheen et al. 2001). In order to (1) model linkage zone functionality; (2) identify potential bottlenecks for carnivores within this potential linkage zone; and (3) test survey methodology, the Wildlife Conservation Society is studying relative densities, area usage, and movement patterns for a suite of large carnivores.

**Survey units:** The entire Centennial Mountain complex and surrounding valleys were overlaid with a 5 × 5 km grid (25 km²), and one transect was sur-

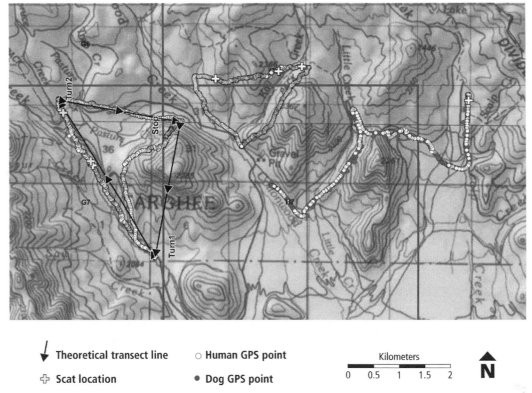

**↓** **Theoretical transect line**   ○ **Human GPS point**

✥ **Scat location**   ● **Dog GPS point**

Kilometers
0   0.5   1   1.5   2

**N**

*Figure 7.16.* These track logs from a pilot study in the Centennial Mountains show the paths of scat detection dogs and handlers during three surveys (note that the two paths generally overlap due to the large spatial extent of the figure). In the first (left) track log, the solid black arrows illustrate the theoretical triangle design. "Turn 1," "Turn 2," and "Stop" points were created by marking the actual turn points with a GPS while conducting the survey. As indicated by the topographical lines, the first triangle was interrupted by a sheer vertical cliff on the third leg. The middle track demonstrates a cleaner 6 km triangle track. Lastly, the far right track exemplifies a track that followed a USDA Forest Service road.

veyed in each of the resulting 100 grids. Transects were 6 km in length and triangle-shaped, circumscribing an area of approximately 67 ha (see figure 7.16).

**Survey method:** Each transect was searched by a team consisting of a detection dog, a handler, and an orienteer.

**Dog details:**

- *Which breeds were used?* Two German shepherds, one Labrador retriever, one Labrador retriever mix.
- *How were dogs acquired for the study?* Owned by the professional handlers.
- *On average, how long did dogs survey per day?* Four to five hours.

- *Were dogs usually worked off-leash or on-leash?* Off-leash.
- *How were the dogs housed during the study?* With handlers in travel trailers and rental cabins.
- *What type of indication were dogs trained to provide?* Sit-alert.
- *How were dogs rewarded for a detection?* Period of praise and play with a favored toy (e.g., tennis ball, ball on a rope).

**Survey design and protocol:** Sites (i.e., transects) were visited one to two times by research teams. Handlers were encouraged to stay on the transect line, but some flexibility was necessary to accommodate the extreme topography of the Centennials and to allow the dogs to effectively locate target scats from a distance. Given the high terrain roughness index and the difficulty of sampling remote, roadless areas, transects were not cleared of scats prior to each field season. Dogs carried a GPS unit to accurately depict and quantify areas of search for density calculation purposes (see figure 7.16).

**Analysis and statistical methods:** Multivariate analyses using Program R and a GIS program were used to model habitat, human use of the landscape, and changes in land-use pattern covariates (e.g., vegetation, topographic complexity, road densities, human population densities, logging history, livestock grazing) to disentangle the effects of human activities from natural habitat limitations on large carnivore use of this potential linkage zone. DNA analyses are currently being conducted on collected scats to determine species, sex, and individual identification.

**Preliminary results:**

- Over summers 2004–6, four detection dog teams surveyed 100 transects. Dogs surveyed a total of 1,183 linear km.
- Teams located a total of 474 scats from all four target species. Dogs also located hair samples, kill sites, wolf rendezvous sites, and den sites.
- Predictive occurrence models and maps are currently being developed. Validity tests of the preliminary models and maps for the suite of four species demonstrate that the models are excellent in their fit when predicting both carnivore occurrence and location of bottlenecks in the region.

## Acknowledgments

This chapter represents the innovation and hard work of numerous people who have helped advance scat detection dog methodology in recent years. In particular, we'd like to thank the pioneers of Working Dogs for Conservation Foundation, PackLeader Detector Dogs, and the Center for Conservation Biology at the University of Washington for their respective efforts. We extend our gratitude to Barbara Davenport, Barbara Dean, Aimee Hurt, Larry Myers, Justina Ray, Sam Wasser, Steve Weigley, Alice Whitelaw, Bill Zielinski, and one anonymous reviewer for commenting on earlier drafts and for sharing their relevant experiences and expertise. Ben Alexander, Allison Bidlack, Gretchen Fowles, Mike Gibeau, Phil Hamilton, Linda Kerley, Jane Packard, Paul Paquet, Kathryn Purcell, Sarah Reed, Sara Reindl-Thompson, and Roz Rolland provided valuable insights from the field, while John Cheyne helped educate us about detection dogs in New Zealand and Rob Gillette supplied us with his veterinary newsletter. We are grateful to Linda Kerley and Jon Beckmann for their assistance with case studies 7.2 and 7.3, respectively. Illustrator Stephen Harrison graciously and patiently brought several figures to life. We offer our sincere appreciation to Jon and Kitty Harvey, who granted us critical financial support. Most importantly, we thank the detection dogs that have dedicated their working lives to wildlife research and conservation. Without their enthusiasm, loyalty—and noses—this chapter would not have been possible.

# Appendix 7.1

*Organizations with expertise in training and deploying scat detection dogs*

PackLeader Detector Dogs
14401 Crews Road KPN
Gig Harbor, WA  98329
253-884-5959
www.packleaderdogtraining.net

University of Washington Center for Conservation
    Biology
Department of Biology
Box 351800
University of Washington
Seattle, WA 98195-1800
206-543-1669
http://depts.washington.edu/conserv/

Working Dogs for Conservation Foundation
52 Eustis Road
Three Forks, MT 59752
406-285-9019
www.workingdogsforconservation.org

# Appendix 7.2

*Typical contents of a canine emergency first-aid kit*

- Activated charcoal
- Adhesive tape
- Antibiotic ointments (topical and eye), sprays, powders, and soap
- Bandage scissors
- Cotton-tipped applicators
- Eyewash
- Disposable razors (to clip hair)
- Ear flush
- Elastic bandage
- First-aid reference manual* (B. Davenport, PackLeader Detector Dogs, pers. comm.)
- Hydrogen peroxide
- Instant ice pack* (A. Hurt, Working Dogs for Conservation Foundation, pers. comm.)
- Styptic powder
- Large butterfly closures
- List of emergency numbers (24-hour veterinarian, poison control centers)
- Milk of magnesia
- Muzzle* (properly fitted; B. Davenport, pers. comm.)
- Nail clippers
- Oral antibiotics
- Oral antihistamines
- Pain relievers
- Pepto-Bismol, Kaolin, Kaopectate
- Pliers with wirecutters
- Roll cotton
- SAM splint(s)
- Snake antivenom
- Sterile pads and compresses
- Stomach tube
- Super Glue* (e.g, for quick pad repairs; B. Davenport, pers. comm.)
- Thermometer (rectal)
- Tourniquet
- Tranquilizers
- Tweezers/hemostat
- Two-inch bandages

Note: Items listed are recommended by the American Rescue Dog Association (2002) unless otherwise noted with an *.

# Appendix 7.3

*Select equipment needs for scat detection dogs*

- Leashes (6-foot lead for walking and possibly a 10- or 20-foot lead for working on-lead)
- Identification collar
- Training collar
- Harness and chest plate
- Crate (for resting and transporting dog)
- First-aid kit
- Reward objects (e.g., tennis balls)
- Dog booties (if necessary to protect paws)

- Feeding and water bowls (including collapsible bowls for field)
- Water bottle
- Bedding
- Tie-out cable
- Climate-dependent specialty equipment (e.g., cooling vest)
- Outdoor kennel

# Appendix 7.4

*Sample scat data form (R. Long, unpubl. data)*

DATE:__ __/__ __ __/__ __          **SCAT DATA FORM**          SCAT ID # <u>3</u>  YR  TRN#  VIS   **S**  SCAT#

**PRESUMED SP:**_____

VISIT #:_____   TRANSECT #:_____

**CONFIRMATION METHOD:**_____  **CONFIRMED SP:**_____  **CONFIRMED BY:**_____
(TO BE COMPLETED BY PROJECT LEADER)

OFF TRANSECT?  Y / N       OFF TRANSECT ID:  # <u>3</u> OFF _____ **S** _____

**MULTIPLE SCAT SITE?  N / Y    OTHERS AT SITE:**_____
(WITHIN 20 METERS)

**SAMPLED BY:** ____, _____, ____               **ELEV:** _____ ft     **FROM: GPS / Topo**

**BEGIN WORKUP:**_____  **END WORKUP:**_____     **RAW COORD N:**  __ __.__ __ __ __ __
   (to the 0:05 minute)                          **RAW COORD W:** <u>0</u> __ __.__ __ __ __ __

**SAMPLE INFO:**

**PRESUMED SP:**_____  **POTENTIAL OTHER SP:**_____  **CONFIDENCE:** HIGH   MEDIUM   LOW

**EVIDENCE:**      SIZE      SHAPE      CONTENTS      DOG RESPONSE      ALL      OTHER:_____

**SCAT SITE DESCRIPTION:**     ON LEAF LITTER    ON NEEDLE LITTER    ON LOG    ON ROCK    IN BRUSH
                              BURIED        BEDSITE    RUB    LATRINE    OTHER:_____

**SCAT FORM:**  LOGS      TWISTED      PLOP      REMNANT      OTHER:_____

**OUTSIDE COND:** MOIST FRESH  MOIST MOLDY  DECAYED  DRY
**INSIDE COND:** MOIST FRESH  MOIST MOLDY  DECAYED  DRY

**SCAT SIZE:**  DECOMPOSED    MAX LOG WIDTH:_____    MAX LOG LENGTH:_____    # OF LOGS:_____ PATTY DIAM_____

**SCAT COLOR:**  BLACK     GREEN/BLACK     BROWN     GREY     OTHER:_____

**PRIMARY CONTENTS:** VEG  HAIR  BONE  PORC  BERRY  HARDMAST  APPLE  BIRDSEED  OTHER:_____
**SECOND. CONTENTS:** VEG  HAIR  BONE  PORC  BERRY  HARDMAST  APPLE  BIRDSEED  OTHER:_____

**LIVE INVERTS?:**  Y / N    **EST. AGE:** 1–2 days  3days–1 week   1 week–1 month  >1month  possibly last season  unknown
               **JUSTIFICATION:**   CONDITION    OTHER:_____

**SAMPLE COLLECTED?** Y / N      **EXTRA COLLECTED?** Y / N   **VIAL COLLECTED?** Y / N

**PHOTO?** Y / N    **CAMERA:**_____    **TYPE:** SL / PR / DG    **ROLL #:**_____    **IMAGE#:**_____
            **TEMP FILE NAME:**_____    **FINAL FILE NAME:**_____

**OTHER SIGN:** TRACK      HAIR      SCENT MARK      OTHER:

**DOG RELATED INFO:**

**DOG RESPONSE:** NONE    AVOID    AMBIGUOUS    LIGHT    STRONG    **DOG LEAD TO AREA?** Y / N
**RESPONSE DESCRIPTION:**    CHANGE OF BEHAVIOR    WORK TO SOURCE    PINPOINT    AUTO-SIT
       SHOW-ME    REDIRECT    WALK    LOOK    OTHER:_____
**LIKELIHOOD OF FINDING SCAT WITHOUT DOG:** VERY LOW    MODERATE    HIGH    UNCERTAIN

**DISTURBANCE INFO:**

**HUMAN DISTURBANCE?** Y / N    **DISTURBANCE TYPE:** ROAD    HARVEST    CAMPING    TRAIL/HIKING
                                          OTHER:

# Appendix 7.5

*Sample dog data form (R. Long, unpubl. data)*

## Daily Dog Report

Date:_____/_____/_____    Handler:_____ Transect:_____

Dog:_____    Orienteer:_____Visit:_____

**Consecutive days worked (including current day):**
1 – 1 day
2 – 2 days
3 – 3 days
4 – 4 days
5 – 5 days
6 – Other _____

**Duration (hrs):**
1 – 0–1
2 – 1–2
3 – 2–3
4 – 3–4
5 – 4–5
6 – 5–6
7 – 6+

**Weather:**
1 – Sunny
2 – Partly Cloudy
3 – Cloudy
4 – Humid
5 – Drizzle
6 – Raining

**Temperature (°F):**
2 – 30–50
3 – 50–60
4 – 60–70
5 – 70–80
6 – 80–90
7 – 90+

**Wind:**
1 – None
2 – Slight
3 – Intermittent
4 – Moderate
5 – Severe

**Predominant terrain:**
1 – Short grass (field)
2 – Tall grass (field)
3 – Dense understory
4 – Open understory
5 – Clear-cut
6 – Other _____

**Elevation change/slope:**
1 – Level
2 – Slight
3 – Medium
4 – Steep

**Insects:**
1 – None
2 – Few
3 – Many

**% On/off-lead:**
1 – On _____
2 – Off _____

**# of finds:**
1 – Independent _____
2 – Assist _____
3 – Aggressive _____
4 – Avoidance _____
5 – Walk _____
6 – Miss _____

**# of aggressive responses:**
1 – Licking _____
2 – Eating _____
3 – Marking _____
4 – Digging _____
5 – Body contact _____

**# of non-target scat indications:**
1 – Deer _____
2 – Moose _____
3 – Coyote _____
4 – Raccoon_____
5 – _____
6 – Unknown_____

**Dog's behavior (in 10% increments):**
1 – Cooperative _____
2 – Unresponsive _____
3 – Frustrated _____
4 – Uninterested _____
5 – Fatigued _____
6 – Injured _____
7 – Distract/insects _____

**# of samples:**

|  | Hit | Collected |
|---|---|---|
| Bear | _____ | _____ |
| Bobcat | _____ | _____ |
| Fisher | _____ | _____ |
| Other | _____ | _____ |
|  |  |  |
| **Total** | _____ | _____ |

Comments:

*Chapter 8*

# Integrating Multiple Methods to Achieve Survey Objectives

*Lori A. Campbell, Robert A. Long, and William J. Zielinski*

The preceding chapters describe a variety of noninvasive methods for surveying carnivores, each with its own strengths and limitations. For surveys with multiple objectives or multiple target species, or when a single method is insufficient for collecting adequate data, the use of multiple detection methods may be warranted. This chapter discusses considerations for employing combinations of detection methods for surveying one or more target species.

## Background

Conducting successful carnivore surveys requires clear articulation of survey objectives and the implementation of the most efficient detection methods to achieve those objectives. For the purposes of this chapter, the term "method" refers to the type of survey approach or device employed (e.g., snow tracking, detection dogs, enclosed track plates, remote cameras, hair snares). A method is distinct from a "survey design," which refers to *how* the methods are employed (e.g., survey sites are located 10 km apart, five track plate stations are visited every other day for ten days). We focus on scenarios in which multiple methods are integrated within a single survey area.

Carnivores are notoriously difficult to study given their generally low population densities, large area requirements, and reclusive habits (Raphael 1994; Minta et al. 1999; Gese 2001). Considerable effort has been dedicated to comparing carnivore detection methods in the field (e.g., Bull et al.1992; Fowler 1995; Foresman and Pearson 1998; Harrison et al. 2002; Schauster et al. 2002; Sargeant et al. 2003a; Gompper et al. 2006; Harrison 2006; Long et al. 2007b), and summarizing their strengths and weaknesses (e.g., Raphael 1994; Zielinski and Kucera 1995a; Gese 2001; Wilson and Delahay 2001; Long et al. 2007b). Much of this work highlights the fact that no single method is appropriate in all environments or under all conditions, or for detecting every individual in a population or all target species present at a sampling location.

Species detection requires that a target animal encounter and interact with a given detection device, or deposit natural sign (e.g., tracks or scat) that can be positively identified by researchers. The ideal survey would employ a single, reliable, efficient, and affordable method capable of detecting the presence of each target species—and indeed, of every individual within each target species. Achieving this goal, however, is almost always unrealistic given the variety of activity patterns and sensory modes that characterize different species, and the diversity of intraspecific behavior among individuals.

For many applications, a single type of survey method is sufficient to satisfy basic inventory objectives (Main et al. 1999; Crooks 2002). Nonetheless, integrating multiple methods can increase the overall probability of detection, resulting in less biased and more precise estimates of occurrence or abundance. Just as the replication of devices within a sample unit (discussed in chapter 2) can lead to a greater probability of detection, using multiple detection methods can improve survey efficiency for one or more species and provide corroborative or complementary information to address multiple objectives.

## Treatment of Objectives

The application of multiple survey methods can be useful for achieving several important objectives discussed in this volume, including occupancy, distribution, relative and absolute abundance, and monitoring. For occupancy, distribution, and monitoring studies, integrating multiple methods will contribute most to community-level efforts addressing multiple species (Campbell 2004; Talancy 2005; Gompper et al. 2006; Long 2006). When targeting single species, multiple methods may benefit the estimation of both relative and absolute abundance. This approach can facilitate the calibration and validation of the least costly method, which can then be used to achieve an acceptable estimate of relative abundance. For example, track station surveys can be calibrated as an abundance index when coupled with the collection of genetic samples from hair or scat (Squires et al. 2004; McKelvey et al. 2006; Zielinski et al. 2006).

Newly developed capture-recapture techniques for estimating absolute abundance can also incorporate multiple noninvasive methods. A capture-recapture study of fishers (*Martes pennanti*) in the southern Sierra Nevada used a combined remote camera and hair collection device to "recapture" marked individuals (Jordan 2007). Although this particular study first captured and ear-tagged individuals that were later "recaptured" with the noninvasive device, this device could be used alone for both capture and recapture (see chapter 6 for further discussion of meeting capture-recapture objectives with hair collection methods).

## Potential Benefits of Multiple Methods

Given the resulting practical and analytical complexities (see *Practical Considerations* later in this chapter), the advantages of integrating multiple methods into a survey may not be immediately apparent. Several such advantages are discussed below.

### Pilot Testing

Multiple methods can be pilot tested during the planning stages of a survey to identify the best taxa- and region-specific method(s) for subsequent data collection. Once identified, several methods can be incorporated into the survey to increase efficiency and reduce cost, provide increased taxonomic resolution, assist in addressing multiple study objectives, and facilitate the detection of multiple species.

While a number of the methods included in this volume are well represented in the literature, studies comparing methods for target taxa in a specific region may not exist. Conducting a comparative pilot study of candidate methods should provide the opportunity to assess the feasibility and cost of implementing these methods, as well as their relative efficiency. Schauster et al. (2002), for instance, gained valuable insights into method selection by field-testing candidate methods for estimating swift fox (*Vulpes velox*) abundance. Numerous other studies (e.g., Gompper et al. 2006; Harrison 2006; O'Connell et al. 2006; Long et al. 2007b) have taken a similar approach.

Sites used for pilot testing should be representative of the areas where the survey is to be conducted

*Figure 8.1.* A gray wolf is photographed (via remote digital camera) navigating a barbed wire hair collection "fence" on a wildlife overpass in Banff National Park, Alberta, Canada. Photos by A. Clevenger.

(i.e., they should encompass the expected range of variation in environmental conditions, species composition, and population densities of target species). If a portion of the intended study area is used for pilot testing, care should be taken to minimize any biases that could be introduced into subsequent surveys. For example, baits or lures used in pilot testing might persist in the environment, potentially affecting the results of later surveys.

In some cases, a secondary survey method can be employed to refine a primary method (see figure 8.1). For instance, remote still or video cameras can be used to record individual behaviors associated with another detection device (e.g., enclosed track plates or hair snares; A. Clevenger, Western Transportation Institute, pers. comm.). Such information can help improve the design or implementation of the primary survey method, thus increasing the probability that it will be successful. Once improvements have been maximized, the secondary method can be discontinued.

## Efficiency and Effectiveness

Given variation in individual behavior, multiple survey methods may facilitate the detection of individuals that would remain undetected by a single method. In a comparison of methods to detect swift foxes, Schauster et al. (2002) found that two methods failed to detect the target species during some surveys, while two other methods confirmed presence. Further, in Vermont, although scat detection dogs were the overall superior method for surveying bobcats (*Lynx rufus*), cameras contributed unique detections at 14% of sites where the species was detected (Long et al. 2007b). Scat collection (via human searchers, not dogs) was also superior for surveying coyotes (*Canis latrans*) in the Adirondacks; had these surveys relied on camera traps for detection of this species, they would have recorded absences in many sites where scats were found (Gompper et al. 2006).

Multiple methods can reduce survey duration by improving detectability and reducing *latency-to-*

*first-detection* (LTD). In Montana, Foresman and Pearson (1998) noted that unenclosed track plates had a shorter LTD for American martens (*Martes americana*) than did enclosed track plates. Although the unenclosed track plates were more vulnerable to precipitation (and hence, device failure) than enclosed track plates, these researchers recommended the incorporation of both types of track plates to reduce survey duration when conditions were dry (Foresman and Pearson 1998). Indeed, most survey methods exhibit optimal performance under specific environmental conditions. For example, snow tracking is most effective in certain snow conditions and within a fairly narrow timeframe (Halfpenny et al 1995; also see chapter 3). McKelvey et al. (2006) suggest that, by collecting hair or scat along snow tracking routes for Canada lynx (*Lynx canadensis*), snow tracking surveys can be applied across a broader array of snow conditions with a reduced penalty of track misidentification.

If one of the survey goals is to gather data across multiple seasons, methods may need to be varied depending upon seasonal conditions. In California, a multiseason study of marten response to motorized recreation (i.e., off-highway vehicles and snowmobiles) required the use of two different methods. Enclosed track plate stations were the default method due to their ease of use, availability, and low cost, but they were replaced by cameras during the winter because they were not effective in deep snow (Zielinski et al. 2007).

Furthermore, certain methods are highly effective in some habitats but not in others. In evaluating methods for detecting bobcats, Harrison (2006) speculated that the success of detection devices might be affected by how habitat influences animal movement. For instance, while hair snares may be effective in areas where habitat features (e.g., dense brush) compel animals to move along well-defined travel routes, they may not function as well in more open landscapes. To address this problem, one could deploy multiple hair snares to try to account for all possible travel routes, or employ a second method such as track or scat surveys to sample open areas.

## Cost-Effectiveness

Another way in which multiple methods can improve efficiency is by acting as a filter. For example, if remote cameras or track plates are deployed at hair collection stations (e.g., Zielinski et al. 2006), photographs or tracks can be used to exclude nontarget hair samples from genetic analysis, resulting in a potentially large cost savings (figure 8.2). Similarly, integrating detection-only methods with techniques

A

B

*Figure 8.2.* A combined camera and hair collection device from (A) the outside and (B) the inside (with ringtail). Photos by M. Jordan.

that provide more extensive information (e.g., sex ratio, individual identity, abundance estimates, habitat use) can reduce costs by excluding nontarget samples from more expensive analysis (e.g., genotyping). This approach may also facilitate the development of abundance indices from cost-effective occurrence data and can provide a more complete spatial and temporal picture of species ecology.

## Resolution

Integrating multiple survey methods can enhance data resolution. In North America, many canids, felids, and mustelids can be difficult to differentiate based on tracks alone (Foresman and Pearson 1998; Ministry of Environment, Lands and Parks 1999; but see Zielinski and Truex 1995 and chapter 3, this volume). Squires et al. (2004) described snow tracking surveys for delineating lynx distributions and argued for augmenting these surveys with the collection of hair samples at daybeds along track-lines as a means of verifying species presence and identity. More generally, incorporating remote cameras or hair or scat collection into track-based surveys can provide corroborative information—either to improve taxonomic resolution or for exclusionary purposes.

## Multiple Objectives

Each detection method employed during a given survey may provide unique information, potentially enhancing the ability to explore multiple objectives. Track or camera surveys often generate only detection data (Zielinski and Kucera 1995a; Gese 2001; but see chapters 3 and 5 for other potentially achievable objectives), and the capacity to assign individual identification will be limited to DNA-based approaches for most species (but see Herzog et al. 2007). The collection of scat or hair in combination with track station surveys, however, can be used to generate estimates of relative or absolute abundance using molecular techniques (Wan et al. 2003; Zielinski et al. 2006). In addition, hair and scat surveys may complement track or camera surveys by provid-

ing information on sex ratio, reproductive status, or stress (e.g., Bonier et al. 2004; Wasser et al. 2004). Such method combinations are particularly effective and synergistic when those methods with the highest resolution (e.g., hair and scat surveys yielding genetic data) result in a lower probability of detection than the coarser methods employed (e.g., track or camera surveys that produce only occurrence data).

Specifically, track or scat surveys may yield different types of temporal and spatial information than camera or hair collection devices (Harrison 2006; Long et al. 2007b). In many cases, searching for tracks or scats is less biased than station-based methods in that no attractant is necessary; the resulting detections reflect retrospective occurrence (potentially over extensive time periods) and habitat use (Long et al. 2007b). In contrast, because most track stations, remote cameras, and hair collection devices use attractants to draw individuals to the detection device (but see chapters 4, 5, and 6 for exceptions), these devices record presence only after deployment and over a much shorter time frame. Further, the use of attractants may result in the collection of biased information and provide misleading inferences about habitat use (see chapter 10 for more discussion).

## Multiple Target Species

Perhaps the most obvious applications for multiple methods are community-level inventories or surveys for which there are more than one target species. Carnivores are frequently evaluated as potential focal species for regional conservation efforts and ecosystem management (Noss 1990; Linnell et al. 2000; Carroll et al. 2001). Diverse life histories and activity patterns may make it difficult to identify a single method that can adequately sample all target species. Thus, the collection of detection-nondetection data at the community level generally requires a variety of methods, which will likely vary in effectiveness across species and environmental conditions (Thompson et al. 1998; Yoccoz et al. 2001). For example, tree-mounted remote cameras (figures 5.8,

5.9) might readily detect species with arboreal or semiarboreal habits but may be essentially inaccessible to most canids. Likewise, certain species may be reluctant to enter enclosed devices such as enclosed track plates but might be readily detected at open devices such as unenclosed track plates or track plots (e.g., many canids and felids; Sargeant et al. 2003a; also see chapter 4).

A number of studies have compared detection rates among species (Campbell 2004; Gompper et al. 2006; O'Connell et al. 2006; Long et al. 2007b). In the central Sierra Nevada of California, three techniques (i.e., enclosed track plates, unenclosed track plates, remote cameras) varied in their effectiveness for detecting target carnivores (Campbell 2004). In this study, each of the three methods contributed unique species' detections at a sample unit (table 8.1), suggesting the need for a multiple-method approach. In the northeastern United States, Gompper et al. (2006) found that five noninvasive detection methods varied considerably in detection efficiency by carnivore species. For instance, black bears (*Ursus americanus*) were readily detected by cameras, but coyotes were not, and scat surveys conducted (by human searchers) along established trails were inefficient at detecting species other than coyotes (Gompper et al. 2006).

## Meeting Assumptions in Capture-Recapture Studies

The integration of multiple methods may be particularly helpful for capture-recapture studies. Given the increased accessibility of genetic techniques that can be used to determine individual identity, "captures" for capture-recapture studies can now be accomplished via the noninvasive collection of hair or scat. As with any capture-recapture approach for estimating population size, heterogeneity in the probability of capturing individual animals leads to less precise and more biased size estimates (see chapter 2 for further discussion). One form of heterogeneity results from behavioral responses to a given type of detection device, when individuals are either more or less likely to be redetected after the initial detection (i.e., trap-happy or trap-shy animals).

Although capture-recapture designs employing three or more sampling occasions with one type of detection device enable the modeling of capture heterogeneity, such surveys require substantial time and labor. Less resource-intensive designs with only two sampling occasions (e.g., the Lincoln-Peterson model) using one detection device, however, cannot detect or adjust for capture heterogeneity. By employing two or more independent, noninvasive techniques—one for captures, one for recaptures—

*Table 8.1.* Statistics for three detection methods used in carnivore surveys in the central and southern Sierra Nevada 1996–99.

| Method | Number of stations | Number of species occurrences[a] | Number of species[b] | Mean number of species | Number of unique detections[c] |
|---|---|---|---|---|---|
| Enclosed track plate | 492 | 88 | 12 | 1.58 | 19 |
| Unenclosed track plate | 164 | 116 | 12 | 1.44 | 46 |
| Remote camera | 164 | 76 | 10 | 0.94 | 25 |
| Total | 820 | 154 | 13 | — | 90 |

Source: Campbell 2004.
Note: Each of eighty-two sample units was composed of six enclosed track plates, two unenclosed track plates, and two remote cameras. The mean number of species detected at a sample unit was significantly different between methods (Chi-square with Yates continuity correction; df = 10, P < 0.01).
[a]Indicates a species detection by at least one station of that type at the sample unit.
[b]Includes species and species groups.
[c]Number of instances in which detections of a species by this method represented the only detections of that species at a sample unit.

the problem of capture heterogeneity may be partially ameliorated in this type of design (as well as in designs with >2 sampling occasions; Boulanger et. al. 2008). Furthermore, because multiple, independent survey techniques (e.g., baited and unbaited hair collection devices) can be operated simultaneously, therefore reducing the time required to collect sufficient data, population closure concerns can also be minimized.

## Outreach and Public Relations

Regardless of the primary survey method used, it may be worthwhile to incorporate a method that produces a visual record of the target species, such as track stations or remote cameras. Tracks are more accessible to most people than hair or scat, and might inspire lay audiences to become interested and engaged in the study. Likewise, remote cameras often produce fascinating, candid photographs of wildlife that can be used in proposals, presentations, and publications to help generate funding and popular support for the research. Consequently, these methods can add value to the project—even if this value is nonscientific in nature.

# Practical Considerations

Surveys integrating multiple methods, like those employing a single detection method, must address sampling effort, configuration within sample units, and survey design (e.g., attractants, duration of survey, number of sampling occasions). Three critical characteristics of successful surveys are that they (1) readily detect the target species when present; (2) are reliable and repeatable; and (3) are not prohibitively expensive. For almost all survey objectives, it is important that the methods selected have a known minimum probability of detecting all target species when they are present, or that the ability to estimate and incorporate detectability is part of the survey design. Although detailed considerations for survey design are addressed in chapter 2, here we specifi-

cally highlight a few issues generated by the integration of multiple survey methods.

## General Survey Design Issues

Identifying the optimal survey design for a single species is far from straightforward, and the exercise becomes significantly more complex if multiple species (Thompson et al. 1998; MacKenzie et al. 2002; Field et al. 2005) or multiple methods are considered. When exploring sample unit design for a single-method survey, one must address the configuration or spatial replication of the method within the sample unit (e.g., six track stations versus three track stations), as well as the probability of detecting one or more target species. If the specific probabilities of detection are known or can be estimated, the appropriate sample unit design can be determined relatively easily by incorporating the increase in the probability of detection resulting from the addition of stations or surveys (figure 8.3; also see chapter 2) and power analysis (Field et al. 2005). Sample unit design becomes more complicated when evaluating possible combinations of methods (e.g., six track plates and two cameras versus three track plates and three cameras)—especially if the devices are deployed simultaneously.

Selection of the preferred design can be based on a number of criteria, including a minimum threshold probability of detecting one or more species, geometric or harmonic mean probabilities of detection averaged across all target species, and cost, time, or equipment requirements. For example, Long et al. (2007b) estimated method-specific probabilities of detection and costs to compare the efficacy of three methods for detecting black bears, fishers, and bobcats, as did Gompper et al. (2006) for five methods used to detect eight species. Campbell (2004) used method-specific probabilities of detection and a minimum threshold detectability of 0.80, and estimated costs for multiple methods and sampling occasions, to evaluate alternative survey designs and to characterize a carnivore assemblage. In each of the above studies, optimal methods based

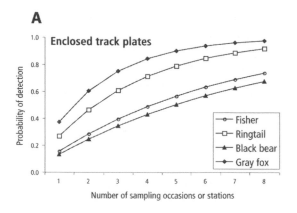

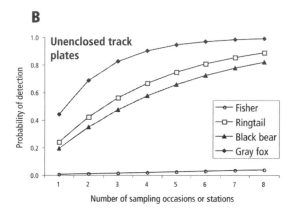

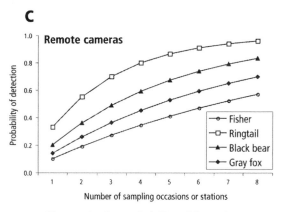

*Figure 8.3.* Changes in the probability of detection at a sample unit resulting from additional stations or sampling occasions. Graphs include four species detected by (A) enclosed track plates; (B) unenclosed track plates; and (C) remote cameras employed during surveys in the central Sierra Nevada, California, 1996–99. Research described in Campbell (2004).

on detectability differed from those based solely on cost and resource considerations.

If the primary goal of a survey is to determine the distribution or relative abundance of an individual species, the single most effective method will probably be the most cost-effective as well. But if the goal is to characterize a community, the most rarely detected species will typically drive method selection. In a post hoc analysis of noninvasive methods, Campbell (2004) found that rarely detected species (i.e., American badgers [*Taxidea taxus*]; raccoons [*Procyon lotor*]; and felids [e.g., *Felis catus, Lynx rufus*, and *Puma concolor*]) necessitated the inclusion of more detection devices, the addition of a second and more expensive method, and a longer survey duration. The resulting survey design cost nearly five times as much as a design sufficient for surveying all other species in the community (Campbell 2004). When possible, the evaluation of multiple methods in initial or pilot surveys and the analysis of resultant detection probabilities should help to determine the minimum number of methods and devices necessary, as well as the most appropriate survey design (Field et al. 2005; O'Connell et al. 2006; Bailey et al. 2007).

## Method Interaction or Conflict

For the purposes of estimating detectability, species detections and devices at a site are generally assumed to be independent (i.e., the detection of an individual or species by a given device should not influence the detection of a conspecific or a second species by the same device). Similarly, in the case of multiple methods, the detection of a species by one method should not influence the detection of a species by a second method. Under field conditions, these assumptions may be difficult to meet (Zielinski and Stauffer 1996; Foresman and Pearson 1998; Sargeant et al. 2003a).

In practice, integrating multiple methods into carnivore surveys may lead to method competition or conflict, which can alter survey results. Target species and potential methods should be carefully reviewed for possible conflicts in space or time. For example, attractants typically used with a particular method or target species may interfere with the detection of the same or another species with other methods (see chapter 10). In Vermont, Long et al. (2007b) had to ensure that camera- and hair collection-based carnivore surveys employing attractants did not interfere with initial surveys conducted with scat detection dogs. By conducting dog surveys first, Long et al. (2007b) guaranteed that target species were not "unnaturally" attracted to the area by the baited methods that followed.

Another concern is that species responding to one type of device may alter the probability of detecting the same or different species at the same device or at a second type of device. For instance, competitive or predatory relationships among target species could potentially influence their respective detections. Foresman and Pearson (1998) found that martens were readily detected by track plates but were never recorded by this device *after* a fisher (a potential marten competitor and predator; Powell and Zielinski 1983) had been detected at the same station—despite camera detections indicating marten presence.

The most extreme example of method competition may occur when multiple methods are physically integrated. Methods can be combined to enable synchronized data collection, such as in a device that records a photograph of the target animal while also collecting hair samples (figure 8.2). But an animal attracted to this device may be adversely affected by the noise or flash from the camera before coming into contact with the hair snare. In tests of a combined track plate/hair collection device, Zielinski et al. (2006) found that the detection of one species could be affected by the subsequent detection of a second species. Although tracks were recorded for both fishers and gray foxes (*Urocyon cinereoargenteus*), the probability that fishers would be identified from a hair sample was dramatically lower when gray foxes were present. In contrast, when fishers were the only species to visit the device, their presence was readily confirmed by DNA analysis of hair

(Zielinski et al. 2006). Fortunately, recent advances in estimating detectability provide the means to explicitly evaluate the influence of one species on the detections of another (MacKenzie et al. 2006).

To minimize violations of the assumptions of detection independence, careful consideration must be given to the combination of detection methods to be used and their spatial and temporal arrangement within the sample unit. If dependencies among methods within a sample unit are high, the data should be collapsed to represent a single detection-nondetection event across multiple devices (see chapter 2). Alternatively, the degree of correlation or redundancy among detections could be explicitly estimated and incorporated (e.g., Sargeant et al. 2003a; Kauffman et al. 2007).

## Occupancy Studies

Assuming that (1) the probability of detecting a target species with each method is known, and (2) there is no dependence or conflict between methods, the overall probability of detecting a species within a sample unit or at a survey site can be estimated for a particular combination of methods (Campbell 2004; Long et al. 2007b; also see chapter 11). To estimate the probability that a species will be detected by at least one of multiple methods during a given survey, the following formula can be used:

$$P = 1 - \prod_{i=1}^{m} (1 - p_i)^{n_i}$$

where $m$ is the number of detection methods with detection probabilities $p_1, p_2, \ldots, p_m$, and the number of method-specific surveys is represented by $n_1$, $n_2, \ldots, n_i$ (Long et al. 2007b; note that a slightly different, station-based version appears in Campbell 2004).

Detections from different methods can be combined into a single encounter history, and individual method-specific probabilities of detection estimated, as long as each encounter represents an independent detection event. For example, Long et al. (2007b) used three different detection methods (i.e., three detection dog surveys, one remote camera survey, and one hair snare survey) to generate a combined encounter history (e.g., *11010* represented detections by the first and second dog surveys and the remote camera survey, but not the third dog survey or the hair snare survey). These researchers then estimated method-specific and overall probabilities of detection via the occupancy option in program MARK (White and Burnham 1999).

If the independence of detection devices or methods cannot be assumed, all detections at a site or within a sample unit can be pooled to create a net detection history (e.g., a single detection or nondetection event for the sample unit) during a given survey interval. Alternately, independence among a number of devices or methods at a site can be formally tested using likelihood ratio tests or other approaches (see Kauffman et al. 2007 and chapter 11), and in some cases device dependence can be modeled during the analysis phase (see chapters 2 and 11). Indeed, a new multimethod approach in program PRESENCE (Proteus Wildlife Research Consultants, Dunedin, New Zealand) may help to incorporate device dependencies into model estimates (L. Bailey, US Geological Survey, pers. comm.).

## Future Directions and Concluding Thoughts

While much of this chapter has described the potential benefits of combining multiple methods to address specific survey objectives, this approach should be taken only when it is necessary and practical. The design of a survey, whether using a single method or multiple methods, should be closely matched with the objectives and scope of the study. The extra costs (e.g., materials, labor, analyses) associated with additional methods should be weighed against the expected increase in the amount or quality of data provided. A given method might be a cost-effective addition for long-term, extensive surveys but may

not be worthwhile for spatially intensive surveys of shorter duration. When incorporated properly, however, multiple methods can reduce the heterogeneity of detection rates and can also result in higher detection probabilities that might otherwise be achieved only with more effort when using single methods—if at all.

Improvements to existing survey methods and the development of new techniques may ultimately diminish the need for integrating multiple methods. The ideal noninvasive survey method for carnivores should not rely on unusual, unnatural, or complex behavior for a detection to be recorded. If a device is used (e.g., track plate, remote camera, barbed rub pad for hair collection), it should be accessible and attractive to a wide array of species. Optimal methods should provide not only detection data, but also information that can be used to help characterize the status of populations (e.g., individual identity, sex, endocrine information). Further, these methods should be useful for both single and multispecies surveys.

Among the noninvasive methods that are currently available, track surveys, scat surveys (without DNA analysis), track stations, and remote cameras are broadly applicable to a diversity of species in North America. Nonetheless, these methods are generally limited to gathering detection-nondetection data due to their inability to distinguish individuals for all but a handful of species, and often under fairly specific circumstances (e.g., Grigione et al. 1999; Kelly 2001; Lewison 2001; Karanth et al. 2004a; Herzog et al. 2007)—though resulting data have been successfully used to generate indices of abundance (Reid et al. 1987; Hayward et al. 2002; Silver et al. 2004). Last, track stations, remote cameras, and hair collection devices typically require the elicitation of certain behaviors to yield detections; specifically, the individual must encounter and interact with the given device.

Hair and scat surveys have the potential to provide both detection and population information if coupled with genetic analysis. But hair collection devices tend to be species-specific and may be difficult to adapt to multispecies survey objectives due to physical constraints. Scat surveys, in contrast, are relatively easy to apply to multiple species and do not require atypical behavior or interaction with a device. The efficiency of scat surveys is usually limited by the ability of the observer to locate and correctly identify scat from the species of interest, and the extent to which useful DNA can be extracted. Scat detection dogs appear to provide an efficient means of locating and identifying samples (Smith et al. 2003; Wasser et al. 2004; Harrison 2006; Long et al. 2007a; also see chapter 7). Further development of this technique, when combined with more efficient methods to extract and amplify DNA from scat (see chapter 9), may offer the best opportunity to reduce the need for multiple method surveys in a variety of applications (see chapter 7 for a detailed discussion of detection dog survey considerations).

With an increasing array of noninvasive methods tailored to effectively detect only one or very few target species in specific environments, it is likely that surveys in the future will require multiple methods only when investigating species diversity and carnivore community structure and dynamics. Analysis techniques for addressing relevant data are also evolving rapidly, and a host of new and improved approaches will no doubt be available in the near future. In addition to using species detectability to identify the best survey design, potential designs could be evaluated based on the proportion of the carnivore assemblage they are likely to detect. Refined assessments of survey design effectiveness could explicitly address the influence of detecting one species on detecting others (MacKenzie et al. 2006), and the degree of redundancy among detections could be evaluated using variance inflation factors (Sargeant et al. 2003a). Species accumulation curves could then be used to estimate total carnivore diversity. Improving the estimates of detectability, and accounting for detection redundancy, should enhance estimates of local and regional carnivore diversity. Finally, new statistical approaches that permit dependencies between species detections or detection devices will expand the applicability of multiple-method surveys.

---

### CASE STUDY 8.1:
### ENCLOSED TRACK STATIONS COMBINED WITH
### HAIR COLLECTION DEVICES

**Source:** Zielinski, W. J., F. V. Schlexer, K. L. Pilgrim, and M. K. Schwartz. 2006. The efficacy of wire and glue hair snares in identifying mesocarnivores. *Wildlife Society Bulletin* 34:1152–1161.

**Location:** Hoopa Valley Indian Reservation (HVRI), Sierra National Forest (SNF), and Lassen National Park (LNP), California.

**Target species:** Fisher and marten.

**Size of survey area:** Variable, from 580 to 4,825 km².

**Purpose of survey:** To test and compare two designs for a combined enclosed track plate and hair collection device.

**Survey units:** The survey unit was a station composed of two hair collection/track plate devices, one containing a barbed wire hair snare and the second, a glue hair snare.

**Survey method:** Collection devices consisted of two enclosures attached end-to-end—a hair snare enclosure, and an entrance enclosure. Each snare enclosure comprised a 25 × 25 × 85 cm box constructed of lightweight, corrugated plastic, with one end blocked by hardware cloth. A hair collection device (either a barbed wire snare or a glue snare) was placed at the open end of the box. Barbed wire snares were constructed from three strands of barbed wire arrayed in a Z pattern across the opening of the snare enclosure (figure 6.10A). Glue snares were constructed from 1.9 × 20 cm strips of a commercially available glue trap for mice. The strips were attached to a wooden slat inserted into the sides of the snare enclosure. Two additional plain slats were respectively inserted above and below the glue strip (figure 6.10B). A second enclosure—the entrance enclosure—was attached to the open end of the snare enclosure. A sooted track plate was placed in both enclosures.

**Survey design and protocol:** Hair snares were baited with chicken and scent lure and checked every other day for two weeks. Collected hairs were placed into an empty film canister with silica desiccant and stored at room temperature. At HVRI, stations ($n =$ 6) were established in January at 2-km intervals along roads. At SNF, twelve stations were placed on an established monitoring grid for fishers and operated September–October. At LNP, ten stations were deployed at locations where martens had been documented by other survey efforts.

**Analysis and statistical methods:** Hair snares were evaluated by four metrics: (1) permeability—the ease with which species would move through the hair collection device to acquire the bait, based on the proportion of times that both the entrance enclosure and hair snare enclosure track plates recorded a species; (2) effectiveness—the proportion of time that both tracks and hair were present; (3) quantity and quality of DNA collected; and (4) cost and relative ease of use.

**Results and conclusions:**

- Five species were detected based on tracks: fisher, marten, gray fox, ringtail (*Bassariscus astutus*) and western spotted skunk (*Spilogale gracilis*). Generally, if a species' tracks appeared at one of the hair snares, its tracks were also recorded at the other snare.
- Hair samples ($n = 96$) were collected from both barbed wire snares ($n = 41$) and glue snares ($n = 55$). A total of 26% of the samples could not be amplified. DNA analysis identified five species: fisher, marten, gray fox, ringtail, and black bear. Sufficient DNA was available to assess individual identity for fishers and martens.
- Barbed wire snares tended to collect fewer hairs than glue snares, but were more permeable to more species. Barbed wire snares reliably collected hair from fishers and gray foxes but not from martens. Glue snares were highly effective (92%–100%) at collecting hair from all five species. The quality of DNA was equivalent for both devices.

- Materials for barbed wire snares were more expensive than for glue snares, but handling the glue strips and the collected hair from glue snares was more difficult and time-consuming in both the field and the laboratory.
- Glue snares were more effective at detecting both martens and fishers, yielding 72% and 75% successful species identification, respectively, for these species.
- On all occasions when both fishers and gray foxes were detected based on tracks, neither species was detected with hair snares. The authors urge the development of a single-visit hair snare device that would become inoperative after a single animal was recorded.
- Species identification from fisher hair was only moderately successful (58%–75%). This shortcoming may be addressed by employing the hair collection device in a multistation array, thus providing increased opportunity for hair collection.
- The authors recommend using track plates and remote cameras when multiple species detection is the only goal. If individual identification is required, hair collection techniques should be utilized.

---

**CASE STUDY 8.2:**

**COMPARISON OF MULTIPLE METHODS FOR CHARACTERIZING A CARNIVORE ASSEMBLAGE**

**Source:** Campbell, L.A. 2004. Distribution and habitat associations of mammalian carnivores in the central and southern Sierra Nevada. PhD dissertation. University of California, Davis. 166 pp.

**Location:** Central and southern Sierra Nevada, California.

**Target species:** Fisher, marten, gray fox, ringtail, western spotted skunk, striped skunk (*Mephitis mephitis*), bobcat, coyote, weasels (*Mustela* spp.), black bear, badger, wolverine (*Gulo gulo*), Sierra Nevada red fox (*Vulpes vulpes necator*).

**Size of survey area:** 9,600 km$^2$.

**Purpose of survey:** To collect information about

carnivore distribution and co-occurrence and describe habitat associations of carnivores. Post hoc comparisons of characteristics of detections by three survey methods were used to develop hypothetical sampling schemes that could be compared based on efficiency (probability of detecting target species) and relative cost.

**Survey units:** Sample units were established at every other cell within a systematic grid of 5 × 5 km cells, resulting in a spacing of 10 km between survey locations.

**Survey method:** Surveys compared data from baited remote cameras and enclosed and unenclosed sooted track plates.

**Survey design and protocol:** Two baited cameras, two unenclosed track plates and an array of six enclosed track plates were established in each sample unit. All three devices were baited with chicken and a commercial scent lure. Stations were visited at two-day intervals over the sixteen-day sampling period ($n = 8$ sampling occasions); scent lure was applied during station establishment and again on the fourth occasion. Surveys were conducted May–November 1996–9.

**Analysis and statistical methods:** Characteristics of detections by each method were compared using: (1) the number of species detected, (2) LTD, (3) proportion of functioning trap nights, (4) proportion of species detections accounted for by each device type, and (5) number of times a species detection by a device represented the only detection of that species at a sample unit. Detection probabilities associated with each species/method combination were used to explore alternative survey designs (including varying combinations of cameras, unenclosed track plates, and enclosed track plates, as well as varying numbers of stations and sampling occasions) at a sample unit with the intent of achieving an 80% probability of detecting the target species. The cost for each hypothetical sample unit was also evaluated.

**Results and conclusions:**

- Each method yielded a set of unique detections, with unenclosed track plates providing the greatest contribution of single method detections at a sample unit (i.e., 7%–50% of the total number of species detections). For certain species, such as those within the Ursidae, Canidae, and Felidae, the detections recorded by unenclosed track plates provided important distribution information that would have otherwise been unavailable.

- Unenclosed track plates recorded detections of twelve species or species groups but failed to detect badgers. Enclosed track plates failed to detect raccoons—as did remote cameras, which also failed to detect striped skunks, badgers, or weasels. Unenclosed track plates were prone to failure due to their sensitivity to precipitation. Remote cameras and enclosed track plates operated properly during >92% of trap nights.

- No species exhibited significant differences in LTD between methods. But each sample unit featured three times the number of enclosed track plates as remote cameras or unenclosed track plates, suggesting that for the relative number of stations, a single unenclosed track plate or remote camera detected species as quickly as three enclosed track plates.

- The probability of detecting a species when present varied by species and method. As the number of sampling opportunities increased, probability of detection approached >80% for all methods after five to eight sampling opportunities. Within a sample unit, this level of detectability could be achieved by repeatedly surveying a single station five to eight times or by surveying five to eight stations once (see figure 2.7), assuming independence of occasions and stations. Felids, raccoons, and badgers were not well represented by any method and failed to achieve 80% probability of detection for the maximum survey effort considered.

- The optimal survey design for surveying multiple species, when cost was not considered, was dictated by the limitations of the one species that was most difficult to detect. Because the remaining species had a higher probability of being detected by any of the methods, these species were still well represented by a sample unit design that favored rarely detected species.

- A combination of four enclosed track plates and four unenclosed track plates run for eight sampling occasions provided an overall probability of detection of >95% for ten of thirteen species. The least expensive combination of methods to produce a probability of detection of ≥80% for all ten species was a sample unit composed of two enclosed track plates run for six occasions.

---

### CASE STUDY 8.3:
### MULTIPLE METHODS USED TO SURVEY
### MULTIPLE SPECIES

**Source:** Gompper, M. E., R. W. Kays, J. C. Ray, S. D. LaPoint, D. A. Bogan, and J. R. Cryan. 2006. A comparison of noninvasive techniques to survey carnivore communities in northeastern North America. *Wildlife Society Bulletin* 34:1142–1151.

**Location:** Adirondack State Park (ADK) and Albany Pine Bush Preserve (APB), New York.

**Target species:** Primary—coyote, marten, fisher, raccoon, weasels, black bear. Secondary—red fox, gray fox, striped skunk, house cat (*Felis catus*), bobcat, American mink (*Neovison vison*).

**Size of survey area:** 25,000 km$^2$ (ADK) and 60 km$^2$ (APB).

**Purpose of survey:** To document the distribution and potential interactions between carnivore species in the Adirondack region of northern New York. With the regional extirpation of the two largest predators—wolves (*Canis lupus*) and cougars (*Puma concolor*)—coyotes have moved into the region as the top carnivore. Because of the coyote's potential to affect the distribution of smaller carnivores, and its nearly ubiquitous distribution, special efforts were made to quantify the abundance of coyotes, while other species were assessed only in terms of detection-nondetection. Scat-based surveys also allowed descriptions of carnivore diet and disease.

Various noninvasive techniques were evaluated and compared.

**Survey units:** At ADK, surveys were conducted along 5-km transects consisting of hiking trails or dirt roads. A total of fifty-four transects were distributed evenly across three broadly defined types of landscapes: wilderness forest, logged forest, and suburban-agricultural. Transects were spaced a minimum of 5 km apart to avoid sampling the same individual carnivores at multiple sites. At APB, surveys were focused on twenty-one independent forested plots embedded in a suburban landscape.

**Survey method:** A total of five techniques were used: baited camera traps (ADK, APB), track plates (ADK), scent stations (APB), snow tracking (APB), and scat surveys (ADK).

**Survey design and protocol:** Surveys were conducted during 2000–2002. With the exception of snow tracking, all sampling took place during summer months (June–September). At ADK, three baited cameras (spaced 2 km apart) and six track plates (spaced 50 m apart) were placed along each transect, 25–50 m off the trail. Cameras were run for thirty days. ADK cameras were rebaited every seven to ten days; track plates were run for twelve days and rebaited every two days. Cameras and track plates were not run concurrently at a trail. Five scent stations were established at each of twenty-one APB sites. Each scent station had a 1 m radius and was baited with a fatty acid scent tablet at the center. Scent stations were open for a single night and monitored by cameras. Snow tracking was conducted by surveying 1 ha plots within twenty-one independent study sites. Surveys were conducted four times over two winters. Scat surveys were conducted once monthly for three months. Transects were cleared of scats, then all putative carnivore scats were collected by researchers walking trails. Scats were identified to the species level via the analysis of fecal mitochondrial DNA, and coyote scats were identified to the individual level via microsatellite DNA fingerprinting (see chapter 9).

**Analysis and statistical methods:** Method efficiency was evaluated by comparing LTD and probability of detection. Indices of abundance (i.e., individuals per trail) were calculated only for coyotes based on fecal DNA.

**Results and conclusions:**

- Although cameras, track plates, and snow tracking all demonstrated utility for communitywide surveys, no single method allowed all species to be simultaneously surveyed.
- Baited cameras and track plates recorded detections of ten species: coyote, black bear, weasels (ermine, *Mustela erminea*, and long-tailed weasel, *M. frenata*), raccoon, marten, fisher, red fox, gray fox, and opossum (*Didelphis virginiana*). Surveys failed to detect bobcat, mink, or North American river otter (*Lontra canadensis*).
- Track plates and cameras were similarly effective for detecting midsized mammals. Cameras appeared to be efficient at detecting black bears, but were very inefficient at detecting coyotes and smaller-bodied species. Scat collection showed coyotes to be present at nearly all sites but to vary in abundance. Cameras detected coyotes at only twelve of twenty-eight sites where genetic evidence confirmed their presence.
- Trail-based scat surveys were very efficient in detecting canids but not other species. The number of coyote scats per 5 km transect was directly related to the number of individual coyotes using the area and was therefore an accurate index of abundance.
- Across all species detected by both cameras and track plates, LTD for cameras tended to be lower than for track plates, but the variance was higher. Although weasels and martens were most rapidly detected by cameras, track plates had a higher probability of detecting these species. The two techniques appeared to

be equivalent (based on LTD) for detecting fishers and raccoons, however, track plates had a higher probability of detecting each species.

- Cameras at APB were more efficient at detecting carnivores than were scent stations. Tracks were missed for 17% of carnivore detections, whereas photos were missed for <2% of detections.
- Snow tracking surveys were more effective for coyotes, fishers, and gray foxes than were other methods, and provided a high probability of detecting coyotes, fishers, and domestic cats.
- No correlation was found between the number of coyote scats and LTD by cameras in ADK.
- Measures of snow track density per unit area could be used to compare local activity across sites.
- Based on extrapolations of detectability, the authors suggest that surveys of approximately two weeks should detect most species present but that approximately a month of surveying may be required for exhaustive inventories.

## Acknowledgments

We are very grateful to the numerous inventive biologists who have used noninvasive survey methods for a variety of purposes, refined the techniques, and created innovative combinations of methods to address multiple objectives. We would like to thank Andy Royle, Justina Ray, and Roland Kays for their thoughtful contributions; Tony Clevenger and Mark Jordan for sharing their research and photographs; and Paula MacKay for keeping us on track and setting a high editorial standard. Rick Truex and Matt Gompper provided insightful comments on earlier versions of this manuscript.

*Chapter 9*

# Genetic and Endocrine Tools for Carnivore Surveys

*Michael K. Schwartz and Steven L. Monfort*

Modern literature and Hollywood proved decades ahead of science in imagining the information that could be obtained from single hairs or feces. Indeed, from Aldous Huxley's *Brave New World* (1932) to the cult movie *GATTACA* (Columbia Pictures Corporation 1997), writers and producers foreshadowed the scientific value of noninvasive samples. In the 1990s, with the advance of both molecular genetics and endocrine biology, forensic scientists developed tools to determine the identity, sex, health, and social status of humans from samples left at crime scenes (e.g., hair, scat, urine, saliva). As with many technological advances in human biology, these developments soon transferred to other disciplines—including wildlife biology.

In this chapter, we review the state-of-the-art tools that molecular and endocrine biologists employ to learn about carnivores and other wildlife through noninvasive means. The chapter is divided into three sections, with the first describing advances in molecular ecology, the second recounting advances in endocrinology, and the final section briefly discussing the synergy obtained by combining DNA and endocrine tools for understanding carnivore ecology. The primary objectives of this chapter are to (1) provide a general overview of laboratory methods and demonstrate their application via examples;

and (2) share practical information with field biologists regarding what and how to sample, and how to treat samples in the field to optimize efficacy in the molecular genetics or endocrinology laboratory.

We don't expect readers to be experts in genetics or endocrinology after reading this chapter, yet we believe that a limited understanding of laboratory methods is helpful—if only to aid in communication with laboratory scientists. Thus, rather than exhaustively describe all existing laboratory techniques, we include material on commonly asked questions and how these questions are typically addressed in the laboratory. Our goal is not to provide a lab manual but to create a useful resource for field biologists. We have strived to balance simplification with precision and to be neither pedantic nor so technically thorough that we fail to convey our meaning. We have also included a glossary of genetic and endocrine terms (appendix 9.1) to assist readers in need of more information and have boldfaced glossary terms upon first use in the chapter. Last, we have chosen to combine genetics and endocrinology into one chapter because there are many cases where a researcher may wish to obtain both endocrine and genetic information from the same sample. In some ideal situations, the information obtained from these disciplines is complementary. The last section of this chapter therefore attempts to integrate genetics and endocrinology.

# Genetic Approaches for Studying and Monitoring Carnivores

Our understanding of the natural world has been dramatically expanded by the field of molecular biology. Yet, modern breakthroughs in technology, the hype of this technology in popular culture, and the remarkable applications of new tools for answering age-old questions have lead to some confusion about the realistic abilities of molecular genetic techniques in the context of wildlife research. While it is true that noninvasive genetic sampling, coupled with molecular biology tools, has proven to be very effective at answering important management, evolutionary, and ecological questions (table 9.1), these tools are not a panacea. Here we try to separate fact from fiction in terms of what can be accomplished with noninvasive genetic sampling.

## A Primer on Molecular Genetic Tools for Studying Wildlife

When used to best effect, molecular data are integrated with information from ecology, observational natural history, ethology, comparative morphology, physiology, historical geology, paleontology, systematics, and

*Table 9.1.* Examples of the use of genetic sampling to address objectives pertinent to carnivore ecology, management, and conservation

| Objective | Species | DNA source material | Reference |
|---|---|---|---|
| Abundance | Brown bear | Feces | Bellemain et al. 2005 |
| | Eurasian badger (*Meles meles*) | Hair | Frantz et al. 2004 |
| Relative abundance | Coyote | Feces | Kohn et al. 1999 |
| | Mountain lion | Hair | Ernest et al. 2000 |
| Occupancy | Fisher | Hair | Zielinski et al. 2006 |
| | Eurasian lynx (*Lynx pardinus*) | Feces | Palomares et al. 2002 |
| Trend in abundance | Brown bear | Hair | Boulanger et al. 2004a |
| | Brush-tailed rock wallaby (*Petrogale penicillata*) | Feces | Piggott et al. 2006 |
| Hybridization | Red wolf | Feces | Adams et al. 2003 |
| | Canada lynx | Hair | Schwartz et al. 2004 |
| Paternity and relatedness | Wombat (*Vombatus ursinus*) | Hair | Banks et al. 2002 |
| | Gray wolf | Feces | Lucchini et al. 2002 |
| Sex identification | Ursids | Feces | Taberlet et al. 1997 |
| | Felids | Hair | Pilgrim et al. 2005 |
| Diet assessment | Felids | Feces | Farrell et al. 2000 |
| Sex-specific movement | Brown bear | Hair | Proctor et al. 2004 |
| | Wolverine | Feces | Flagstad et al. 2004 |
| Turnover rates and survival | Coyote | Feces | Prugh et al. 2005 |
| | Wolverine | Hair | Squires et al. 2007 |
| Phylogeography and population genetics | Dhole (*Cuon alpinus*) | Feces | Iyengar et al. 2005 |
| | Louisiana black bear (*Ursus americanus luteolus*) | Hair | Triant et al. 2004 |
| Spatial organization | Eurasian otter (*Lutra lutra*) | Feces | Hung et al. 2004 |
| Landscape genetics | American black bear | Hair | Cushman et al. 2006 |
| | Eurasian otter | Feces | Hobbs et al. 2006 |

other time-honored disciplines. Each of these traditional areas of science has been enriched, if not rejuvenated by contact with the field of molecular genetics. (Avise 2004)

The majority of noninvasive genetics studies have used DNA as a diagnostic **marker** to acquire information about difficult-to-study species. For instance, we can determine species identification, sex, and individual identification from a hair sample using diagnostic molecular genetic tools. In this context, noninvasive genetic sampling has been used to address questions of occupancy, abundance, and geographic range (see table 9.1), and when these metrics are collected over time, for genetic monitoring purposes (Schwartz et al. 2007). Bellemain et al. (2005), for example, collected brown bear (*Ursus arctos*) feces throughout south-central Sweden over two consecutive years. Using individual identification information from diagnostic DNA markers, along with four approaches to estimating abundance (two rarefaction indices, a Lincoln-Peterson estimate, and a closed capture model in program MARK), the authors were able to arrive at estimates of abundance for each sex per year.

Noninvasive genetic samples can also be used in a population genetic framework. Population genetics is the study of the distribution and frequency of genes. In this context, noninvasive genetic sampling has been used to investigate effective population size, gene flow, mating systems, genetic diversity, and relationships between populations of many species (Schwartz et al. 1998; Manel et al. 2003; Miller et al. 2003; Wisely et al. 2004; Schwartz et al. 2004; Leonard et al. 2005). Specifically, noninvasive genetic sampling has provided new means for collecting population genetic samples from species that are otherwise difficult to study. Cushman et al. (2006), for instance, used DNA from noninvasive hair snares, coupled with population and landscape genetic analyses, to determine the effects of roads, forest cover, slope, and elevation on black bear (*Ursus americanus*) movement.

Contemporary biologists may take for granted the ability to obtain either diagnostic or population genetic information from genetic samples, but it wasn't until Sir Alec Jeffreys began studying DNA variation and the evolution of gene families through the use of "hypervariable" regions of human DNA that molecular biologists were able to produce a genetic fingerprint (Jeffreys 1985a, b; see Avise 2004 for a review of earlier isozyme research). Jeffreys discovered that particular regions of the human genome, which consist of short sequences repeated multiple times, also contain a "core sequence" that could be developed into a tool called a **probe**. Further, Jeffreys recognized that these probes could be used to explore multiple regions of the human genome that also contain tandem repeats (two or more nucleotides sequentially repeated (e.g., CATG, CATG, CATG)—ultimately producing something biologically akin to a barcode—such that each individual has a unique genetic signature.

This technique became known as DNA fingerprinting, and instantly became a staple of forensic and paternity work worldwide. Two years after the development of DNA fingerprinting, Wetton et al. (1987) found these same Jeffreys probes to be useful for studying house sparrows (*Passer domesticus*) in the United Kingdom. Within a decade, DNA fingerprinting was common in many wildlife studies, rewriting conventional wisdom on mating systems and gene flow among populations (Wildt et al. 1987; Lynch 1988; Burke et al. 1989).

Initially, there were two major impediments to DNA fingerprinting for addressing wildlife issues. First, although Jeffreys probes provided a DNA "barcode," it was typically not possible to know which bars were associated with which **locus**. Thus, population genetic models that relied on locus-specific information needed to be adapted or abandoned. Second, high quantities of high-quality DNA were required for these barcodes to be visible on a standard electrophoresis gel. DNA fingerprinting from hair and other samples containing minimal or degraded DNA was therefore unreliable or impossible.

The next two breakthroughs, which have led to the recent boom in genetic techniques, were the development of the **polymerase chain reaction** (**PCR**) and the discovery of new classes of genetic markers (Saiki et al. 1988; Tautz 1989; Weber and May 1989). PCR is the "amplification" of a gene, part of a gene, or part of any section of the genome. The process can be likened to a molecular photocopy machine, where a few short DNA fragments are copied many times, ultimately allowing visualization of the PCR product on a standard electrophoresis gel. PCR is conducted in a thermal cycler, which heats and subsequently cools a chemical process to precise temperatures during multiple steps (figure 9.1). Origi-

nally, during the PCR process, a critical enzyme (called a polymerase) would break down when DNA strands were heated, making PCR extremely labor intensive, as more polymerase needed to be added during every cycle in the process. The process was greatly facilitated by the discovery of a thermostable enzyme called *Taq* **polymerase** (derived from the organism *Thermus aquaticus*, discovered in hot springs in Yellowstone National Park, USA), which allows the polymerase chain reaction to be subject to extreme temperature increases and decreases without disintegrating the polymerase.

**Primers** are critical components of the PCR reaction. A forward and a reverse primer together act as

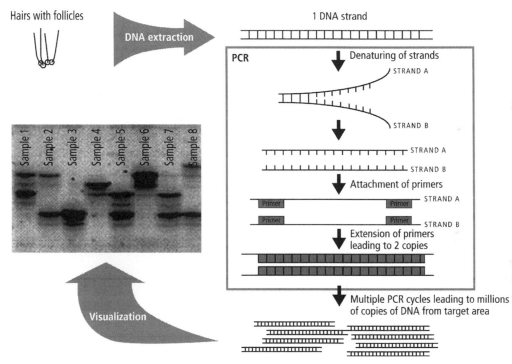

*Figure 9.1.* Schematic illustrating the process of deriving individual identification from hair samples. This process begins with DNA extraction, which produces DNA strands. Next, a particular region of the DNA is amplified using (in this case) microsatellite primers, and the polymerase chain reaction (PCR). The PCR process involves three major steps: (1) the denaturing of the double-stranded DNA molecule; (2) the attachment of primers at a particular locus in the genome; and (3) the extension of these primers to produce a copy of the original locus. After mulitple PCR cycles, millions—if not billions—of copies of the region are created, and can then be visualized on an electrophoresis gel. This gel image shows eight Canada lynx samples evaluated at one microsatellite locus. Even with one locus, multiple individuals can already be discerned, but samples 5 and 7 produce the same banding pattern at this locus. Ultimately, when additional loci were run, these two samples were determined to be from different individuals.

bookends denoting the section of the genome to be copied. These primers can originate from either the mitochondrial or nuclear genomes of an organism. Deciding whether to examine sections of the nuclear or mitochondrial genome will depend largely on the goals of the study. For instance, if the goal is to determine species from hair or fecal samples, the mitochondrial genome is typically used. **Mitochondrial DNA** (mtDNA) is often less variable within a species than **nuclear DNA** but variable between species (see *Wildlife Genetics in Practice: A Hypothetical Example*). If the goal is to produce individual identification or fine-scaled population genetic information, however, nuclear DNA is often preferred.

Currently, **microsatellites** are one of the most common genetic tools for producing individual identification from noninvasive genetic samples and for conducting population genetic analyses. Microsatellites belong to a class of primers that contain variable numbers of tandem repeats—in general, these repeats are two to five base pairs in length. They are highly variable in nearly all vertebrates, which ultimately allows the differentiation of individuals within a population.

Microsatellites have several advantages over Jeffreys DNA fingerprinting probes. First, microsatellite **loci** are codominant markers, meaning that **alleles** from both of the chromosome pairs in diploid organisms are observed. Second, when used for individual identification—genotyping in genetic terms—each pair of bars on the barcode is a separate microsatellite locus. Thus, a **heterozygous** individual (an individual with two different alleles) would have two bars (called bands or fragments) from a single microsatellite (see samples 1, 2, 5, 7 and 8 in figure 9.1). Alternatively, a **homozygous** individual has only one fragment at the microsatellite locus (see samples 3, 4, and 6 in figure 9.1). The ability to distinguish loci from one another enables traditional population genetic models to estimate phenomena such as gene flow or relatedness (Wright 1969). Third, microsatellites are believed to be selectively **neutral**, conforming to many population genetic

models. These properties, plus the ability to either inexpensively develop microsatellites for a particular species or use those already developed for related taxa, have made microsatellites a popular tool for molecular ecologists studying wildlife.

In summary, the coupling of PCR and microsatellite or mtDNA primers allows small amounts of DNA (e.g., from cells attached to the follicle of a single hair) to be transformed into a diagnostic identifier of individuals, species, and populations, and makes noninvasive genetic sampling feasible.

## Wildlife Genetics in Practice: A Hypothetical Example

Imagine a genetic sampling survey with the goal of estimating carnivore species diversity in a western forest. Samples in this hypothetical survey consist of feces (also called scat, pellets, dung, or turds, depending on the publication) located by scat detection dogs (chapter 7) and hair snared at bait stations (chapter 6). In this section, we walk through the different analyses that are commonly conducted on such samples. It is important to note, however, that the particular molecular genetic techniques applied will depend on the objectives and species under study, as well as the expertise of the laboratory. One size doesn't fit all in molecular ecology.

### Species Identification

The first question of interest to wildlife researchers is often, what species were detected by my survey? There are several ways in which a laboratory can ascertain species identification, but one of the most common is to **sequence** a region of the mitochondrial genome. For identifying carnivore species in particular, a standard approach is to use the PCR reaction with primers for the 16S rRNA region of the mitochondrial genome (following the protocols in Hoezel and Green 1992; Mills et al. 2000). Most North American carnivores have a distinct sequence at the 16S rRNA region, and the majority of these species' sequences are entered into a national database (GenBank; National Center for Biotechnology Information 2007), which facilitates identification by matching sequences. For carnivores outside of North America, however, and for other taxa, reference sequences may not be available—although this is changing rapidly.

DNA barcoding is a new trend in molecular biology for species identification. With this approach, short, standardized DNA sequences—typically from a mitochondrial gene—are used to identify known species and to discover new species quickly and easily (Herbert et al. 2004; Savolainen et al. 2005). The initial goal of barcoding was to use a standardized region of the mitochondrial genome to uniquely identify all species, although this is proving difficult. Regardless, the barcoding databases established for many taxa will aid in developing noninvasive surveys designed for carnivore species' detection worldwide.

DNA sequencing can be expensive. In some cases, however, we can reduce expenses by approximately 35% by using a restriction enzyme test to ascertain species identification. Here, as in sequencing, we amplify the 16S rRNA region, but we then immerse the DNA in particular enzymes which cut the DNA at diagnostic "restriction sites." We can identify species by examining the patterns of restriction enzyme-digested, PCR-amplified mtDNA (figure 9.2). This was the approach taken with the thousands of samples collected in the USDA Forest Service's National Lynx Survey (McKelvey et al. 1999; see chapter 6). The downside is that research is required to develop such assays. Further, when nontarget species are encountered, they often cannot be identified to the species level and may even confound the identification of the target species. Given our hypothetical survey, with the goal of identifying every species that deposited a sample, we would likely sequence either 16S rRNA or another region of the mitochondrial genome and compare these sequences to known species sequences in a genetic database (e.g., GenBank). But if we only wanted to know whether or not the sample was deposited by a

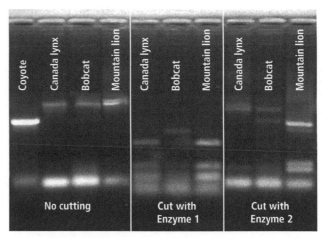

*Figure 9.2.* An electrophoresis gel showing the results of a restriction digest test to determine species identification. Without deploying any restriction enzymes, separation of felids and canids is possible. Using two different enzymes, Canada lynx, bobcats, and mountain lions can be discerned.

given target species, we might choose a restriction enzyme test (e.g., Paxinos et al. 1997; Mills et al. 2000; Dalen et al. 2004b).

## Gender Determination

Let's suppose that laboratory results show our hypothetical survey to have detected ten gray wolves (*Canis lupus*), eight Canada lynx (*Lynx canadensis*), four fishers (*Martes pennanti*), one elk (*Cervus elaphus*), and one bushy-tailed woodrat (*Neotoma cinerea*). The forest manager may want to know if there are any female lynx or gray wolves present in the forest (i.e., to assess whether these might be breeding populations). One of three genes is typically used to identify gender in carnivores. The first is the SRY gene (the *testis determining factor*), present only on the male Y chromosome. When a sample from a male is analyzed with SRY-specific primers, one band appears on an electrophoresis gel. If the sample is from a female, no bands appear. Unfortunately, a negative result (i.e., no band) can mean either that the sample originated from a female, or that it was of low quality and did not contain adequate amounts of DNA. Therefore, it is common for researchers to coamplify a microsatellite locus, which amplifies regardless of gender. Failure of the microsatellite to amplify signals that the DNA was of poor quality and the results should be discarded. Multiple repeats of this process are recommended for accuracy.

A second method for identifying sex is to sequence a gene in the *zinc-finger region* (*ZF*) of the X and Y chromosomes. In felids, the *ZFY* (male) band has a three-base pair deletion compared to the *ZFX*. Thus, a male lynx—and males of most other mammal species—will show two bands on an electrophoresis gel (i.e., a band for the X chromosome and a band for the Y chromosome, which vary in length because of the deletion on Y), whereas a female will only show one band (i.e., females have two X chromosomes, with no length variants; Pilgrim et al. 2005). Similar tests have been published for canids, cetaceans, and bovids. Last, a similar gene, called the amelogenin gene (which codes for proteins found in tooth enamel) has a twenty-base pair deletion on the Y chromosome of felids (and some other species), providing another gender determination test that works for felids (Pilgrim et al. 2005) as well as ursids (Poole et al. 2001).

### Individual Identification

Our hypothetical research reveals that, of the eight lynx samples, five were produced by males and three by females. The next common question might be, how many individuals are represented by these samples? While several tools exist to provide individual identification, the most common are microsatellites (see figure 9.1). For lynx, a panel of six microsatellites is frequently used to determine individual identification (Schwartz et al. 2004). In a lynx study in Minnesota, six microsatellites provided a **probability of identity** of $1.55 \times 10^{-06}$ (M. Schwartz, unpubl. data), which translates to the probability of two randomly chosen lynx in the Minnesota population having identical **genotypes** as being 1 in 645,161. Given that surveys to date have detected fewer than two hundred lynx in Minnesota, the survey was deemed to have had sufficient power to distinguish

individuals with six microsatellites. The number of microsatellites necessary for individual identification depends on the amount and distribution of genetic variation in the species (characterized by the probability of identity; Waits et al. 2001). In other work, as few as four microsatellites, or as many as ten, have been required to achieve a reasonable probability of identity, depending on the population and its history (e.g., small and inbred populations tend to have little variability).

In cases where existing microsatellites have low variability, one solution can be to develop microsatellites specifically for the population of interest. Given that microsatellites show an ascertainment bias (i.e., they are more variable in the species and, in some cases, the population for which they are developed), this approach can result in variable microsatellites for the target population, thus requiring fewer microsatellites for individual identification. Today, a number of commercial companies can quickly develop variable microsatellites for a target population at a reasonable cost (e.g., $10,000–$15,000 USD).

The decision to develop microsatellites for a particular species, versus assessing whether suitable primers have already been developed from a closely related species, lies in the initial costs of developing markers, the availability of markers for a related species, and the purpose of the project. For instance, a recent study seeking a panel of microsatellites for sampling mountain beavers (*Aplodotia rufa*) found that the lack of congenerics and the requirement for short microsatellites to be used with individual hairs dictated the development of microsatellites (Pilgrim et al. 2006).

Once sufficient power to discriminate between individuals is achieved, the resulting microsatellite genotypes can be compared to determine the number of unique individuals (figure 9.1). When employing microsatellites to identify individuals with noninvasively collected genetic samples, it is important that some method be used to ensure that the resultant data are error free (see box 9.1). It should be noted that other genetic tools can be used to

## Box 9.1

### A cautionary note on field and laboratory errors

Errors occur in all scientific disciplines; the critical issue is whether and how they are detected and reported. Researchers conducting noninvasive genetic sampling have openly acknowledged, and attempted to address, errors that occur in the laboratory. Such errors are of two general types: those associated with labeling samples, misreading labels, and misscoring electrophoresis gels—deemed *human errors*—and those that are inherent when using low-quality or low-quantity DNA samples, called *genotyping errors*. The two most common types of genotyping errors are **allelic dropout**—the preferential amplification of one of two alleles from a codominant marker—and **false alleles**, amplification products that mimic true alleles.

Genotyping errors can dramatically affect survey results, especially when estimating abundance via genetic sampling (Waits and Leberg 2000; McKelvey and Schwartz 2004a, b). For example, Creel et al. (2003) demonstrated that, without addressing genetic errors, the population size of gray wolves in Yellowstone National Park would be overestimated by 550%. Similarly, before error-checking, Schwartz et al. (2006) found a 28.1% overestimate in the number of unique genotypes of black bears in Northern Idaho. This bias is in contrast to the underestimation that can occur when researchers use too few microsatellites to determine individuals and thus mistakenly infer a single individual from samples obtained from two unique individuals (deemed the "shadow effect"; Mills et al. 2000).

There are four primary methods used to identify and remove genetic errors. The first is called the multitube approach (Taberlet et al. 1996). In this approach, each sample at each locus is run up to seven times to ensure genotype consistency. While multitubing will detect genotyping errors, it has the disadvantage of being expensive in terms of cost and sample use. That is, if a ten-locus genotype needs to be rerun seven times at each locus, seventy runs of the sample are required. Sometimes there is not enough DNA for this approach (depending on the initial sample), or no DNA remains for future analyses. Thus, a small cottage industry has developed with the aim of finding less time-consuming, costly, and DNA-intensive methods to detect and remove errors.

The second approach is to quantify the amount of target DNA in the sample. Morin et al. (2001) developed an assay designed to measure the amount of amplifiable nuclear DNA in low DNA-concentration extracts. This method provides an indication of the concentration of DNA in the sample prior to any analyses and allows researchers to discard the sample or estimate the optimal number of times each sample should be analyzed to eliminate errors (see Morin et al. 2001 for details). While this approach undoubtedly reduces errors, it has two shortcomings. First, even if a relatively large amount of DNA is present, not all genotyping errors are caught. Second, the equipment required to measure target DNA remains uncommon in many laboratories, although this is changing rapidly.

A third method for addressing genotyping errors uses computer algorithms to detect them. Depending on the data and objective, various algorithms have been suggested. For capture-recapture data, McKelvey and Schwartz (2004a, b) developed two tests that were incorporated into program DROPOUT (www.fs.fed.us/rm/wildlife/genetics/software.php; McKelvey and Schwartz 2005) to detect both human and genotyping errors. The first test utilizes a longer genetic tag (e.g., a genotype that produces individual identification) than is typically used in capture-recapture studies and examines the distribution of differences in loci between all samples. The second test proposed by McKelvey and Schwartz (2004a, 2005) assesses the number of loci required to provide enough power to distinguish individuals (e.g., five loci), employs a greater number than this number (e.g., nine loci), and then runs through different combinations of five-locus genotypes to determine the number of unique individuals. If a locus has errors, its use in creating a genotype will result in an inflation of the number of unique individuals.

Another commonly used algorithm for detecting genotyping errors is the maximum likelihood-based method contained in program RELIOTYPE (Miller et al. 2002). This approach minimizes errors by estimating genotype reliability and directs which samples should be multitubed to remove errors. In other words, RELIOTYPE uses the allelic frequencies from a population and creates maximum likelihood estimates of the allelic dropout rate per locus (Miller et al. 2002). Those loci estimated to have the most errors can then be reanalyzed. Additional programs include GIMLET (Valiere et al.

---

### Box 9.1 (Continued)

2002), which is useful for evaluating errors when samples have been multitubed (Taberlet et al. 1996); PED-MANAGER (Ewen et al. 2000), applicable when pedigree information is available; MICRO-CHECKER (Van Oosterhout et al. 2004), which compares randomly constructed genotypes to observed genotypes in order to determine scoring errors due to stutter and short allele dominance; and HW-QUICKCHECK (Kalinowski 2006), which uses exact tests to detect departures from Hardy-Weinberg proportions—a sign of genotyping error. Relatively little research has been conducted to evaluate the effectiveness of one approach versus another (but see Smith et al. 2006).

The final approach to handling genotyping errors is used solely in capture-recapture studies (Lukacs and Burnham 2005a). This method incorporates the probability of genotyping error into the closed-population models of Otis et al. (1978), Huggins (1989), and Pledger (2000) by analyzing the disproportionate number of genotypes collected once relative to genotypes collected more frequently. While approaches that incorporate error into estimates have potential value, they have only recently been developed and have not been widely used or thoroughly evaluated with actual datasets (for further discussion of genotyping error, see chapter 11).

---

distinguish individuals (see Avise 2004 for a description of these techniques)—with each having its own benefits and limitations.

*Inference to Populations*

Our hypothetical forest manager might need information about the population composition (e.g., origin of a reintroduced population, population membership) of fishers found in the forest. For example, fishers in the Rocky Mountains are either descended from animals reintroduced between the 1950s and the 1990s or from a remnant population that escaped extinction (Vinkey et al. 2006). Reintroduced fishers have a unique mitochondrial DNA signature at both the **cytochrome**-*b* region (mtDNA) and the **control region** (mtDNA), compared to native individuals. We could thus analyze the four fisher samples from our hypothetical survey to determine the historical maternal origins of the animals from which the samples were taken (remember that mitochondrial DNA is transmitted only through mothers, thus nothing can be inferred about the paternal lineage with this approach). Alternatively, we might be able to use molecular markers, such as microsatellites, to assign individuals to a population (if reference databases are available), or even to classify and subsequently assign subspecies or other taxo-

nomic designations (for a review of genetic tools and additional types of analyses for determining subspecies, see Baker 2000; Avise 2004; Allendorf and Luikart 2007; Palsbøll et al. 2007).

## Types of Genetic Samples

DNA analysis has been attempted on biological samples ranging from historical pelts to regurgitates found on the side of a trail (table 9.2), but hair (Morin and Woodruff 1992; Taylor et al. 1997) and feces (Taberlet et al. 1997; Prugh et al. 2005) are the most common sources of noninvasive genetic material collected for wildlife research and monitoring (table 9.1). In one of the first applications of noninvasive genetic sampling, Taberlet et al. (1997) used a combination of hair and feces collected in the field to track a small population of Pyrenean brown bears. After extracting DNA from the samples, the authors were able to amplify six variable microsatellite loci and obtain individual and gender identification information that showed that the population consisted of at least five bears: four males and one female.

The target for sampling hair is the follicle located at the end of the hair shaft; follicles are larger on guard hairs than on underfur. Occasionally, hair

*Table 9.2.* Source material typically used for genetic sampling

| Source Material | Species | Purpose | Reference |
|---|---|---|---|
| Hair | Chimpanzee (*Pan troglodytes*)<br>Brown bear | Paternity and gene flow estimates<br>Abundance | Morin et al. 1994<br>Mowat and Strobeck 2000 |
| Feces | Coyote<br>Eurasian badger | Population size estimation | Kohn et al. 1999 ; Prugh et al. 2005<br>Frantz et al. 2004 |
| Regurgitates | Gray wolf | Documentation of dispersal | Valiere 2003 |
| Saliva | Coyote<br>Wolf | Predator identification | Williams et al. 2003; Blejwas et al. 2006<br>Sundqvist et al. 2008 |
| Urine | Wolverine | Methodological study | Hedmark et al. 2004 |
| Menstrual bleeding | Taiwan macaque (*Macaca cyclopis*) | Microsatellite development | Chu et al. 1999 |
| Sloughed skin | Humpback whale<br>(*Megaptera novaeangliae*)<br>Ringed seal (*Phoca hispida*) | Abundance estimation | Palsbøll et al. 1997<br><br>Swanson et al. 2006 |
| Blood in snow | Wolf<br>Multiple felids | Species identification<br>Species identification | Scandura 2005<br>M. Schwartz, unpubl. data |
| Museum specimens | Wolverine<br>Brown bear | Evolutionary significant units | Schwartz et al. 2007<br>Leonard et al. 2000 |
| Prey amplification | Prey from carnivores | Diet | Farrell et al. 2000 |

samples without follicles provide positive DNA—but not reliably. As a source of DNA, one advantage of hair over scat is that hair contains fewer chemical inhibitors. Furthermore, contamination from other DNA sources (e.g., prey DNA found in scat) are minimized with hair (although allogrooming and other social behaviors may cause cross-contamination). There is often a high rate of success in determining individual identification from hair (e.g., Frantz et al. [2004] had a 93% success rate with European badgers)—although this varies by study. Published success rates vary from 15% to greater than 90%.

Alternatively, fewer cells—and less total DNA—are generally available in a hair sample than a scat sample. Thus, unless a clump of hair is obtained, hair samples are often expended after a single DNA extraction; failure to obtain DNA leaves no opportunity for a second extraction attempt. Prior to launching a survey, it is highly advisable to conduct a pilot study to determine the rate of success for obtaining DNA from the hair of the target species under normal survey conditions (Goosens et al. 1998).

The large variation in success is due to such factors as the morphological characteristics of the species' hair, the social characteristics of the species, the environmental conditions under which the sample is collected, storage and laboratory methods, the goals of the study, and the quality of results accepted by the researcher and the laboratory.

One practical consideration is whether to consider a clump of hair as having originated from a single or multiple individuals. The answer will likely depend on many variables, including the life history characteristics of the species and the goal of the study. If the goal is species detection, the laboratory can often identify which species are represented in a mixed sample. Laboratories can often detect if multiple individuals were sampled, but they can do little to recover information regarding *which* individuals were present in a mixed sample (Alpers et al. 2003; Roon et al. 2005). Alpers et al. (2003), however, note that if there are few alleles in a population, there will be times when mixed samples will not be accurately identified as having come from multiple individuals. Instead, "new" individuals will be erroneously "created" by the combined genotype profile of multiple individuals. Thus, if individual identification is indeed a goal, researchers should attempt to minimize collecting samples from multiple individuals. This might be accomplished by frequently revisiting hair collection devices, or by using single-catch hair collection methods (e.g., Belant 2003a; Bremner-Harrison et al. 2006; also see chapter 6).

Advantageously, feces contain many sloughed epithelial cells. In addition, most carnivore fecal samples are large enough to allow multiple attempts at DNA recovery. Last, there is usually relatively little ambiguity as to the number of individuals that deposited fecal samples, although overmarking by conspecifics or sampling from latrines can potentially produce cross-individual contamination.

The greatest constraints to fecal analysis are the chemical inhibitors present in feces that restrict the amplification of DNA. Also, amounts and quality of fecal DNA are known to vary by species, temperature at time of collection, age, season, preservation method, diet, storage time, and extraction protocol (Murphy et al. 2002; Piggott and Taylor 2003; Maudet et al. 2004; Nsubuga et al. 2004; also see table 9.3). As a result, rates of species and individual identification from feces are extremely variable. McKelvey et al. (2006), for example, were 100% successful in identifying species from lynx feces collected in Washington, although individual identification rates were significantly lower (K. Pilgrim, USDA Forest Service, pers. comm.). Bellemain et al. (2005) report a 70% individual identification success rate with brown bear fecal samples at six to seven microsatellite loci, including a locus diagnostic for gender.

There has been relatively little study of the success of obtaining DNA from urine, regurgitates, saliva, or menstrual blood. In our experience, these materials have proven suboptimal sources of DNA compared to hair and feces, are difficult to systematically sample, and are better left to opportunistic collecting (but see Hedmark et al. 2004, who reported 40% success in determining individual identification with wolverine [*Gulo gulo*] urine as compared to a

65% success rate with feces). Regardless of the sample type, the way in which it is treated in the field will drastically affect the effectiveness of the survey.

## Sample Treatment and Preservation

Almost all wildlife genetic studies require complex and expensive field operations to obtain samples. In fact, once field personnel, transportation, equipment, housing, communications, and other field costs are accounted for, laboratory costs usually pale in comparison. In many instances, field data are meticulously collected, yet samples are treated improperly or are inadvertently contaminated. Below we discuss the handling and treatment of field samples.

### Contamination

Contamination is a major concern for many noninvasive studies (depending on the objective), and can occur in the field or laboratory. In the field, for example, contamination can be caused by baits, lures, previously handled animals, accompanying pets, or field personnel. Considering that the target sample may comprise only a few cells at the end of a hair, it is important to limit contact with material that can mask the target sample. We recommend the use of new latex gloves and sterile mechanical devices (e.g., tweezers, wooden picks) for handling all samples in the field. Gloves should be changed between the handling of different samples, and mechanical devices can be sterilized with ethanol and a lighter, washed in a weak bleach solution, or replaced between samples. Some of these safeguards can be diminished depending on the research question at hand. For instance, if microsatellites are to be used to determine the individual identification of nonprimates, contamination from field personnel is less of a concern—although cross-sample contamination must still be guarded against. But contamination can be an issue for studies that use universal genetic tools (e.g., mitochondrial DNA to identify species for estimating occupancy, or gender-specific markers to identify sex). Given that the goals of many studies

*Table 9.3.* Summary of fecal preservation methods used in seven recent DNA-based studies

| | | | Method | | | | | | | | | | | | |
| | | | DET[a] solution/ buffers | 70% EtOH[b] | 90%–100% EtOH[b] | Frozen –20°C | Air dry | Freeze dry | Oven dry | Micro-wave | Silica desiccant | Oven/ silica | Oven/ stored –20°C | EtOH/ silica | Drierite |
| Reference | Species | DNA type | | | | | | | | | | | | | |
|---|---|---|---|---|---|---|---|---|---|---|---|---|---|---|---|
| Frantzen et al. 1998 | Baboon[c] (w) | Mt | 2 | 4 | | 3 | 1 | | | | | | | | |
| Frantzen et al. 1998 | Baboon (w) | N | 1 | 3 | | 3 | 2 | | | | | | | | |
| Murphy et al. 2000 | Brown bear (c) | Mt | | | | | | 1 | 2 | 3 | 4 | | | | |
| Murphy et al. 2000 | Brown bear (c) | N | | | | | | 1 | 2 | | | | | | |
| Murphy et al. 2002 | Brown bear (c) | Mt | 1 | | 1 | | | | | | 5 | 4 | 3 | | |
| Murphy et al. 2002 | Brown bear (c) | N | 2 | | 1 | | | | | | 5 | 4 | 3 | | |
| Wasser et al. 1997 | Black bear (w) | Mt | 4 | 6 | 5 | | | 2 | | | 1 | | | | 3 |
| Wasser et al. 1997 | Black bear (w) | N | 4 | 5 | 5 | | | 2 | | | 1 | | | | 3 |
| Nsubuga et al. 2004 | Mountain gorilla[d] (w) | N | 3 | | | | | | | | 2 | | | 1 | |
| Roeder et al. 2004 | Mountain gorilla (c) | N | | | 2 | | | | | | 3 | 1 | | | |
| Roeder et al. 2004 | Mountain gorilla (c) | Mt | | | 2 | | | | | | 2 | 1 | | | |
| Piggot and Taylor 2003b | Tasmanian pademelon[e] (c) | N | 4 | 2 | | 2 | 1 | | | | | | | | |

Note: Numbers in cells are rank orders (with 1 being highest) of the effectiveness of each method within each study based on our interpretation of tables and discussions in the original work. Studies that examined the influence of preservation methods on both mitochondrial DNA (Mt) and single-copy nuclear DNA (N) are featured in two respective rows. Each paper included caveats for field conditions, duration of storage, and discussions as to why the particular method was likely superior—see original publications for specific details. In the species column, (w) and (c) denote studies of samples collected from wild and captive animals, respectively.

[a] DET is a DMSO salt solution comprised of 20% DMSO, 0.25 M sodium-EDTA, 100mM Tris, pH 7.5, and NaCl to saturation (Seutin et al. 1991).

[b] Ethanol

[c] *Papio cynocephalus ursinus*

[d] *Gorilla gorilla*

[e] *Thylogale billardierii*

change over time, we recommend implementing protocols that minimize all types of contamination.

Every laboratory has protocols in place for detecting contamination (see box 9.2 for factors to consider when choosing a laboratory), and laboratories that routinely process noninvasive genetic samples have separate facilities for receiving and extracting DNA. In addition, the bleaching or UV irradiation of laboratory surfaces is routine. Furthermore, laboratories specializing in noninvasive genetic sampling may limit access to areas where noninvasive samples are analyzed and discourage technicians from entering a main laboratory before processing samples in the satellite facility. Finally, all laboratories will routinely run both positive and negative controls (e.g., samples comprising simply distilled water) to detect laboratory contamination. While such measures minimize contamination from other samples or PCR products found within the laboratory, even the most stringent lab will unlikely be able to discern field contamination (e.g., the cross-contamination of samples between hair snares as a result of improper handling).

### Preventing Sample Deterioration

Given proper storage conditions, DNA is a robust and stable molecule that can persist for thousands of years (e.g., Hofreiter et al. 2001; Leonard et al. 2005). The main adversaries of DNA are **hydrolysis**, oxidation, physical cleavage through freeze-thaw cycles, alkylation, and UV radiation. Most storage techniques are designed to halt the principal enemy of DNA—hydrolysis—by eliminating water from the sample either through chemical or physical drying. Placing a sample in a silica desiccant or in an oven mechanically dries the sample, thus minimizing degradation. Alternatively, depositing a sample in ethanol or a buffer solution chemically dries the sample.

There have been a multitude of studies to examine the best way to minimize deterioration of fecal samples (see table 9.3). These studies have compared the integrity of samples preserved by the following methods:

1. Drying at room temperature or in a warm room, oven, or microwave, and storing dry.
2. Drying and storing in 70%–100% ethyl alcohol.
3. Freezing at −20°C.
4. Saturating and storing in a buffer solution.
5. Drying and storing in a silica- or Drierite-based desiccant.
6. Drying in a lyophilizer (i.e., a freeze dryer).
7. Drying with an oven or ethanol, then storing with silica desiccant.

Most of these studies have been limited to only a few methods applied to samples from one species. The most striking finding from our comparison of results (see table 9.3) is the lack of consistency between studies. For example, silica desiccant proved to be the best storage mechanism for black bears (Wasser et al. 1997), and the worst for brown bears (Murphy et al. 2002). Similarly, storage in ethanol performed poorly for Frantzen et al. (1998) and Wasser et al. (1997) but was the second best storage system for Piggot and Taylor (2003). These discrepancies are likely due to factors relating to the species (e.g., omnivores versus carnivores, species with high-lipid versus low-lipid diets), environmental conditions (e.g., mesic versus xeric, many freeze-thaw cycles versus constant cold), field and laboratory protocols (e.g., duration of storage, speed of sample drying, laboratory extraction technique, dessication protocols), and study objectives (e.g., individual versus species identification). Piggott and Taylor (2003) noted an interaction between storage method and extraction technique in the laboratory (i.e., certain extraction techniques performed better with certain storage methods, and vice versa). Again, these results strongly support conducting a pilot study to explore the performance of various storage and extraction techniques.

Given such varied study results, it is difficult to make sweeping recommendations as to the best way to store fecal samples. Here are some general rules of thumb:

## Box 9.2

### Choosing a DNA laboratory

We recommend selecting a laboratory at the beginning of a survey, and working closely with this lab for the duration. While much of the equipment used in various labs is the same, the interests and expertise of each lab is slightly different. Some labs are well equipped to conduct noninvasive genetic sampling projects, while others are not. And while some labs have collaborated with dozens of noninvasive projects and employ experienced technicians to handle the anomalies that arise during such efforts, others are less experienced. Finally, certain labs will conduct or assist with post-genotyping statistical analyses (e.g., assignment tests, tests for genetic structuring).

Although there are many factors to consider when choosing a lab as a partner, most project managers overemphasize cost at the expense of other considerations. We propose ten questions that a researcher should address prior to choosing a lab:

1. Has the lab worked with the type of samples (e.g. hair, scat) that will be used in this particular survey? Many laboratories have little or no experience with noninvasive genetic samples, nor do they have a separate lab to conduct DNA extractions, which will reduce the risk of contamination.
2. Has the lab conducted analyses for surveys of similar size and scale?
3. Has the lab worked with your target species and employed the particular genetic tools you prefer?
4. How experienced are the technicians? Commercial labs employ technicians with many years of experience, whereas universities sometimes rely on relatively untrained students. While there are benefits to training students, there also may be costs in terms of quality. Further, because of academic calendars and the demands of a student's own work, timelines for the delivery of genetic results may be difficult to predict. Agency labs are another option, but they often focus solely on projects central to the agency's mission.
5. Are you looking only for lab results, or are analysis and interpretation also important?
6. Over what time frame is the project scheduled? As genetic monitoring approaches become more common, long-term studies will also increase in number. Using one lab consistently prevents errors that may result from changing labs (and therefore protocols and technicians). Furthermore, many data types (e.g., those produced by microsatellites) are relative—versus absolute—measures. Changing labs will require both the former lab and the new lab to calibrate initial results if data are to be analyzed over time.
7. Can the lab store your samples over time? This may be important if you need to run additional analyses in the future.
8. How does the lab check for errors? Is the lab willing to re-run samples that contain potential errors? Error-checking has become an important aspect of genetic analyses, and different labs are likely to approach this topic in different ways (see box 9.1).
9. Are lab costs competitive given the services and quality offered?
10. If the samples or results are contentious and could end up presented in court, the following questions may also apply:
    • Does the lab have forensic certification or follow forensic protocols?
    • Has anyone from the lab served as an expert witness in a trial?
    • How many people have access to the lab?
    • Are samples secured? Is the lab secured?

Finally, we recommend confirming who will own the resulting data, and to whom and by whom they can be disseminated. Posing these questions early on can eliminate contention later in the research process.

1. *Extract early.* Almost all studies that have examined sample quality in relation to time have demonstrated a deterioration of DNA (Roon et al. 2003). It may be useful to send samples to the laboratory (see box 9.3 for instructions on labeling, tracking, and shipping samples) and to have DNA extracted throughout the duration of the survey—even if the survey's exact objectives are still undetermined. DNA should persist longer in a laboratory buffer than in feces.

2. *Pilot studies.* Whenever possible, a pilot study should be conducted to test storage methods and extraction procedures. For example, feces collected in a captive setting can be subjected to various conditions for varying lengths of time and stored using several different methods to establish species-specific protocols.

3. *Imitate success.* If a pilot study is not possible, consider the species, its diet, the size of the sample, the environment, the laboratory, and laboratory extraction methods. Choose a storage technique that has been successful in other studies with similar conditions.

4. *The devil is in the details.* When investigating a storage protocol with a proven track record, research the specific products used (e.g., Fisher brand silica desiccant, mesh size 10–18, part number S161212) and the precise details of the protocol. Ethanol varies in concentration and contains contaminants added to prevent human consumption. Similarly, silica desiccant varies by mesh size, and the results of air drying differ with field technician accommodations. It is also important to understand protocol details in terms of absolute amounts of sample collected, ratios of sample to ethanol or silica desiccant (e.g., 5 ml ethanol/1 g feces), container sizes (e.g., surface area exposed), and an approximate rate at which the sample will dry.

5. *Field conditions matter.* Even the *best* storage system is useful only if it can be effectively implemented in the field. Asking a technician who lives in a tent to air dry a sample may not be realistic, even if air drying is deemed the most effective protocol. Silica desiccant is often used in these situations, as it doesn't leak, it is easily portable, and it often contains an indicator chemical that changes color when it is saturated with moisture.

Less is known about hair preservation than fecal preservation. Roon et al. (2003) compared storing hair samples in silica desiccant with freezing them at –20°C. Although these researchers found no significant difference between methods used to preserve hair for mitochondrial DNA work, freezing was slightly—but consistently—better for microsatellite (nuclear) DNA tests. It should be noted, however, that this was a study of captive animals, thus enabling the freezing of samples upon collection. In some field studies, freezing samples at –20°C may not be an option until well after they are collected. If freezing at the field site is possible, it is important to avoid subsequent thawing in transit to the laboratory, as this process can mechanically cleave DNA—thus diminishing its quality. In field situations, we have found that storing hair either directly in silica desiccant or in paper envelopes in a silica desiccant dryer produces adequate results. To our knowledge, there is no information available to permit the comparison of storage methods for other types of noninvasively collected samples.

## Pitfalls, Cautions, and Future Perspectives

No technology is a panacea; this certainly holds true for molecular markers. Although molecular geneticists can currently identify individuals, sex, and sometimes population membership, little information can be acquired about an individual's age or life-stage (i.e., young-of-the year, juvenile, adult) from a noninvasive sample (although see Nakagawa et al. 2004 for future possibilities). To obtain some data for population demographic analyses, tags may need to be placed on animals that don't possess naturally unique identifying markings—often requiring physical capture of the animal.

---

### Box 9.3

### Labeling, tracking, and shipping genetic samples

It is critical to work with the genetics laboratory to accurately label and track samples, especially for larger studies. Each sample requires a unique and obvious identifier. While this may seem trivial, it is not uncommon to end up with vials having similar labels if multiple field crews are working concurrently (e.g., "sample collected 9/27/04, hair #1"). If samples arrive at the lab without clear and accurate documentation, confusion can ensue. Thus, we recommend that a designated field coordinator be assigned to organize all samples, ensure that each sample is assigned a unique identifier, compile a master list, and send all this information to the lab. Our second recommendation is to use a barcode system; a number of labs that process many samples have purchased bar code readers in recent years (see figure 6.16 for a description of barcode labeling).

From the perspective of the lab, the following information is helpful to include in a shipment of samples:

1. *Sample list.* Many researchers include an electronic sample list and a print-out with their samples. Often, this list includes the field data associated with a sample (e.g., location, collector, comments). After genotyping a sample, genetic data is entered into the initial electronic sample list. This reduces transcription errors and errors associated with manipulating spreadsheet files at the lab.
2. *Copies of the necessary permits.* Many samples are collected under state, national, and international permits. Most labs maintain a file of these permits. Additional permits (e.g., CITES permits) may be necessary if samples are sent to a lab outside the country of collection.
3. *Chain of custody form.* If samples are potentially contentious, a chain of custody form should be completed to track access to each sample.

---

There are many other cautions that need to be heeded before conducting a molecular genetic study. A frequent mistake made by many researchers is to assume that simply sending noninvasive samples to a laboratory will yield answers to all questions of interest. It is not unusual for someone to send samples to a laboratory and to expect a report without ever having posed a question or explicitly described the desired data. It is even more common for wildlife researchers to underestimate the effort required to conduct an analysis for a given project; after all, on television, human forensic samples are analyzed between commercials. For instance, most biologists are aware that molecular markers can determine parentage (to estimate the relative abundance of offspring in a sample), yet it is frequently assumed that this is a trivial exercise. Often, numerous molecular markers are required to provide adequate power for assigning paternity and maternity simultaneously—sometimes more than are readily available or affordable. By comparison, determining paternity given known maternity (or other information acquired in the field) is far less intensive. Thus, combining field data with genetic data can save analysis time and money. These same caveats hold true for a suite of other questions, including those related to estimating absolute abundance and distribution.

On the positive side, molecular genetic methods are advancing quickly. In the foreseeable future, additional molecular tools such as microarrays—which allow the examination of hundreds of loci at one time—and **single nucleotide polymorphisms** (SNPs; see Luikart et al. 2003), another type of molecular marker, may enable more information to be obtained from genetic samples. In fact, SNPs may ultimately replace microsatellites, as they have the advantage of better conforming to well-characterized models of evolution and are more common throughout the genome (Aitken et al. 2004; Seddon et al. 2005). Furthermore, unlike microsatellites, which yield relative scores that require standards to be used for comparing results between laboratories,

SNPs are believed to provide data with absolute scores—thus facilitating collaboration between researchers studying the same species. To date, the expense of developing SNPs, and questions regarding error rates, ascertainment biases, their effectiveness with noninvasive samples, and within-population variability, have limited their use in conservation genetics (Morin et al. 2004). These issues will likely soon fade, however (Kohn et al. 2006; Morin and McCarthy 2007).

Rapid developments in the field of molecular ecology will continue to advance how noninvasive genetic sampling can be used to estimate abundance and occurrence. To maximize the utility of the approaches used, close collaboration between laboratories and field biologists must continue and improve. Field biologists should understand the limits of their data, while laboratory biologists must develop new tools with field applications in mind. Genetic sampling—although simple in principle—is actually complex in its execution, with attention to detail required from survey design through data analysis. Fortunately, there has been significant interest in this area, and the resulting research has demonstrated our ability to use noninvasive genetic sampling to monitor and study wild carnivore populations.

# Endocrine Approaches for Studying and Monitoring Carnivores

Yalow and Berson (1959) were awarded the Nobel Prize for developing the first **immunoassay** for assessing minute concentrations of hormones (i.e., $10^{-12}$ gm/ml) in blood circulation. These methods were initially adapted for wildlife to measure **steroid hormones** in urine from diverse primate species (Hodges et al. 1979). **Steroid hormone metabolites**, quantified "noninvasively" in excreta (urine or feces), permit wildlife biologists to study reproduction and stress physiology in individuals, populations, or species, without disturbing animals (Lasley and Kirkpatrick 1991; Monfort 2003).

Urinary (in nonhuman primates; Hodges et al. 1979; Andelman et al. 1985) and fecal (in domestic livestock; Bamberg et al. 1984; Möstl et al. 1984) steroid monitoring techniques, pioneered with captive animals, have been adapted to study diverse biological phenomena in free-ranging animals, including reproductive seasonality, gonadal status, pregnancy rates and age-specific fecundity, and the endocrine mechanisms controlling reproductive fitness in social mammals. Adrenal glucocorticoid metabolites (GCs) were first used as a proxy for "stress" in wildlife studies of bighorn sheep (*Ovis canadensis*; Miller et al. 1991). Urinary and fecal GC metabolites have since been used to evaluate physiological stress associated with social status, and the effects of environmental disturbance on animal well-being and fitness (Goymann et al. 1999; Creel et al. 2002; Sands and Creel 2004; Young et al. 2006).

## A Primer on Endocrine Tools: Measures for Assessing Reproduction and Stress

Noninvasive endocrine measures avert the physiological stress resulting from animal capture, restraint, and/or anesthesia, allowing normal underlying hormonal patterns to remain undisturbed. Additionally, unlike blood samples, which yield a single "point in time" measure of endocrine status, urine and feces provide an integrated measure of hormone production (i.e., from hours to days) due to the preexcretion "pooling" that occurs in the urinary bladder—or, in the case of feces, in the intestinal tract (Millspaugh and Washburn 2004). Pooled hormone measures dampen episodic secretory patterns that normally occur in blood circulation—a potential benefit when seeking to evaluate overall patterns of hormone production (Monfort 2003). Endocrine methods permit repeated longitudinal sampling in individuals, as well as sampling across populations to facilitate population-scale studies.

With the exception of estrogens (i.e., estradiol, estrone), little unmetabolized steroid (e.g., progesterone, testosterone, corticosterone) is excreted in urine and feces. This is because gonadal and adrenal steroids circulating in the bloodstream are metabolized in the liver and/or kidney before excretion into urine or bile (which is delivered to the gastrointestinal tract and eliminated via defecation), and rendered biologically impotent during metabolism through subtle molecular changes and/or through the attachment (conjugation) of highly charged side chain molecules that increase water solubility to facilitate excretion (Taylor 1971).

The proportion of metabolites derived from any given class of steroids (i.e., estrogens, androgens, progestagens, or corticosteroids) excreted in urine or feces is species- or taxon-specific, ranging, for example, from >90% fecal excretion in felid species (Brown et al. 1994) to >80% steroid excretion in baboon (*Papio cynocephalus cynocephalus*) urine (Wasser et al. 1994). To complicate matters further, some species excrete one class of steroid molecule into feces (e.g., progesterone metabolites), whereas another class (e.g., estrogen metabolites) may be excreted predominantly in urine (e.g., Sumatran rhinoceros [*Dicerorhinus sumatrensis*], Heistermann et al. 1998; African elephant [*Loxodonta Africana*], Wasser et al. 1996). In all species, there is a variable **excretion lag-time** between when steroid production and secretion occur in the bloodstream and byproducts appear in excreta, ranging from <12 hours (e.g., African wild dog [*Lycaon pictus*]; Monfort et al. 1997); 12–24 hours (e.g., scimitar-horned oryx [*Oryx dammah*], Morrow and Monfort 1998), and 24–48 hours in primates (e.g., baboons; Wasser et al. 1994) and colon or hindgut fermenters (e.g., elephants; Wasser et al. 1996).

Steroid metabolites can be quantified using a variety of immunoassays (e.g., radioimmunoassay, enzyme immunoassay, fluorescent immunoassay) techniques. In general, immunoassays that employ a broad spectrum or "group specific" antibody that cross-reacts with a host of similarly structured steroid metabolites within a given class of steroid (e.g., estrogens, progestagens, androgens, corticosteroids) are preferred for assessing steroid metabolites in excreta, and literally dozens of different commercially available or custom-made immunoassays have provided suitable results (Lasley and Kirkpatrick 1991; Brown et al. 1994; Wasser et al. 1994, 2000; Schwarzenberger et al. 1996; Monfort 2003; Young et al. 2004; Heistermann et al. 2006).

Regardless of the immunoassay technique employed, each assay must be validated for each new species or biological fluid to demonstrate that it yields reliable and consistent estimates of hormone production (Niswender et al. 1975; Reimers et al. 1981). It is particularly important to demonstrate that measured hormone concentrations provide physiologically relevant information. For example, an ovarian cycle might be validated by (1) comparing two independent measures of the same hormone in matched samples (i.e., fecal versus urinary estrogen); (2) comparing temporal hormone excretion patterns with external signs of reproductive status (e.g., sex skin swelling, copulatory or rutting behavior); (3) demonstrating cause-and-effect patterns, such as a rise and fall in hormone concentrations coincident with pregnancy onset and parturition, respectively; or (4) analyzing gonadal or adrenal responsiveness to a challenge with *gonadotropin releasing hormone* [GnRH] or *adrenocorticotrophic hormone* [ACTH], respectively (Monfort 2003).

## Endocrine Monitoring in Practice

Here we discuss two common applications for endocrine monitoring in the context of noninvasive wildlife research—those pertaining to reproduction and stress.

### Reproductive Life History Strategies and Pregnancy Determination

In the DNA section above, we discussed a hypothetical survey that used molecular markers to individually identify eight lynx. One question posed was whether or not this was a breeding population. Through the use of molecular markers, we were able to determine that there were three females in this population, but very little could be ascertained about their reproductive status. Here we can now turn to endocrine biology, assuming a proper validation of hormone levels has been conducted for the target species.

Endocrine monitoring has been extensively used to characterize reproductive status in free-ranging wildlife, including carnivores (Creel et al. 1992, 1995, 1997, 1998; Goymann et al. 1999, 2001; Clutton-Brock et al. 2001; Moss et al. 2001; Dloniak et al. 2004; von der Ohe et al. 2004; Wasser et al. 2004; Young et al. 2006), rodents (Billiti et al. 1998; Harper and Austad 2000; Touma et al. 2003), ungulates (Chaudhuri and Ginsberg 1990; Kirkpatrick et al. 1993; Foley et al. 2001; Pelletier et al. 2003), and especially primates (Wasser 1996; Hodges and Heistermann 2003; Ziegler and Wittwer 2005). In the case of primates, investigations have focused on establishing the basic interrelationships between hormones and behavior in both sexes (Brockman and Whitten 1996; Brockman et al. 1998; Strier and Ziegler 1997; Ziegler et al. 1997, 2000; Cavigelli 1999; Curtis et al. 2000; Herrick et al. 2000; Fujita et al. 2001; Lynch et al. 2002; French et al. 2003; Harris and Monfort 2003, 2005; Campbell 2004; Muller and Wrangham 2004a, b).

Furthermore, fecal progesterone metabolites have been used to assess pregnancy status in a variety of wildlife species (e.g., elk, Garrott et al. 1998; Stoops et al. 1999; moose [*Alces alces*], Monfort et al. 1993; Schwartz et al. 1995; Berger et al. 1999; horses [*Equus caballus*], Kirkpatrick et al. 1991b; bison [*Bison bison*], Kirkpatrick et al. 1993, 1996; bighorn sheep, Schoenecker et al. 2004; black rhinoceros [*Diceros bicornis*], Garnier et al. 1998); and meerkats [*Suricata suricatta*], Moss et al. 2001; Young et al. 2006). For example, fecal progesterone metabolites (Berger et al. 1999) were assessed to determine whether a reduction in moose numbers was the result of decreased fecundity associated with habitat degradation or of increased neonate predation by reintroduced wolves and grizzly bears in the Greater Yellowstone Ecosystem. Hormones confirmed that pregnancy rates were among the lowest of any moose population in North America, enabling biologists to conclude that neonate predation could not explain the observed decline in moose populations.

## Stress Related to Social, Environmental, and Human Disturbance Factors

Urinary and fecal GC metabolites are being used to evaluate adrenal status in a growing number of wildlife species (Goymann et al. 1999; Wasser et al. 2000; Hunt and Wasser 2003; Millspaugh et al. 2003; Monfort 2003; Harper and Austad 2004; Kuznetsov et al. 2004; Cavigelli et al. 2005; Gould et al. 2005; Mateo and Cavigelli 2005; Heistermann et al. 2006; Young et al. 2006). This application is based on the premise that stress hormone production—mediated by neural and psychosocial inputs from higher brain centers that stimulate the production and secretion of GCs from the adrenal cortex (Moberg 1985; McEwen 1998, 2000)—can be approximated by assessing GC metabolites in the excreta of free-ranging wildlife (Wasser et al. 2000). Glucocorticoid excretion patterns have been especially useful for evaluating the interrelationships between hormonal measures, dominance rank, age, genetic relatedness, reproductive status, and rates of aggression among members of each social group (Creel et al. 1992, 1996, 1997; Muller and Wrangham 2004a, b; Sands and Creel 2004; Creel 2005). In one study, it was shown that dominant African wild dogs (Creel et al. 1996, 1997) excreted elevated GC concentrations compared to subordinate pack members. Because dominant wild dogs were involved in more aggressive interactions—presumably to maintain dominance—it was hypothesized that stress may be a cost of social dominance (Creel et al. 1996).

Additionally, GC excretion has been used to investigate the impact of human disturbance on wildlife, including, for example, the influence of logging activities on Northern spotted owls (*Strix occidentalis caurina*; Wasser et al. 1997; Tempel and Gutierrez 2003, 2004); the effect of snowmobiling activity on wolves and elk (Creel et al. 2002); the stress of radio-collaring wild dogs (Creel et al. 1997); and the stress physiology of prerelease conditioning and reintroduction on whooping cranes (*Grus americana*; Hartup et al. 2005). These studies show the promise of GC assessments for studying the relationships between environmental stressors and animal well-being. It is critical to emphasize, however, that there is a serious risk of oversimplifying and/or overinterpreting GC data by suggesting that elevations in fecal GCs, in and of themselves, signal physiological or psycho-social "stress." In some cases, elevated or reduced GCs may be completely normal, reflecting changes in metabolic or nutritional status, reproductive life history, pregnancy, seasonality, prehibernatory preparations, and myriad other adaptive physiological states (von der Ohe and Servheen 2002). Thus, elevated GCs alone are not necessarily an index of stress, and these measures should be assessed in concert with other prospective indicators—activity patterns, behaviors, body condition, disease status, immunocompetence—the sum total of which may signal that an animal is experiencing physiological or psycho-social stress.

## Types of Samples for Endocrine Evaluations

Various types of samples can be used to conduct endocrine evaluations. Here we discuss urine and fecal samples.

### Urine Samples for Endocrine Evaluations

Small amounts of urine can be rapidly absorbed into natural substrates, making this a challenging medium for the wildlife biologist to use in conducting endocrine evaluations. Nevertheless, urinary gonadal (indicators of reproduction) and adrenal steroid (putative indicators of stress) metabolites have been evaluated in a diversity of free-ranging primate, carnivore, and ungulate species (see review, Monfort 2003). For example, hormones have been assessed in urine collected directly off the ground (elephants, Poole et al. 1984; Przewalski's horses [*Equus przewalskii*], Monfort et al. 1990; bison, Kirkpatrick et al. 1991a; gorillas [*Gorilla gorilla*], Robbins and Czekala 1997); from urine-soaked snow (feral horses, Kirkpatrick et al. 1991b) and Kalahari sand (meerkats, Clutton-Brock et al. 2001; Moss et al. 2001); by positioning an observer under arboreal

primates until they urinated onto a piece of aluminum foil attached to the end or stick (Harris and Monfort 2003, 2005); and even from rubber 'flip-flop' sandals that were urinated on as part of gregarious scent-marking behavior (dwarf mongooses [*Helogale parvula*], Creel et al. 1992, 1995).

A major advantage of urinary hormone monitoring versus fecal steroid monitoring is that samples can be assayed without further processing, thus minimizing labor costs. Another advantage is that excretion lag-times are generally short relative to hormone secretion in blood circulation, and day-to-day fluctuations in fluid balance can be easily calibrated to creatinine excretion (Taussky 1954). In summary, urine samples are generally difficult to collect under field conditions, but the tradeoff is that laboratory procedures are simplified and relatively inexpensive.

### Fecal Samples for Endocrine Evaluations

Reproductive and adrenal steroids have been assessed in feces collected from a range of free-living mammal and bird species (Monfort 2003). Feces are generally easy to collect under field conditions, which is one of the main reasons that this approach has become so popular over the past decade. Furthermore, new scat detection dog methods increase the likelihood of locating feces in the field (see chapter 7). But feces contain large numbers of bacteria that produce enzymes that can alter the structural integrity of steroid metabolites postdefecation (Millspaugh and Washburn 2004). To minimize these postdefecation impacts, feces should be collected as soon as possible after defecation, followed immediately by treatment to minimize continued **bacterial degradation**. A disadvantage of fecal steroid monitoring, relative to urinary methods, is the need for extensive processing before immunoassay. The associated increase in time, labor, and overall cost depends, in part, on the fecal extraction method employed, including whether samples are extracted wet or dry, procedural losses are documented, and heat is used during the extraction process (Monfort 2003).

## Sample Treatment and Preservation

Once collected, urine can be preserved indefinitely through frozen storage (e.g., household or propane freezer, liquid nitrogen tank) or with "field-friendly" storage methods, including absorbing urine onto filter paper (Harris and Monfort 2003) or storing it in 10% ethanol at room temperature for up to twelve weeks (Whitten et al. 1998). In contrast, if feces cannot be collected and preserved within one to two hours postdefecation, one needs to systematically evaluate the potential impact of bacterial degradation on hormone metabolite concentrations, and develop a sample storage strategy to mitigate this impact (Whitten et al. 1998; Khan et al. 2002; Terio et al. 2002; Washburn and Millspaugh 2002; Hunt and Wasser 2003; Lynch et al. 2003; Millspaugh et al. 2003; Beehner and Whitten 2004; Galama et al. 2004; Palme 2005; Palme et al. 2005). This effect must be further evaluated if the samples are treated to kill potential pathogens, as required by the US Department of Agriculture. For instance, Millspaugh et al. (2003) assessed the effect of chemical and heat treatment on glucocorticoid concentrations from fresh and frozen white-tailed deer (*Odocoileus virginianus*) and elk fecal samples stored for six days. These researchers found that fecal glucocorticoid concentrations were significantly altered by the chemical and heat treatments, although treatment in a 2% acetic acid solution followed by freezing had the least impact on the sample.

Frozen storage is generally considered the gold standard of fecal preservation methods (Hunt and Wasser 2003), but fecal steroid metabolite concentrations may vary over time even in frozen specimens (Khan et al. 2002). Adopting standardized collection and storage methods is a crucial consideration when designing any field study given that inappropriate sample storage can invalidate hormone measures. In general, it is prudent to collect feces as soon after defecation as possible (i.e., again, within one to two hours), and to maintain the specimens in a portable insulated container cooled with frozen ice packs during transfer to the field station, where samples can be frozen, treated, or processed in preparation for analysis. Remote field sites without freezers present a special challenge. In such situations, prospective research is essential to validate alternative sample preservation methods such as drying in portable ovens (Stoops et al. 1999; Lynch et al. 2003; Galama et al. 2004; Pettitt et al. 2007) or fixing samples in alcohol or other chemical-bactericidal media (Wasser et al. 1994; Khan et al. 2002; Millspaugh et al. 2003). Alternately, other field-friendly extraction methods (Whitten et al. 1998; Beehner and Whitten 2004) may be available.

Simply put, investigators should presume that each species is potentially unique with respect to how reproductive and adrenal steroid hormones are metabolized, excreted, and degraded postdefecation. Descriptions of the myriad tests or experimental designs that might be employed to validate fecal sampling and storage procedures is beyond the scope of this chapter, but the burden is squarely on the investigator to demonstrate that sampling regimens employed—including postdefecation metabolization—have been tested to ensure the physiological validity of the resulting endocrine data. Endocrine data derived without controlling for the elapsed time from defecation to fecal storage, and for which the impact of storage method and the duration of fecal sample storage on hormonal measures has not been tested, should be interpreted with extreme caution. A useful option for validating fecal sample collection and storage methods is to use freshly collected feces from captive subjects maintained in zoos or wildlife centers to conduct controlled experiments.

For international field studies, specimen exportation can be avoided completely if sample processing and immunoassays are conducted in the countries where specimens are collected. This is now feasible given the advent of nonradiometric immunoassays, and even noninstrumented immunoassays (Kirkpatrick et al. 1993), which are increasingly portable and transferable to remote field sites.

## Pitfalls, Cautions, and Future Perspectives

As Cervantes said, "A word to the wise is enough." This applies to ensuring that the necessary validations are conducted for each hormone assay and species of interest. Failure to do so can completely invalidate the usefulness of endocrine measures. For example, accurate pregnancy diagnosis requires that steroid measures be initially confirmed independently using alternate methods (e.g., rectal palpation, ultrasonography, pregnancy-specific proteins), as well as direct visual confirmation of neonate status. Additionally, a priori knowledge of normal endocrine excretion dynamics is essential for determining the optimal time for sampling (i.e., early- versus mid- or late-pregnancy), and the degree of individual, seasonal, and age-related effects on fecal hormone production.

Likewise, hormone tests of adrenal status require that one demonstrate a cause-and-effect relationship between a stressor and its ability to induce an associated temporal increase in excreted GC concentrations (Wasser et al. 2000; Millspaugh et al. 2003). Special caution is necessary for assessing GCs in the context of other potentially relevant biological factors, such as seasonal and diurnal rhythms; body condition; nutritional status; and the social or reproductive status, age, and sex of the animal being sampled (von der Ohe and Serveen 2002; Millspaugh and Washburn 2004; Palme 2005; Touma and Palme 2005). Further, treatment effects (e.g., human-induced stress) must be considered against the background variation from these potentially confounding factors to demonstrate that measured endocrine changes provide physiologically relevant information.

Wildlife biologists should be aware that no two immunoassays are created equal: each employs a unique antibody with a characteristic specificity or ability to recognize minute, three-dimensional structural differences among similar classes of hormones. Thus, two different assays for the same hormone (e.g., progesterone)—sold by different manufacturers or developed by separate labs—may not necessarily recognize the same exact hormone metabolite. This is because downstream hormone metabolites in excreta are diverse, and each antibody may recognize one or more metabolites that are unique for that particular assay. Results from two different labs and/or assays for the same hormone may therefore not be directly comparable, which reinforces the need for the validations emphasized above. Additionally, the storage or processing method used (e.g., ethanol boiling versus cold buffer solubilization for feces) may affect the diversity and overall concentration of metabolites quantified by any particular assay. In short, it is essential to carefully evaluate the endocrine methods employed, even when using extraction procedures or immunoassays that have been previously reported to be effective for documenting steroid metabolites in a closely related species.

Urinary and fecal steroid monitoring have now been employed in dozens of species of mammals (Monfort 2003; Palme et al. 2005) and birds (Goymann 2005; Touma and Palme 2005), although fewer studies have been conducted under field conditions (Monfort 2003). Despite progress, we still know strikingly little about the reproductive biology and stress physiology of most species of mammals and birds. Excellent new examples of field applications are being published each year, reflecting this rapidly emerging field of investigation. It is becoming increasingly common for field investigations to merge behavioral, genetic, and hormone measures to define life history requirements of individuals, populations, and species, as well as complex ecological relationships and the evolution of mating systems— including phenomena such as dominance, social stress, and reproductive suppression. Such methods are a boon to enhancing our fundamental physiological knowledge of reproductive status, health, and the effect of human disturbance on animal well-being. Increasingly, these methods are proving helpful to wildlife managers and decision makers in their attempts to ensure the survival of viable populations of carnivores and other species.

## Synergy of Genetic and Endocrine Data

Advances in both molecular biology and endocrinology, coupled with noninvasive sampling, are providing new insights into the population dynamics of many species. From a single fecal sample we can now identify not only the species, but the individual and its sex, population of origin, reproductive status, and social status by amalgamating molecular and endocrine approaches. Yet, to obtain this synergy, appropriate pilot work must be conducted and samples must be treated in a manner acceptable to both disciplines.

Pilot molecular ecology work will provide information about the power of the molecular markers being used, the efficacy of the storage and extraction techniques, and the success rate of sampling from the specific species and location. Given species- and location-specific variation, results of pilot work should then guide future sampling efforts. In the field of endocrinology, pilot work is critical to providing baseline data and to ensuring that changes in assayed hormone levels are biologically meaningful and provide reliable and consistent estimates of hormone production. Such validation is often achieved by independently confirming that physiological and behavioral changes induce predictable changes in steroid levels. In addition, pilot work can determine the optimal timing for sampling, and the degree of individual, seasonal, and age-related effects on hormone production.

If one hopes to obtain both DNA-based (e.g., species, individual identification, sex, population genetic) and endocrine-based (e.g., social, physiological, reproductive status) information from the same sample, a priori planning is absolutely critical. Sampling hair, while effective for molecular genetics, is not as useful for endocrinology. Urine has proven effective and inexpensive for endocrine studies and can provide molecular genetic information as well—but probably at a diminished success rate when compared to other source material. The cost of failed DNA samples must be balanced against the savings generated in the endocrinology laboratory. Fecal samples may be the best medium for obtaining both DNA and adrenal or gonadal steroids. Freezing samples is ideal for endocrine work and is adequate for many DNA studies (although this approach never ranked highest in table 9.3 and is known to present difficulties for obtaining sufficient DNA from some species). If only one sample can be acquired, freezing should enable it to provide maximal information. Alternatively, we recommend dividing a freshly collected fecal sample into two subsamples for best results. The first subsample should be immediately frozen and sent to the endocrine lab, and the second freeze dried or placed in silica desiccant, buffer, or ethanol (depending on results from preliminary studies) and sent to the DNA lab. The use of a barcode system with multiple labels can help researchers track identical samples shipped to different labs (see box 9.3 and figure 6.16 for more details).

Given pilot studies and careful sample collection, treatment, and storage, there is much information to be gained from noninvasively collected samples toward the study of carnivore abundance, reproductive and social status, occupancy, and geographic range. Furthermore, as both molecular genetics and endocrinology are rapidly advancing fields, we expect to see new developments that will allow even more information to be obtained from noninvasive sampling.

## Acknowledgments

We would like to thank Paula MacKay, Robert Long, Justina Ray, Bill Zielinski, Lisette Waits, and an anonymous reviewer for comments on earlier drafts. We also thank Kristy Pilgrim and Cory Engkjer for contributing ideas to the DNA portion of the chapter. Mike Schwartz was supported by a Presidential Early Career Award for Scientists and Engineers and the USDA Forest Service Rocky Mountain Research Station during the writing of this manuscript.

# Appendix 9.1

*Glossary of select genetic and endocrine terms*

**allele.** Form or variable DNA coding of a gene or nongene sequence.

**allelic dropout.** Failure of an allele to be amplified during the PCR process due to low-quality or a low quantity of DNA.

**bacterial degradation.** Alteration of the structure of steroid hormones or their metabolites as a result of enzymatic activity exerted by fecal bacteria during the postdefecation interval, resulting in quantitative differences in immunoreactive hormone measures over time.

**control region.** Segment of the mitochondrial genome, approximately 1,000 base pairs in length, which initiates DNA replication and transcription. It has been found to be highly variable in many organisms and is often used to build phylogenetic trees.

**cytochrome.** Gene of the mitochondrial genome used for electron transport or to catalyze redox reactions. These genes are commonly used to build phylogenetic trees.

**dNTPs.** Free T, A, C, and Gs (deoxyribonucleotides) used in a PCR reaction to build new strands of DNA.

**excretion lag-time.** Lag-time from steroid production and secretion in the blood stream to the appearance of their metabolites in urine (generally <12 hours) or feces (12–48 hours).

**false allele.** PCR error that leads to the appearance of an allele when, in fact, one does not exist.

**genotype.** Combination of alleles that makes up an individual's genetic profile.

**heterozygous.** State of carrying different alleles of the same gene. In diploid individuals (e.g., most mammals and all carnivores), this is the state of having two alleles at a locus—one on each of the corresponding chromosomes.

**homozygous.** State of carrying the same alleles of the same gene.

**hydrolysis.** Chemical process in which DNA is cleaved into sections by the addition of a water molecule.

**immunoassay.** Immunological techniques, such as radioimmunoassays (RIA) and enzyme immunoassays (EIA), are sensitive laboratory methods used to quantify molecules (e.g., steroid hormones) in biological fluids.

**locus** (plural: **loci**). Segment of DNA at a particular place in the genome.

**marker.** Segment of DNA with a known location on a chromosome. Often used interchangeably with *locus* and *primer*, although they all have subtle differences in meaning.

**microsatellite.** Short blocks of tandem DNA repeated many times within the genome. Due to the cell's inefficiency in copying tandem repeats, mutations are frequent at microsatellite sites, producing genetic variability. Also known as *SSTRs*.

**mitochondrial DNA.** DNA found in the mitochondria of a cell. The mitochondrial genome is circular, and in humans contains thirty-seven genes plus a control region. Mitochondrial DNA is unique in its inheritance pattern as it is only passed on to offspring from mothers.

**neutral.** Genes or loci that are believed to be "junk DNA" and not subject to selection. Neutral DNA tends to be highly variable and useful for many population genetic analyses.

**nuclear DNA.** DNA from the nucleus of a cell.

**PCR** or **polymerase chain reaction.** A technique used to amplify (increase) the number of copies of a particular segment of DNA for the purpose of being able to visualize this area (locus). Critical components of the PCR include *Taq* polymerase, **dNTPs,** and primers, as well as coenzymes and buffers.

**Primer.** Loosely synonymous with *marker* or *probe.* Short, single-stranded segment of DNA which is used to "prime" a *PCR* reaction. In microsatellite analyses, a primer pair is used to "bookend" the area of interest. These primers are often labeled with a fluorescent dye for visualization purposes.

**probability of identity.** Probability that two randomly chosen individuals in a population have identical genotypes

**probe.** See *primer.*

**single nucleotide polymorphism** (SNP). Point mutations in the DNA sequence of an individual. This form of genetic variation holds promise to be useful in future wildlife and conservation studies.

**sequence.** Process to determine the nucleotide composition of DNA (verb) or the composition itself (noun).

**steroid hormone.** Hormones synthesized mainly by endocrine glands, such as the testis (e.g., testos-terone), ovary (e.g., estrogen, progesterone), the adrenals (e.g., cortisol, corticosterone) and, during gestation, by the fetoplacental unit (e.g., estrogens, progesterone). Once released into blood circulation, steroids exert their impact in target tissues.

**steroid hormone metabolites.** Circulating steroid hormones are extensively metabolized by the liver and kidney before excretion as urinary or fecal steroid metabolites or byproducts. The proportion of steroid excreted in urine or feces is usually species- or taxon-specific. Adrenal steroid metabolites are often termed *corticoids* or *glucocorticoids;* testosterone metabolites are generically referred to as *androgen metabolites;* and ovarian steroid metabolites are typically referred to as *estrogens, progestins, progestagens,* or *progesterone metabolites.*

***Taq* polymerase.** A polymerase is an enzyme whose function is to catalyze the production of new DNA from an existing DNA template. *Taq* polymerase is a thermostable polymerase central to PCR, as it does not degrade during the reaction.

# Chapter 10

# Attracting Animals to Detection Devices

*Fredrick V. Schlexer*

For wildlife research purposes, an attractant is any substance, material, device, or technique used to attract a target species. Attractants are used with most of the survey methods described in this book, excluding natural sign surveys (chapter 3), some track stations (chapter 4), remote cameras on trails (chapter 5), hair collection from natural rub objects or along travel routes (chapter 6), and scat detection dogs (chapter 7). Indeed, the selection of an attractant is often an integral part of the survey-planning process. This chapter describes the various substances and methods used to draw North American carnivores to noninvasive sampling devices—from historical, scientific, and traditional perspectives. Further, it provides practical recommendations on how to acquire, apply, and store baits, lures, and other attractants and describes scientific efforts to test their efficacy.

Although the terms *bait* and *lure* are often used interchangeably, each has a unique meaning in the context of surveying wildlife:

- *Bait* is a food item or other substance that attracts an animal by appealing to its sense of taste and smell. Baits are typically intended to be consumed by the target species, although nonreward baits (discussed later in the chapter) may preclude consumption.
- *Lures* include scent lures, visual lures, and sound lures. A *scent lure* is any substance that draws animals closer via their sense of smell. *Visual lures* engage an animal's sense of sight, while *sound lures* elicit a curiosity approach by simulating noises made by prey species or conspecifics.
- *Natural attractants* are objects in the existing environment (e.g., trees, snags, or latrine sites) that are regularly used by target animals as part of their behavioral repertoire.

## Background

Over thousands of years, humans developed various trapping methods to capture animals for food and hides, and to protect themselves and their property from predators. Through trial and error, trap effectiveness was increased by the refinement of methods to entice animals into traps. Many historical fur trappers had their own "secret formula" for attracting target species, and were reluctant to share the lists of ingredients with others because of competition and the potential loss of income (Geary 1984).

As a result, multiple baits and scent lures were developed for each furbearing species.

This traditional knowledge base—accumulated from the combined experience of indigenous peoples, hunters, trappers, and naturalists—has been incorporated into modern efforts to attract animals for wildlife research. Unfortunately, most attractants have not been scientifically tested and are used on the basis of tradition rather than proven effectiveness. Numerous researchers have endeavored to evaluate and standardize traditional attractants (e.g., Graves and Boddicker 1987; McDaniel et al. 2000; Stanley and Royle 2005), but the predominant reliance on unverified methods to draw animals to survey devices underscores the need for additional and rigorous scientific testing (see *Evaluating the Effectiveness of Baits and Lures* later in this chapter).

The use of attractants in carnivore surveys has a long history (e.g., Cook 1949; Wood 1959). Early attempts to evaluate attractants were directed at the development of a reliable method to estimate coyote (*Canis latrans*) abundance using scented track stations (see chapter 4). Natural scent lures were tested with captive animals (Roughton 1979) and in the field (Linhart and Knowlton 1975; Linhart et al. 1977; Roughton and Bowden 1979), and efforts were soon expanded to include synthetic scents (Turkowski et al. 1979; Martin and Fagre 1988). A synthetic fatty acid scent (FAS) was ultimately selected as a standard lure for coyotes by the US Fish and Wildlife Service (Roughton 1982), and a standardized delivery method was developed in the form of an inexpensive plaster disk saturated with this scent (Roughton and Sweeny 1982). FAS continues to be used today, primarily for canid and felid scent station surveys (e.g., Harris and Knowlton 2001; Zoellick et al. 2004). More often, however, researchers employ commercially available scent lures (e.g., Caven's Gusto, Carman's MegaMusk) for noninvasive carnivore surveys (Romain-Bondi et al. 2004; Zielinski et al. 2005; Gompper et al. 2006; also see appendix 10.1). Although many such lures are created based on traditional recipes—and at least some yield positive results—most have not been rigorously tested (see *Evaluating the Effectiveness of Baits and Lures*).

Valuable information about attractants can be found in unpublished reports produced by fish and wildlife agencies at the national, provincial, state, and local levels. Private wildlife groups (e.g., World Wildlife Fund, Wildlife Conservation Society) are also rich sources of relevant research. Many of these unpublished reports can be accessed via the internet (e.g., Henschel and Ray 2003; Uresk et al. 2003; Kendall et al. 2004). Traditional attractants are further discussed in furbearer trapping "how-to" books (e.g., Carman 1975; Wyshinski 2001) and popular outdoor magazines (e.g., *Fish and Fur, Field and Stream, Outdoor Life*). Last, trapping supply distributors usually include information on attractants both in print catalogs and on their websites (see appendix 10.2).

## Description of Attractants

This section describes various types of attractants that can be used individually or in combinations. A list of recommended attractants for each target species or group is presented in table 10.1.

### Baits

Baits are typically composed of food, and fall into several general categories, including both natural dietary items and less customary consumables. Fresh or decomposed meat, poultry, and fish are often used as bait, as are canned fish and canned or dried pet foods. Live animals are also occasionally deployed as bait or lures (e.g., Zezulak and Schwab 1979; Caso 1994; Dillon 2005); researchers wishing to use live animals should follow Institutional Animal Care and Use Committee guidelines (ACUC 1998; IACUC 2006). Some carnivores respond to nonmeat baits such as fruits or vegetables, fruit jams, seeds and nuts, baked goods, and cheese (table 10.1; appendices 10.1, 10.3). It is also possible to combine several types of bait at a single detection device to

*Table 10.1.* Recommended attractants for carnivore surveys, in order of preference, listed by target species or group

| Species or group | Baits | Scent lures | Visual lures | Sound lures |
|---|---|---|---|---|
| **Canids (except foxes)** | Raw chicken (pieces or whole)<br>Meat or whole carcasses[c]<br>Fish (whole or canned) | FAS[a]<br>Canid glands or urine<br>Catnip[d] oil<br>Raw wool<br>Commercial lures[e, f, g, h] | | VR[b] |
| **Foxes**<br>(except arctic fox) | Raw chicken (pieces or whole)<br>Fish (whole or canned)<br>Dog or cat food (dry or canned)<br>Meat or whole carcasses[c]<br>Nuts, raisins, other fruits | FAS[a]<br>Fox glands or urine<br>Catnip[d] oil<br>Commercial lures[e, f, g, h, i] | | |
| Tropical felids[j] | Live animals[k]<br>Fish (whole or canned)<br>Raw chicken (pieces or whole) | Felid glands or urine<br>Commercial lures[f, g, l, m]<br>FAS[a]<br>Catnip[d] (oil, dried, or fresh) | Flashers | VR[b] |
| Temperate felids[n] | Meat or whole carcasses[c]<br>Fish (whole or canned)<br>Raw chicken (pieces or whole) | Catnip[d] (oil, dried, or fresh)<br>Commercial lures[f, g, l, m]<br>Beaver castoreum<br>FAS[a]<br>Felid glands or urine | Flashers | VR[b] |
| **Mephitids** | Raw chicken (pieces or whole)<br>Fish (whole or canned)<br>Rabbit or beaver meat<br>Chicken eggs | Commercial skunk-scented lure[o]<br>FAS[a]<br>Fish oil | | |
| **Mustelids**<br>  Wolverine | Meat or whole carcasses[c]<br>Fish (whole or canned)<br>Raw chicken (pieces or whole)<br>Rotten meat | Commercial skunk-scented lure[o]<br>Fish oil<br>Beaver castoreum | Flashers | |
|   North American<br>  river otter | Fresh, whole fish | | | |
|   American marten,<br>  fisher, weasels<br>  (*Mustela* spp.) | Raw chicken (pieces or whole)<br>Fish (whole or canned)<br>Rabbit or beaver meat | Commercial skunk-scented lure[o]<br>Fish oil | Flashers | |
|   American mink | Fresh, whole fish<br>Fresh meat[p] (rabbit, beaver, muskrat, birds) | Mink glands and urine<br>Fish oil | | |
|   American badger | Raw chicken (pieces or whole)<br>Fresh meat (rabbit, beaver, muskrat, birds) | Commercial skunk-scented lure[o] | | |
| Procyonids<br>  Ringtail | Raw chicken (pieces or whole)<br>Dog or cat food (dry or canned)<br>Fish (whole or canned)<br>Rabbit or beaver meat<br>Fruit jam | Commercial skunk-scented lure[o]<br>Ringtail glands or urine<br>FAS[a]<br>Fish oil | | |
|   White-nosed coati | Dog or cat food (dry or canned)<br>Fish (whole or canned)<br>Live animals[k]<br>Marshmallows | FAS[a]<br>Fish oil<br>Commercial lure[g] | | |

*Table 10.1.* (Continued)

| Species or group | Baits | Scent lures | Visual lures | Sound lures |
|---|---|---|---|---|
| Raccoon | Raw chicken (pieces or whole) | Commercial skunk-scented lure[o] | | |
| | Fish (whole or canned) | Fish oil | | |
| | Dog or cat food (dry or canned) | FAS[a] | | |
| | Rabbit or beaver meat | | | |
| | Fruit jam | | | |
| Ursids | Raw chicken (pieces or whole) | Commercial skunk-scented lure[o] | | |
| | Fish (whole or canned) | Liquid fish fertilizer | | |
| | Meat or whole carcasses[c] | Fish oil | | |
| | Fish food pellets | Anise oil or vanilla extract | | |
| | Molasses, maple syrup, or honey (diluted with water) | | | |
| | Livestock blood | | | |
| | Fruit jam | | | |
| | Fruits and vegetables (apples, corn) | | | |
| | Stale pastries (e.g., bagels, donuts, cookies) | | | |
| | Rotten meat | | | |

Note: Attractants were selected based on a synthesis of those used in the surveys included in appendix 10.1, and on the author's experience and professional opinion. Actual attractant(s) chosen should depend on survey goals, season, and availability.
[a]Synthetic fatty-acid tablets.
[b]Vocalization recording.
[c]E.g., wild ungulate, domestic livestock, beaver.
[d]*Nepeta cataria.*
[e]E.g., any commercial liquid fox lure, liquid coyote lure, or fox gland lure, such as Caven's Fox #1, Caven's Fox #2, Caven's Canine Force.
[f]E.g., Marak's Bobcat Lure, Marak's Coyote Lure, Marak's Gray Fox Lure, Marak's Raccoon Lure.
[g]E.g., Carman's Canine Call, Pro's Choice, Bobcat Gland Lure, Trophy Deer Lure, and Mega Musk.
[h]E.g., Carman's Canine Call.
[i]E.g., Trailing Scent.
[j]*Leopardus* spp., *Puma yagouaroundi, Panthera onca.*
[k]E.g., chickens or chicks, rabbits, quail, pigeons.
[l]E.g., Hawbaker's Wildcat #2.
[m]E.g., Weaver's Cat Call.
[n]*Lynx* spp., *Puma concolor.*
[o]E.g., Caven's Gusto.
[p]Do not use rotted fish or meat.

increase the probability of detecting a given species or to attract multiple species (see *Target Species*). Proprietary commercial baits are available, but their superiority to commonly available meat or fish baits has not been demonstrated.

The attraction capabilities of meat or fish bait decline over time due to decomposition. At high concentrations, the wide variety of amines and sulfur compounds characteristic of microbial activity serves as a cue to the target animal, allowing it to identify a piece of meat as rotten and inedible

(Janzen 1977). At lower concentrations, however, these same compounds signal the presence of edible bait (Stager 1964). Thus, the products of decay are both attractive and repulsive, depending on their concentration. The optimal condition of bait (a function of detectability and desirability) is reached when the carcass is odorous enough to be detected at a distance, but not so rotten as to discourage investigation. Because carnivores possess a more sensitive olfactory system than do humans, and are thus able to detect odors at lower concentrations (Hepper and

Wells 2005), it is impossible for researchers to accurately assess where a given bait falls along the attraction-repulsion scale for a given target species.

Bait deployed such that it can be consumed by the target species is considered a reward bait. This type of bait presentation can limit sampling to the first animal that reaches the site—a potentially desirable outcome in some instances (e.g., if genetic methods can only utilize samples collected from one individual at a time; see chapter 6)—but may contribute to repeated sampling of the same individual if the bait is regularly replaced and the animal becomes habituated to obtaining food (Brongo et al. 2005). Inaccessible or nonreward baits alleviate this problem and will continue to draw additional individuals to the site until the bait becomes unattractive. Nonreward baits also serve well as scent lures.

Carnivores usually respond best to baits comprising potential prey species (Schemnitz 1996; Cypher and Spencer 1998; Kamler et al. 2002). Ethical considerations and animal care and use protocols prohibit the harvest of prey animals for baiting purposes (Powell and Proulx 2003), but effective substitutes (e.g., commercially available meat and fish) are widely available (appendix 10.1).

## Scent Lures

Scent lures (also known as long-distance lures or call lures) exploit an animal's hunger or curiosity or convey social or territorial signals. Scent lures are available in a variety of forms (e.g., solid, viscous, liquid, granulated, or powdered), and can be animal-based, vegetable/fruit-based, inorganic, or synthetic. For many carnivore species, attraction to a survey location may be maximized by using scent lures in combination with bait (Kucera et al. 1995a; Zielinski 1995). Further, some baits, such as rotten meat or fish, can effectively serve as scent lures because they release volatile compounds (Bullard 1982).

Scent lures sometimes contain plants or plant extracts, such as catnip (*Nepeta cataria*), for example (McDaniel et al. 2000; Weaver et al. 2005). Fresh or dried catnip attracts a variety of carnivores (appendices 10.1, 10.3) but is primarily used for felids (Tucker and Tucker 1988). Other ingredients used in traditional scent lure manufacture include fixatives (i.e., stabilizing agents), essential oils, and seafood essences (appendix 10.3).

Commercial scent lures are proprietary mixtures of animal blood, organs, urine, glands or other items (some trappers even add small amounts of cheap perfume to their mixtures [Schemnitz 1996]), often fermented for weeks or months. Lures may include scents from prey or nonprey species, such as American beaver (*Castor canadensis*) castoreum and muskrat (*Ondatra zibethicus*) scent glands. Every trapper or animal damage control agent has a favorite lure, and these lures work with varying degrees of success (Baker and Dwyer 1987; Graves and Boddicker 1987; Dobbins 2004). Although many lure manufacturers advertise "proven results" or that their lures have been "trapline tested," details of such tests are usually unavailable. Several commercial lures have been scientifically evaluated (e.g., Martin and Fagre 1988; Stapper et al. 1992), and a few brands have consistently been used in carnivore surveys (appendix 10.1).

Species-specific scent lures that stimulate social or territorial responses usually include urine, musk, and/or macerated scent glands from the target species (Wyshinski 2001; Dobbins 2004). These lures are often called matrix lures by trappers (Hanson 1989). Although many proprietary lure mixtures are derived from such substances, the basic ingredients can also be acquired from trapping supply distributors or from zoos and game ranches.

Most scent lures are combined with a base material or an extending medium that assists in distributing the scent and acts as an antifreeze, diluent, evaporative retardant, additional attractant, or preservative (table 10.2). Examples include lanolin, which allows a concentrated lure to be easily spread over multiple sites, and molasses, which supplements the attractant qualities of the lure. Blood lures require the use of an anticoagulant (e.g., sodium cit-

*Table 10.2.* Common scent lure bases and their uses

| Base | Use |
|---|---|
| Glycerine | Antifreeze, evaporative retardant, preservative |
| Honey | Antifreeze, attractant, diluent, evaporative retardant |
| Lanolin (anhydrous) | Antifreeze, evaporative retardant |
| Molasses | Antifreeze, attractant, diluent, evaporative retardant |
| Propylene glycol[a] | Antifreeze, preservative |
| Sodium benzoate | Preservative, antifungal |
| Sodium citrate[b] | Anticoagulant for blood |
| Tallow fat | Antifreeze, attractant, evaporative retardant |
| Vegetable oil or shortening | Antifreeze, evaporative retardant |
| Zinc valerate | Preservative |

[a]Similar to glycerine but not as viscous.
[b]Use a solution of 1:7 sodium citrate to water in a 1:9 ratio of anti-coagulant to blood.

rate) to be effective. Due to their physical character (e.g., liquid, powder), most scent lures must be deployed using absorbent materials or containers (table 10.2; see *Deployment of Attractants*).

## Visual Lures

Commercial trappers and wildlife researchers frequently use visual lures (collectively known as flashers or flags; Young 1958; Geary 1984; Baker and Dwyer 1987), sometimes in concert with scent lures or baits. Flashers typically consist of a lightweight object—for example, a piece of aluminum foil or a pie pan (figure 10.1A), a whole dried bird wing or a large feather (figure 10.1B), a patch of fur, a piece of light-colored cloth, or an old cassette tape or compact disk (figure 10.1C)—suspended above the detection device with string or fishing line, and in some cases a swivel (figure 10.1D). In a slight variation, an opaque piece of cloth or burlap hung across the front of track plate stations has been shown to attract mustelids and raccoons (*Procyon lotor*; Loukmas et al. 2003). Flashers are generally designed to flutter or move in a breeze, and are effective at attracting the attention of numerous carnivore species (Zielinski 1995). Visual lures are most commonly used with felids (Mowat et al. 1999, Weaver et al. 2005), which are more responsive to visual stimuli than to scents (Kitchener 1991). In areas where dense vegetation limits visibility, scent lures can help

draw target animals close enough to notice the flasher (Kucera et al. 1995a). It is not known whether any carnivores are repelled by flashers.

## Sound Lures

Imitating the vocalizations of conspecifics or distress calls of prey animals will often attract predators (Wise et al. 1999; Shivik 2006). This attraction method employs mechanical or electronic sounds to engage the target species and stimulate exploration or a territorial approach. While such predator calls are often used by hunters, their application in carnivore surveys is limited because all age and sex classes are not necessarily attracted equally (Windberg and Knowlton 1990). Sound lures have been identified as a potentially effective technique for surveying felids in tropical habitats (Kitchener 1991).

## Natural Attractants

Some objects in the landscape (e.g., trees, posts) naturally attract certain carnivore species. Brown bears (*Ursus arctos*) and American black bears (*Ursus americanus*), for example, are especially likely to rub on trees or other objects as they travel through an area (Kendall et al. 2004; chapter 6), leaving behind hair samples that can be easily collected and used to meet various survey objectives (Kendall et al. 1992; Kendall and Waits 2003). If natural attractants can

*Figure 10.1.* Examples of visual attractants (also known as flashers). (A) Researcher R. Long hangs an aluminum pie pan bent into an S shape to promote spinning. The pan is suspended from a branch with baling wire, a swivel, and monofilament fishing line (photo by P. MacKay). (B) Bird feathers suspended from a branch with monofilament fishing line (photo by F. Schlexer). (C) Compact disk suspended from a branch with the same setup as in figure 10.1A (photo by P. MacKay). (D) Close-up of the swivel used in figure 10.1A and 10.1C (after Weaver et al. 2005; photo by P. MacKay).

be identified for a given target species, these objects can be integrated into carnivore survey methods (e.g., barbed wire-wrapped trees for sampling bears; see chapter 6).

## Practical Considerations

The success of a given survey depends on the selection of an effective and appropriate attractant for the target species, the detection method, and the survey area. For example, the ease with which survey stations can be accessed by researchers should be carefully evaluated when selecting an attractant. Stations located in remote areas far from roads restrict the use of large, heavy baits such as ungulate carcasses, which are often employed for remote camera surveys. Track stations and hair collection devices typically use smaller amounts of bait, providing more leeway for site placement. The replenishment of baits and lures is also constrained by difficult site access. Snowmobiles, all-terrain vehicles, pack animals, or helicopters should be considered where appropriate, although these methods of transport can add considerably to the cost of a project. A number of additional practical considerations for integrating attractants into a survey protocol are discussed here.

### Target Species

Knowledge of the natural history, ecology, and behavior of the target species is essential when selecting attractants for a survey. For example, does the species prefer fresh or rotted bait, and in the form of small pieces or whole carcasses? Is it attracted to scent lures, or are flashers a better choice? Is the species less active in winter? What age and sex classes will likely be drawn to the attractant? A solid understanding of these and other species-related questions should help researchers design effective surveys.

Surveys focusing on multiple species may experience greater success if several attractants are used and might also benefit from a combination of baits and lures. Researchers should keep in mind that the suite of detectable species may change over time as bait decomposes. Care should be taken to select target species–specific attractants to prevent nontarget species from being drawn to (and potentially compromising) the detection device. Further, it is important to avoid scenarios in which an attractant for one target species repels another (Doty 1986). For example, scent from fisher (*Martes pennanti*) glands placed at a device may deter American martens (*Martes americana*), which are preyed upon by fishers (Raine 1983). Other such examples of interspecific predation among carnivores are described by Palomares and Caro (1999).

For some target species, particularly those with large home ranges, detectability (see chapter 2) can be improved by prebaiting. Prebaiting involves placing consumable bait in the prospective survey area a few days to several months before the survey begins. This allows individual animals to discover and become habituated to the presence of bait. Prebaiting is a common practice for furbearer trapping (Baker and Dwyer 1987) and is effective for noninvasive surveys when time, site access, and personnel availability permit (Mace et al. 1994; Way et al. 2002; Shivik et al. 2005).

### Deployment of Attractants

Once attractants have been selected, the next step is to determine the presentation method. Reward bait stations are easier to set up than nonreward bait stations (which require additional wire and other materials to isolate the bait), but reward baits must be replenished frequently and should be used with scent lures in case the bait is consumed early in the sampling occasion (see *Survey Design Issues*).

Detection methods involving attractants have specific requirements for positioning bait or scent lures. The position and amount of attractant will vary by method and target species, but it is always critical to configure the survey station such that animals must contact or otherwise trigger the detection device to investigate the attractant. Placing bait above a barbed wire-wrapped post or tree bole, for example, entices the target animal to climb across the wire—thus depositing hair (see chapter 6; figure 6.6). Chapters 4, 5, and 6 discuss method-specific considerations for locating attractants at detection stations.

When deploying baits and scent lures, care must be taken to avoid transferring odors to detection devices, which could potentially be disturbed or destroyed by curious or hungry animals. This is especially true for costly remote cameras, camera sensors, and sensor wires that are easily contaminated if the same person handles both the attractants and the device (chapter 5). To avoid loss of data and damage to equipment, two-person crews should be used during setups involving attractants, with one person installing the detection device and the other handling the bait and/or scent lure. Bears are particularly notorious for destroying cameras and track stations when this protocol is not followed (see chapters 4 and 5).

### Baits

Bait placement can be as simple as laying a piece of chicken on a track plate (see figure 6.8A). Such a reward presentation allows the animal to remove the bait, which must then be regularly replenished until the survey is terminated. In contrast, the presentation of nonreward bait must preclude animals from stealing the bait. One common technique entails puncturing a can of fish several times and nailing it to a tree above the detection device (figure 10.2A); fish odor can escape, but the can itself cannot be removed for consumption. Frozen meat baits can be nailed directly to a tree (figure 10.2B) or wrapped against the trunk with wire (figure 10.2C; but see *Wildlife Heath and Safety*).

Another deployment strategy, especially suitable for larger pieces of bait, is to hang the bait—unprotected or inside a breathable, cloth bag that limits insect damage—from an overhanging branch above the detection device. If hanging branches are unavailable, a catenary system can be constructed using steel cable (figure 10.3), but be aware that baits presented in this way may become accessible to animals after a snowfall. Care should be taken to prevent bait removal by nontarget animals. For example, whenever possible, place large baits under a dense forest canopy or cover them to minimize visits by avian scavengers (Bortolotti 1984; Baker and Dwyer 1987; Aubry et al. 1997).

### Scent Lures

Scent lures can be used in their original formulation or mixed with a viscous substance (see table 10.2) to dilute and extend the service life of concentrated lures and allow them to be spread easily on vegetation (figure 10.4A). Various materials and containers can also be used to facilitate the dispersal of lures over time (see box 10.1). Naturally occurring applicators or vehicles for dispersal, such as sticks or branches, may be found at the survey site—thus reducing material costs and the amount of supplies that must be carried into the field.

Liquid or powdered lures are often poured into containers, which are then perforated and suspended above the detection device. Some containers (e.g., film canisters, cans, bottles) can be acquired at no cost from photo labs or recycling centers. Containers can also be filled with absorbent material, such as wool or cotton, to limit evaporation (figure 10.4B). Prepared containers can be sealed for transport and then opened or perforated in the field. Additional cotton balls, pipe cleaners, or rags saturated with lure can be hung directly from vegetation using lightweight string or fishing line (figure 10.4C). Pelleted lures (e.g., fish meal) are best dispersed in breathable or mesh bags (figure 10.3), and can be mixed with liquid lures (e.g., molasses). Lures spread on vegetation should be applied at sufficient heights to prevent inadvertent contact with field personnel, and lures dispersed in containers should be placed out of reach of animals.

The effective distance of a scent lure changes with variables that can be difficult to control (e.g.,

*Figure 10.2.* Various types of nonreward bait presentations that prevent target species from immediately removing the bait. (A) Punctured can of cat food nailed to a tree above the detection device. Canned fish may also be used (photo by F. Schlexer). (B) American marten seizing frozen, raw chicken drumsticks nailed to a tree above the detection device (photo by USDA Forest Service). (C) Whole, frozen raw chicken carcass nailed to a tree above the detection device and further secured to the tree trunk with multiple wraps of baling wire (photo by F. Schlexer).

temperature, precipitation, humidity, wind speed and direction, topography, and vegetation). Such confounding factors can affect visitation rates independent of target species density (Rice et al. 2001) but can often be managed by lure placement. Generally, scent lures should be positioned to allow for maximum diffusion of the scent plume while still being close enough to the survey station to lure animals to the detection device (Carman 1975). Scent lures can be applied to tree branches or to stakes to elevate odor. Topography also affects local air flow, and can be exploited to maximize scent dispersal

(see chapter 7 for a brief introduction to the movement of scent across the landscape).

The amount of scent lure required depends largely on lure viscosity and weather conditions. As the volatile molecules produced by lures form a more concentrated and localized odor signal in cool, dry, and calm air than in warm, moist and turbulent air (Vickers 2000), additional lure should be used when the former conditions prevail. Small amounts (approximately 5 cc) of liquid lure can be splashed or smeared directly onto trees and vegetation near the detection device, but it is not

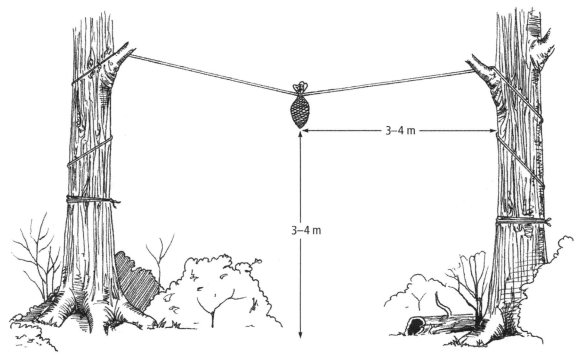

*Figure 10.3.* Nonreward bait presentation. A breathable mesh bag contains bait or scent lure and is suspended above the detection device—out of reach of the target species and potential scavengers. Illustration by S. Harrison.

---

## Box 10.1

### Materials commonly used to disperse scent lures

Containers

  cloth bags
  cotton stockings
  empty paint cans
  film canisters
  microcentrifuge tubes
  nylon stockings or panty hose
  plaster disks (also available pre-scented with lure)
  plastic bottles
  plastic vials or capsules
  poultry egg shells

Absorbent materials

  corn cobs
  cotton balls

cotton lamp wicks
cotton or felt cloth
gauze pads
natural sheeps' wool pads
paper towels
pipe cleaners
tampons

Other materials

  naturally available sticks
  cotton-tipped swabs
  fence posts
  tongue depressors
  tree or shrub branches
  wooden or bamboo stakes

*Figure 10.4.* Methods of dispersing scent lures. (A) Paste lure applied directly to a branch (photo by F. Schlexer). (B) Perforated film canister containing cotton balls saturated with liquid lure and suspended from a branch using monofilament fishing line (photo by F. Schlexer). (C) Gauze pad saturated with liquid lure and suspended from a branch using monofilament fishing line (photo by F. Schlexer). Scent lures in figures 10.4B and 10.4C should be hung out of reach of the target species and potential scavengers.

necessary to saturate the bark. In some cases, over-application of lures may have a repellent effect (Carman 1975; Dobbins 2004).

### Visual Lures

The most important factor to consider when installing visual lures is sight distance. Vegetation may hinder both the visibility of the lure and breezes to provide motion. The lure should thus be suspended (with string or monofilament fishing line) from a branch in an opening above the detection device, at a height of 1–3 m. If string or twine is used, laid (twisted) line provides more motion than braided line. Attaching the line to a tree limb via heavy gauge wire and a fishing swivel can help to maximize lure movement and minimize twisting and entanglement with tree limbs (figure 10.1D; Weaver et al. 2005). Scent lures can be used to draw an animal within range of the visual lure and the detection device. This may not be effective or necessary for felids, however, which primarily rely on vision during foraging (Kitchener 1991) and can be readily attracted to visual lures without additional scent lures (Mowat et al. 1999). In areas of high human use, care should be taken to conceal visual lures from human view in order to minimize vandalism or theft of detection devices.

## Acquisition and Storage

Baits and lures can be an expensive component of a carnivore survey. Thus, it is important to seek out low-cost sources and to employ effective storage methods. Appendix 10.4 provides cost information for some commercial baits and lures.

### Baits

Chicken is a good choice of bait because it is readily available, relatively inexpensive, and can be obtained in convenient sizes (Zielinski 1995). Chicken necks, backs, and wings, and other types of bait (e.g., canned meat or fish) can usually be purchased in bulk at a discount. Grocery stores or butcher shops can be excellent sources of free meat or fish that is outdated or spoiled. Butcher shops and meat packing plants may also be able to supply meat scraps or organs that can't be sold for human consumption, and slaughterhouses can provide livestock blood that would otherwise be discarded. Similarly, fish markets and fish packing plants will often provide free trimmings, fish heads, viscera, or rancid whole fish. These can either be used as is or rendered into fish oil. Whole fish are sometimes available from fish hatcheries or commercial fishermen. Nonmeat baits, such as rotten fruit or vegetables and stale baked goods, are often available at no charge.

Many carnivores are opportunistic and can be attracted with ungulate carcasses (Hornocker and Hash 1981) or those of other species, such as beaver. Two potential sources of carcasses are trappers/hunters and road-killed animals. Deer (*Odocoileus* sp.) are the most commonly available roadkill, but elk (*Cervus elaphus*) and moose (*Alces alces*) are obtainable in some areas. As it is often illegal to handle or transport road-killed game without permission, it is important to contact the local game agency before pursuing this type of bait. Trappers may be able to provide carcasses representing the target species' typical prey (note that trappers are occasionally paid a small fee for this service). Kucera et al. (1995a) recommend using whole carcasses when available, but hindquarters can be more manageable. Whole carcasses can also be cut into smaller pieces and frozen for future use.

Researchers should be prepared to take advantage of opportunistic sources of large amounts of bait, particularly outside of the field season (e.g., roadkill, meat sales at the local market). If storage space is limited, it may be cost-effective to rent freezer space. Bait should be cut into single-use portions and individually wrapped before freezing, thus allowing the appropriate amount of bait to be removed during the survey with minimal handling.

Whether fresh or rancid bait is ultimately chosen, storage and disposal methods should be carefully considered in advance. Meat, blood, and fish baits require refrigeration or freezing. If appropriate facilities are not available or convenient, canned baits should be explored as alternatives. Provisions should be made to safely and lawfully dispose of unused bait. To avoid confounding survey results, uneaten or nonreward baits should be removed from the survey area and discarded in a manner compliant with local waste-disposal laws.

### Scent Lures

Although certain scent lures, such as fish emulsion and cod-liver oil, can be obtained from a variety of sources (e.g., garden and farm supply stores), some researchers prefer to use commercial products or to mix their own lures using ingredients available from trapping supply companies (appendices 10.2, 10.3). Lure recipes and manufacturing methods are available from traditional trapping sources (e.g., Carman 1975; Hanson 1989), and descriptions of how to prepare livestock blood and fish oil for use as lures can also be readily found (e.g., Wyshinski 2001; USDI 2003). A few substances used in lures, such as honey and molasses, are available in bulk from discount grocery stores, canned food warehouses, and bakery suppliers—in quantities ranging from 1 gal. bottles to 55 gal. drums. Matrix lures can be acquired from trapping suppliers, and potentially from hunters, zoos, or game ranches. FAS and catnip oil can be acquired from the USDA Pocatello Supply Depot (appendix 10.2).

Given that scent lures contain volatile compounds, they should be stored in airtight containers in a dark, dry place. Sealed bottles should be stored at room temperature and can have a shelf life of up to two years (Wyshinski 2001). Opened bottles should be frozen for long-term storage but may be kept at room temperature when use is pending.

## Health Concerns

Baits have the potential to cause disease—not only in wildlife, but in researchers conducting surveys. The possibility of infection in both humans and wildlife can be mitigated by the careful selection of attractants and safe handling methods. Some methods of bait presentation may also put animals at risk and should be avoided.

### Safe Handling of Baits and Scent Lures

The potential risks of handling raw or rotted meat or fish are a legitimate concern, and all survey protocols involving bait should include instructions for safe bait handling. Table 10.3 lists the most common pathogens that can cause illness in humans who handle contaminated meat or fish. Some of these agents are found in the intestines of animals, but others are ubiquitous in the environment and can contaminate fresh bait after it has been deployed at the survey station, particularly in warm weather. Indeed, bait can become contaminated in as little as four hours at 20°C (USDA 2005). In a volunteer study with humans (Black et al. 1988), *Campylobacter* infection occurred in subjects who ingested as few as 800 organisms—an amount that can be present in just one drop of juice from raw chicken.

Bait should be carried into the field in containers to protect researchers from contamination. One safe transport method is to place individual pieces of bait in plastic Ziplock bags and freeze them until needed. To further prevent infection, personal protective equipment such as latex gloves or kitchen tongs should be used when handling fresh, old, or rancid bait. Hands should always be washed with soap and (preferably warm) water after handling bait, particularly before touching one's face or consuming food. If hands are not visibly soiled, disposable antiseptic wipes or waterless disinfectant may be liberally applied as an alternative. These alcohol-based hand sanitizers should contain at least 60% alcohol to be effective (Reynolds et al. 2006). Researchers with recent skin abrasions should avoid direct contact with bait. Finally, to prevent cross-contamination, clothing or other gear should not come into contact with hands or gloves used to handle bait.

In study areas where bears occur, researchers should exercise caution when carrying and handling bait to reduce the likelihood of human-bear interactions. Bait containers should be completely sealed to

*Table 10.3.* Common pathogens that may contaminate meat or fish baits used in noninvasive surveys

| Pathogenic agent | Source | Potential bait reservoirs | Method of transmission | Infective dose* |
|---|---|---|---|---|
| *Brucella* spp. | Urine, blood, and tissues of infected animals | Ruminants[a], swine[b], canids | Aspiration, ingestion, mucosal contact, dermal abrasions | Very small |
| *Campylobacter jejuni* | Feces and intestinal tracts of animals and birds | Ruminants, swine, fowl[c], rodents | Aspiration, ingestion | Small |
| *Campylobacter coli* | Feces and intestinal tracts of animals and birds | Ruminants, swine, fowl, rodents | Aspiration, ingestion | Small |
| *Clostridium perfringens* | Soil, feces, and intestinal tracts of animals | Ruminants, swine, fowl, fish | Aspiration, ingestion, dermal abrasions | Very large |
| *Escherichia coli* O157:H7 | Water, feces, and intestinal tracts of mammals | Any domestic or wild mammal | Aspiration, ingestion, dermal abrasions | Unknown, but assumed to be very small |
| *Francisella tularensis* | Soil, water, blood, and tissues of infected animals | Many domestic and wild mammals and birds | Aspiration, ingestion, dermal contact | Aspiration—very small, ingestion and dermal contact—very large |
| *Leptospira interrogans* | Urine, blood, and tissues of infected animals | Ruminants, swine, rodents, reptiles, amphibians | Aspiration, ingestion, dermal abrasions | Very small |
| *Listeria monocytogenes* | Soil, water, blood, feces and intestinal tracts of animals | Any domestic or wild mammal or bird | Aspiration, ingestion, dermal abrasions | Unknown, but assumed to be small |
| *Salmonella* spp. (over 2,300 species) | Water, feces and intestinal tracts of animals and fish | Ruminants, swine, fowl, rodents, reptiles, fish | Aspiration, ingestion | Very small |

Source: USDA 2005; FDA 2006; PHAC 2006.
*Infective dose is the number of organisms needed to cause disease in average healthy individuals. *Very small* indicates as few as 10–100 organisms; *Small* indicates 500–1,000 organisms; *Very large* indicates $\geq 10^8$ organisms.
[a]Ruminants include deer, elk, moose, caribou, wild sheep and goats, and domestic livestock (i.e., cattle, sheep, goats, horses).
[b]Swine include wild and domestic pigs.
[c]Fowl include wild birds and domestic poultry.

minimize external odor. In brown bear habitat, field personnel should never hike alone, be aware of their surroundings, and make noise to alert bears of their presence—particularly in dense brush. Researchers should also be prepared to quickly surrender the bait container if a bear charges. Additional safety tips are available from the American Bear Association (ABA 2006).

Due to the potency and disagreeable odor of many scent lures—and in order to avoid attracting animals to anything but the detection device—care should be taken to prevent contamination of field personnel (i.e., skin and clothing), gear, and vehicles. This can be accomplished by sealing the lure in a plastic Ziplock bag or container (e.g., Loukmas et al. 2003). Military surplus ammunition cans, 5 gal. plastic tubs, or airtight plastic or aluminum camera cases are ideal for transporting both scent lures and baits—as long as they don't need to be carried a long distance.

### Wildlife Health and Safety

Given that rotten meat is commonly used to attract carnivores (Bullard 1982), questions sometimes arise regarding the potential effects of such baits on the health of target species. Many carnivores regularly consume carrion or are at least occasional scavengers; most can safely tolerate the high bacterial load in rotten meat due to having short digestive tracts and appropriate digestive enzymes and acids (DeVault et al. 2003). Harrison et al. (2006) tested for bacterial contamination of carcass meat (including deer and elk) donated to a zoo and concluded that such meat appears to be reasonably safe for carnivores.

A more serious health threat for some carnivores occurs when raw fish is used as bait. Salmon poisoning disease (SPD) and Elokomin fluke fever (EFF) are acute, infectious diseases, primarily affecting canids. Animals become infected by ingesting salmon, steelhead, or trout that contain a rickettsia-infected fluke. SPD can kill up to 90% of infected animals, while EFF usually manifests in a milder form (Aiello 1998). SPD has been commonly seen in coyotes (Foreyt et al. 1987), foxes (Cordy and Gorham 1950), and gray wolves (*Canis lupus*; Darimont et al. 2003), and has been reported in cougars (*Puma concolor*; Kistner et al. 1979) and American black bears (Farrell et al. 1973) as well. SPD microorganisms are also transmittable to domestic animals and humans (Aiello 1998). EFF has been reported in canids, ursids, procyonids, and mustelids (Aiello 1998). Infected fish are found along the northern Pacific coast and in rivers used for migration. Because the encysted flukes are resistant to freezing (Aiello 1998), fresh or frozen salmonids should only be used as bait if they are cooked or canned, or if they originate from outside infected areas.

An additional safety consideration for wildlife lies in bait presentation. For some survey methods (e.g., remote cameras), nonreward meat baits are routinely wrapped and fastened to trees in woven wire mesh (e.g., chicken wire) or hardware cloth to increase the duration of attractiveness. There is increasing concern among researchers that portions of wire could be incidentally consumed with the bait, posing a health risk from metal poisoning or intestinal perforation. This method of bait presentation, therefore, should be avoided. The preferred alternative is to nail small frozen bait directly to a tree (figure 10.2B)—or to wrap large bait to a tree with thin-gauge wire (figure 10.2C)—within the target area of the detection device.

## Survey Design Issues

Survey objectives may constrain attractant selection. Detection-nondetection surveys might require a specific scent lure to attract a target species within a sample unit. Other types of surveys, such as those focusing on foraging behavior or habitat use, might be confounded by a strong lure if the effective sampling distance is great and animals deviate from their natural paths to investigate (Zielinski et al. 2005). Hence, the use of strong scent lures in such situations is not recommended (Gese 2001). Caution should also be applied in scat-based diet studies, which may yield unreliable results if commercial foods or atypical bait items are consumed.

### Habituation and Avoidance

Some canids—particularly coyotes—are susceptible to trap-shyness and learn to recognize and avoid traps and associated attractants (Conner et al. 1998). Coyotes that have been trapped appear to make fewer visits to noninvasive scent stations (Andelt et al. 1985). Reciprocally, recent or nearby trapping efforts (either for recreational, control, or research purposes) may inflate survey detection rates if animals become conditioned to bait as a food source (Brongo et al. 2005). The use of novel attractants (i.e., those not widely used by trappers or animal control personnel) can potentially mitigate these problems. Reward-based attractants (e.g., the coyote lure operative device or CLOD; Marsh et al. 1982) can be used to attract trap-shy animals (Berentsen et al. 2006).

Reward baits can have both ethical and sampling implications. In terms of the former, some animals become reliant on the food value of bait, potentially resulting in a caloric deficit when the survey station is removed (Brongo et al. 2005). Sampling bias is also a concern in this situation. Habituated animals may remove bait early in the sampling occasion, reducing the attractiveness of the device and thus causing undersampling. Conversely, a habituated individual can cause oversampling by repeatedly visiting the same device in hopes of obtaining food. Nonreward baits likely reduce return visits by the same individual, but can attract nontarget scavenger species when baits decompose. Bait presentation should strive to maximize the probability of detecting the target species while simultaneously minimizing multiple detections of the same individual (Zielinski et al. 1995b).

Attractant effectiveness can vary with survey duration. Martin and Fagre (1988) determined that coyote visitation rates at scented track stations were significantly lower at the end of a six-day survey period than at the beginning. In contrast, Stapper et al. (1992) found that visitation rates did not change over the course of three-day surveys, suggesting that some carnivores neither avoided nor were attracted back to a lure after their initial visit when the survey period was relatively short.

Results from studies of captive animals (Harrison 1997) and repeated scent surveys conducted over a short period (Robson and Humphrey 1985) suggest that a given population's response to scent-based attractants may decline over time. Free-ranging carnivores, however, are less likely to become accustomed to scents that they encounter only a few times each year. The concern of habituation should thus not deter the use of scent lures for long-term monitoring of carnivore populations (Harrison 1997).

## Standardization of Attractants

Switching attractant types, or employing multiple attractant types, during a survey can create attraction biases, including variations in effective sampling distance, unequal detection probability, and lack of spatial independence (see chapter 2). For example, sampling distance might change depending on the strength of the odor associated with a scent lure, and switching to a bait that is less attractive to the target species could violate the assumption of equal detection probability. To minimize such issues, attractant type and quantity, and the protocol for deploying attractants, should be standardized for most surveys, particularly those comparing estimates of absolute abundance (Buckland et al. 2006) or relative abundance (Romain-Bondi et al. 2004; Gompper et al. 2006) over geographic areas or among years (Raphael 1994).

The use of standardized attractants for relative abundance surveys increases the probability that observed detection rates reflect differences in population size versus differences in methodology (Raphael 1994). Even species presence cannot be reliably inferred using nonuniform methods (McKelvey et al. 1999). A standardized, reliable set of attractants applied with consistent protocols will help to generate statistically valid data and facilitate repeatability. Standardized attractants were used in the National Lynx Detection Protocol, for example (McKelvey et al. 1999; see chapter 6). This rigorous protocol stipulated the type, proportions, and placement of lures used to attract Canada lynx (*Lynx canadensis*), allowing the pooling of data collected by a large number of agencies and administrative jurisdictions over a broad geographic area.

Attempts have been made to develop standardized attractants for particular species. The development of FAS arose from the testing and field evaluation of standardized lures intended to attract depredating coyotes (Roughton and Bowden 1979), and standardized attractants have also been proposed for some species of felids (Clapperton et al. 1994a; McDaniel et al. 2000), foxes (Steelman et al. 1998), mustelids (Clapperton et al. 1994b; Zielinski et al. 2005), and ursids (Mowat and Strobeck 2000). Attractant standardization methods for remote camera and track station surveys are respectively recommended by Kucera et al. (1995a) and Zielinski (1995).

## Seasonal Issues

Baits and lures can be used in any season, but researchers should select attractants and associated protocols based on the expected temperature and humidity during the survey season (see *Frequency of Reapplication*). The effective sampling distance will typically be greater in warm versus cold weather. Large baits, which resist decay and desiccation, may be more appropriate in warm weather. Wind and temperature not only affect scent dispersion but can influence animal behavior as well. In general, carnivores are more likely to investigate baits and scents during winter when prey availability is more limited and less diverse (Carman 1975). Conducting surveys in winter also prevents conflicts with bears, which can inflict damage to equipment (see *Safe Handling of Baits and Scent Lures*) and alter the behavior of the target species.

Some attractants are limited by seasonal availability. Whole fish or road-killed carcasses may be sporadically accessible, for example, and certain lures (e.g., cow blood) need to be aged or premixed under specific environmental conditions. Most liquid scent lures require the addition of an antifreeze agent if they are to be used in below-freezing temperatures (table 10.2). Finally, commercial trapping lures may be in high demand and difficult to obtain in quantity immediately before a trapping season.

## Frequency of Reapplication

Weather conditions dictate how often bait must be replenished. Baits can be washed out by rain or desiccated by heat, leaving them odorless and ineffective. Given that baits decompose most rapidly in warm weather, summer field personnel should carry extra bait during station checks in case replacement is warranted. Zielinski (1995) recommends that reward baits be replaced at enclosed track plates every visit (i.e., every two days), although detections may

occur as long as some bait remains (Raphael 1994). Frozen baits deployed in subfreezing conditions are resistant to decomposition and therefore require less frequent replacement. Such baits, however, may not be as effective as a distance lure due to the reduced release of aromatic compounds. Thus, under these conditions, rotted bait is preferred to fresh bait.

Researchers should not rely on their own sense of smell to determine if scent lures are in need of reapplication. A lure reapplication schedule should be based on scientific literature or on experimental testing and should address environmental variables such as topography, climate, and season. The reluring interval can vary from several days to several weeks (Dobbins 2004), depending on survey duration, lure type, and weather conditions. Zielinski (1995) recommends that scent lures be applied at enclosed track plates at least twice during a twelve-day survey period. As many scent lures are oil-based and therefore are not seriously diluted by rain or snow, reapplication after every weather event is unnecessary. Lures with a skunk-based scent are more effective at low temperatures (Carman 1975), but some scent lure base materials (e.g., lanolin) become unusable at temperatures below freezing.

# Evaluating the Effectiveness of Baits and Lures

The majority of baits and lures used by commercial and recreational trappers and hunters are founded on tradition and time-tested success. Many of these attractants may be valid for noninvasive carnivore surveys as well and should be scientifically evaluated using rigorous, repeatable protocols. Researchers have generally used attractants based on their history of effectiveness (appendix 10.1), and the scientific testing of attractants didn't begin until the last few decades (e.g., Linhart and Knowlton 1975). Some such testing continues today, following the systematic approach of separating out the components of a given attractant and assessing each component individually (e.g., Kimball et al. 2000)—

often in collaboration with local trappers, animal control agents, or analytical chemists (Turkowski et al. 1979; Wood et al. 2005).

One common testing method involves presenting a captive animal with a variety of attractants and measuring its relative interest by recording behavioral responses. This method can quantitatively evaluate such behaviors as sniffing, scent-marking, scraping, rubbing/rolling, licking/biting, and defecating, as well as response enthusiasm (Fagre et al. 1981, Harrison 1997). Various attractants can thus be ranked according to behavioral response.

Field testing is more appropriate for assessing the effectiveness of attractants in wildlife research, as it incorporates environmental factors and population density. In such tests, visitation rate (or detection rate)—as opposed to behavioral response—is often used as a means of evaluation (Graves and Boddicker 1987). Scent stations provide an effective venue for assessing visitation. Bullard et al. (1983), for example, found that free-ranging coyote visits increased with lure type and intensity, and that widely different odors elicited similar visitation rates. Andelt and Woolly (1996) used experimental manipulation to determine the responses of urban carnivores to a variety of natural and proprietary lures at scent stations (see appendix 10.1). The randomization of attractants and the rotation of lures at a given location allow for statistical comparison with a control lure (e.g., water). Combining captive animal behavioral trials and field evaluations is another successful approach to assessing baits (e.g., Fowler and Golightly 1993).

When evaluating attractants for a noninvasive survey, it is important to consider a number of factors beyond attractiveness. These include, for instance, the survey season, study area, target species, and duration. Martin and Fagre (1988) found that such variables significantly affected outcome when testing natural and synthetic lures.

Clapperton et al. (1994a) assessed the effect of a variety of odors on captive wild and domestic cats (*Felis* spp.) and on feral cats (*Felis catus*) in field trials. These researchers identified catnip and matatabi

(*Actinidia polygama*, otherwise known as Japanese catnip) as the most successful candidate lures for attracting cats. Scent station visits and behavioral responses to scent lures in captive and free-ranging Central American felids were evaluated by Harrison (1997), who found that behavioral scores were more effective at evaluating lures than were investigation times. A randomized test of natural and proprietary lures found that beaver castorium and catnip oil were most effective at attracting Canada lynx (McDaniel et al. 2000). And the USDA Forest Service is evaluating a broad spectrum of scent lures to assess their potential for attracting wolverines (*Gulo gulo;* Copeland et al. 2004). Nearly thirty individual compounds have been tested, and wolverine urine and anal gland secretions show promise (Wood et al. 2005).

Much effort has been expended to develop palatable baits for delivering poison or fertility control drugs to "pest" (e.g., coyotes; Robinson 1962) and nonnative species (e.g., stoats [*Mustela erminea*]; Clapperton et al. 1994b). Similar research has been aimed at developing bait-based methods for administering rabies vaccines to Arctic foxes (*Vulpes lagopus;* Follmann 1988), gray foxes (*Urocyon cinereoargenteus;* Steelman et al. 1998), and raccoons (Wolf et al. 2003). This category of research has employed rigorous methods for testing the efficacy of baits and lures (see also Turkowski et al. 1979; Graves and Boddicker 1987; Mason et al. 1999).

Advanced statistical methods can validate experimental manipulations of attractants. Stanley and Royle (2005) used Poisson and negative binomial models to evaluate retrospective data quantifying the effect of bait supplementation at scent stations (Hein and Andelt 1994). Both studies showed that coyotes used scent stations baited with a supplemental deer carcass more often than stations without supplemental bait.

Among the many salient questions pertaining to the use and evaluation of attractants for noninvasive surveys, three stand out: Why are such a wide variety of carnivores attracted to skunk-based scent lures? Which species prefer rotten bait to fresh bait? What is the sampling radius (effective distance) over which specific lures are able to attract particular species?

Addressing these questions would do much to enhance the reliability and repeatability of carnivore survey efforts. Meanwhile, the studies presented here illustrate how carnivore surveys can benefit from the systematic testing of attractants. Although folk tradition should not be ignored, this field will be handicapped until quantifiable and repeatable testing of traditional attractants supplants anecdotal conjecture. The identification of scientifically valid and effective baits and lures will conserve scarce research funds and provide standardized and defensible results for surveys designed to inform the conservation of carnivores in a changing world.

## Acknowledgments

I thank the USDA Forest Service Pacific Southwest Research Station for supporting this work. Russ Carman and Paul J. Dobbins assisted with research and shared some "inside secrets" from the lure business. I am also grateful to Roland Kays and the late Eric York for providing helpful reviews of early drafts of this chapter.

# Appendix 10.1

*Baits and lures (scent, visual, and sound) that have been used in carnivore surveys, by target species*

| Bait or lure by species | Reference |
|---|---|
| **Coyote** | |
| Baits | |
| unplucked chickens | Aubry et al. 1997 |
| fish (salmon and steelhead) | Aubry et al. 1997 |
| deer carcasses | Aubry et al. 1997 |
| black-tailed prairie dogs | Kamler et al. 2002 |
| black-tailed jackrabbits | Kamler et al. 2002 |
| cottontail rabbits | Kamler et al. 2002; Way et al. 2002 |
| gray squirrels | Way et al. 2002 |
| woodchucks | Way et al. 2002 |
| supermarket meat scraps | Way et al. 2002 |
| lamb meat | Shivik et al. 2005 |
| jackrabbit meat | Shivik et al. 2005 |
| deer meat | Shivik et al. 2005 |
| raw chicken | Way et al. 2002; Gompper et al. 2006 |
| deer meat (2–5 kg) | Gompper et al. 2006 |
| beaver meat (2–5 kg) | Aubry et al. 1997; Gompper et al. 2006 |
| Scent lures | |
| FAS[a] | Harrison 1997*; Sargeant et al. 1998 |
| catnip[b] oil | Harrison 1997* |
| bobcat urine | Harrison 1997* |
| commercial lure[c] | Harrison 1997* |
| wool | Shivik et al. 2005 |
| unspecified commercial lure | Shivik et al. 2005 |
| commercial lure[d] | Gompper et al. 2006 |
| Sound lures | |
| vocalization recordings | Knowlton and Stoddart 1984 |
| **Gray wolf** | |
| Baits | |
| meat | Van Ballenberghe 1984 |
| Scent lures | |
| wolf urine | Van Ballenberghe 1984 |
| unspecified commercial lure | Van Ballenberghe 1984 |
| FAS[a] | Sargeant et al. 1998 |
| **Gray fox** | |
| Baits | |
| raisins and other fruits | Fuller 1978; Trapp 1978; Hallberg and Trapp 1984 |
| honey-based commercial bait | Berchielli and Leubner 1981 |

| Bait or lure by species | Reference |
|---|---|
| fish | Smith and Brisbin 1984 |
| dog food | Weston and Brisbin 2003 |
| raw chicken | Zielinski et al. 2005; Gompper et al. 2006 |
| deer meat (2–5 kg) | Gompper et al. 2006 |
| beaver meat (2–5 kg) | Smith and Brisbin 1984; Gompper et al. 2006 |
| Scent lures | |
| fox gland lure | Berchielli and Leubner 1981 |
| fox urine | Berchielli and Leubner 1981 |
| FAS[a] | Harrison 1997* |
| catnip[b] oil | Harrison 1997* |
| bobcat urine | Conner et al. 1983; Harrison 1997* |
| commercial lure[c] | Harrison 1997* |
| commercial lure[e] | Weston and Brisbin 2003 |
| commercial lure[d] | Zielinski et al. 2005; Gompper et al. 2006 |
| **Island fox** | |
| Baits | |
| dry cat food | Kohlmann et al. 2005 |
| canned cat food | Kohlmann et al. 2005 |
| Scent lures | |
| loganberry paste commercial lure | Kohlmann et al. 2005 |
| **Arctic fox** | |
| Baits | |
| fish | Garrott and Eberhardt 1987 |
| **Kit fox** | |
| Baits | |
| carrion (especially lagomorphs) | O'Farrell 1987 |
| birds[f] | O'Farrell 1987 |
| small mammals[f] | O'Farrell 1987 |
| sardines | O'Farrell 1987 |
| cooked chicken parts | O'Farrell 1987 |
| cheese | O'Farrell 1987 |
| canned mackerel | O'Farrell 1987; Cypher and Spencer 1998; Koopman et al. 2000; Warrick and Harris 2001 |
| black-tailed jackrabbits | Zoellick and Smith 1992 |
| leporids | Cypher and Spencer 1998; Koopman et al. 2000 |
| Scent lures | |
| FAS[a] | Warrick and Harris 2001 |

| Bait or lure by species | Reference |
|---|---|
| **Swift fox** | |
| Baits | |
| chicks[f] | Scott-Brown et al. 1987 |
| rabbits[f] | Scott-Brown et al. 1987 |
| deer | Scott-Brown et al. 1987 |
| raw chicken | Covell 1992 |
| beef scraps | Harrison et al. 2002 |
| black-tailed prairie dogs | Kamler et al. 2002 |
| black-tailed jackrabbits | Kamler et al. 2002 |
| desert cottontails | Kamler et al. 2002 |
| canned mackerel in oil | Uresk et al. 2003 |
| Scent lures | |
| cod-liver oil-mackerel commercial lure[g] | Harrison et al. 2002; Harrison 2003 |
| | |
| **Red fox** | |
| Baits | |
| honey-based commercial bait | Berchielli and Leubner 1981 |
| raw chicken | Zielinski et al. 2005; Gompper et al. 2006 |
| deer meat (2–5 kg) | Gompper et al. 2006 |
| beaver meat (2–5 kg) | Gompper et al. 2006 |
| Scent lures | |
| fox gland lure | Berchielli and Leubner 1981 |
| fox urine | Berchielli and Leubner 1981 |
| FAS[a] | Sargeant et al. 1998 |
| commercial lure[d] | Gompper et al. 2006 |
| | |
| **Ocelot** | |
| Baits | |
| live chickens or chicks | Tewes 1986; Emmons 1988; Crawshaw and Quigley 1989; Laack 1991; Caso 1994; Horne 1998; Harveson et al. 2004; Dillon 2005 |
| live rabbits | Tewes 1986; Caso 1994 |
| live quail | Caso 1994 |
| live pigeons | Horne 1998 |
| sardines in oil | Trolle 2003; Trolle and Kery 2003; Dillon 2005 |
| chicken parts | Dillon 2005 |
| Scent lures | |
| ocelot, bobcat, and fox urine | Laack 1991 |
| FAS[a] | Harrison 1997* |
| commercial lure[h] | Boddicker et al. 2002 |
| catnip[b] | Shinn 2002; Weaver et al. 2005 |
| cod-liver oil | Trolle 2003 |
| commercial lure[j] | Shinn 2002; Weaver et al. 2003; Weaver et al. 2005 |
| commercial lure[i] | Dillon 2005 |
| Visual lures | |
| pie plate flasher | Shinn 2002; Weaver et al. 2005 |

| Bait or lure by species | Reference |
|---|---|
| **Margay** | |
| Scent lures | |
| FAS[a] | Harrison 1997* |
| catnip[b] oil | Harrison 1997* |
| bobcat urine | Harrison 1997* |
| commercial lure[c] | Harrison 1997* |
| commercial lure[h] | Boddicker et al. 2002 |
| | |
| **Canada lynx** | |
| Baits | |
| chicken | Zielinski 1995 |
| carrion | Kucera et al. 1995a |
| deer (> 5 kg) | Kucera et al. 1995a |
| fish | Kucera et al. 1995a |
| rabbit | Shenk 2001 |
| Scent lures | |
| unspecified commercial lure | Zielinski 1995; Kucera et al. 1995a |
| skunk musk/essence/ tincture | Kucera et al. 1995a |
| beaver castoreum | McDaniel et al. 2000* |
| catnip[b] oil | McDaniel et al. 2000* |
| Visual lures | |
| flasher | Young 1958; Baker and Dwyer 1987; Zielinski 1995; Kucera et al. 1995a |
| | |
| **Bobcat** | |
| Baits | |
| fresh meat | Kitchings and Story 1979; Zezulak and Schwab 1979; Smith and Brisbin 1984 |
| live chickens | Kitchings and Story 1979; Zezulak and Schwab 1979; Fischer 1998; Horne 1998 |
| live rabbits | Kitchings and Story 1979; Zezulak and Schwab 1979 |
| fish | Smith and Brisbin 1984 |
| unplucked chickens | Aubry et al. 1997 |
| fish (salmon and steelhead) | Aubry et al. 1997 |
| beaver meat | Aubry et al. 1997 |
| deer carcasses | Aubry et al. 1997 |
| live pigeons | Horne 1998 |
| raw chicken | Long et al. 2007b |
| commercial lure[j] | Long et al. 2007b |
| Scent lures | |
| FAS[a] | Roughton 1979*; Diefenbach et al. 1994; Sargeant et al. 1998 |
| bobcat urine | Morrison et al. 1981; Conner et al. 1983 |
| commercial lure[j] | Shinn 2002 |
| commercial lure[d] | Long et al. 2007b |

| Bait or lure by species | Reference |
|---|---|
| Visual lures | |
| flasher | Young 1958; Baker and Dwyer 1987; Shinn 2002 |
| **Jaguar** | |
| Baits | |
| sardines in oil | Trolle 2003 |
| Scent lures | |
| catnip[b] | Kitchener 1991 |
| commercial lure[h] | Boddicker et al. 2002 |
| cod-liver oil | Trolle 2003 |
| **Cougar** | |
| Baits | |
| unplucked chickens | Aubry et al. 1997 |
| fish (salmon and steelhead) | Aubry et al. 1997 |
| beaver meat | Aubry et al. 1997 |
| deer carcasses | Aubry et al. 1997 |
| sardines in oil | Trolle 2003 |
| Scent lures | |
| cod-liver oil | Trolle 2003 |
| **Jaguarundi** | |
| Baits | |
| live chickens | Caso 1994 |
| live rabbits | Caso 1994 |
| live quail | Caso 1994 |
| Scent lures | |
| FAS[a] | Harrison 1997* |
| bobcat urine | Harrison 1997* |
| commercial lure[c] | Harrison 1997* |
| **Striped skunk** | |
| Baits | |
| smoked herring | Bailey 1971 |
| fish | Smith and Brisbin 1984; Greenwood et al. 1997 |
| deer carcasses | Smith and Brisbin 1984; Aubry et al. 1997 |
| sardines | Rosatte 1987; Bartelt et al. 2001 |
| chicken entrails | Rosatte 1987 |
| dog food | Rosatte 1987 |
| unplucked chickens | Aubry et al. 1997 |
| fish (salmon and steelhead) | Aubry et al. 1997 |
| beaver meat | Aubry et al. 1997 |
| chicken eggs | Greenwood et al. 1997 |
| dry dog food | Greenwood et al. 1997 |
| sunflower seeds | Greenwood et al. 1997 |
| canned cat food | Baldwin et al. 2004 |
| raw chicken | Zielinski et al. 2005 |

| Bait or lure by species | Reference |
|---|---|
| Scent lures | |
| various chemical attractants | Rosatte 1987 |
| FAS[a] | Greenwood et al. 1997; Sargeant et al. 1998 |
| mink gland and salmon oil (1:1) | Loukmas et al. 2003 |
| commercial lure[d] | Zielinski et al. 2005 |
| **Western spotted skunk** | |
| Baits | |
| unplucked chickens | Aubry et al. 1997 |
| fish (salmon and steelhead) | Aubry et al. 1997 |
| beaver meat | Aubry et al. 1997 |
| deer carcasses | Aubry et al. 1997 |
| raw chicken | Zielinski et al. 2005 |
| Scent lures | |
| FAS[a] | Sargeant et al. 1998 |
| commercial lure[d] | Zielinski et al. 2005 |
| **Wolverine** | |
| Baits | |
| fresh meat (1 kg) | Hash and Hornocker 1980 |
| carrion | Kucera et al. 1995a |
| deer (> 5 kg) | Kucera et al. 1995a; Copeland et al. 1995 |
| fish | Kucera et al. 1995a; Copeland et al. 1995 |
| chicken | Zielinski 1995 |
| beaver carcasses | Copeland 1996; Fisher 2005 |
| rotten meat | Mowat 2001 |
| Scent lures | |
| fish oil | Mowat 2001 |
| beaver castor | Mowat 2001 |
| unspecified commercial lure | Zielinski 1995; Kucera et al. 1995a |
| commercial lure[k] | Fisher 2005 |
| Visual lures | |
| cloth flasher | Hash and Hornocker 1980 |
| flasher | Zielinski 1995; Kucera et al. 1995a |
| **North American river otter** | |
| Baits | |
| whole fish | Melquist and Dronkert 1987 |
| **American marten** | |
| Baits | |
| beaver carcasses | Strickland and Douglas 1987 |
| canned sardines | Strickland and Douglas 1987; Gosse et al. 2005 |
| strawberry or raspberry jam | Strickland and Douglas 1987 |

| Bait or lure by species | Reference |
|---|---|
| beaver meat (2–5 kg) | Baker and Dwyer 1987; Aubry et al. 1997; Gompper et al. 2006 |
| carrion | Kucera et al. 1995a |
| fish | Kucera et al. 1995a |
| deer (> 5 kg) | Kucera et al. 1995a; Aubry et al. 1997 |
| raw chicken | Zielinski 1995; Zielinski et al. 2005; Gompper et al. 2006 |
| unplucked chickens | Aubry et al. 1997 |
| fish (salmon and steelhead) | Aubry et al. 1997 |
| partially decomposed chicken wings | Mowat et al. 2001 |
| deer meat (2–5 kg) | Gompper et al. 2006 |
| Scent lures | |
| beaver fat | Baker and Dwyer 1987 |
| anise oil | Strickland and Douglas 1987 |
| fish oil | Strickland and Douglas 1987 |
| unspecified commercial lure | Zielinski 1995; Kucera et al. 1995a; Mowat et al. 2001 |
| rendered fish oil | Mowat et al. 2001 |
| skunk scent commercial lure | Gosse et al. 2005 |
| commercial lure[d] | Zielinski et al. 2005; Gompper et al. 2006 |
| Visual lures | |
| flasher | Zielinski 1995; Kucera et al. 1995a |
| **Fisher** | |
| Baits | |
| beaver carcasses | Douglas and Strickland 1987 |
| canned sardines | Douglas and Strickland 1987 |
| beaver meat | Baker and Dwyer 1987; Aubry et al. 1997 |
| meat scraps | Jones and Garton 1994 |
| carrion | Kucera et al. 1995a |
| deer (> 5 kg) | Kucera et al. 1995a; Aubry et al. 1997 |
| fish | Kucera et al. 1995a |
| chicken | Zielinski 1995 |
| unplucked chickens | Aubry et al. 1997 |
| fish (salmon and steelhead) | Aubry et al. 1997 |
| raw chicken | Zielinski et al. 2005; Long et al. 2007b; Gompper et al. 2006 |
| deer meat (2–5 kg) | Gompper et al. 2006 |
| beaver meat (2–5 kg) | Gompper et al. 2006 |
| Scent lures | |
| beaver fat | Baker and Dwyer 1987 |
| anise oil | Douglas and Strickland 1987 |
| unspecified commercial lure | Jones and Garton 1994; Zielinski 1995; Kucera et al. 1995a |

| Bait or lure by species | Reference |
|---|---|
| mink gland and salmon oil (1:1) | Loukmas et al. 2003 |
| commercial lure[d] | Zielinski et al. 2005; Long et al. 2007b; Gompper et al. 2006 |
| Visual lures | |
| flasher | Zielinski 1995; Kucera et al. 1995a |
| **Ermine** | |
| Baits | |
| unplucked chickens | Aubry et al. 1997 |
| fish (salmon and steelhead) | Aubry et al. 1997 |
| deer carcasses | Aubry et al. 1997 |
| beaver meat (2–5 kg) | Aubry et al. 1997; Gompper et al. 2006 |
| fresh meat | Gonzales 1997 |
| partially decomposed chicken wings | Mowat et al. 2001 |
| raw chicken | Gompper et al. 2006 |
| deer meat (2–5 kg) | Gompper et al. 2006 |
| Scent lures | |
| rendered fish oil | Mowat et al. 2001 |
| unspecified commercial lure | Mowat et al. 2001 |
| mink gland and salmon oil (1:1) | Loukmas et al. 2003 |
| commercial lure[d] | Gompper et al. 2006 |
| **Long-tailed weasel** | |
| Baits | |
| dead domestic mice | DeVan 1982; Gehring and Swihart 2004 |
| unplucked chickens | Aubry et al. 1997 |
| fish (salmon and steelhead) | Aubry et al. 1997 |
| deer carcasses | Aubry et al. 1997 |
| beaver meat (2–5 kg) | Aubry et al. 1997; Gompper et al. 2006 |
| fresh meat | Gonzales 1997 |
| raw chicken | Zielinski et al. 2005; Gompper et al. 2006 |
| deer meat (2–5 kg) | Gompper et al. 2006 |
| Scent lures | |
| mink gland and salmon oil (1:1) | Loukmas et al. 2003 |
| unspecified commercial lure | Gehring and Swihart 2004 |
| commercial lure[d] | Gompper et al. 2006 |
| **Least weasel** | |
| Baits | |
| live mice | Fagerstone 1987 |

| Bait or lure by species | Reference |
|---|---|
| fresh meat | Henderson 1994; Gonzales 1997 |
| **American mink** | |
| Baits | |
| fresh rabbit, muskrat, birds | NDFA 1997 |
| fresh fish | NDFA 1997 |
| Scent lures | |
| fish oil | NDFA 1997 |
| mink glands and urine | NDFA 1997 |
| unspecified commercial mink lure | Loukmas and Halbrook 2001 |
| ranch mink scat | Loukmas and Halbrook 2001 |
| mink gland and salmon oil (1:1) | Loukmas et al. 2003 |
| **American badger** | |
| Baits | |
| raw chicken | Minta and Marsh 1988; Zielinski et al. 2005 |
| chicken carcass | Gonzales 1997 |
| ground squirrels | Newhouse and Kinley 2000; Apps et al. 2002 |
| rabbits[f] | Newhouse and Kinley 2000; Apps et al. 2002 |
| beef liver | Newhouse and Kinley 2000; Apps et al. 2002 |
| Scent lures | |
| commercial lure[l] | Newhouse and Kinley 2000 |
| unspecified commercial lure | Apps et al. 2002 |
| commercial lure[d] | Zielinski et al. 2005 |
| **Ringtail** | |
| Baits | |
| raisins | Hallberg and Trapp 1984; Kaufmann 1987 |
| fruit jam | Kaufmann 1987 |
| fish | Kaufmann 1987 |
| unplucked chickens | Aubry et al. 1997 |
| fish (salmon and steelhead) | Aubry et al. 1997 |
| beaver meat | Aubry et al. 1997 |
| deer carcasses | Aubry et al. 1997 |
| raw chicken | Zielinski et al. 2005 |
| Scent lures | |
| ringtail urine | Kaufmann 1987 |
| ringtail musk | Kaufmann 1987 |
| commercial lure[d] | Zielinski et al. 2005 |
| **White-nosed coati** | |
| Baits | |
| bananas | Kaufmann 1987 |

| Bait or lure by species | Reference |
|---|---|
| marshmallows | Kaufmann 1987 |
| canned and dry pet food | Kaufmann 1987 |
| live chickens | Caso 1994 |
| live rabbits | Caso 1994 |
| live quail | Caso 1994 |
| sardines | Valenzuela and Ceballos 2000 |
| Scent lures | |
| commercial lure[h] | Boddicker et al. 2002 |
| **Raccoon** | |
| Baits | |
| fresh fish | Smith and Brisbin 1984; Sanderson 1987 |
| deer meat (2–5 kg) | Smith and Brisbin 1984; Gompper et al. 2006 |
| dry, chunk-style dog food | Sanderson 1987 |
| canned fish | Sanderson 1987 |
| unplucked chickens | Aubry et al. 1997 |
| fish (salmon and steelhead) | Aubry et al. 1997 |
| deer carcasses | Aubry et al. 1997 |
| beaver meat (2–5 kg) | Aubry et al. 1997; Gompper et al. 2006 |
| sardines | Bartelt et al. 2001 |
| marshmallows | Bartelt et al. 2001 |
| strawberry jam | Bartelt et al. 2001 |
| raw chicken | Zielinski et al. 2005; Gompper et al. 2006 |
| Scent lures | |
| bobcat urine | Conner et al. 1983; Rucker 1983; Leberg and Kennedy 1987 |
| FAS[a] | Smith et al. 1994; Sargeant et al. 1998 |
| mink gland and salmon oil (1:1) | Loukmas et al. 2003 |
| commercial lure[d] | Zielinski et al. 2005; Gompper et al. 2006 |
| **American black bear** | |
| Baits | |
| apples | Baker and Dwyer 1987 |
| fish | Baker and Dwyer 1987 |
| rotten meat (2 kg) | Woods et al. 1999 |
| corn | Brown 2004 |
| honey (diluted with water) | Brown 2004 |
| maple syrup (diluted with water) | Brown 2004 |
| stale pastries (e.g., bagels, donuts, cookies) | Brown 2004; Knorr 2004 |
| canned sardines | Brongo et al. 2005 |
| raw chicken | Zielinski et al. 2005; Long et al. 2007b; Gompper et al. 2006 |

| Bait or lure by species | Reference |
|---|---|
| fish food pellets | Long et al. 2007b |
| molasses | Long et al. 2007b |
| deer meat (2–5 kg) | Gompper et al. 2006 |
| beaver meat (2–5 kg) | Gompper et al. 2006 |
| Scent lures | |
| liquid fish fertilizer | Woods et al. 1999 |
| commercial lure[d] | Zielinski et al. 2005; Long et al. 2007b; Gompper et al. 2006 |

**Grizzly bear**

| | |
|---|---|
| Baits | |
| raw meat (wild ungulate, domestic livestock) | Mace et al. 1994 |
| livestock blood | Mace et al. 1994; Boulanger et al. 2004c; Proctor et al. 2004; Romain-Bondi et al. 2004 |
| rotten meat (2 kg) | Woods et al. 1999; Proctor et al. 2004; Romain-Bondi et al. 2004 |
| Scent lures | |
| canned blueberries | Mace et al. 1994 |
| anise extract | Mace et al. 1994 |
| vanilla extract | Mace et al. 1994 |
| commercial skunk scent | Mace et al. 1994 |

| Bait or lure by species | Reference |
|---|---|
| liquid fish fertilizer | Woods et al. 1999; Proctor et al. 2004; Romain-Bondi et al. 2004 |
| fish oil | Boulanger et al. 2004c |

Note: Asterisk (*) indicates studies that have empirically tested and evaluated specific lures for the target species.

† Lures available from multiple trapping supply distributors. See appendix 10.2 for names and addresses.

[a]Synthetic fatty-acid tablets (USDA, Pocatello Supply Depot, Pocatello, ID).

[b]*Nepeta cataria* (fresh and dried catnip leaves are available from pet stores and multiple trapping supply distributors; catnip oil is available from USDA, Pocatello Supply Depot, Pocatello, ID).

[c]Hawbaker's Wildcat #2 †.

[d]Caven's Gusto †.

[e]Liquid Fox and Coyote Lure, Fox Gland Lure (On Target A.D.C., Cortland, IL); Caven's Fox #1, Caven's Fox #2, Caven's Canine Force †.

[f]Bait is presumed to be dead (author did not state)

[g]Trailing Scent (On Target A.D.C., Cortland, IL).

[h]Carman's Canine Call, Pro's Choice, Bobcat Gland Lure, Trophy Deer Lure, and Mega Musk †.

[i]Marak's Bobcat Lure, Marak's Coyote Lure, Marak's Gray Fox Lure, Marak's Raccoon Lure †.

[j]Weaver's Cat Call (John L. Weaver, Wildlife Conservation Society, St. Ignatius, MT).

[k]O'Gorman's LDC Extra †.

[l]Carman's Canine Call †.

# Appendix 10.2

*Select commercial suppliers of baits and scent lures, lure ingredients, and other attractants*

| Item | Supplier | Item | Supplier |
|------|----------|------|----------|
| FAS (fatty acid scent)<br>Plaster predator survey<br>    disks<br>Catnip oil | USDA, APHIS, WS<br>Pocatello Supply Depot<br>238 East Dillon Street<br>Pocatello, ID 83201<br>208-236-6920<br>psdusda@qwest.net | | Knob Mountain Fur Company<br>430 Monroe Street<br>Berwick, PA 18603<br>570-759-7035<br>www.knobmountainfur.com/<br>    index.php |
| Carnivore urines<br>Glands, musks, and<br>    proprietary lures<br>Botanical oils and<br>    extracts<br>Vocalization recordings | AllPredatorCalls.com<br>PO Box 90163<br>Tucson, AZ 85752<br>520-293-2972<br>www.allpredatorcalls.com/ | | M & M Furs, Inc.<br>PO Box 15<br>26445 435th Avenue<br>Bridgewater, SD 57319-0015<br>605-729-2535<br>www.mandmfurs.com/ |
| | Adirondack Outdoor Company<br>PO Box 86<br>Elizabethtown, NY 12932<br>518-873-6806<br>www.adirondackoutdoor.com/<br>    trapping.htm | | Minnesota Trapline Products<br>6699 156th Avenue N.W.<br>Pennock, MN 56279<br>320-599-4176<br>www.minntrapprod.com/ |
| | Cumberland's Northwest<br>    Trappers Supply<br>PO Box 408<br>Owatonna, MN 55060<br>507-451-7607<br>www.nwtrappers.com/default.asp | | On Target A.D.C.<br>PO Box 480<br>Cortland, IL 60112<br>815286-3073<br>www.wctech.com/ontarget/ |
| | Dobbins' Products<br>208 Earl Drive<br>Goldsboro, NC 27530<br>919-580-0621<br>www.trapperman.com/catalog.html | | The Snare Shop<br>858 East U.S. Highway 30<br>Carroll, IA 51401<br>712-792-0601 |
| | Funke Trap Tags & Supplies<br>2151 Eastman Ave.<br>State Center, IA 50247<br>641-483-2597<br>www.funketraptags.com/ | | Sterling Fur and Tool Company<br>11268 Frick Road<br>Sterling, OH 44276<br>330-939-3763 |
| | S. Stanley Hawbaker & Sons<br>PO Box 309<br>Fort Louden, PA 17224 | | Sullivan's Scents and Supplies<br>429 Upper Twin<br>Blue Creek, OH 45616<br>740-858-4416<br>www.sullivansline.com/sline/<br>    slhome.htm |
| | Kishel's Quality Animal Scents &<br>    Lures, Inc.<br>c/o Rettig's Outdoor Supplies<br>107 Harvey Lane<br>Saxonburg, PA 16056<br>724-352-7121<br>www.kishelscents.com/index.asp | | Wasatch Wildlife Products<br>PO Box 753<br>Magna, UT 84044<br>801-250-9308<br>www.wasatchwild.com |

# Appendix 10.3

*Scents and oils used in traditional and commercial lure manufacture*

| Scent | Use | Characteristic | Canidae | Felidae | Mephitidae | Mustelidae | Procyonidae | Ursidae |
|---|---|---|---|---|---|---|---|---|
| Acorn oil | attractant | herbal | | | | | | x |
| Almond extract | attractant | sweet | | | x | x | x | x |
| Ambergris oil (synthetic) | fixative | musky | x | | x | x | x | x |
| Amber oil | fixative | minty | x | | | x | | |
| Ambrette musk | attractant | musky, sweet | x | | x | x | x | |
| Anise oil | attractant | sweet, licorice | x | | x | x | x | x |
| Apple oil | attractant | sweet | | | | | x | x |
| Asafoetida gum | attractant | pungent | x | x | | | | |
| Asfoetida tincture | attractant | pungent | x | x | | | | |
| Banana essence oil | attractant, additive | floral | | | x | x | x | |
| Bergamot oil | attractant, additive | minty | x | x | | | x | x |
| Balsam oil | attractant, additive | herbal | | | | | x | |
| Birch oil | attractant, additive | sweet | | | | | x | |
| Black prune oil | additive | fruity | | | | | x | |
| Bleach | additive | pungent | | x | | | | |
| Blue cheese oil | attractant | sharp | x | | x | x | | |
| Blueberry essence | attractant | fruity | x | | x | x | x | x |
| Calamus oil | attractant | sweet | | | | | x | x |
| Calamus powder | attractant | sweet | | | | | x | x |
| Catnip oil | attractant | herbal | x | x | | | x | |
| Catnip, dried | attractant | herbal | x | x | | | x | |
| Catnip, fresh | attractant | herbal | x | x | | | x | |
| Caramel essence | additive | sweet | | | | | | x |
| Canton musk | fixative | musky | x | x | | | | |
| Chenopodium oil | fixative | musky | x | | | | | |
| Cherry oil | attractant, additive | sweet | | | | | x | |
| Cheese essence | attractant | pungent | x | | | | x | x |
| Civet oil | attractant | musky | x | | | | | |
| Cod liver oil | attractant | fishy | | | | x | x | |
| Cumin | fixative | pungent | x | | | | | |
| FAS (fatty acid scent) | attractant | pungent | x | x | | x | x | |
| Fennel oil | attractant | herbal | | | | | x | |
| Fig extract oil | additive | sweet | | | | | x | |
| Fish oil | attractant | fishy | x | | | x | x | x |
| Fish extract | attractant | fishy | x | | | x | x | x |
| Garlic essence | attractant | pungent | | | x | x | x | x |
| Grape essence | attractant | fruity | x | | | | x | |
| Honey essence oil | attractant, additive | sweet | x | | | | x | x |
| Honeysuckle oil | attractant | sweet, floral | x | | | | x | |
| Lavender oil | attractant | floral | | | | | x | |
| Liquid smoke | attractant | pungent | x | | | | x | x |
| Loganberry oil | attractant | fruity | x | | x | | x | x |
| Lovage oil | attractant | herbal | x | | | | | |
| Lovage root powder | attractant | herbal | x | | | | | |
| Melon oil | attractant | fruity | x | | | | x | |
| Muscaro musk | attractant | musky | x | | | | x | |
| Orange oil | attractant | citrus | | | | | x | |

| Scent | Use | Characteristic | Canidae | Felidae | Mephitidae | Mustelidae | Procyonidae | Ursidae |
|---|---|---|---|---|---|---|---|---|
| Pennyroyal oil | attractant | minty | | | | | x | |
| Peppermint oil | attractant | minty | | | | | x | |
| Persimmon oil | attractant, additive | fruity | x | | | | x | |
| Phenyl acetic, crystals | attractant, additive | sweet | | | | | x | |
| Phenyl acetic, liquid | attractant, additive | sweet | | | | | x | |
| Prune oil | attractant | sweet | | | | | x | x |
| Raspberry oil | attractant | fruity | x | | | | x | x |
| Rhodium oil | attractant | minty | | | | | x | |
| Rue oil | attractant, fixative | herbal | x | | | | | |
| Salmon oil | attractant | fishy | x | | | x | x | x |
| Spearmint oil | attractant, additive | sweet, minty | x | | | | x | |
| Shellfish oil | attractant | fishy | x | | | x | x | x |
| Shrimp essence | attractant | fishy | x | | | x | x | |
| Strawberry oil | attractant | fruity | x | | x | | x | x |
| Sweetcorn oil | attractant, additive | herbal | x | | x | | x | |
| Synthetic fermented egg | attractant | pungent | x | | x | x | x | |
| Tabasco | attractant | pungent | | | | | | x |
| Tonka bean extract | additive | vanilla | | | | | x | x |
| Tonquin musk, synthetic | attractant | musky | x | | | | | |
| Trout oil | attractant | fishy | x | | | | x | x |
| Valerian root extract | attractant | pungent | x | x | | | x | |
| Vanilla oil | additive | vanilla | | | | | x | x |
| Watermelon oil | attractant | fruity | x | | | | | |
| White thyme oil | additive, fixative | minty | x | | | | | |
| Wintergreen oil | attractant, additive | sweet, minty | | | | | x | |
| Ylang ylang oil | attractant, additive | floral, sweet | x | x | x | x | x | |

Note: Musk tibetine and musk ketone, synthetic substances with a typical musky scent that are widely used as fixatives in lure manufacture and in the cosmetics industry, are priority-listed Persistent, Bioaccumulative and Toxic (PBT) chemicals (OSPAR 2004) and also cannot be recommended due to their potential carcinogenic effects (Schmeiser et al. 2001; Apostolidis et al. 2002).

Source: Trapping supply catalogs; see appendix 10.2 for names and addresses of commercial lure suppliers.

# Appendix 10.4

*Approximate cost of select baits, lures, and lure bases*

| Item | Approximate cost* and units |
|---|---|
| Baits | Per pound |
| Chicken, whole, fresh or frozen | $0.50–$1.50 |
| Chicken quarters, fresh or frozen | $0.80–$1.30 |
| Chicken thighs, frozen (4 lb. bag) | $0.70– $0.90 |
| Chicken legs, frozen (4 lb. bag) | $0.70– $0.90 |
| Chicken drumettes (wings), frozen (4 lb. bag) | $0.70–$0.90 |
| Beef liver, heart, or other organ meat | $0.50– $1.50 |
| Canned fish (mackerel, sardines, salmon, tuna) | $2.00–$3.00 |
| Canned pet food (cat or dog) | $0.60–$0.80 |
| Dry pet food (cat or dog) | $0.20–$0.40 |
| Proprietary baits (ground animal meat) | $10.00–$20.00 |
| | |
| Lures | Per fluid ounce |
| Beaver castor | $3.50–$5.00 |
| Botanical oils | $3.00–$5.00 |
| Carnivore glands | $3.50–$4.50 |
| Carnivore urine | $0.10–$0.25 |
| Catnip, dried | $2.00–$4.00 |
| Catnip, oil | $4.00–$23.50 |
| Cod-liver oil | $1.30–$1.50 |
| Fatty acid scent (FAS), diluted | $9.00 |
| Fatty acid scent (FAS), undiluted | $5.25 |
| Fish fertilizer, liquid | $0.10– $0.20 |
| Fish oil | $0.15–$0.30 |
| Musk oils, natural or synthetic | $4.00–$18.00 |
| Proprietary scent lures | $3.50–$5.00 |
| Skunk scent, tincture | $3.50–$5.50 |
| Skunk scent, pure | $18.00–$20.00 |
| | |
| Lure bases | Various units |
| Glycerine | $22–$35/gal. |
| Honey | $25–$30/gal. |
| Lanolin, anhydrous | $10–$15/pt. |
| Molasses | $35–$40/gal. |
| Predator survey disks, scented with FAS | $0.43 ea. |
| Predator survey disks, unscented | $0.21 ea. |
| Propylene glycol | $20–$30/gal. |
| Sodium benzoate, powder | $35–$40/gal. |
| Sodium citrate | $0.30– $0.40/oz. |
| Vegetable oil | $5–$6/gal. |
| Zinc valerate, powder | $7–$15/oz. |

*U.S. dollars as of July 2006. Prices may be lower if bought in quantity.

# Chapter 11

# Statistical Modeling and Inference from Carnivore Survey Data

*J. Andrew Royle, Thomas R. Stanley, and Paul M. Lukacs*

Much of contemporary ecological theory and conservation biology is concerned with variation in species abundance or occurrence, or the dynamics of abundance and occurrence (e.g., survival probability, local extinction, colonization). Unfortunately, in ecology, it is not always possible to directly observe the state variable that is the object of inference. Often, ecological studies are observational studies that yield an incomplete characterization of population state or demographic processes.

Species nondetection is one important source of observational error or bias. That is, because many carnivores are secretive by nature, some animals present at sample units may not be detected by the survey. Inference methods for animal survey data have been strongly influenced by the issue of imperfect detection (or nondetection bias), and most of the methods discussed here address imperfect detection explicitly. This issue has catalyzed considerable research on animal sampling and quantitative methods in ecology, with many books and monographs dealing extensively with the topic of estimating detection probabilities (e.g., Seber 1982; Williams et al. 2002; MacKenzie et al. 2006). How one chooses to address—or the extent to which one can address—the issue of nondetection or detectability largely de-

termines the inferential framework that can be applied to any particular problem.

The objective of this chapter is to provide an overview of strategies for modeling and drawing inferences from data that arise from noninvasive carnivore sampling. We describe methods that apply to two broad sampling situations. The first scenario is when detection-nondetection data are recorded but it is not possible to distinguish individuals. Examples include surveys employing hair snares (Foran et al. 1997b; Belant 2003a), track plates (Barrett 1983; Orloff et al. 1993; Zielinski and Stauffer 1996; Olson and Werner 1999), snow tracking (Stanley and Bart 1991; Hayward et al. 2002), and remote cameras that allow identification only to the species level (Moruzzi et al. 2002; Gompper et al. 2006). The second scenario is the classical animal sampling situation in which a closed population of size $N$ is sampled repeatedly and the identification of individuals is achievable. Thus, capture-recapture approaches can be used for inference. This latter scenario would pertain to remote camera approaches in which the photo identification of individuals is feasible or to surveys in which genetic material can be obtained from hair or scat and used to discriminate individuals. Before proceeding further, we offer a few philosophical thoughts on the development of study objectives and sampling considerations.

## Establishing Objectives

The establishment of objectives is usually regarded as the first step in sound study design (see chapter 2). Indeed, scientific studies of animal populations should always be driven by a clear statement of objectives prior to the collection of data. In principle, when coupled with a concise method of inference, the a priori formulation of objectives allows explicit consideration of the target population such that "optimal" (or at least adequate) sampling designs are chosen. By optimal, we mean with respect to some variance-based criterion for the quantity that is to be estimated (e.g., Field et al. 2005; MacKenzie and Royle 2005).

Most studies of carnivore populations focus on understanding or characterizing some aspect of population state or dynamics. Common metrics of population state include extent of occurrence (i.e., distribution), relative abundance, abundance, density, or the response of these measures to habitat structure, landscape structure, or some other environmental covariate. The population dynamic processes of survival and recruitment are also fundamental to many animal population studies.

While an explicit a priori statement of objectives is typical in most scientific investigations, many monitoring programs are established with only an implicit, vague, and ill-defined objective to monitor population status. The assumption is that whatever data are being collected are relevant to at least some important aspect of population state. Generally speaking, many of the basic strategies that are relevant to modeling and inference under carefully designed research studies are also relevant to data obtained as part of a monitoring initiative. The utility of such strategies typically depend more on the particular details of design than on the methods used for data acquisition. In the case of monitoring programs, sound design with respect to particular objectives may arise out of happenstance—or it may not.

## Sampling Considerations

In a conceptual discussion of study design, Yoccoz et al. (2001) identify three key questions that need to be considered when designing a monitoring program: (1) *why* collect the data; (2) *what* kind of data should be collected; and (3) *how* should data be collected. Indeed, these considerations are important to any data collection enterprise. The first question is concerned at some level with study objectives, whereas the second and third largely determine the tangible aspects of the study design (e.g., how much money will be spent and what are the field sampling methods to be used). Previous chapters in this book deal with questions two and three in great detail. A literal interpretation of these key questions alone, however, is insufficient for developing a sampling design. Instead, we should broaden our interpretation (especially of how to collect data) to include elements of when to sample, where to sample, and how many spatial or temporal replicates to conduct.

Even so, the conceptual framework of Yoccoz et al. (2001) does not easily yield a sufficiently detailed and concise framework for developing a sampling design for any particular problem; it does not provide a cookbook or instruction manual for designing carnivore surveys. Rather, it presents a view that is conceptually too broad, and that can only be made useful when combined with a specific method or framework of inference. That is, *how will the data be analyzed to help us achieve our objective?* This requires a rendering in statistical terms of the why, what and how questions. Some of the subsequent sections of this chapter, as well the material in chapter 2, are helpful in formalizing this transition. For example, if the survey objective is based on assessing distribution or occurrence, then the section on *Estimating Occupancy and Distribution* in this chapter provides a reasonably concise rendering of the problem in statistical terms (and chapter 2 discusses considerations for designing such surveys). It should then be fairly straightforward to develop a formal analysis of the statistical design problem (MacKenzie and Royle 2005). Conventional views of power analysis (Link and Hatfield 1990) also provide an example of this formalization (i.e, the rendering of objectives and methods of inference in statistical terms for the sake of informing the design of a study).

The practical difficulty of providing general guidance on sampling design is that, once a framework for analysis is selected, it might well affect the "what" or "how" questions presented by Yoccoz et al. (2001). In this sense, the design of studies becomes a classic chicken-and-egg problem. A more holistic view of sampling design is thus required, as some important design elements arise directly from the statistical framework we impose, and others arise out of the biology of the species in question. In order to develop objectives (the why), one would consider target species' biology and management issues (and consult chapter 2). One might read chapters 2–10 of this book to further explore matters of "what" and "how". Finally, one should learn some statistics to put it all together, come up with a sampling design, and ultimately permit inferences to be drawn from the survey data.

## Statistical Methods

To develop a framework for inference from observational survey data, it is necessary to formalize the relation between the data and the state or process of interest (e.g., abundance, occurrence). Thus, we construct a model, typically mathematical, that makes explicit the dependence of the data on the parameter of interest.

We will deal exclusively with probability models, or models in which data, $y_1, y_2, \ldots, y_n$, are realizations of a random variable (i.e., the outcome of a stochastic process) having some probability distribution which we will denote by the expression $f(y|\theta)$ (read as "the distribution of $y$ given $\theta$"), where $\theta$ is an unknown parameter (and, typically, the object of

inference). As an example, suppose that we survey ten spatial sample units and record whether or not a given species is detected at each unit. Further suppose that the random variable of interest is the number of occupied sample units, denoted by $y$. This random variable can take on any integer value between 0 and 10. A natural probability distribution for use in describing this random variable is the binomial distribution, having sample size 10 and parameter $\theta$, usually referred to as the *probability of success* (e.g., the number of successful detections out of ten tries). We will repeatedly encounter the binomial distribution (or its multivariate generalization, the multinomial) in this chapter.

Other specific models for various types of data and sampling designs are described in subsequent sections. We will use these models to derive inference about the parameter $\theta$—for example, to estimate it or to provide a confidence interval around it. One particularly powerful approach for accomplishing such goals is the method of maximum likelihood (ML), in which estimators of parameters are obtained by maximizing a function—the likelihood—which is related directly to the probability model used to describe the data. The likelihood is usually denoted by the expression $L(\theta|y_1, y_2, \ldots, y_n)$ and read as "the likelihood of the parameter $\theta$ given the data." The maximum likelihood estimator (MLE) of $\theta$ is found by maximizing this function $L$, typically using conventional numerical optimization methods common in most statistical software packages. A number of software programs for estimating parameters under many of the models discussed in this chapter are described later in *Software*.

## Onward

We've presented very cursory treatments of various topics, including objectives, design, statistical methods, and general philosophical considerations underlying animal sampling. The remainder of this chapter is concerned with certain aspects of the question, *how will the data be analyzed to help us achieve our objectives?* As mentioned previously, the answer to this question is not only relevant for its own sake, but is central to a formal assessment of objectives and to developing an appropriate survey design.

It is difficult to cover even one methodological topic exhaustively in a single chapter. Fortunately, there are some excellent synthetic treatments of animal sampling that should be consulted for more details on most of the methods outlined below. Williams et al. (2002), for example, present a modern view of many animal sampling concepts, and MacKenzie et al. (2006) deal with occupancy models exclusively. Two references that focus specifically on carnivore monitoring are Karanth and Nichols (2002) and Henschel and Ray (2003). Finally, we suggest that readers also consult chapter 2 for a more in-depth overview of objectives and design considerations for noninvasive carnivore surveys.

## Estimating Occupancy and Distribution

Occurrence-based measures of population status, such as site occupancy, the proportion of area occupied, probability of occurrence, and distribution of species, are widely used as a basis for the monitoring and assessment of animal populations (Bayley and Peterson 2001; Nichols and Karanth 2002; Tyre et al. 2003; Wintle et al. 2004; Pellet and Schmidt 2005; Weir et al. 2005). Such metrics involve a binary state variable (i.e., present or not present) characterizing the occupancy state of a number of spatial sampling units.

The characterization of sample units by a binary state variable seems particularly appealing for large, territorial carnivores, as any particular sample unit is unlikely to be occupied by more than one or a small number of individuals. Indeed, this approach is being readily adopted for monitoring tigers in India (Karanth and Nichols 2002) and other carnivores (O'Connell et al. 2005; Sargeant et al. 2005; Long 2006). Practical interest in occurrence-based summaries is motivated, at least in part, by the generality,

simplicity, and operational efficiency of implementing survey designs that yield information about occurrence. Models for occurrence require only detection-nondetection data and, consequently, sampling is efficient in many field situations and across a variety of taxa. Conversely, estimation of more detailed descriptors of population state such as abundance, survival, and recruitment typically require considerably more sampling effort in order to obtain information at the level of the individual animal.

The issue of nondetection bias is important when objectives involve occurrence-based summaries of population status (MacKenzie et al. 2002; Nichols and Karanth 2002). In the presence of imperfect detection, an observed zero (i.e., putative absence) is ambiguous because it can result for one of two reasons: (1) the species was not present at the sample unit, or (2) the species was present but was undetected during sampling. Thus, the inference problem, when confronted with ambiguous observations of detection-nondetection, is to properly partition observed zeros into *sampling zeros* (i.e., presence but nondetection) and fixed or *structural zeros* (i.e, nonpresence). The issue of detection bias in detection-nondetection surveys of animal populations has received considerable attention in recent years, with much effort focused on evaluating its effects (Moilanen 2002; Gu and Swihart 2004) and the development of methods to accommodate it (Zielinski and Stauffer 1996; Bayley and Peterson 2001; MacKenzie et al. 2002; Royle and Nichols 2003; Tyre et al. 2003; Wintle et al. 2004).

Explicit interest in detection probability when the object of inference is occupancy state can be motivated by a simple power analysis in the context of an ecological assessment problem. For example, suppose a proposal exists to convert a patch of landscape from forest to houses, and an assessment is required as to whether this patch is presently providing habitat for a carnivore species of conservation interest. Further, using an appropriate sampling method for the target species, suppose unambiguous evidence of its presence is obtained (e.g., DNA from scat, a photo from a remote camera). At this point, the role of the biological assessment is complete, and society must make value judgments relating to population status of rare carnivores and proposed land use.

Suppose, on the other hand, that the species was *not* detected by the sampling activity. Is it safe to conclude that the species does not exist there? Obviously, this depends on the adequacy of the sampling apparatus, and the issue of detection probability is directly relevant—and indeed critical—to this assessment. To formalize the problem, consider the null hypothesis to be that a site is unoccupied by the species in question, and the alternative to be that the site is occupied. We might write $H_0 : z = 0$ versus $H_a : z = 1$, where $z$ is the binary occupancy status of the site. In keeping with conventional views of experimental design in statistics, we would like to design a study to achieve some specified power of rejecting the null *if* the alternative holds. That is, we specify $\Pr(\text{reject } H_0 | H_a)$—read as "the probability of rejecting $H_0$ given $H_a$— a priori, and then set the sample size to achieve this probability. In classical statistics, $\Pr(\text{reject } H_0 | H_a)$ is referred to as power. And, as it turns out, this probability is a function of the detection probability parameter, $p = \Pr(\text{species is detected} | \text{species is present})$, the probability of detecting the species when it is present. The precise nature of the relationship between power and detection probability depends on the sampling protocol.

As an example, suppose sampling is conducted via scat detection dogs (see chapter 7), and the site is surveyed in a systematic manner (let's say [dog effort]/area is held constant). Further assume that the site is surveyed $T$ times such that the $T$ replicate samples are independent of one another. In this case, the power to detect the species is given by the expression $1 - (1 - p)^T$, where $p$ is the per survey probability of detection. Obviously, power increases rapidly with both $T$ and $p$, and this allows one to formally evaluate various methods and designs in order to achieve desired levels of power (see chapter 2 for additional discussion of this topic).

The practical problem is that one will not generally know the value of $p$ that applies to any particular survey method, target species, or scenario. Heuristically, we might seek sites that are known to be occupied and subject them to repeated sampling in order to estimate detection probability. Such approaches, however, are generally not practical for most carnivore survey scenarios. Here we formalize the estimation of detection and occupancy probability parameters for data collected at sites for which the true occupancy state is unknown.

## Estimating and Modeling Site Occupancy in the Presence of Imperfect Detection

We introduce the technical framework of this section by considering the problem of estimating site occupancy, or the proportion of occupied sites, based on a sample of $i = 1, 2, \ldots, R$ sites. Suppose that each site is characterized by its unobserved binary *occupancy status*, $z_i$, which takes on a value 1 if the site is occupied, and 0 if it is not. The parameter of interest is $\psi = \Pr(z = 1)$, which is the average value of $z$ over some hypothetical collection of replicate sites. The parameter $\psi$ is often referred to as site occupancy, or the proportion of sites occupied, or—as we will often say here—the probability of occurrence or the probability that a site is occupied.[a] The key concept underlying subsequent development is that we are unable to observe the occupancy status; rather, we make an imperfect observation of occupancy state $y_i$ which may be 0 or 1 for an occupied site ($z = 1$). Whereas, for an unoccupied site ($z = 0$), $y_i$ must take on the value 0 with probability 1[b].

---

[a] As commonly used, site occupancy is usually understood to be synonymous with both the proportion of occupied sites in a sample of size $R$ and also the probability that a site is occupied, $\psi$. Strictly speaking, however, the former is a realization of random variables, whereas $\psi$ is a population parameter. See MacKenzie et al. (2006) and also Royle and Kéry (2007) for a discussion of the distinction.

[b] This is in keeping with the convention that false positives are not possible. While this assumption is debatable in some instances, it yields a conceptually clear and concise development of this class of models. Moreover, some recent work suggests that this assumption may not be required. See Royle and Link (2006).

In practice, models are frequently developed for $y$ under the assertion that $y$ is a perfect measure of occupancy status. For example, logistic regression is commonly applied to binary "presence-absence" data for the purpose of evaluating the relationship between occurrence and habitat. Note, however, that $E[y] = p\psi$. Thus, unless $p$ is sufficiently large, substantial errors can result from the naïve use of data to estimate occupancy rates or factors that influence probability of occurrence because $p$ is confounded with $\psi$. The problem addressed next is the formal estimation of both $p$ and $\psi$ in the presence of imperfect detection.

One way to obtain information about $p$ is to obtain replicate observations for each spatial sampling unit (MacKenzie et al. 2002; Nichols and Karanth 2002). Suppose that each site is surveyed $t = 1, 2, \ldots, T$ times, yielding the encounter history (alternately known as a detection history) for site i: $\mathbf{y}_i = (y_{i1}, y_{i2}, \ldots, y_{iT})$. For example, a hypothetical site is surveyed $T = 3$ times and yields the encounter history $(1, 1, 0)$, indicating that the species was detected during the first and second surveys but not the third. For clarity, we will assume that detection probability, $p$, is constant over the $T$ replicate surveys, and that the population is closed. In this case, for purposes of estimating $p$ and $\psi$, we can summarize the data by the site-specific totals $y_i = \sum_{t=1}^{T} y_{it}$. The probability distribution of the total number of detections, $y_i$, can be specified as a function of parameters $p$ and $\psi$. This is closely related to the simple binomial probability distribution (with success probability $p$), but has the additional parameter $\psi$, which can be thought of as controlling the number of extrabinomial zeros.

Under this design, in which replicate data are taken at each of the $R$ sites, maximum likelihood estimates of the parameters $p$ and $\psi$ can be obtained by numerically maximizing the likelihood function. Numerical maximization can be achieved using software packages such as R (Ihaka and Gentleman 1996; R Development Core Team 2005), S-PLUS (Insightful Corporation, Seattle, WA), SAS (SAS Institute, Cary, NC), or WinBUGS (Medical Research Council Biostatistics Unit, Cambridge, UK; Gilks et al. 1994), or specialized software such as PRESENCE (Proteus Wildlife Research Consultants, Dunedin, New Zealand) or Program MARK (White and Burnham 1999). See *Software* for further discussion.

Nichols and Karanth (2002) noted the duality between the problem of estimating $\psi$ and estimating the size of a closed population. In both cases, the data are sequences of Bernoulli observations (i.e., independent binary random variables—"coin flips") with parameter $p$ representing the probability of detection. By equating the detection history obtained by replicate detection-nondetection surveys to the capture history in classical capture-recapture analyses, many useful generalizations of the basic site occupancy model described above can be developed. For example, there may be covariates that influence detection probability, such as extant environmental conditions during sampling, or features of the sampling protocol (e.g., effort, method). In addition, a common focus of ecological studies is the assessment of factors thought to influence occupancy, such as summaries of habitat or landscape structure. To accommodate these possibilities, we require a more general formulation of the likelihood in terms of the full encounter histories (instead of only the total number of detections).

As a rule, the probability of observing any particular encounter history can be represented as a function of sample-specific detection parameters, $p_i$ and the occurrence probability parameter $\psi$. For instance, if observations are independent across samples, then the probability of observing detection history $\mathbf{y}_i = (1, 1, 0)$ for a site that is occupied is $\Pr[(1, 1, 0)|z = p_1 p_2 (1 - p_3)$. That is, the probability of this history equals the joint probability of detecting the target species given that it occurs there on occasions 1 and 2, and not detecting it (despite its presence) on occasion 3. The corresponding probability for an unoccupied site is, assuming that false-positives cannot occur, $\Pr[(1, 1, 0)|z = 0] = 0$. The likelihood is constructed as an average of these two pieces, weighted by the corresponding probabilities of each piece. That is, $\Pr[(1, 1, 0)] = p_1 p_2 (1 - p_3) \psi + 0 \times (1 - \psi)$. In general, there are $2^T$ possible encounter histories, and the likelihood can be described in terms of parameters $p_1$, $p_2$, and $p_3$ and $\psi$ (see MacKenzie et al. 2006).

The detection history formulation of the likelihood can also be used for developing extensions of the detection probability model, for example, when site-specific or sampling occasion-specific covariates that influence detection probability can be recorded or measured. While this is only slightly more cumbersome, we leave the technical details aside and instead refer the interested reader to MacKenzie et al. (2006).

Another very useful extension of the site occupancy modeling framework described here applies to the case where $\psi$ varies among sites as a function of measurable covariates (e.g., those that describe landscape or habitat structure). A common way to incorporate such covariates is to use the logit link (e.g., Agresti 1990):

$$u_i \equiv \mathrm{logit}(\psi_i) = \beta_0 + \beta_1 x_{i1} + \ldots + \beta_k x_{ik},$$

where $x_{ij}$ is the value of the $j$th covariate ($j = 1, \ldots, k$) at site $i$ and $\{\beta\}$ are the parameters to be estimated. Such models would typically be considered in the development of maps of range and distribution (e.g., Long 2006). See box 11.1 for a worked example of this approach.

## Temporally Dynamic (*Multiple-Season*) Occupancy Models

MacKenzie et al. (2003) extended the basic site-occupancy model described in the previous section to an open system, allowing for local extinction and colonization as important demographic processes for modeling temporal dynamics of species occurrence. These authors considered a situation with multiple primary sampling periods (which they referred to as seasons), within which there were multiple secondary sampling periods akin to the robust

design of Pollock (1982; described in *More Flexible Models* later in this chapter). Closure is assumed within a primary period but is not a necessary assumption between primary periods. MacKenzie et al. (2003) refer to this as a "multiple-season" model and provide a likelihood-based framework for making inferences about model parameters. We sidestep the overly technical details of likelihood-based analysis of this model but instead attempt to provide a simplified and concise description of the model structure using the formulation of the model described by Royle and Kéry (2006).

Denote the occupancy state of site $i$ during season $t$ as $z(i, t)$. We assume that if a site is occupied at any point in time, its probability of "surviving" (i.e., remaining occupied) until the next time period is $\phi$. Note that MacKenzie et al. (2003) describe the model in terms of local extinction, $\varepsilon = (1 - \phi)$, rather than survival rate. Secondly, we suppose that if a site is unoccupied at any point in time, then the probability that it becomes occupied (i.e., is "colonized") by the next time period is $\gamma$—which is referred to as *local colonization* probability and analogous to a recruitment rate in conventional population ecology. These processes are manifest in the following model for occupancy: The initial occupancy states $z(i, 1)$, are assumed to be independent and identically distributed Bernoulli random variables with parameter $\psi_1$:

$$z(i, 1) \sim \mathrm{Bernoulli}(\psi_1).$$

Then, the occupancy state transitions from time $t - 1$ to time $t$. Thus, for $t = 2, 3, \ldots, T$ are governed by a Bernoulli distribution in which the success probability is conditional on the previous state:

$$z(i, t) | z(i, t - 1) \sim \mathrm{Bernoulli}(\phi[z(i, t - 1)] + \gamma[1 - z(i, t - 1)]).$$

This expression is understood as follows: for a site that is occupied at time $t - 1$ (i.e., $z(i, t - 1) = 1$), the survival component of the model (the first component of the Bernoulli success probability) is

---

## Box 11.1

### Estimating occupancy with fisher survey data from California

To demonstrate occupancy estimation, we consider forest carnivore survey data described in Zielinski (1995) and Zielinski et al. (2005). This survey focused on determining the current distribution of mesocarnivores in forests of northern and central California, and comparing this distribution to mapped records of occurrence collected approximately 100 years prior (Grinnell et al. 1937). The distribution of fishers (*Martes pennanti*), in particular, had been reported to have declined on the basis of an earlier interpretation of preexisting survey data (Zielinski et al. 1995a).

Over a seven-year period, researchers surveyed 464 sample units that were systematically located across much of the forested area of this region. Surveys comprised an array of six track plates (Zielinski 1995) and one remote camera at each unit (Zielinski et al. 2005). Sample units were located by selecting every other grid location from the nationwide Forest Inventory and Analysis (USDA Forest Service) survey grid, resulting in a distance of 7–10 km between the centers of adjacent units. The track plates and cameras were checked for evidence of detections every two days for a total of sixteen days (i.e., eight occasions). The surveys confirmed a previously described 400-km gap in the distribution of fishers in the Sierra Nevada, and failed to detect two rare species: the wolverine (*Gulo gulo*) and the red fox (*Vulpes vulpes*). Data also described a fragmented distribution of American martens (*Martes americana*) in the Cascades of northeastern California, which did not appear to exist in historical accounts. The distributions of most of the generalist predators remained unchanged (Zielinski et al. 2005).

To estimate site occupancy, we will begin by assigning a detection at a given unit if any of the six track plate stations registered a detection (remote camera data were not included in this analysis). A total of 464 stations were visited $T = 8$ times, yielding the following frequencies of detection (corresponding to the number of sites with 0, 1, 2, . . . , 8 detections): (400, 16, 12, 12, 5, 10, 3, 4, 2). Thus, for example, fishers were not detected at four hundred sites and were detected during all eight visits at two sites. Application of the simple occupancy model described in *Estimating and Modeling Site Occupancy in the Presence of Imperfect Detection* (this chapter) yields parameter estimates $\hat{p} = 0.40$, and $\hat{\psi} = 0.14$. Note that the naïve (observed) occupancy rate was 0.138, and thus there is little adjustment in (estimated) occupancy due to imperfect detection. This is because the (estimated) power to detect fishers with the given survey method and design (assuming the track plates were detecting animals independently) was $1 - (1 - 0.40)^8 = 0.98$.

We note that the six track plates can be viewed as subsamples of the primary spatial sample units. As the subsamples are relatively close together in space, subsample detections may not be independent. Pooling the subsamples is one valid treatment of such data, but other formulations such as the newer multiple-device approaches described in the chapter are also possible.

---

operative, and $z(i, t)$ is a Bernoulli trial with probability $\phi$. Conversely, if a site is *not* occupied at time $t - 1$ (i.e., if $z(i, t - 1) = 0$), then the recruitment component of the model is operative, in which case $z(i, t)$ is a Bernoulli random variable with probability $\gamma$. Thus, occupied sites survive with probability $\phi$, and unoccupied sites become "colonized" with probability $\gamma$.

The observation model, specified as conditional on the occupancy state variables $\{z(i, t); \forall i, t\}$, is a Bernoulli process with parameter $z(i, t)p_{it}$ and is denoted by:

$$y(i, t)|z(i, t) \sim \text{Bernoulli}(z(i, t)p_{it}).$$

Thus, if the site is occupied at time $t$, the data are Bernoulli trials with parameter $p$, whereas if a site is unoccupied at time $t$, then the data are Bernoulli trials with $\Pr(y(i, t) = 1) = 0$.

A number of strategies are available for fitting this model. Barbraud et al. (2003) suggested a method of analysis based on analogy with classical capture-recapture models (Cormack-Jolly-Seber models), whereas MacKenzie et al. (2003) developed a formal likelihood-based procedure for estimating model parameters. Royle and Kéry (2007) note that the hierarchical structure of the model—that is, the state-space formulation just described—is directly amenable to Bayesian analysis using simulation-based methods known as Gibbs sampling and Markov chain Monte Carlo (see discussion of Win-BUGS in *Software*).

## Assumptions and Potential Effects of Assumption Violation

The occupancy models described in the previous sections rely on two classes of assumptions. First are those assumptions pertaining to the observations, conditional on the true occupancy state of each site (i.e., assumptions related to data collection), and second are those related to the underlying occupancy state both within a site (i.e., among sampling occasions) and among sites (i.e., assumptions of the process under investigation).

Regarding the first type of assumption, these models assume that the binary observations made at an occupied site are independent, meaning that detection probability does not change in response to previous detection or nondetection. In principle, this assumption could be modified in specific ways (e.g., to allow for an increase in detection probability subsequent to initial detection) by the development of more general models. Such generalizations might have application to certain carnivore sampling problems. Consider surveys conducted by scat detection dogs (chapter 7). If the scat is removed during the first sampling occasion (e.g., for analysis in the laboratory), this may lead to a decrease in the probability of detection in subsequent samples. Conversely, the dog, or handler, might develop an efficient search strategy during early sampling occasions, thereby increasing the probability of detection during subsequent occasions. Other instances that might result in sampling dependence include situations in which previous detections cause an observer to sample with increased effort because they know the target species to be present, or in which sampling devices are baited such that, once animals discover the device, they are more likely to return in later periods or to visit adjacent devices at the same sample unit. Such models have been developed for modeling behavioral responses in classical capture-recapture settings (Yang and Chao 2005) but they have not been applied to occupancy surveys to the best of our knowledge.

Two assumptions are also made regarding the occupancy state variables, $z_i$. The first assumption is analogous to the closure assumption in classical capture-recapture models. In the context of occupancy models, the equivalent assumption is that the occupancy status for each site, $z_i$, is constant across sampling periods (or within secondary sampling periods in the case of the model described in *Temporally Dynamic [Multiple-Season] Occupancy Models*). If occupancy status is not constant, however, and changes in occupancy are random, then the method still gives an unbiased estimate—although the estimate should be considered to represent site *use* rather than site *occupancy*.

Another important assumption pertaining to $z_i$ is that the occupancy states are independent among sites, so the $z_i$ also constitute an independent sample of Bernoulli random variables. That is, occupancy status at each site is independent of occupancy status at other sites. In fact, this assumption is not critical to the estimation of site occupancy, as it can be violated and still yield unbiased estimates of site occupancy. In general, though, estimated variances will be negatively biased such that, for example, confidence intervals will be too narrow. Alternatively, the model may be extended to explicitly permit certain types of dependencies (e.g., see Sargeant et al. 2005; *Predicting Occurrence without Site Covariates* in chapter 2).

## Model Extensions and Other Topics

Thus far we have dealt largely with estimating site occupancy from basic detection-nondetection data. In some cases, however, researchers may be interested in questions that are more complex than simple occupancy, or may wish to use data that were collected in ways inconsistent with the assumptions of the basic occupancy model. In such cases it may be possible to use extensions of the simple occupancy model.

*Multiple-Device Sampling*

It is common for carnivore surveys to employ multiple detection devices at the same site, either in an array of similar devices (e.g., five track plates with prescribed spacing) or as a mix of different devices such as track plates and remote cameras. The main methodological consideration in such cases is that of dependence among devices in the sense that if detection occurs via one, during a particular interval, then it is probably more likely to occur via others. This issue is primarily due to the movement of carnivores about their range. An individual animal may not be exposed to an array of devices during certain sampling periods, but once it moves into the vicinity and encounters a single device, it will likely be exposed to the entire array.

A relatively new conceptual framework for dealing with such dependency issues has been developed and is equivalent to the "robust design" (Pollock 1982) in which surveys are repeated over T primary periods comprised of K secondary periods (see *More Flexible Models*). In the multiple device scenario, the primary periods are the sampling occasions, and the secondary periods are the multiple devices (J. D. Nichols, USGS Patuxent Wildlife Research Center, pers. comm.). If an individual is available for detection during a given primary period, then it is available for detection by all subsample devices (even though it may not actually be detected). During other sampling occasions, the animal may not be available and, therefore, not exposed to any of the devices. This situation is analogous to temporary emigration (Kendall et al. 1997) in classical capture-recapture theory, and the robust design framework provides a method to deal with such differences in availability. This approach will soon be implemented in the multimethod option of program PRESENCE (L. Bailey, US Geological Survey, pers. comm.).

While the basic formulation of the problem as an application of the robust design seems to resolve the treatment of multiple devices, we believe that the further development of models and inference in these problems is an important area of research. In particular, the spatial and temporal scale of sampling is fundamental to the interpretation of the detection and availability parameters, and the dependence structure within primary samples.

*Spatial Models*

The site occupancy modeling framework generalizes to accommodate explicit spatial structure, such as that representing contagion or similarity in occupancy status among sites. Such models are commonly used for image restoration and are widely applied in many disciplines. A recent paper by Sargeant et al. (2005) considers one class of models modified to allow for imperfect observation of the state variable. These models are similar to the classic autologistic model (Augustin et al. 1996), but also allow for false negative errors (see further discussion in chapter 2 under *Predicting Occurrence Without Site Covariates*).

*Multispecies Models*

The extension of the occupancy modeling framework to multiple species has been considered in several different contexts. MacKenzie et al. (2006) developed models that allow for the estimation of co-occupancy among species. As the number of parameters in these models increases rapidly with the number of species, they are most useful when interest is focused on a small number of species (say, two to five). Dorazio and Royle (2003, 2005) developed models of animal community structure and species richness based on species-specific models of occupancy.

*Relationship Between Abundance and Occurrence*

One rationale for choosing to focus on occupancy as a summary of demographic state is that it is believed to be related to abundance or density of a species in a spatially averaged sense. That is, common species that are more abundant will tend to occupy more of the available habitat, and the reverse is true of uncommon species. This issue has been addressed in some detail by He and Gaston (2003), Royle and Nichols (2003), Stanley and Royle (2005), and Royle et al. (2005).

An intuitive method for obtaining abundance estimates from estimated occupancy is to equate $\hat{\psi}$ to the proportion of area occupied by the species. Then, given an independent estimate of the typical home range size of the species (say, $H$), an estimate of total population size is $\hat{N} = \dfrac{A\hat{\psi}}{H}$ where $A$ is the total survey area. This approach, however, requires that a number of assumptions be met (e.g., one individual per home range consistently across the survey area)—some of which may be unrealistic for carnivores.

A formal, model-based approach was described by Royle and Nichols (2003). This approach recognizes that variation in abundance among sample units is manifested as heterogeneity in detection probability (e.g., of the type described by Dorazio and Royle 2003). Specifically, if $p_i$ is the probability of detecting a species at site $i$, $N_i$ is the number of animals exposed to sampling (i.e., the local abundance), and $r$ is the probability of detection of *individual* animals, then $p_i = 1 - (1 - r)^{N_i}$—assuming individuals are detected independently. Under certain model assumptions, parameters of the local abundance distribution ($N_i$) can be estimated using conventional methods (e.g., MLE). In addition, estimates of individual local abundance parameters (i.e., values of $N_i$ for specific locations) can also be obtained (Royle 2004b).

## Inference about Relative Abundance

Much of the previous discussion in this chapter has focused on estimating state variables from sample data subject to imperfect detection using methods that explicitly model detectability and incorporate other demographic parameters of interest (e.g., local extinction and colonization). Application of these procedures requires specific sampling designs that yield information about both types of parameters. Unfortunately, in practical sampling situations, it is not always possible to employ such designs or to be sufficiently satisfied with the validity of the assumptions (e.g., independence of observations) to have confidence in resulting inferences. In contrast to approaches where the detection process is made explicit, there are many survey and analytical methods that ignore detection bias altogether—because the survey design is insufficient, the assumptions are questionable, or the aims and goals of the study do not require an explicit accounting of detectability. In these cases, the object of inference is often asserted to be relative abundance.

Unfortunately, relative abundance is not a precise term, and it is difficult to describe general classes of methods for making relevant inferences. Indeed, to the best of our knowledge, no synthetic treatments of inferential methods for relative abundance exist. Unlike abundance, occupancy, or dynamic attributes of populations (e.g., survival, recruitment), relative abundance is not a quantity inherent to an animal population. Rather, the definition of relative abundance is in the eye of the analyst.

Lacking explicit interest in population size, $N$, one definition that facilitates a reasonably general development of an inferential framework is that relative abundance is the expected value of the observations at some point in space and time. That is, we let $y_{it}$ denote some count statistic for a sample unit $i$ during time $t$. Then, we define $E[y_{it}] = \mu_{it}$ as the relative abundance (i.e., an index) and focus on developing models of $\mu_{it}$. This strategy seems reasonable when certain constraints are imposed on detectability. In the presence of imperfect detection, where $y_{it}$ is a random sample from a population of size $N_t$ with detection probability $p_{it}$, then $E[y_{it}] = p_{it}N_t$. Thus, for example, relative changes in population size over time may be estimated from observed counts without bias, provided that $p_{it} = p$ (i.e., that detectability is constant among replicates and across time, so that the expected value of the observed counts is proportional to abundance). This is unlikely in most practical settings because detectability can be influenced by a number of factors, including observer ability, observer effort, time of day, weather, and habitat.

A slightly more generic view relaxes the strict equality assumption regarding detection probability in favor of an assumption in which $E[p_{it}]$ = constant. That is, there may be variation in detection probability (spatially or temporally), but the mean detection probabilities are equal over suitable spatial units or temporal periods. A minor modification of this assumption occurs when factors that influence detectability can be identified and modeled explicitly. Such notions underlie the trend modeling efforts of Link and Sauer (1998a, 1999). The validity of such treatments seems most reasonable under rigorous standardization of sampling protocols.

Modeling and inference from count survey data can be carried out using standard statistical methods for generalized linear models (GLMs; McCullagh and Nelder 1986). As there are very few special considerations for the application of such methods to animal survey data, we refer readers to Agresti (1990) for a detailed discussion of GLMs. Below we discuss two common situations in which GLMs might be applied to animal survey data.

## Poisson Regression

Count data are widely used in studies of avian populations and avian monitoring—for example, the North American Breeding Bird Survey (BBS; Robbins et al. 1986; Peterjohn 1994; Peterjohn et al. 1994) and the annual Christmas Bird Count (CBC; Butcher 1990; Link and Sauer 1999)—and numerous models have been developed to analyze such data for patterns of change through time and space (e.g., Sauer and Droege 1990; Link and Sauer 1998a, b, 1999). Despite the fact that such models have been developed primarily for bird count data, the relevant issues confronting carnivore biologists are similar to those faced by avian biologists, and many of the models developed for birds can be applied directly to

carnivore count data. For example, similar to point counts from songbird surveys, it may be possible to use detections of individual carnivores obtained via remote camera surveys (for individually identifiable species) or DNA analysis from collected hairs or scats. The important assumption is that the count statistic (or its expectation) is proportional to the number of individuals in the sampled population. Although scat or track counts could also be analyzed with Poisson methods, it is difficult to determine under what circumstances such statistics would be related to population size (see chapters 2, 3, and 4 for pitfalls of using sign counts as relative abundance indices).

As discussed earlier, count data are not censuses and cannot typically be used to estimate carnivore abundance because the counts represent an unknown fraction ($p$) of the true population. Furthermore, patterns of change in count data may provide a biased view of patterns of change in population size, as the process of collecting count data introduces numerous nuisance factors that can create patterns in counts unrelated to abundance (Link and Sauer 1998b, 1999). Such nuisance factors might include effort expended by observers, observer ability or the sensitivity of the detection device, and the particular method of observation or detection. Consequently, the use of count data to estimate relative abundance requires robust models with realistic assumptions that link the count data with a reasonable measure of relative abundance. Link and Sauer (1998a) provide a model for inference about changes in relative abundance that was developed for CBC count data, but that is also applicable to carnivore count data. For a more thorough treatment of the estimation of relative abundance in space under such loglinear models, see Link and Sauer (1998b), as well as publications describing extensions of the basic idea that have arisen subsequently (Link and Sauer 2002; Thogmartin et al. 2004).

## Binomial (Logistic) Regression

Frequently, animal surveys yield binary indicators of detection or nondetection, or aggregations of detections over some fixed number of sample units. In either case, the binomial distribution is a natural choice for describing variation in such data. The binomial GLM is most often referred to as *logistic regression* and is probably one of the most common classes of models applied to ecological data. Such models are widely adopted, indeed almost universally, for modeling the distribution and range of carnivores. The occupancy models described earlier in the chapter can be thought of as an extension of the class of logistic regression models to account for imperfect detection.

For binary data, we have a binomial distribution with sample size one (typically referred to as a Bernoulli random variable), and in the more general case (for aggregated Bernoulli random variables), we have a binomial distribution with sample size $K$—the number of "trials" that were aggregated. As an example, suppose a network of scent stations is established along a roadside route (see Sargeant et al. 1998 for an example), where a route comprises ten distinct stations. We might consider pooling the data from these ten stations and asserting that the total number of detections represents the outcome of a binomial random variable with sample size $K = 10$ and parameter $\theta$.

Without replication, it is not possible to disentangle variation due to the detection process from that due to occurrence (unlike with the class of models considered in *Estimating Occupancy and Distribution*). Thus, the binomial parameter $\theta$ in this problem represents the aggregation of detection and occurrence probabilities. Nevertheless, we may develop models for $\theta$ in precisely the same manner as for loglinear models under *Poisson Regression*, except that here, instead of the log-link, the more natural link function for the binomial distribution is the logit link. As an example, let $y_i$ denote the observed number of detections out of $K$ trials for route $i$, and let $x_i$ denote the value of some measured covariate (e.g., describing landscape structure) relevant to route $i$. We suppose then that $y_i$ has a binomial distribution with parameter $\theta_i$, where:

$$\log(\theta_i/(1 - \theta_i)) = \alpha + \beta x_i$$

Within this framework, very general models describing variation in the parameter $\theta_i$ may be developed, including models possessing random effects, spatial variation, trend models, and so forth. We caution, however, that because it is not possible to estimate and incorporate false-negative errors, logistic regression approaches should be limited to situations where overall detectability is known to be high (e.g., the fisher example provided in box 11.1).

# Estimating Abundance

Estimating population size from noninvasive survey data shares many concepts with other abundance estimation problems that use information taken from actual animal captures. Here we consider data analysis issues for methods where individual identification of animals can be obtained. Noninvasive survey data also add some unique components to abundance estimation because identification is not always certain.

## Basic Ideas and Model Structure

Similar to true capture-based methods, noninvasive capture-recapture approaches for abundance estimation entail the detection (capture) and redetection (recapture) of animals that can be individually identified (i.e., effectively "marked"). Samples (e.g., hair, feces, photos) are collected at several points in time and individuals are matched across sampling occasions, enabling the construction of a detection or encounter history. Similar to encounter histories recorded for occupancy estimation, histories for abundance estimation comprise a vector of 1s and 0s indicating sampling occasions during which the animal was or was not detected (as opposed to the record of detections and nondetections of species at sites). For example, suppose a four-occasion survey is conducted for mountain lions (*Puma concolor*), and a given lion is detected on occasions 1, 3, and 4. Its encounter history is thus (1, 0, 1, 1). This history—the elements of which are random variables—serves as the data structure for the method.

Capture-recapture estimators of abundance ($N$) follow a common theme. The estimators take the form of a count of the number of individuals detected ($C$) divided by a function of detection probability ($p^*$):

$$\hat{N} = \frac{C}{p^*}.$$

A large body of work presents various ways to estimate abundance from capture-recapture data given a few basic assumptions, as well as specific assumptions about detectability (Seber 1982; Williams et al. 2002). As for the basic assumptions, the population is first assumed to be demographically and geographically closed as previously described in *Sampling Considerations*. Second, all individuals are assumed to have some chance of being detected. Thus, there cannot be a segment of the population that is unavailable for some reason (e.g., a fraction of the population is hibernating). Third, it is assumed that all individuals have an equal probability of being detected. Finally, all individuals are assumed to be correctly identified. These assumptions allow us to build a statistical model relating data to parameters such that abundance can be estimated from capture-recapture data. These assumptions may be relaxed in more general models to better accommodate realities of field conditions.

Much of the work in abundance estimation builds off of the multinomial likelihood models discussed in Otis et al. (1978). These models allow detection probability to vary across sampling occasions and to change in response to detection. In general terms, such models recognize that encounter histories can take on only a discrete number of possible values. Thus, the encounter history frequencies have a multinomial distribution, with cell probabilities that are functions of detection probability parameters, and a total sample size, $N$, which is the population size to be estimated. The cell probabilities are, precisely, the probability of obtaining each potential encounter history. Suppose $h_i$ is an encounter history (e.g., [1, 0, 1, 1] from above), and $n_{hi}$ is the number of

animals with encounter history $h_i$. The probability of the encounter history is developed by describing the events that are required for the encounter history to be observed. Returning to the sample mountain lion encounter history (1, 0, 1, 1) above, the probability of the history is:

$$p_1(1 - c_2)c_3c_4,$$

where $p_1$ is the probability of capture in period 1, and $c_t$ are the corresponding recapture probabilities in periods $t = 2, 3, 4$. The terms *capture probability* and *detection probability* are often used synonymously in noninvasive sampling because the relevant statistical methods were originally developed for trapping studies. Various approaches to describing the probability of an encounter history allow the analyst flexibility in model building. For example, if detection probability is nearly equal across sampling periods, the probability of the encounter history could be simplified by setting the $p$'s and $c$'s constant:

$$p(1 - c)cc.$$

We will use this ability to model the encounter history as our framework for dealing with assumption violations.

Abundance and capture probability can be estimated using maximum likelihood. To more easily estimate population size, it can be defined as the sum of the number of animals detected ($M_{t+1}$) and the number of animals not detected ($f_0$). Given this definition, only $f_0$ needs to be estimated in order to estimate population size. The log-likelihood function is defined as:

$$\log L(f_0, \mathbf{p}, \mathbf{c}) \propto \log((f_0 + M_{t+1})!) - \log(f_0!) + \sum_{i=1}^{M_{t+1}} \log(\Pr[h_i]).$$

The values of the parameters that result in $\log L$ achieving its maximum value are the best estimates of the parameters. The maximum value of $\log L$ is also important for model selection methods such as the information-theoretic approaches of Burnham and Anderson (2002).

One common assumption violation is the presence of individual variation in detection probability, commonly referred to as individual heterogeneity. When individual animals have different detection probabilities, animals with higher detection probabilities are more likely to be detected first and more often. If this variation is not taken into account, detection probability is overestimated and population size is thus underestimated.

Many attempts have been made to deal with individual heterogeneity in detection probability during data analysis. For example, Burnham and Overton (1979) proposed a jackknife estimator to correct for bias caused by individual heterogeneity. The jackknife estimator essentially estimates population size repeatedly with a different subset of animals removed from the analysis each time, then averages all of the estimates. This is a useful estimator, but it is outside of the likelihood framework and limits our ability to use contemporary model selection procedures (e.g., Burnham and Anderson 2002). Alternatively, Norris and Pollock (1996) and Pledger (2000) proposed modeling detection probability as a finite mixture. Finite mixture models probabilistically split the population into "groups" having equal detection probability, but allow these probabilities to differ. This approximates continuous variation and tends to produce good estimates of abundance in the face of individual heterogeneity. In other approaches, Dorazio and Royle (2003) proposed continuous mixture models to account for heterogeneity, and Royle (2006) provided a summary of different heterogeneity models relevant to occupancy-type models. No single method, however, has completely resolved the problems that arise with individual heterogeneity. One basic design guideline that helps to reduce bias resulting from individual heterogeneity is to design the survey such that probability of detection is quite high (see chapter 2). When this is the case, most animals are detected and there is little room for bias. Indeed, well thought-out survey designs can help to alleviate many causes of detection heterogeneity before

they affect the data and subsequent analysis (see chapter 2 as well as individual survey method chapters for further discussion).

## Identifying Individuals via Genetic Data

The requirement that detected individuals be individually identifiable typically limits the application of noninvasive capture-recapture approaches to methods that either collect genetic samples (i.e., hair or scat-based methods) or enable individual identification via photographs taken with remote cameras (although see chapters 3 and 4 for some exceptions based on tracks). A unique set of issues arise when using noninvasive genetic data to identify individuals, and in turn to estimate abundance. More specifically, two types of constraints must be addressed: (1) there is no information available regarding marks possessed by the population as a whole, and (2) sampling occasions may not be well defined. Each of these issues can be overcome with suitable attention during design and analysis. In addition, other sources of data, such as from telemetered or harvested animals, are often available to supplement information from noninvasive samples.

When genotypes are used to identify individuals from hair, scat, or other relatively low-quality DNA sources, uncertainty exists in individual identification. While it is commonly argued that physical tags placed on animals during true capture-recapture surveys are sometimes misread—just as genotypes can occasionally be erroneous—there is a fundamental difference between misread tags and incorrect genotypes. When researchers mark animals in a tagging study, they record marks on a list and refer to this list when animals are recaptured. If a given tag number is not found on the list, the observation is known to be wrong and is immediately corrected or discarded. But when genotypes are used to identify individuals, no such list exists. All animals have a genotype, of course, but because it is not known which genotypes exist in the population as a whole, this information must be inferred. Thus, when an incorrect genotype is produced, there is no obvious error-checking mechanism beyond that of careful laboratory protocol. Incorrect genotypes may be unknowingly added to the data set, causing an underestimation of detection probability, the count of individuals to be too large, and a resulting overestimation of abundance.

Genotyping error rates can be estimated from capture-recapture data, however. Lukacs and Burnham (2005a) present an extension to the closed-population capture-recapture models that directly estimates genotyping error and produces unbiased estimates of abundance. The model assumes that genotyping errors result in genotypes that do not exist in the actual population and will therefore never be detected twice. The probability of an encounter history is then expanded to consider the possibility that genotypes detected only once may be incorrect. The abundance estimator is adjusted to account for genotyping errors such that:

$$\hat{N} = \hat{\alpha}(\hat{f}_0 + M_{t+1}),$$

where $\alpha$ is the probability of correctly genotyping an individual, $f_0$ is the number of genotypes not detected, and $M_{t+1}$ is the number of genotypes detected. If $\hat{\alpha} = 1$, the estimator simplifies to the standard closed-population abundance estimator. See chapter 9 for additional methods to detect and address genotyping errors.

Sampling occasions for collecting noninvasive genetic data may not be discrete. For example, let's suppose that scat samples are collected from throughout a defined survey area. Some individuals will likely deposit more scat than other individuals. It is reasonable to assume that animals that deposit more scats could be more easily detected than those that, for whatever reason, deposit fewer scats. This scenario causes additional heterogeneity in detection probability. Two methods have been developed to deal with multiple samples from a single individual collected within one sampling occasion. For example, Miller et al. (2005) presented an abundance estimator called *capwire*, based on counts of the number of times each individual is encountered, assuming sampling with replacement. Lukacs et al.

(2008b) modeled multiple detections as a mixture distribution to estimate both the mean number of samples available per individual and abundance. Given that a fixed number of samples are to be genotyped, using multiple samples from a single individual is relatively inefficient compared to using samples from multiple individuals. Therefore, it is advantageous to design surveys in a way that minimizes the number of repeated samples from the same individual collected within one sampling occasion (e.g., see chapter 6 for a description of some single-catch hair collection methods).

Often, noninvasive sampling data and laboratory tests can be combined to more accurately estimate abundance from genetic data. If laboratory tests of genotyping error are performed (see chapter 9), the test results can be analyzed together with the capture-recapture data to get a more robust estimate of genotyping error than can be achieved with either method individually (Lukacs 2005)—which should ultimately result in a more precise estimate of abundance. One efficient way to obtain multiple sources of data is to combine samples from animals killed by hunters or trappers with hair or scat collected via noninvasive methods. For example, Dreher et al. (2007) used hair collected from barbed wire at scent stations along with hair and tissue samples collected from harvested black bears (*Ursus americanus*) to estimate bear abundance in the lower peninsular of Michigan. The harvested animals provided both a high-quality DNA source (tissue) to serve as a known genotype and a low-quality DNA source (hair) to be tested against it—thus enabling the direct calculation of error rates. Bear abundance was much more precisely estimated with this approach than would have otherwise been possible.

## Identifying Individuals via Photographic Data

For some species, such as tigers (*Panthera tigris*), it may be possible to uniquely identify individuals based on characteristics such as spotting patterns, stripes, or scars captured with photographs from remote cameras (Karanth and Nichols 1998; also see chapter 5). Photographic data presents another set of challenges for capture-recapture analyses. Photo misidentification can be a potentially serious problem that varies with photo quality (Stevick et al. 2001). Stevick et al. (2001) developed a method that accounts for misidentification in two-sample capture-recapture surveys. These authors presented a modification to the Petersen estimator that allows error rate to be estimated. Given that photo quality is an important factor affecting matching ability, error rate is estimated by strata of similar photographic quality.

For studies of longer duration, the method of Lukacs and Burnham (2005a; described earlier) for detecting errors in genetic data may be useful *if* an error is not made in the same way more than once. This assumption is likely to be less justified in photographic studies than in genetic studies. Both the methods of Stevick et al. (2001) and Lukacs and Burnham (2005a) assume that false negatives (e.g., two photographs of the same individual failing to be matched) are far more likely to occur than false positives (e.g., erroneously matching two different individuals). Well-thought-out field protocols, such as photographing both sides of the body of an animal simultaneously (Karanth and Nichols 1998), may help to minimize both types of errors.

If a species' identifying characteristics change throughout its life, photo identification may not be an appropriate method for distinguishing individuals (Jefferson 2000). Indeed, changes in such characteristics are subtly analogous to tag loss in traditional capture-mark-recapture approaches. Although changes in frequently detected individuals can potentially be taken into account, those individuals that are not detected for long periods of time are likely to be incorrectly identified as new individuals.

## More Flexible Models

Numerous extensions to capture-recapture models exist. These extensions allow more detailed biological questions to be explored and enable the management of a wider array of assumption violations. There are currently capture-recapture models to estimate survival, animal movement, state-transitions,

and population growth rate, among other parameters. Lukacs and Burnham (2005b) provide a review of capture-recapture methods applicable to noninvasive genetic data.

One extension of particular interest is the robust design (Pollock 1982; Kendall et al. 1997). This design combines closed- and open-population capture-recapture models to estimate parameters that could not be estimated by either method alone. The robust design consists of two levels of sampling (figure 11.1). The primary sampling periods are separated by longer time periods (relative to the species' biology)—perhaps a year for a carnivore species. Between primary periods, the population is assumed to be open and animals may move in or out of the population, or may die. At the start of each primary sampling period is a set of secondary sampling occasions. Secondary sampling occurs over a biologically short period of time—a few days or weeks, for example. Across secondary sampling occasions, the population is assumed to be closed (note that these are analogous to a standard set of closed-population sampling occasions). The secondary sampling periods allow detection probability and abundance to be estimated. The robust design has been extended to handle genotyping error as well (Lukacs et al. 2008a).

Capture-recapture models have also been extended to a GLM framework such that parameters can be modeled as functions of covariates through a link function. Link functions allow a linear model to be built on a scale of $(-\infty$ to $\infty)$ and maps the function back to the scale of the parameter, such as $(0–1)$ for a probability. The capture-recapture models of Otis et al. (1978) have been expanded to handle environmental covariates related to capture and recapture probability, which can help deal with data exhibiting capture heterogeneity (also see chapter 2). Environmental covariates are measurable attributes that apply equally to all animals in the population (e.g., air temperature, sampling effort). Covariates relating to individual animal characteristics (e.g., sex, age) are more difficult to handle when estimating abundance because the value of the covariate is not known for animals that have not been detected.

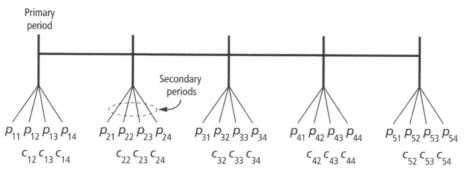

*Figure 11.1.* The robust design framework depicting primary sampling periods (between which the population may be open), secondary sampling periods (between which the population is considered closed), and closed population model parameters including capture probability ($p$) and recapture probability ($c$).

To address this problem, Huggins (1989) proposed a conditional likelihood formulation of the closed-population capture-recapture model to estimate abundance. Huggins (1989) proposed dividing all the probabilities of observed encounter histories by the probability of observing an animal at least once. This effectively removes the undetected animals from the analysis, or "conditions" them out of the likelihood. With undetected animals removed, individual covariates can be utilized and abundance estimated as the sum of one over the probability of detecting each individual:

$$\hat{N} = \sum_{i=1}^{M_{t+1}} \frac{1}{p_i}$$

where $p_i$ is the probability of detection for animal $i$ (itself a function of covariates) and $M$ is the total number of detected individuals. The GLM framework allows more detailed biological questions to be addressed, as well as better handling of individual heterogeneity in capture probability.

## Software

The data structures and analysis approaches presented in this chapter are so diverse that identifying a single software package to address all possible scenarios is not feasible. We've thus summarized a few of the more common and useful packages.

### R: An Environment You Can Grow With

The R programming environment (Ihaka and Gentleman 1996; R Development Core Team 2005), is a very generic and flexible option for data analysis that is available as a free download (www.r-project.org/). R features a broad user base, many add-on packages, and a wide array of built-in procedures for carrying out many standard operations (e.g., nonlinear optimization). This approach requires substantial technical expertise but may pay off in the long run if analysts find themselves working on a large number of different types of problems, or if methodological modifications that have not been adopted by existing packages are necessary.

### Bayesian Analysis in *WinBUGS*

The Bayesian analysis of most models described in this chapter can be implemented using the software package WinBUGS (Medical Research Council Biostatistics Unit, Cambridge, UK; Gilks et al. 1994). Examples demonstrating the use of WinBUGS can be found in MacKenzie et al. (2006) and Royle and Dorazio (2003). Further, an R package (*R2WinBUGS*;

www.stat.columbia.edu/~gelman/bugsR/) is available that permits the user to carry out data organization, manipulation, and related computation in R, as well as to call WinBUGS in batch mode to fit models.

## Program MARK

Program MARK (White and Burnham 1999) is freely available on the internet (www.warnercnr .colostate.edu/~gwhite/mark/mark.htm), and offers extensive capabilities for analyzing capture-recapture and occupancy data. MARK is capable of fitting both closed- and open-population models to estimate parameters such as abundance, survival, population growth rate, and state transitions, and includes models specifically designed for the analysis of genetic data. MARK uses the individual animal's encounter history as the basic data unit (or, for occupancy analysis, the species' detection history at individual sites), which provides a flexible framework for analysis and allows for the inclusion of environmental covariates, individual covariates, and grouping. MARK also allows random effects models to be fit with a moment estimator, or simulated with MCMC. A MARK-specific user's forum and detailed documentation about the program and its myriad applications can be found at www.phidot .org/software/mark/. Like WinBUGS, R can also be used to execute MARK (see www.phidot.org/ software/mark/rmark).

## Other Options

PRESENCE (Proteus Wildlife Research Consultants, Dunedin, New Zealand) is a specialized, stand-alone software package designed for fitting all classes of site occupancy models (e.g., single- and multiseason, two-species, abundance-induced heterogeneity, multimethod). It is similar to Program MARK in that it provides comprehensive analysis capabilities, including the ability to incorporate covariates, and also outputs statistics useful for model selection. It can be obtained at no cost from the website www .mbr-pwrc.usgs.gov/software/presence.html.

The USGS Vermont Cooperative Fish and Wildlife Research Unit's spreadsheet project (www.uvm .edu/envnr/vtcfwru/spreadsheets/) provides a number of downloadable Excel spreadsheets designed to demonstrate and teach the fundamentals of analyzing animal survey data. Specifically, detailed modules are available for both occupancy (Donovan and Hines 2007) and abundance (Donovan and Alldredge 2007) estimation. In addition, the site contains detailed instructions and links to pertinent literature on a variety of analytical topics. Each spreadsheet consists of a number of different peer-reviewed worksheets, with new material building on materials from previous sheets. Many of the applications are tutorials for analyses that are available within Program MARK or PRESENCE. The exercises were developed in Microsoft Excel for Windows, although Mac users should be able to complete the exercises with minor modifications.

# Future Directions and Concluding Thoughts

Myriad types of data arise from carnivore surveys. Intensive studies of wildlife populations using capture-recapture methods (see *Estimating Abundance* earlier) are capable of providing the most detailed demographic information and are preferred whenever possible. These methods yield estimates of population size, and can be extended in numerous ways to estimate other demographic parameters, including survival and recruitment rates. Indeed, capture-recapture methods can be ideal for inference about carnivore populations because several very practical, noninvasive techniques exist for uniquely identifying individuals (e.g., remote cameras, DNA analysis of hair or scat samples).

Many carnivore survey scenarios, however, do not easily fit into a capture-recapture framework. Instead, they may comprise methods (e.g., snow tracking) or spatial networks of sampling devices (e.g., track stations, remote cameras) that do not yield data amenable to individual identification or even permit the delineation or definition of physical populations of individuals. There are a number of inferential frameworks available for the treatment of such data, including conventional GLM or regression approaches for which the putative objective is to assess relative variation in some summary of observed data. Certain modifications to survey design (i.e., replicate surveys) allow for the consideration of occupancy modeling approaches that permit the estimation and incorporation of detectability into estimates.

We see a number of areas for advancement in the development of analytic tools for modeling and inference from carnivore survey data. First is the collection and analysis of DNA data. The diversity of species for which genetic data are available is growing rapidly, and methods that combine multiple sources of data, such as hair sampling and hunter kills or radio-telemetry, can greatly improve the precision of parameter estimates. The extension of analysis methods that utilize genetic data to address objectives beyond abundance estimation is also possible and could be fruitful. Second, additional information related to abundance is contained in occupancy metrics and could be more efficiently exploited to yield estimates of abundance or density.

While certain possibilities have already been described (Royle and Nichols 2003), additional work on this topic seems warranted—especially for rare or elusive species for which capture-recapture is difficult or impossible. Third, a deficiency of many of the inference methods described in this chapter is that the area effectively sampled with certain sampling methods—especially those that utilize attractants (chapter 10)—is unknown. This impairs the interpretation of parameters in some models and, when devices are too close together, yields nonindependence of both occupancy status and detection data. Advances in modeling approaches designed to address each of these issues would provide valuable tools for those studying carnivores.

## Acknowledgments

We would like to thank Bill Zielinski for providing fisher survey data and the text describing the survey program. We also thank Jim Nichols for his input on certain technical issues and Andrew Tyre, Terri Donovan, and Glen Sargeant for helpful comments and suggestions.

## Chapter 12

# Synthesis and Future Research Needs

*Robert A. Long, Paula MacKay, Justina C. Ray, and William J. Zielinski*

Only two decades ago, the survey and monitoring of carnivores was a formidable task, owing largely to their elusiveness as a group and the extensive terrain they inhabit. Few methods, save the fitting of relatively small numbers of animals with radio-telemetry collars, reliably provided sufficient data for even rudimentary analyses. Stimulated in part by this scenario, and boosted by technological advances, noninvasive techniques appropriate for surveying carnivores have now blossomed into a viable and widely available toolset. These techniques have expanded from a somewhat limited suite of tracking methods to the highly technical detection devices, laboratory analyses, and analytical approaches presented in this volume. While today's ground-based snow tracking surveys (chapter 3) are quite similar to those carried out by renowned naturalist Olaus Murie, for example, other methods (e.g., scat detection dogs; chapter 7) and materials, such as lightweight folding enclosures for collecting hair samples (chapter 6), have emerged only in the last several years. Indeed, a decade ago, there would have been little practical use for hairs sampled with these latter devices, as most hair analysis was limited to laborious and often imprecise and subjective morphometric or chromatographic approaches.

Each of the chapters in this book has dealt with a particular noninvasive survey method, or a topic relevant to a number of methods (e.g., genetic analysis, attractants). In our final chapter, we highlight the critical advances that we believe have contributed most to the field and discuss some of the techniques and materials that currently define the state of the art. We also compare the various methods with respect to their suitability for surveying North American carnivores and discuss their relative strengths and weaknesses. Finally, we share our thoughts on where the field may be headed.

## Critical Advances and Today's State of the Art

New technologies and innovations have allowed noninvasive survey methods to grow by leaps and bounds since 1995—the year the first comprehensive manual on this topic was published (Zielinski and Kucera 1995a). A number of specific advances stand out as having contributed disproportionately to this growth. Here we briefly describe each of these advances (and their current limitations).

## The Rise of Genetic Methods

The genetic methods discussed in chapter 9 have yielded possibilities that many of our predecessors could never have dreamed of. Over the past decade, genetic advances have propelled the survey of carnivores from an endeavor largely based on (sometimes uncertain) species identification to one where the identification of species, individual, and sex can often be ascertained from a scat or single hair. The two most important advances in this regard were the polymerase chain reaction—which allows a few copies of DNA to be amplified into the many copies required for analysis—and the development of new classes of genetic markers (e.g., microsatellites). These genetic breakthroughs have enabled DNA-based survey methods (e.g., hair and scat collection) to flourish.

Regardless of the survey objective, many contemporary carnivore surveys include a genetic component. Even surveys employing tracking (chapter 3) and track stations (chapter 4)—methods which once relied solely on track identification to assess and confirm species presence—may now integrate genetic data. In many cases, these data are used to simply corroborate species detections. But increasingly, data pertaining to individual animals can supplement standard detection data, allowing population genetic and demographic questions to be explored. Given that DNA can be extracted and stored with relatively little effort, it is not uncommon for projects to stockpile samples until time and money permit further analysis. Indeed, one of the current constraints for DNA-based projects is that the demand for high-quality analyses has created a backlog of samples and thus bottlenecks at many DNA laboratories. Researchers must often wait months (or even years) to receive results. Another limitation is that DNA from some noninvasive samples, especially scats, is present in low quantity or quality—thus making extraction and amplification difficult, or potentially generating excessive genotyping errors (Taberlet et al. 1999). Nevertheless, we cannot overstate the positive effects of modern molecular techniques on carnivore field research, particularly given continuing efforts to address these very challenges (see *The Future*).

## Remote Camera Technology Comes of Age

Remote cameras have been the poster child of wildlife detection technology for years. Their slow but steady evolution from the heavy, cumbersome, expensive, and unreliable novelties of yesteryear to their current status as light, fairly reliable, and affordable detection devices capable of being deployed almost anywhere has been remarkable (see chapter 5 for details). Nonetheless, most remote film cameras are still plagued by certain shortcomings, including reliability issues in certain climates (e.g., regions characterized by high humidity or extreme rainfall or cold), film and battery costs, a limited number of exposures, troublesome external wires, audible shutter mechanisms, long shutter "wake-up" times, and disruptive flash technology.

With the recent advent of digital technology and dramatic improvements in engineering, remote camera designs are experiencing ongoing, radical changes. The high capacity memory cards used to store information in digital models allow a practically limitless number of photos. In some models, shutter mechanisms are completely silent, wake-up times are almost instantaneous, and infrared illumination permits flashless photography—even in complete darkness. These advances help maximize performance and minimize animal disturbance and the detection of cameras by human passersby. Contemporary camera-sensor units are also more rugged, reliable, and versatile, often utilizing highly waterproof housings, requiring less battery power, permitting the imprinting of custom information (e.g., location identifiers) directly onto photos, and, in certain cases, allowing both still photos and short video clips to be recorded.

Unfortunately, the most sophisticated cameras remain quite expensive and will likely be beyond the budgets of many survey projects for the foreseeable future (although highly rated and less costly alternative models are becoming increasingly available). And while emerging advances are primarily good news for camera users, rapidly changing technologies and the concomitant lack of consistency in equipment and methods can make it difficult to maintain and service a standard set of camera and sensor units over time and to compare results between studies. Further, researchers who are not consistently involved in field research may find it daunting to keep abreast of camera advances.

## New Materials and Innovative Device Design

Wildlife researchers habitually explore the use of newly available materials, and have demonstrated amazing ingenuity when designing animal detection devices—often finding relatively simple solutions to seemingly complex problems. For example, only a dozen years ago, track plate enclosures (chapter 4) were primarily made from heavy plywood and required extensive effort for construction and deployment. New, fluted plastic materials that are both lightweight and rugged now allow enclosures to be built and deployed with a fraction of the effort and time. Similarly, genetic approaches that permit the extraction of DNA from hairs have spurred the search for species-specific hair collection devices (chapter 6), yielding numerous, creative hair-snagging mechanisms and designs. These include gun-cleaning brushes and various adhesives, as well as the positioning of attractants and snagging devices to maximize sampling. By mounting hair-snagging devices in track plate enclosures—and potentially incorporating remote cameras—researchers can collect substantial information from a single visit by target species (see chapter 8). The Scentinel, an automated detection tunnel that incorporates several recording and measurement devices (King et al. 2007; Technology Transfer, Wellington, New Zealand, www.scentinel.co.nz), may represent the pinnacle of multiple-device innovation (see chapter 4).

## Detection Dogs Open Up New Doorways

The collection of hairs for DNA sampling typically requires a target individual to enter or interact with a collection device (chapter 6). Alternately, scat is a readily available by-product that not only yields DNA-based information such as species identification, individual identification, and sex, but also permits researchers to evaluate growth, reproductive, and stress hormones (chapter 9), and diet. The problem is that scats are generally difficult to find via visual searches.

Enter scat detection dogs (chapter 7), which have revolutionized scat surveys. Proven highly effective in numerous studies (Smith et al. 2003; Wasser et al. 2004; Harrison 2006; Long et al. 2007a), detection dogs are able to work in most climates and under extremely variable weather conditions, can be trained to locate scats from multiple species, and circumvent much of the bias associated with either attractant- or trail-based survey methods. Scat detection dogs have already been employed in dozens of surveys across the globe and are of rapidly increasing interest to wildlife researchers and agencies. Among the hindrances to their widespread adoption are accessibility, cost, and logistical constraints—these dogs require careful selection, training, testing, and maintenance, and there are currently limited options for professional dog trainers or consultants specializing in their deployment.

## Analytical Methods and Software Meet Modern Detection Data

Biostatisticians have experienced their own breakthroughs in recent years. New developments in analytical methods and software permit much more powerful, accurate, and precise data analyses than would have been possible ten years ago, and have provided fertile ground for the evaluation of wildlife populations (Williams et al. 2002) and animal occurrence (Scott et al. 2002; MacKenzie et al. 2006). Parallel advances in field and analytical methods have helped to drive one another. For example, modern occupancy approaches (chapters 2 and 11) allow the estimation and explicit incorporation of detectability, and have created a new framework for analyzing the types of data generated by many non-invasive survey methods. Occupancy methods also permit the use of multiple detection methods and provide a standardized metric (i.e., probability of detection) that can be used to evaluate and compare methods. Furthermore, modern occupancy approaches require only detection-nondetection data (as opposed to data representing the detection of specific individuals), yet still yield meaningful parameter and precision estimates. Due to such attributes, occupancy methods enable the analysis of data that was, until recently, often considered insufficient for making solid inferences.

Advances in analyses have not been limited to assessments of occupancy and distribution. Indeed, much of the theory behind occupancy estimation borrows from modern developments in the analysis of capture-recapture data (chapters 2 and 11). This specialized field continues to progress, with newer models able to accommodate departures from certain assumptions (e.g., various forms of detection heterogeneity) that would have crippled earlier models. In addition, models have been developed to deal specifically with some of the challenges of non-invasively collected data, such as errors in genotyping (chapters 9 and 11) and photo identification (chapters 5 and 11). Finally, the advent of powerful, affordable, and (relatively) user-friendly software packages have enabled researchers—with a bit of guidance from biostatisticians—to conduct fairly sophisticated data analyses on their own.

Today's analytical methods and software benefit the design of carnivore surveys as well. Prior to having such accessible software at their disposal, researchers were limited to fairly rudimentary tools for determining which survey designs would be the most precise, unbiased, and efficient for meeting their objectives. Presumably, this limitation often resulted in insufficient data or excessive survey effort. Cutting-edge programs (e.g., GENPRESS [Bailey et al. 2007]; chapter 2) now permit researchers to test

various survey designs a priori to decide which may be the most efficient for meeting survey objectives.

## Sorting Through the Toolbox

For most carnivore species, there now exist one or more noninvasive methods to choose from that enable simple species detection (table 12.1), with some also allowing the collection of additional information (e.g., from DNA). While no method is clearly superior for detecting carnivores across the board, we refer to table 12.1 and make a few general observations:

- Remote cameras, scat detection dogs, and the various types of unenclosed track stations all appear to be suitable or potentially suitable options for detecting every species included in table 12.1. Remote cameras have been effectively used to survey more species than detection dogs, but this will no doubt change as dogs are employed to survey an increasing number of new target species. Of the three methods, only dogs provide a source of DNA and other biological material that can be used in further analyses. Of course, scat surveys conducted by human searchers also allow for the collection of scat samples, but low detection rates and certain biases that typify visual searches are limiting.
- Like scat surveys, hair collection methods yield a valuable source of DNA. But there are a fairly large number of species for which efficient hair-snaring devices have not yet been developed or tested (resulting in a number of "unsuitable" values in table 12.1). While some hair collection devices (e.g., corrals) are highly effective for certain species (ursids, in this case), behavioral and morphological variations in how individual species interact with hair snares hinder the development of snares capable of collecting samples from multiple species.
- Enclosed track plates, while producing the best quality results of all the track station methods,

restrict the diversity of species that can be detected because the typical enclosure size is relatively small. Scent stations and track plots have the advantage of being potentially suitable for all species, but are somewhat limited by the challenges of species identification. And although unenclosed track plates can register high quality tracks and do not inherently exclude larger-bodied species, they are severely constrained by weather conditions.

- Snow tracking on the ground is suitable or potentially suitable for surveying almost as many species as detection dogs, cameras, and unenclosed track stations, with the exception of hibernating species or those that occur in regions that do not routinely receive snow. Snow track-based surveys, however, have limited ability to achieve unambiguous species identification. Aerial snow tracking further restricts the number of species for which unambiguous species identification is possible, primarily because such identification is only reliable for species with relatively large body sizes and in areas characterized by a sustained period of snow and habitats with suitable visibility.

- In relation to carnivore families, canids and ursids appear to be suitably detectable via the greatest number of methods, and mephitids and felids via the fewest. Those seeking to make a contribution to this field may wish to focus their attention on skunks because, as reflected in table 12.1, there has been very little effort to evaluate methods to detect them. Similarly, we encourage more research on methods to survey low density and secretive felids, as members of this group seem to vary dramatically and regionally in their response to attractants and in their willingness to interact with hair and track collection devices.

Despite recent advances—or perhaps because of them—selecting the most appropriate survey method(s) for a particular target species (or especially for multiple target species) can be daunting.

No single method can be considered universally effective for all situations, and every method has its relative strengths and weaknesses (table 12.2). Often, the characteristics of a given survey or survey area (e.g., objective, land cover, topography, climate, budget) will help dictate which method(s) to employ. Even if a given survey objective can theoretically be achieved via noninvasive methods, circumstances may preclude success. A scat-based capture-recapture survey for a species that is very rare, for instance, may require too much effort to detect a sufficient number of individuals (although note that conducting a similar survey with live-capture approaches may also be unrealistic). Such considerations are discussed in detail within the method-specific chapters of this book.

The amount of labor, personnel training, and specialized equipment required for a particular method will also affect its suitability for a given project. While such considerations often correlate with overall cost, in some cases, the *partitioning* of cost may be the important factor. For example, scat detection dogs typically require high up-front costs as well as advanced training for surveyors (unless professional detection dog teams are to be hired). In addition, money must ultimately be budgeted for the DNA analysis of scats. Remote camera surveys, while entailing substantial up-front equipment costs (unless cameras are already available), can be executed with less skilled personnel. Some methods that have little or no equipment or supply costs (e.g., natural sign surveys, track stations) may nonetheless require extensive travel and revisits to stations—both of which translate to high personnel costs. Tracking surveys also demand skilled personnel to accurately assess animal sign in the field. Finally, it is critical to consider a given method's probability of detecting the target species when it is present. Methods that are inexpensive in terms of equipment or training may also have a very low probability of detecting the target species. Reciprocally, certain methods are expensive in terms of up-front costs (e.g., detection dogs), but may result in much higher detectability.

*Table 12.1.* Suitability of select[a] noninvasive survey methods for detecting North American carnivores. Rankings reflect the opinions of the editors, which are based on field experience, input from chapter authors and other experts, and the scientific literature. As suitability will ultimately depend on survey-specific considerations (e.g., objectives, location, season) discussed throughout the book, readers should refer to relevant chapters to conduct a more detailed evaluation.

| Species | Natural sign | | | Track station | | | | | Hair collection | | | Remote camera | Scat detection dog |
|---|---|---|---|---|---|---|---|---|---|---|---|---|---|
| | Ground snow tracking | Aerial snow tracking | Scat collection | Track plot | Scent station | Unenclosed track plate | Enclosed track plate[b] | Corral | Tree/post snare | Rub station | Cubby[b] | | |
| Coyote | ● | ● | ● | ● | ● | ● | x | ○ | x | ○ | x | ● | ● |
| Gray wolf | ● | ● | ● | ● | ● | ● | x | ○ | x | ○ | x | ● | ● |
| Red wolf | x | x | ○ | ● | ● | ● | x | ○ | x | ○ | x | ● | ○ |
| Gray fox | ○ | x | ○ | ● | ● | ● | ● | ○ | x | ○ | ● | ● | ● |
| Island fox | x | x | ○ | ● | ○ | ● | ○ | ○ | x | ○ | ○ | ● | ○ |
| Arctic fox | ● | ● | ○ | ● | ● | ● | ○ | ○ | x | ○ | ○ | ● | ○ |
| Kit fox | x | x | ● | ● | ● | ● | ○ | ○ | x | ○ | ● | ● | ● |
| Swift fox | ○ | ○ | ● | ● | ● | ● | ○ | ○ | x | ○ | ● | ● | ● |
| Red fox | ● | ● | ● | ● | ● | ● | ● | ○ | x | ○ | ○ | ● | ○ |
| Ocelot | x | x | ○ | ○ | ○ | ○ | x | x | x | ● | x | ● | ○ |
| Margay | x | x | x | ○ | ○ | ○ | x | x | x | ○ | x | ● | ○ |
| Canada lynx | ● | ● | x | ○ | ● | ○ | x | x | x | ● | x | ● | ● |
| Bobcat | ● | x | x | ● | ○ | ○ | x | x | x | ● | x | ● | ● |
| Jaguar | x | x | ○ | ● | ● | ○ | x | x | x | ○ | x | ● | ● |
| Cougar | ● | ● | ○ | ● | ● | ○ | x | x | x | ● | x | ● | ● |
| Jaguarundi | x | x | x | ○ | ○ | ○ | x | x | x | ○ | x | ● | ○ |
| American hog-nosed skunk | x | x | x | ○ | ○ | ○ | ○ | x | x | x | ○ | ● | |
| Hooded skunk | x | x | x | ○ | ○ | ○ | ○ | x | x | x | ○ | ○ | ○ |
| Striped skunk | ○ | x | x | ● | ● | ● | ● | x | x | x | ● | ● | ○ |
| Western spotted skunk | ○ | x | x | ○ | ○ | ● | ● | x | x | x | ● | ● | ● |
| Eastern spotted skunk | ○ | x | x | ○ | ○ | ○ | ○ | x | x | x | ○ | ● | |
| Wolverine | ● | ● | x | ○ | ○ | ○ | x | x | ● | x | x | ● | ○ |
| North American river otter | ● | ● | ● | ● | ○ | ● | x | x | x | x | ● | ● | ○ |
| American marten | ● | ○ | x | ○ | ○ | ● | ● | x | ○ | x | x | ● | ○ |
| Fisher | ● | ● | x | ○ | ○ | ● | ● | x | ○ | x | ● | ● | ● |
| Ermine | ● | x | x | ○ | ○ | ● | ● | x | x | x | ● | ● | ○ |
| Long-tailed weasel | ● | x | x | ○ | ○ | ● | ● | x | x | x | ○ | ● | ○ |
| Black-footed ferret | ● | x | x | ● | ○ | ○ | ○ | x | x | x | ○ | ○ | ● |

*Table 12.1* (Continued)

| Species | | | | | | | | | | | | |
|---|---|---|---|---|---|---|---|---|---|---|---|---|
| Least weasel | ● | x | x | ○ | ○ | ○ | ○ | x | x | x | x | ○ |
| American mink | ● | x | x | ● | ● | ● | ● | ● | x | x | x | ○ |
| American badger | ○ | x | ○ | ● | x | ● | x | ● | x | x | x | x |
| Ringtail | x | x | x | ○ | ○ | ○ | ● | x | x | x | x | ○ |
| Raccoon | ● | x | x | ● | ● | ● | ● | x | ○ | x | x | ○ |
| White-nosed coati | x | x | x | ● | ○ | ○ | x | x | x | x | x | x |
| American black bear | x | x | ○ | ● | ● | x | x | ● | ○ | ○ | x | ● |
| Grizzly bear | x | x | ● | ● | ○ | x | x | ● | ○ | ○ | x | ● |
| Polar bear | ● | ● | ○ | ● | ○ | x | x | ○ | x | x | x | ● |

● = suitable—the method has been effectively used to detect the target species.

○ = potentially suitable—although the method has not (to our knowledge) been specifically employed to detect the target species, results from other surveys, coupled with the target species' characteristics, suggest that it could be suitable for detecting this species under appropriate conditions.

x = unsuitable—given our relevant experience with and knowledge of the method, combined with the characteristics of the target species, we consider the method currently unsuitable for reliably detecting this species. In some cases, we simply have too little information to rank the method as "potentially suitable."

[a] Dust/mud track surveys and passive hair collection methods (i.e., natural rub objects and travel route snares) are excluded from this table due to limited applications to date.

[b] Assumes standard-sized enclosure, as discussed in chapters 4 and 6.

Table 12.2. A comparison of noninvasive survey methods

| Attribute | Ground snow tracking | Aerial snow tracking | Mud/dust tracking | Scat collection | Track plots | Scent stations | Unenclosed track plates | Enclosed track plates | Remote cameras | Hair collection | Scat detection dogs |
|---|---|---|---|---|---|---|---|---|---|---|---|
| **Target species** | | | | | | | | | | | |
| Potential for surveying multiple species | High | High | High | High | High | High | High | Moderate | High | Low | High |
| Is a behavioral response required? | No | No | No | No | No | Yes | Yes | Yes | Yes[a] | Yes[b] | No |
| **Nature and quality of information** | | | | | | | | | | | |
| Ease of verifying species | Low | Low | Low | Moderate[c] | Low | Low | Moderate | Moderate | High | Moderate[c] | Moderate[c] |
| Potential for high quality, permanent record | Low–Moderate[d] | Low–Moderate[d] | Low–Moderate[d] | High | Low–Moderate[d] | Low–Moderate[d] | High | High | High | High | High |
| Can evidence be collected at first visit? | Yes | Yes | Yes | Yes | No | No | No | No | No | No[e] | Yes |
| Sign persistence in environment | Short–Moderate | Short–Moderate | Short–Moderate | Long | Short–Moderate | Short–Moderate | Short–Moderate | Moderate–Long | N/A | Long | Long |
| Ability to collect behavioral information | High | High | Low | Moderate | Low | Low | Low | Low | High[f] | Low | Moderate |
| Potential for collecting DNA samples | Moderate | Low[g] | Low | High | No[h] | No[h] | No[h] | No[h] | No[h] | High | High |
| **Labor and equipment needs** | | | | | | | | | | | |
| Special equipment requirements (excluding attractants)[i] | Low | High | Low | Low | Low | Low | Moderate | Moderate | High | Moderate | High |
| Is an attractant required? | No | No | No | No | No | Yes | Yes | Yes | Sometimes | Sometimes | No |
| Required level of training | High | High | High | Moderate | Moderate | Moderate | Low | Low | Moderate | Low | High |
| Cost of materials, equipment, or preparation[j] | Low | High | Low | Low[j] | Low | Low | Low | Moderate | High | Low[j] | High[j] |
| Labor intensity required[k] | High | Moderate | High | High | Moderate | Moderate | Moderate | Moderate | Moderate | Moderate | Moderate |
| Are revisits to sites required to apply attractants or maintain devices? | No | No | No | No | Yes | Yes | Yes | Yes | Yes | Yes | No |

Table 12.2 (Continued)

| | | | | | | | | | |
|---|---|---|---|---|---|---|---|---|---|
| **Potential for variation between observers** | High | High | High | High | Moderate | Moderate | Low | Low | Low | Moderate |
| **Seasonal and substrate requirements** | | | | | | | | | | |
|   Substrate conditions required | Snow | Relatively open canopy | Mud/dust | None | Suitable substrate | Suitable substrate | None | None | None | None |
|   Seasons of use | Winter | Winter | Minimal rain | None | Non-snow | Non-snow | Non-snow | All | All | All |
| **Public interface** | | | | | | | | | | |
|   Potential for incorporating citizen science | High | Low | Low | Low | Moderate | Moderate | Moderate | High | Moderate | Low |
|   Potential for theft of deployed equipment | N/A | N/A | N/A | N/A | N/A | Low | Moderate | High | Low | N/A |

a Except when unbaited sets are used.
b Except when sampling travel routes.
c Typically requires DNA confirmation.
d Photos or plaster casts of tracks can be acquired if tracks are of sufficient quality.
e Except in the case of some natural rub object surveys.
f With video or near-video options.
g Requires landing aircraft.
h Except with some combined methods (e.g., hair snare attached to enclosed track plate).
i Excludes ground transportation.
j DNA analysis costs may be high.
k Defined as "labor per detection" or "labor per area searched."

In closing, we are compelled to reiterate the admonition of MacKenzie and Royle (2005), who remind us that "... the lack of clear objectives will often lead to endless debate about design issues as there has been no specification for how the collected data will be used in relation to science and/or management; hence judgments about the 'right' data to be collected cannot be made" (1107). Indeed, the most important message of this book may be to encourage the early and specific identification of survey objectives, regardless of the survey methods to be employed.

## The Future

Each of the exciting advances described above constitute today's state of the art in survey methodology. Future developments will no doubt be equally impressive and useful. Thompson (2004a: 395) summed up the future of sampling rare species with three words: "innovation, technology and software." Although he was largely referring to sampling designs, we presume that these entities will also be pivotal to the success and continued growth of noninvasive survey methods in general. And while we are reluctant to try to predict exactly how noninvasive survey methods will evolve, we'll briefly speculate on a few advances that might realistically emerge soon. Many of these thoughts are discussed further in relevant method-specific chapters.

### Genetic Methods: Moving Toward Increased Reliability, Convenience, and Affordability

As instrumental as genetic techniques have been in advancing the noninvasive methods showcased in this book, we foresee that they will become even more important to the field in the future. DNA laboratories are becoming increasingly adept at extracting and amplifying DNA from scats, and new analytical methods that permit the direct assessment of population size from molecular data (see chapter 9)

will make the extraction of such data all the more valuable. Virtually instantaneous assays to confirm species from noninvasive samples in the field are clearly on the horizon. In fact, affordable bear detection "kits," which allow researchers to confirm that a sample of hair, bone, meat or dried material originated from a bear (www.carnivoreconservation .org/files/issues/bear_detection_kit.pdf), have already been developed. While these kits are currently being tested primarily as an anticontraband tool, the ramifications for wildlife surveys are profound. One advance that has yet to be realized is a method for accurately estimating animal age from noninvasively acquired biological samples. Such a breakthrough would truly transform noninvasive survey methods by providing critical demographic information.

### Detection Dogs: A "One-Stop" Survey Solution?

Detection dogs have the potential to provide a one-stop solution for carnivore surveys under many circumstances. Further testing and comparisons of dog performance will increase confidence in the dependability of this method. In addition, standardized training protocols and certification programs for dogs and handlers should be developed. Potential advances in genetic methods, such as more effective DNA extraction from scats, the ability to age target individuals noninvasively, and, especially, rapid "in-field" species confirmation kits, would provide even more opportunities for field-testing dogs and utilizing dog-collected samples to meet carnivore survey objectives. A reliable technique for aging scat samples would also help address potential closure issues, since dogs can detect scat long after an animal has left the area. Lastly, as detection dogs continue to prove effective with an ever-wider variety of species, we envision the increased availability of professional detection dog trainers and handlers, as well as natural resource agencies devoting special departments or branches entirely to detection dog methods.

## Remote Cameras: Wireless Technology and Sensor Networks

The current generation of remote cameras, while clearly more convenient and reliable than their predecessors, are still individual operating units that require relatively high amounts of effort to deploy and maintain. Technological advances toward increased ruggedness and weatherproofing, battery life, digital memory, and image quality, coupled with lighter-weight equipment and lower costs, will further propel this method. But the ability to download images remotely and wirelessly would revolutionize remote camera surveys. Wireless camera networks deployed with automated baiting mechanisms could enable the long-term monitoring of a given target species without the need for costly revisits to the site. Indeed, if expenses could be sufficiently minimized, entire camera networks (see chapter 5) could be deployed across a broad region, providing real-time detection and monitoring data for carnivore populations.

## Survey Methods Old and New

Although newer methods that exploit modern DNA techniques and other technologies can clearly be used to meet many carnivore survey objectives, we do not envision that they will altogether supplant some of the more traditional noninvasive survey methods—which have continued to make advancements in their own right. Notably, track surveys and track stations, which were the basis for much of the robust methodology we have at our disposal today (chapter 1), will remain methods of choice under certain conditions. For example, in snow-covered terrain that is largely inaccessible by ground, only aerial snow tracking can provide sufficient survey coverage for some wide-ranging species (Magoun et al. 2007). Ground-based snow tracking will continue to perform an altogether different role by affording opportunities for public engagement and citizen science, in addition to scientific data.

Considering the widespread adoption of scent stations (chapter 4)—the precursor to modern track plates—we do not anticipate that this simple, low-technology method will decline in use either. A recent study identified scent stations as one of the most efficient of four methods tested in relation to cost, performance, and logistics, having simultaneously detected more species over large areas than live-trapping or remote cameras (Barea-Azcón et al. 2007). Innovative track plate enclosures (chapter 4) have allowed hair collection devices and camera technology to be incorporated into these structures (Zielinski et al. 2006; King et al. 2007), thereby broadening their potential applications as well.

Meanwhile, hair snares illustrate the notion that the best technique is not always that which targets the widest array of species. When a single species is of interest, techniques tailored to its behavioral proclivities often stand to capture the greatest amount of high-quality information. Hair collection methodologies are being increasingly adapted in this regard, resulting in a proliferation of devices to snag hair from passing animals (chapter 6). In light of this discussion, we encourage studies that compare the efficacy of multiple methods for achieving given objectives (e.g., Campbell 2004; Gompper et al. 2006; Harrison 2006; Barea-Azcón et al. 2007; Long et al. 2007b).

## More Flexible Analytical Methods and Design Tools

Methods for analyzing animal survey data are likewise evolving with breathtaking speed. New methods for modeling device dependence are already being tested, as are Bayesian approaches that permit the relaxation of some existing method assumptions (e.g., site dependence; chapters 2 and 11). Such advances will continue to provide ever-increasing options for analyzing noninvasively collected carnivore data, and software designed specifically for such data (e.g., estimation and incorporation of genotyping errors, abundance estimation from continuous data; chapters 9 and 11, respectively) will also continue to emerge. We suspect that occupancy methods will largely supplant more traditional relative abundance indices that are based on either visitation or sign counts (chapters 2, 3, and 4) and may also serve as suitable alternatives to abundance estimation for many carnivore survey and monitoring needs. Occupancy estimation holds particular promise for monitoring populations of rare and elusive carnivores over large areas—populations that are often of conservation concern and that require accurate monitoring in the context of recovery efforts. Such populations may be too small—or declining too quickly—to allow for the long-term surveys necessary to collect sufficient data for precise population estimates.

Finally, the further development of specialty software and analytical approaches to guide the design of carnivore surveys (chapter 2) will substantially contribute to the accuracy of estimates derived from survey data. We encourage researchers, biostatisticians, and software programmers to strive for survey designs and analysis methods that are as simple and user-friendly as possible. Sufficiently realistic designs and accessible analysis tools will allow the greatest number of researchers to employ such methods and will maximize benefits to carnivore conservation.

## Exploring Other Objectives with Noninvasive Surveys

In addition to the key survey objectives discussed throughout the book, noninvasive survey methods can be used to explore many other important objectives related to carnivore demography and ecology. These objectives first rely on the ability to effectively detect the species of interest or to locate relevant samples—topics that are addressed in method-specific chapters. Here, we briefly describe three of the more promising and creative uses for noninvasively collected data.

### Assessing Genetic Structure

With the advent of accessible and affordable genetic techniques, material collected during noninvasive surveys can be used to genetically characterize populations (see Waits and Paetkau 2005; Schwartz et al. 2007). In addition to using such data for detecting presence or estimating population size, genetic material can be employed for evaluating social structure (Utami et al. 2002), detecting hybridization (Adams et al. 2003) and monitoring change in genetic structure both prospectively and retrospectively (Schwartz et al. 2007).

### Evaluating Animal Movement and Habitat Connectivity

Approaches for analyzing relationships between individuals via measures of DNA similarity have lead to a new discipline, landscape genetics, which links genetic structure to landscape features (Manel et al. 2003; Carr 2005; Manel et al. 2005; Cushman et al. 2006). Such approaches enable the testing of hypotheses relating to population fragmentation, functional connectivity, and the movement of carnivores (Cushman et al. 2006). Survey designs appropriate for landscape genetic applications generally include few assumptions or requirements beyond the collection of sufficient samples from well-distributed locations across relatively large areas.

### Development and Evaluation of Other Methods

Some noninvasive survey methods permit researchers to assess the design and performance of other methods. Remotely triggered video cameras, for example, have been employed to evaluate (and ultimately improve upon the design of) hair collection devices and track surfaces for monitoring the use of road crossing structures by carnivores in Alberta (A. Clevenger, Western Transportation Institute, pers. comm.). As remote camera and video technologies advance (e.g., the newest digital remote cameras can photograph at near-video speeds of five frames per second; see chapter 5), the use of remote photography and videography to evaluate other survey methods will likely become more common.

## Final Thoughts: Noninvasive Methods and Carnivore Conservation

The methods and techniques described throughout this book allow us to closely examine the lives of species that are typically unseen by humans. Given the ongoing changes in human demography and land use, the increasing conversion of wildlife habitat to anthropogenic cover types, and global climate change, the ability to assess and monitor carnivore populations will be paramount if we are to ensure a future that includes this remarkable group of animals. Indeed, Laliberte and Ripple (2004) estimated that eight of the thirty-two North American carnivores they reviewed had experienced range contrac-

tions across more than 20% of their historic range. Ironically, the conservation of carnivores may actually help to mitigate larger-scale ecological perturbations resulting from global climate change by regulating community structure and dampening variability in prey populations (Ray et al. 2005; Sala 2007). While more traditional capture-based methods can provide high resolution data on relatively few individuals, noninvasive survey methods have the capacity to be deployed over very large areas and thus to monitor entire populations. We hope that this volume helps bring these methods within reach of as many carnivore researchers, natural resource managers, and conservationists as possible. We also hope to inspire the development of companion protocols, designed to standardize surveys for the collection of data for specific regions or taxa. Such standardized field procedures will be required to maximize the effectiveness of carnivore research and conservation efforts in North America and beyond.

# References

ABA (American Bear Association). 2006. www.american-bear.org/main.htm (accessed 30 November 2007).

Abbott, H. G. B., and A.W. Coombs. 1964. A photoelectric 35-MM camera device for recording animal behavior. *Journal of Mammalogy* 45:327–330.

Acker, R., and J. Fergus. 1994. *Field guide to dog first aid: emergency care for the outdoor dog.* Wilderness Adventures Press, Belgrade, MT.

ACUC (Animal Care and Use Committee). 1998. Guidelines for the capture, handling, and care of mammals as approved by the American Society of Mammalogists. *Journal of Mammalogy* 79:1416–1431.

Adams, J. R., B. T. Kelly, and L. P. Waits. 2003. Using fecal DNA sampling and GIS to monitor hybridization between red wolves (*Canis rufus*) and coyotes (*Canis latrans*). *Molecular Ecology* 12:2175–2186.

Agresti, A. 1990. *Categorical data analysis.* John Wiley & Sons, New York.

Aiello, S. E., editor. 1998. *The Merck veterinary manual.* 8th ed. Merck and Company, Whitehouse Station, NJ.

Aitken, N., S. Smith, C. Schwarz, and P. Morin. 2004. Single nucleotide polymorphism (SNP) discovery in mammals: a targeted-gene approach. *Molecular Ecology* 13:1423–1431.

Akenson, J. J., M. G. Henjum, T. L. Wertz, and T. J. Craddock. 2001. Use of dogs and mark-recapture techniques to estimate American black bear density in northeastern Oregon. *Ursus* 12:203–210.

Alexander, S. M., and N. M. Waters. 2000. The effects of highway transportation corridors on wildlife: a case study of Banff National Park. *Transportation Research Part C-Emerging Technologies* 8:307–320.

Allen, L., R. Engeman, and H. Krupa. 1996. Evaluation of three relative abundance indices for assessing dingo populations. *Wildlife Research* 23:197–206.

Allendorf, F. W., and G. Luikart. 2007. *Conservation and Genetics of Populations.* Blackwell, London, UK.

Alpers, D. L., A. C. Taylor, P. Sunnucks, S. A. Bellman, and W. B. Sherwin. 2003. Pooling hair samples to increase DNA yield for PCR. *Conservation Genetics* 4:779–788.

Altom, E. K., G. M. Davenport, L. J. Myers, and K. A. Cummins. 2003. Effect of dietary fat source and exercise on odorant-detecting ability of canine athletes. *Research in Veterinary Science* 75:149–155.

American Rescue Dog Association. 2002. *Search and rescue dogs: training the k-9 hero,* 2nd ed. Howell Book House, New York.

Amstrup, S. C., T. L. McDonald, and B. F. J. Manly, editors. 2005. *Handbook of capture-recapture analysis.* Princeton University Press, Princeton, NJ.

Andelman, S., J. G. Else, J. P. Hearn, and J. K. Hodges. 1985. The monitoring of reproductive events on wild Vervet monkeys (*Cercopithecus aethiops*) using urinary pregnanediol-3-glucuronide and its correlation with behavioural observations. *Journal of Zoology, London* 205:467–477.

Andelt, W. F., and S. H. Andelt. 1984. Diet bias in scat deposition-rate surveys of coyote density. *Wildlife Society Bulletin* 12:74–77.

Andelt, W. F., C. E. Harris, and F. F. Knowlton. 1985. Prior trap experience might bias coyote responses to scent stations. *Southwestern Naturalist* 30:317–318.

Andelt, W. F., and T. P. Woolley. 1996. Responses of urban mammals to odor attractants and a bait-dispensing device. *Wildlife Society Bulletin* 24:111–118.

Anderson, D. R. 2001. The need to get the basics right in wildlife field studies. *Wildlife Society Bulletin* 29:1294–1297.

———. 2003. Response to Engeman: index values rarely constitute reliable information. *Wildlife Society Bulletin* 31:288–291.

Anderson, D. R., K. P. Burnham, W. R. Gould, and S. Cherry. 2001. Concerns about finding effects that are actually spurious. *Wildlife Society Bulletin* 29:311–316.

Angelstam, P. 1986. Predation on ground-nesting birds' nests in relation to predator densities and habitat edge. *Oikos* 47:365–373.

Anthony, R. M. 1996. Den use by arctic foxes (*Alopex lagopus*) in a subarctic region of western Alaska. *Canadian Journal of Zoology* 74:627–631.

Apostolidis, S., T. Chandra, I. Demirhan, J. Cinatl, H. W. Doerr, and A. Chandra. 2002. Evaluation of carcinogenic potential of two nitro-musk derivatives, musk xylene and musk tibetene in a host-mediated in vivo/in vitro assay system. *Anticancer Research* 22:2657–2662.

Apps, C. D., B. N. McLellan, J. G. Woods, and M. F. Proctor. 2004. Estimating grizzly bear distribution and abundance relative to habitat and human influence. *Journal of Wildlife Management* 68:138–152.

Apps, C. D., N. J. Newhouse, and T. A. Kinley. 2002. Habitat associations of American badgers in southeastern British Columbia. *Canadian Journal of Zoology* 80:1228–1239.

Arnett, E. B. 2006. A preliminary evaluation on the use of dogs to recover bat fatalities at wind energy facilities. *Wildlife Society Bulletin* 34:1440–1445.

Arnold, H. R. 1993. *Atlas of mammals in Britain*. HMO, London, UK.

Arzberger, P. 2004. *Sensors for environmental observatories: report of the NSF sponsored workshop*. WTEC, Baltimore, MD.

Ascensão, F., and A. Mira. 2007. Factors affecting culvert use by vertebrates along two stretches of road in southern Portugal. *Ecological Research* 22:57–66.

Ashbrenner, J. E. 1994. Distribution of marten in the Nicolet National Forest, 1994. *Wisconsin Wildlife Surveys* 4:96–102.

Aubry, K. B., and D. B. Houston. 1992. Distribution and status of the fisher (*Martes pennanti*) in Washington. *Northwestern Naturalist* 73:69–79.

Aubry, K. B., and L. A. Jagger. 2006. The importance of obtaining verifiable occurrence data on forest carnivores and an interactive website for archiving results from standardized surveys. Pages 159–176 in Santos-Reis, M., G. Proulx, J. D. S. Birks, and E. C. O'Doherty, editors. *Martes in carnivore communities*. Alpha Wildlife Publications, Sherwood Park, Alberta, Canada.

Aubry, K. B. and J. C. Lewis. 2003. Extirpation and reintroduction of fishers (*Martes pennanti*) in Oregon: implications for their conservation in the Pacific states. *Biological Conservation* 114:79–90.

Aubry, K. B., F. E. Wahl, J. V. Von Kienast, T. J. Catton, and S. G. Armentrout. 1997. Use of remote video cameras for the detection of forest carnivores and in radio-telemetry studies of fishers. Pages 350–361 in G. Proulx, H. N. Bryant, and P. M. Woodard, editors. *Martes taxonomy, ecology, techniques and management*. Provincial Museum of Alberta, Edmonton, Alberta, Canada.

Augustin, N. H., M. A. Mugglestone, and S. T. Buckland. 1996. An Autologistic Model for the Spatial Distribution of Wildlife. *Journal of Applied Ecology* 33:339–347.

Avise, J. 2004. *Molecular Markers, Natural History, and Evolution*. 2nd ed. Sinauer, Sunderland, MA.

Axel, R. 1995. The molecular logic of smell. *Scientific American* 273:154–159.

Azuma, D. L., J. A. Baldwin, and B. R. Noon. 1990. *Estimating the occupancy of spotted owl habitat areas by sampling and adjusting for bias*. USDA Forest Service Pacific Southwest Research Station General Technical Report PSW-GTR-124, Albany, CA.

Bailey, G. N. A. 1969. A device for tracking small mammals. *Journal of Zoology* 159:513–540.

Bailey, L. L., J. E. Hines, J. D. Nichols, and D. I. MacKenzie. 2007. Sampling design trade-offs in occupancy studies with imperfect detection: examples and software. *Journal of Applied Ecology* 17:281–290.

Bailey, T. N. 1971. Biology of striped skunks on a southwestern Lake Erie marsh. *American Midland Naturalist* 85:196–207.

Baker, A. J. 2000. *Molecular methods in ecology*. Blackwell Science, London, UK.

Baker, J. A., and P. M. Dwyer. 1987. Techniques for commercially harvesting furbearers. Pages 970–995 in M. Novak, J. A. Baker, M. E. Obbard, and B. Malloch, editors. *Wild furbearer management and conservation in North America*. Ontario Ministry of Natural Resources, Toronto, Ontario, Canada.

Baldwin, R. A., A. E. Houston, M. L. Kennedy, and P. S. Liu. 2004. An assessment of microhabitat variables and capture success of striped skunks (*Mephitis mephitis*). *Journal of Mammalogy* 85:1068–1076.

Bamberg, E., H. S. Choi, E. Möstl, W. Wurm, D. Lorin, and K. Arbeiter. 1984. Enzymatic determination of unconjugated oestrogens in faeces for pregnancy diagnosis in mares. *Equine Veterinary Journal* 16:537–539.

Banks, S. C., S. D. Hoyle, A. Horsup, P. Sunnucks, and A. C. Taylor. 2003. Demographic monitoring of an entire species (the northern hairy-nosed wombat, *Lasiorhinus krefftii*) by genetic analysis of noninvasively collected material. *Animal Conservation* 6:101–107.

Banks, S. C., L. F. Skerratt, and A. C. Taylor. 2002. Female dispersal and relatedness structure in common wombats (*Vombatus ursinus*). *Journal of Zoology*, London 256:389–399.

Barbraud, C., J. D. Nichols, J. E. Hines, and H. Hafner. 2003. Estimating rates of local extinction and colonization in colonial species and an extension to the metapopulation and community levels. *Oikos* 101:113–126.

Barea-Azcón, J. M., E. Virgós, E. Ballesteros-Duperón, M. Moleón, and M. Chirosa. 2007. Surveying carnivores at large spatial scales: a comparison of four broad-applied methods. *Biodiversity Conservation* 16:1213–1230.

Barrett, R. H. 1983. Smoked aluminum track plots for determining furbearer distribution and relative abundance. *California Fish and Game* 69:188–190.

Bart, J., S. Droege, P. Geissler, B. Peterjohn, and C. J. Ralph. 2004. Density estimation in wildlife surveys. *Wildlife Society Bulletin* 12:1242–1247.

Bartelt, G. A., R. E Rolley, and L. E. Vine. 2001. *Evaluation of abundance indices for striped skunks, common raccoons and Virginia opossums in southern Wisconsin.* Wisconsin Department of Natural Resources, Research Report 185, Madison.

Bateman, M. C. 1986. Winter habitat use, food habits and home range size of the marten, *Martes americana*, in western Newfoundland (Canada). *Canadian Field-Naturalist* 100:58–62.

Bayley, P. B., and J. T. Peterson 2001. An approach to estimate probability of presence and richness of fish species. *Transactions of the American Fisheries Society* 130:620–633.

Bayne, E., R. Moses, and S. Boutin. 2005. *Evaluation of winter tracking protocols as a method for monitoring mammals in the Alberta Biodiversity Monitoring Program.* Integrated Landscape Management Group, Department of Biological Sciences, University of Alberta, Edmonton, Alberta, Canada.

Beauvais, G. P., and S. W. Buskirk. 1999. An improved estimate of trail detectability for snow-trail surveys. *Wildlife Society Bulletin* 27:32–38.

Becker, E. F. 1991. A terrestrial furbearer estimator based on probability-sampling. *Journal of Wildlife Management* 55:730–737.

Becker, E. F., H. N. Golden, and C. Gardner. 2004. Using probability sampling of animal tracks in snow to estimate population size. Pages 248–270 *in* W. L. Thompson, editor. *Sampling rare or elusive species: concepts, designs and techniques for estimating population parameters.* Island Press, Washington, DC.

Becker, E. F., M. A. Spindler, and T. O. Osborne. 1998. A population estimator based on network sampling of tracks in the snow. *Journal of Wildlife Management* 62:968–977.

Beckmann, J. P. 2006. Carnivore conservation and search dogs: the value of a novel, non-invasive technique in the Greater Yellowstone Ecosystem. Pages 28–34 *in* A. Wondrak Biel, editor. *Greater Yellowstone public lands: a century of discovery, hard lessons, and bright prospects.* Proceedings of the 8th Biennial Scientific Conference on the Greater Yellowstone Ecosystem. Yellowstone Center for Resources Yellowstone National Park, WY.

Beehner, J. C., and P. L. Whitten. 2004. Modifications of a field method for fecal steroid analysis in baboons. *Physiology and Behavior* 82:269–277.

Beier, L. R., S. B. Lewis, R. W. Flynn, G. Pendleton, and T. V. Schumacher. 2005. From the field: a single-catch snare to collect brown bear hair for genetic mark-recapture studies. *Wildlife Society Bulletin* 33:766–773.

Beier, P., and S. C. Cunningham. 1996. Power of track surveys to detect changes in cougar populations. *Wildlfe Society Bulletin* 24:540–546.

Bekoff, M. T., J. Daniels, and J. Gittleman. 1984. Life history patterns and the comparative social ecology of carnivores. *Annual Review of Ecology and Systematics* 15:191–232.

Belant, J. L. 2003a. A hairsnare for forest carnivores. *Wildlife Society Bulletin* 31:482–485.

———. 2003b. Comparison of 3 tracking mediums for detecting forest carnivores. *Wildlife Society Bulletin* 31:744–747.

Bellemain, E., J. E. Swenson, D. Tallmon, S. Brunberg, and P. Taberlet. 2005. Estimating population size of elusive animals with DNA from hunter-collected feces: four methods for brown bears. *Conservation Biology* 19:150–161.

Berchielli, L. T., and A. B. Leubner. 1981. A technique for capturing red and gray foxes. Pages 1555–1559 *in* J. A. Chapman and D. Pursley, editors. *Worldwide furbearer conference proceedings*. R. R. Donnelley and Sons, Falls Church, VA.

Berentsen, A. R., R. H. Schmidt, and R. M. Timm. 2006. Repeated exposure of coyotes to the Coyote Lure Operative Device. *Wildlife Society Bulletin* 34:809–814.

Berger, J., J. W. Testa, T. Roffe, and S. L. Monfort. 1999. Conservation endocrinology: a noninvasive tool to understand relationships between carnivore colonization and ecological carrying capacity. *Conservation Biology* 13:980–989.

Best, M. S., and R. M. Whiting Jr. 1990. Trapping induced changes in scent-station visits by opossums and raccoons. Pages 94–99 *in* P. R. Krausman and N. S. Smith, editors. *Managing wildlife in the Southwest*. Arizona Chapter, Wildlife Society, Tuscon, AZ.

Bider, J. R. 1968. Animal activity in uncontrolled terrestrial communities as determined by a sand transect technique. *Ecological Monographs* 38:269–308.

Billitti, J. E., B. L. Lasley, and B. W. Wilson. 1998. Development and validation of a fecal testosterone biomarker in *Mus musculus* and *Peromyscus maniculatus*. *Biology of Reproduction* 59:1023–1028.

Black, R. E., M. M. Levine, M. L. Clements, T. P. Hughes, and M. J. Blaser. 1988. Experimental *Campylobacter jejuni* infection in humans. *Journal of Infectious Diseases* 157:472–479.

Blejwas, K. M., C. L. Williams, G. T. Shin, D. R. McCullough, and M. M. Jaeger. 2006. Salivary DNA evidence convicts breeding male coyotes of killing sheep. *Journal of Wildlife Management* 70:1087–1093.

Boddicker, M., J. J. Rodriguez, and J. Amanzo. 2002. Indices for assessment and monitoring of large mammals within an adaptive management framework. *Environmental Monitoring and Assessment* 76:105–123.

Boersen, M. R., J. D. Clark, and T. L. King. 2003. Estimating black bear population density and genetic diversity at Tensas River, Louisiana, using microsatellite DNA markers. *Wildlife Society Bulletin* 31:197–207.

Bollinger, E. K., and R. G. Peak. 1995. Depredation of artificial avian nests: a comparison of forest-field and forest-lake edges. *American Midland Naturalist* 134:200–203.

Bonesi, L., and D. W. Macdonald. 2004. Evaluation of sign surveys as a way to estimate the relative abundance of American mink (*Mustela vison*). *Journal of Zoology* 262:65–72.

Bonier, F., H. Quigley, and S. N. Austad. 2004. A technique for non-invasively detecting stress response in cougars. *Wildlife Society Bulletin* 32:711–717.

Bortolotti, G. R. 1984. Trap and poison mortality of golden and bald eagles. *Journal of Wildlife Management* 48:1173–1179.

Boulanger, J., S. Himmer, and C. Swan. 2004a. Monitoring of grizzly bear population trends and demography using DNA mark-recapture methods in the Owikeno Lake area of British Columbia. *Canadian Journal of Zoology* 82:1267–1277.

Boulanger, J., K. C. Kendall, J. B. Stetz, D. A. Roon, L. P. Waits, and D. Paetkau. 2008. Multiple data sources improve DNA-based mark-recapture population estimates of grizzly bears. *Ecological Applications* 18:in press.

Boulanger, J., B. N. McLellan, J. G. Woods, M. F. Proctor, and C. Strobeck. 2004b. Sampling design and bias in DNA-based capture-mark-recapture population and density estimates of grizzly bears. *Journal of Wildlife Management* 68:457–469.

Boulanger, J., M. Proctor, S. Himmer, G. Stenhouse, D. Paetkau, and J. Cranston. 2006. An empirical test of DNA mark-recapture sampling strategies for grizzly bears. *Ursus* 17:149–158.

Boulanger, J., G. Stenhouse, G. MacHutchon, M. Proctor, S. Himmer, D. Paetkau, and J. Cranston. 2005a. *Grizzly bear population and density estimate for the 2005 Alberta (proposed) Unit 4 Management Area Inventory*. Alberta Sustainable Resource Development, Fish and Wildlife Division, Alberta, Canada.

Boulanger, J., G. Stenhouse, and R. Munro. 2004c. Sources of heterogeneity bias when DNA mark-recapture sampling methods are applied to grizzly bear (*Ursus arctos*) populations. *Journal of Mammalogy* 85:618–624.

Boulanger, J., G. Stenhouse, M. Proctor, S. Himmer, D. Paetkau, and J. Cranston. 2005b. *2004 population in-*

*ventory and density estimates for the Alberta 3B and 4B Grizzly Bear Management Area.* Alberta Sustainable Resource Development, Fish and Wildlife Division, Alberta, Canada.

Boulanger, J., G. C. White, B. N. McLellan, J. G. Woods, M. F. Proctor, and S. Himmer. 2002. A meta-analysis of grizzly bear DNA mark-recapture projects in British Columbia. *Ursus* 13:137–152.

Bowers, M. A. 1996. Species as indicators of large-scale environmental change: a computer simulation model of population decline. *Ecoscience* 3:502–511.

Boyce, M. S., P. R. Vernier, S. E. Nielsen, and F. K. A. Schmiegelow. 2002. Evaluating resource selection functions. *Ecological Modelling* 157:281–300.

Boydston, E. E. 2005. *Behavior, ecology, and detection surveys of mammalian carnivores in the Presidio.* Final Report. U.S. Geological Survey, Sacramento, CA.

Bremner-Harrison, S., S. W. R. Harrison, B. L. Cypher, J. D. Murdock, and J. Maldonado. 2006. Development of a single-sampling noninvasive hair snare. *Wildlife Society Bulletin* 34:456–461.

Bridges, A. S., S. Klenzendorf, and M. R. Vaughan. 2004. Seasonal variation in American black bear *Ursus americanus* activity patterns: quantification via remote photography. *Wildlife Biology* 10:277–284.

Brink, H., J. E. Topp-Jørgensen, and A. R. Marshall. 2002. First record in 68 years of Lowe's servaline genet. *Oryx* 36:323–327.

British Columbia Ministry of Environment, Lands and Parks. 1998. *Inventory methods for wolf and cougar.* Terrestrial Ecosystem Task Force, Resources Inventory Committee, Victoria, Canada.

——. 1999. *Inventory methods for medium-sized territorial carnivores: coyote, red fox, lynx, bobcat, wolverine, fisher and badger.* Terrestrial Ecosystems Task Force, Resources Inventory Committee. Victoria, Canada.

Brockman, D. K. and P. L. Whitten. 1996. Reproduction in free-ranging *Propithecus verreauxi*: Estrus and the relationship between multiple partner matings and fertilization. *American Journal of Physiological Anthropology* 100:57–69.

Brockman, D. K., P. L. Whitten, A. F. Richard, and A. Schneider. 1998. Reproduction in free-ranging male *Propithecus verreauxi*: the hormonal correlates of mating. *American Journal of Physiological Anthropology* 105:137–151.

Brody, A. J., and M. R. Pelton. 1989. Effects of roads on black bear movements in western North Carolina. *Wildlife Society Bulletin* 17:5–10.

Brongo, L. L., M. S. Mitchell, and J. B. Grand. 2005. Effects of trapping with bait on bait-station indices to black bear abundance. *Wildlife Society Bulletin* 33:1357–1361.

Brooks, R. T. 1996. Assessment of two camera-based systems for monitoring arboreal wildlife. *Wildlife Society Bulletin* 24:298–300.

Brown, J. A., and C. M. Miller. 1998. Monitoring stoat (*Mustela erminea*) control operations: power analysis and design. *Science for Conservation*: 96. Department of Conservation, Wellington, NZ.

Brown, J. H. 2004. Challenges in estimating size and conservation of black bear in west-central Kentucky. MS thesis. University of Kentucky, Lexington, KY.

Brown, J. L., S. K. Wasser, D. E. Wildt, and L. H. Graham. 1994. Comparative aspects of steroid-hormone metabolism and ovarian activity in felids, measured noninvasively in feces. *Biology of Reproduction* 51:776–786.

Brown, K. P., N. Alterio, and H. Moller. 1998. Secondary poisoning of stoats (*Mustela erminea*) at low mouse (*Mus musculus*) abundance in a New Zealand *Nothofagus* forest. *Wildlife Research* 25:419–426.

Brown, T. J. 1983. *Tom Brown's field guide to nature observation and tracking.* Berkley Trade, New York.

Bryson, S. 1984. *Search dog training.* Boxwood Press, Pacific Grove, CA.

Buckland, S. T., D. R. Anderson, K. P. Burnham, and J. L. Laake. 1993. *Distance sampling: estimating abundance of biological populations.* Chapman and Hall, New York.

Buckland, S. T., R. W. Summers, D. L. Borchers, and L. Thomas. 2006. Point transect sampling with traps and lures. *Journal of Applied Ecology* 43:377–384.

Bulinski, J., and C. McArthur. 2000. Observer error in counts of macropod scats. *Wildlife Research* 27:277–282.

Bull, E. L., R. S. Holthausen, and L. R. Bright. 1992. Comparison of three techniques to monitor marten. *Wildlife Society Bulletin* 20:406–410.

Bullard, R. W. 1982. Wild canid associations with fermentation products. *Industrial and Engineering Chemical Products Research and Development* 21:646–655.

Bullard, R. W., F. J. Turkowski, and S. R. Kilburn. 1983. Responses of free-ranging coyotes to lures and their modifications. *Journal of Chemical Ecology* 9:877–888.

Burke, T., N. B. Davies, W. Bruford, and B. J. Hatchwell. 1989. Parental care and mating behavior of polyandrous dunnocks Prunella modularis related to paternity by DNA fingerprinting. *Nature* 338:249–251.

Burnham, K. P., and D. R. Anderson. 2002. *Model selection and multimodel inference: a practical information-theoretic approach.* 2nd ed. Springer-Verlag, New York.

Burnham, K. P., D. R. Anderson, G. C. White, C. Brownie, and K. H. Pollock. 1987. Design and analysis methods for fish survival experiments based on release-recapture. *American Fisheries Society Monograph* 5.

Burnham, K. P., and W. S. Overton. 1979. Robust estimation of population size when capture probabilities vary among individuals. *Ecology* 60:927–936.

Burst, T. L., and M. R. Pelton. 1983. Black bear mark trees in the Smoky Mountains. *International Conference Bear Research and Management* 5:45–53.

Butcher, G. S. 1990. Audubon Christmas Bird Counts. Pages 5–13 *in* J. R. Sauer and S. Droege, editors. *Survey designs and statistical methods for the estimation of avian population trends.* U.S. Fish and Wildlife Service Biological. Report 90(1).

Cablk, M. E., and J. S. Heaton. 2006. Accuracy and reliability of dogs in surveying for desert tortoise (*Gopherus agassizii*). *Ecological Applications* 16:1926–1935.

Cain, A. T., V. R. Tuovila, D. G. Hewitt, and M. E. Tewes. 2003. Effects of a highway and mitigation projects on bobcats in Southern Texas. *Biological Conservation* 114:189–197.

Campbell, C. J. 2004. Patterns of behavior across reproductive states of free-ranging female black-handed spider monkeys (*Ateles geoffroyi*). *American Journal of Physiological Anthropology* 124:166–176.

Campbell, L. A. 2004. Distribution and habitat associations of mammalian carnivores. PhD dissertation. University of California, Davis.

Carbone, C., S. Christie, K. Conforti, T. Coulson, N. Franklin, J. R. Ginsberg, M. Griffiths, et al. 2001. The use of photographic rates to estimate densities of tigers and other cryptic mammals. *Animal Conservation* 4:75–79.

———. 2002. The use of photographic rates to estimate densities of cryptic mammals: response to Jennelle et al. *Animal Conservation* 5:121–123.

Carey, H. R. 1926. Camera-trapping: a novel device for wild animal photography. *Journal of Mammalogy* 7:278–281.

Carman, R. 1975. *The complete guide to lures and baits.* Spearman Publishing and Printing, Sutton, NE.

Carr, D. 2005. Genetic structure of a recolonizing population of fishers (*Martes pennanti*). MS thesis. Trent University, Peterborough, Ontario, Canada.

Carroll, C. 1997. Predicting the distribution of the fisher (*Martes pennanti*) in northwestern California, U.S.A., using survey data and GIS modeling. MS thesis. Oregon State University, Corvallis.

Carroll, C., R. F. Noss, and P. C. Paquet. 2001. Carnivores a focal species for conservation planning in the Rocky Mountain region. *Ecological Applications* 11:961–980.

Carroll, C., W. J. Zielinski, and R. F. Noss. 1999. Using presence-absence data to build and test spatial habitat models for the fisher in the Klamath region, U.S.A. *Conservation Biology* 13:1344–1359.

Carthew, S. M., and E. Slater. 1991. Monitoring animal activity with automated photography. *Journal of Wildlife Management.* 55:689–692.

Case, L. P. 2005. *The dog: its behavior, nutrition, and health.* Blackwell Publishing, Ames, IA.

Caso, A. 1994. Home range and habitat use of three Neotropical carnivores in northeast Mexico. MS thesis. Texas A&M University, Kingsville.

Cavallini, P. 1994. Feces count as an index of fox abundance. *Acta Theriologica* 39:417–424.

Cavigelli, S. A. 1999. Behavioural patterns associated with faecal cortisol levels in free-ranging female ring-tailed lemurs, *Lemur catta. Animal Behavior* 57:935–944.

Cavigelli, S. A., S. L. Monfort, T. K. Whitney, Y. S. Mechref, M. Novotny, and M. K. McClintock. 2005. Frequent serial fecal corticoid measures from rats reflect circadian and ovarian corticosterone rhythms. *Journal of Endocrinology* 184:153–163.

Cazon-Naravaez, A. V., and S. S. Suhring. 1999. A technique for extraction and thin layer chromatography visualization of fecal bile acids applied to neotropical felid scats. *Revista de Biologia Tropical* 47:245–249.

Chamberlain, M. J., J. W. Mangrum, B. D. Leopold, and E. P. Hill. 1999. A comparison of attractants used for carnivore track surveys. *Proceedings of the Annual Conference of the Southeastern Association of Fish and Wildlife Agencies* 53:297–304.

Chame, M. 2003. Terrestrial mammal feces: a morphometric summary and description. *Memórias do Instituto Oswaldo Cruz* (Rio de Janeiro) 98:71–94.

Chapman, F. M. 1927. Who treads our trails? *National Geographic Magazine* 52:341–345.

ChasingGame. 2006. Trail and scouting camera reviews and performance ratings. www.chasingame.com (accessed 19 December 2007).

Chaudhuri, M., and J. R. Ginsberg. 1990. Urinary androgen concentrations and social status in 2 species of free-ranging zebra (*Equus burchelli* and *E. grevyi*). *Journal of Reproduction and Fertility* 88:127–133.

Chautan, M., D. Pontier, and M. Artois. 2000. Role of rabies in recent demographic changes in red fox (*Vulpes vulpes*) populations in Europe. *Mammalia* 64:391–410.

Choate, D. M., M. L. Wolfe, and D. C. Stoner. 2006. Evaluation of cougar population estimators in Utah. *Wildlife Society Bulletin* 34:782–799.

Choquenot, D., W. A. Ruscoe, and E. Murphy. 2001. Colonisation of new areas by stoats: time to establishment and requirements for detection. *New Zealand Journal of Ecology* 25:83–88.

Chu, J. H., H. Y. Wu, Y. J. Yang, O. Takenaka, and Y. S. Lin. 1999. Polymorphic microsatellite loci and low-invasive DNA sampling in Macaca cyclopis. *Primates* 40:573–580.

Clapperton, B. K., C. T. Eason, R. J. Weston, A. D. Woolhouse, and D. R. Morgan. 1994a. Development and testing of attractants for feral cats, *Felis catus* L. *Wildlife Research* 21:389–399.

Clapperton, B. K., S. M. Phillipson and A. D. Woolhouse. 1994b. Field trials of slow-release synthetic lures for stoats (*Mustela erminea*) and ferrets (*M. furo*). *New Zealand Journal of Zoology* 21:279–284.

Clark, F. W. 1972. Influence of jackrabbit density on coyote population change. *Journal of Wildlife Management* 36:343–356.

Clevenger, A. P., B. Chruszcz, and K. Gunson. 2001. Drainage culverts as habitat linkages and factors affecting passage by mammals. *Journal of Applied Ecology* 38:1340–1349.

Clevenger, A. P., J. Dionne, C. Townsend, and B. Chruszcz. 2005. *Long-term monitoring and DNA-based approaches for restoring landscape connectivity across transportation corridors*. Mid-year report to Parks Canada Agency. Western Transportation Institute, Montana State University, Bozeman.

Clevenger, A. P. and N. Waltho. 2000. Factors influencing the effectiveness of wildlife underpasses in Banff National Park, Alberta, Canada. *Conservation Biology* 14:47–56.

———. 2005. Performance indices to identify attributes of highway crossing structures facilitating movement of large mammals. *Biological Conservation* 121:453–464.

Clevenger, T., M. Gibeau, and M. Sawaya. 2006. *Environmental screening report for "Conservation benefits of wildlife crossings on Banff National Park bear populations using non-invasive DNA methods."* Report prepared for Banff National Park, Parks Canada. Western Transportation Institute, Montana State University, Bozeman.

Clifford, D. L. 2006. Health threats affecting island fox (*Urocyon littoralis*) population recovery on the California Channel Islands. PhD dissertation. University of California, Davis.

Clutton-Brock, T. H., P. N. M. Brotherton, A. F. Russell, M. J. O'Riain, D. Gaynor, R. Kansky, A. Griffin, et al. 2001. Reproductive skew in cooperative mammals. *Science* 291:478–481.

Clyde, M. E., D. Covell, and M. Tarr. 2004. *A landowner's guide to inventorying and monitoring wildlife in New Hampshire*. University of New Hampshire Cooperative Extension, Durham.

Cochran, W. G. 1977. *Sampling theory*. John Wiley & Sons, New York.

Cohen, J. 1988. *Statistical power analysis for the behavioral sciences*. Lawrence Erlebaum Associates, Hillsdale, NJ.

Conner, M. C., R. F. Labisky, and D. R. Progulske Jr. 1983. Scent-station indices as measures of population abundance for bobcats, raccoons, gray foxes, and opossums. *Wildlife Society Bulletin* 11:146–152.

———. 1984. Response to comment on scent-station indices. *Wildlife Society Bulletin* 12:329.

Conner, M. M., M. M. Jaeger, T. J. Weller, and D. R. McCullough. 1998. Effect of coyote removal on sheep depredation in northern California. *Journal of Wildlife Management* 62:690–699.

Connors, M. J., E. M. Schauber, A. Forbes, C. G. Jones, B. J. Goodwin, and R. S. Ostfeld. 2005. Use of track plates to quantify predation risk at small spatial scales. *Journal of Mammalogy* 86:991–996.

Connovation 2006. Blue Prints Trakka. www.connovation.co.nz/mainsite/ProductGroup.TheTrakkaKit.html (accessed 19 December 2007).

Cook, A. H. 1949. *Fur-bearer investigation*. New York State

Conservation Department, Federal Aid in Wildlife Restoration Project 10R, Supplement G, Final Report, Albany, NY.

Copeland, J. 1993. Assessment of snow-tracking and remote camera systems to document presence of wolverines at carrion. Unpublished report. Nongame and Endangered Wildlife Program, Bureau of Wildlife, Idaho Dept. of Fish and Game.

Copeland, J. P. 1996. Biology of the wolverine in central Idaho. MS thesis. University of Idaho, Moscow.

Copeland, J. P., E. Cesar, J. M. Peek, C. E. Harris, C. D. Long and D. L. Hunter. 1995. A live trap for wolverine and other forest carnivores. *Wildlife Society Bulletin* 23:535–538.

Copeland, J. P., L. F. Ruggiero, J. Claar, K. McKelvey, J. R. Squires, and C.D. Long. 2004. *Efficacy of olfactory-based lures for the detection of wolverine.* www .wolverinefoundation.org/research/surveycensus.htm (accessed 30 November 2007).

Copeland, J. P., and J. S. Whitman. 2003. Wolverine (*Gulo gulo*). Pages 672–682 in G. A. Feldhamer, B. C. Thompson, and J. A. Chapman, editors. *Wild mammals of North America: biology, management, and conservation.* 2nd ed. John Hopkins University Press, Baltimore, MD.

Coppinger, R., and L. Coppinger. 2001. *Dogs: a startling new understanding of canine origin, behavior and evolution.* Scribner, New York.

Cordy, D. R., and J. R. Gorham. 1950. The pathology and etiology of salmon disease in the dog and fox. *American Journal of Pathology* 26:617–637.

Covell, D. F. 1992. Ecology of swift fox (*Vulpes velox*) in southeastern Colorado. MS thesis. University of Wisconsin, Madison.

Crawshaw, P. G. Jr., and H. B. Quigley. 1989. Notes on ocelot movement and activity in the Pantanal region, Brazil. *Biotropica* 21:377–379.

Creel, S. 2005. Dominance, aggression, and glucocorticoid levels in social carnivores. *Journal of Mammalogy* 86:255–264.

Creel, S., N. M. Creel, M. G. L. Mills, and S. L. Monfort. 1997. Rank and reproduction in cooperatively breeding African wild dogs: behavioral and endocrine correlates. *Behavioral Ecology* 8:298–306.

Creel, S., N. M. Creel, and S. L. Monfort. 1996. Social stress and dominance. *Nature* 379:212.

Creel, S., N. Marusha-Creel, and S. L. Monfort. 1998. Birth order, estrogens and sex-ratio adaptation in African wild dogs (*Lycaon pictus*). *Animal Reproduction Science* 53:315–320.

Creel, S., N. M. Creel, D. E. Wildt, and S. L.Monfort. 1992. Behavioural and endocrine mechanisms of reproductive suppression in Serengeti dwarf mongooses. *Animal Behavior* 43:231–245.

Creel, S., J. E. Fox, A. Hardy, J. Sands, B. Garrott, and R. O. Peterson. 2002. Snowmobile activity and glucocorticoid stress responses in wolves and elk. *Conservation Biology* 16:809–814.

Creel, S, S. L. Monfort, N. Marushka-Creel, D. E. Wildt, and P. M. Waser. 1995. Pregnancy increases future reproductive success in subordinate dwarf mongooses. *Animal Behavior* 50:1132–1135.

Creel, S., G. Spong, J. L. Sands, J. Rotella, J. Zeigle, L. Joe, K. M. Murphy, and D. Smith. 2003. Population size estimation in Yellowstone wolves with error-prone non-invasive microsatellite genotypes. *Molecular Ecology* 12:2003–2009.

Crooks, K. R. 2002. Relative sensitivities of mammalian carnivores to habitat fragmentation. *Conservation Biology* 16:488–502.

Cunningham, S. C., L. A. Haynes, C. Gustavson, and D. D. Haywood. 1995. Evaluation of the interaction between mountain lions and cattle in the Aravaipa-Klondyke area of southeast Arizona. Arizona Game and Fish Dept. Technical. Report 17. Phoenix, AZ.

Cunningham, S. C., L. Kirkendall, and W. Ballard. 2006. Gray fox and coyote abundance and diet responses after a wildfire in central Arizona. *Western North American Naturalist* 66:169–180.

Curtis, D. J., A. Zaramody, D. I. Green, and A. R. Pickard. 2000. Monitoring of reproductive status in wild mongoose lemurs (*Eulemur mongoz*). *Reproduction, Fertility and Development* 12:21–29.

Cushman, S. A., K. S. McKelvey, J. Hayden, and M. K. Schwartz. 2006. Gene flow in complex landscapes: testing multiple hypotheses with causal modeling. *American Naturalist* 168:486–499.

Cutler, T. L., and D. E. Swann. 1999. Using remote photography in wildlife ecology: a review. *Wildlife Society Bulletin* 27:571–581.

Cypher, B. L., and K. A. Spencer. 1998. Competitive interactions between coyotes and San Joaquin kit foxes. *Journal of Mammalogy* 79:204–214.

Dalen, L., B. Elmhagen, and A. Angerbjörn. 2004a. DNA

analysis on fox faeces and competition induced niche shifts. *Molecular Ecology* 13:2389–2392.

Dalen, L., A. Götherström, and A. Angerbjörn. 2004b. Identifying species from pieces of faeces. *Conservation Genetics* 5:109–111.

Danilov, P., P. Helle, V. Annenkov, V. Belkin, L. Bljudnik, E. Helle, V. Kanshiev, H. Linden, and V. Markovsky. 1996. Status of game animal populations in Karelia and Finland according to winter track count data. *Finnish Game Research* 49:18–25.

Danner, D. A., and N. Dodd. 1982. Comparison of coyote and gray fox scat diameters. *Journal of Wildlife Management* 46:240–241.

Darimont, C. T., T. E. Reimchen, and P. C. Paquet. 2003. Foraging behaviour by gray wolves on salmon streams in coastal British Columbia. *Canadian Journal of Zoology* 81:349–353.

Dark, S. J. 1997. A landscape-scale analysis of mammalian carnivore distribution and habitat use by fisher. MS thesis. Humboldt State University, Arcata, CA.

Davis, F., C. Seo, and W. Zielinski. 2007. Regional variation in home-range-scale habitat models for fisher (*Martes pennanti*) in California. *Ecological Applications* 17:2195–2213.

Davison, A., J. D. S. Birks, R. C. Brookes, T. C. Braithwaite, and J. E. Messenger. 2002. On the origin of faeces: morphological versus molecular methods for surveying rare carnivores from their scats. *Journal of Zoology* 257:141–143.

De Vos, A. 1951. Tracking of fisher and marten. *Sylva* 77:14–18.

Dean, E. E. 1979. *Training of dogs to detect black-footed ferrets*. Southwestern Research Institute, Santa Fe, NM.

DePue, J. E., and M. Ben-David. 2007. Hair sampling techniques for river otters. *Journal of Wildlife Management* 71:671–674.

Desimone, R., and B. Semmens. 2004. *Garnet Mountains mountain lion research: progress report January 2003–December 2004*. Montana Fish, Wildlife and Parks, Helena.

DeVan, R. 1982. The ecology and life history of the long-tailed weasel (*Mustela frenata*). PhD dissertation. University of Cincinnati, Cincinnati, OH.

DeVault, T. L., O. E. Rhodes Jr., and J. A. Shivik. 2003. Scavenging by vertebrates: behavioral, ecological, and evolutionary perspectives on an important energy transfer pathway in terrestrial ecosystems. *Oikos* 102:225–234.

Di Bitetti, M. S., A. Paviolo, and C. De Angelo. 2006. Density, habitat use and activity patterns of ocelots (*Leopardus pardalis*) in the Atlantic Forest of Misiones, Argentina. *Journal of Zoology* 270:153–163.

Dice, L. R. 1941. Methods for estimating populations of mammals. *Journal of Wildlife Management* 5:398–407.

Diefenbach, D. R., M. J. Conroy, R. J. Warren, W. E. James, L. A. Baker, and T. Hon. 1994. A test of the scent-station survey technique for bobcats. *Journal of Wildlife Management* 58:10–17.

Dijak, W., and F. I. Thompson. 2000. Landscape and edge effects on the distribution of mammalian predators in Missouri. *Journal of Wildlife Management* 64:209–216.

Dillon, A. 2005. Ocelot density and home range in Belize, Central America: camera-trapping and radio telemetry. MS thesis. Virginia Polytechnic Institute and State University, Blacksburg.

Dillon, A., and M. J. Kelly. 2007. Ocelot activity, trap success, and density in Belize: the impact of trap spacing and distance moved on density estimates. *Oryx* 41:469–477.

Dloniak, S. M., J. A. French, N. J. Place, M. L. Weldele, S. E. Glickman, and K. E. Holekamp. 2004. Monitoring of fecal androgens in spotted hyenas (*Crocuta crocuta*). *General and Comparative Endocrinology* 135:51–61.

Dobbins, C. L. 2004. *Evaluation of lures, baits, and urines*. 2nd printing. Bullet Printing, Murfreesboro, NC.

Donovan, T. M., and M. Alldredge. 2007. *Exercises in estimating and monitoring abundance*. www.uvm.edu/envnr/vtcfwru/spreadsheets/abundance/abundance.htm (accessed 19 December 2007).

Donovan, T. M., and J. Hines. 2007. *Exercises in occupancy modeling and estimation*. www.uvm.edu/envnr/vtcfwru/spreadsheets/occupancy/occupancy.htm (accessed 19 December 2007).

Dorazio, R. M., and J. A. Royle. 2003. Mixture models for estimating the size of a closed population when capture probabilities vary among individuals. *Biometrics* 59:351–364.

———. 2005. Estimating size and composition of biological communities by modeling the occurrence of species. *Journal of the American Statistical Association* 100:389–398.

Doty, R. L. 1986. Odor-guided behavior in mammals. *Experientia* 42:257–271.

Douglas, C. W., and M. A. Strickland. 1987. Fisher. Pages 511–529 *in* M. Novak, J. A. Baker, and M. E Obbard, editors. *Wild furbearer management and conservation in North America.* Ontario Ministry of Natural Resources, Toronto, Canada.

Downey, P. J. 2005. Hair-snare survey to assess distribution of Margay (*Leopardus wiedii*) inhabiting El Cielo Biosphere Reserve, Tamaulipas, Mexico. MS thesis, Oklahoma State University, Stillwater.

Downey, P. J., E. C. Hellgren, A. Caso, S. Carvajal, and K. Frangioso. 2007. Hair snares for noninvasive sampling of felids in North America: do gray foxes affect success? *Journal of Wildlife Management* 71:2090–2094.

Dreher, B. P., S. R. Winterstein, K. T. Scribner, P. M. Lukacs, D. R. Etter, G. J. M. Rosa, V. A. Lopez, S. Libants, and K. B. Filcek. 2007. Non-invasive estimation of black bear abundance in Michigan incorporating genotyping errors and harvested bears. *Journal of Wildlife Management* 71:2684–2693.

Drew, R. E., J. G. Hallett, K. B. Aubry, K. W. Cullings, S. M. Koepfs, and W. J. Zielinski. 2003. Conservation genetics of the fisher (*Martes pennanti*) based on mitochondrial DNA sequencing. *Molecular Ecology* 12:51–62.

Dumond, M. 2005. *Species at risk populations' monitoring: grizzly bear and wolverine test of a promising method.* Progress Report for the Hunters and Trappers Association. Kugluktuk Angoniatit Association. Nunavut Wildlife Research Group, Department of Environment, Kugluktuk, Nunavut, Canada.

Earle, R. D., and V. R. Tuovila. 2003. Furbearer winter track count survey of 2002. Wildlife Report No. 3397. Michigan Department of Natural Resources, Lansing.

Edwards, G. P., N. D. de Preu, B. J. Shakeshaft, and I. V. Crealy. 2000. An evaluation of two methods of assessing feral cat and dingo abundance in central Australia. *Wildlife Research* 27:143–149.

Eggert, L. S., J. A. Eggert, and D. S. Woodruff. 2003. Estimating population sizes for elusive animals: the forest elephants of Kakum National Park, Ghana. *Molecular Ecology* 12:1389–1402.

Elbroch, M. 2003. *Mammal tracks and sign: a guide to North American species.* Stackpole Books, Mechanicsburg, PA.

Ellingson, A. R., and P. M. Lukacs. 2003. Improving methods for regional landbird monitoring: a reply to Hutto and Young. *Wildlife Society Bulletin* 31:896–902.

Elzinga, C. L., D. W. Salzer, J. W. Willoughby, and J. P. Gibbs. 2001. *Monitoring Plant and Animal Populations.* Blackwell Science, Malden, MA.

Emmons, L. H. 1988. A field study of ocelots (*Felis pardalis*) in Peru. *Revue d'Ecologie (la Terre et la Vie)* 43:133–157.

Engeman, R. M. 2003. More on the need to get the basics right: population indices. *Wildlife Society Bulletin* 31:286–287.

Engeman, R. M., K. L. Christensen, M. J. Pipas, and D. L. Bergman. 2003a. Population monitoring in support of rabies vaccination program for skunks in Arizona. *Journal of Wildlife Diseases* 39:746–750.

Engeman, R. M., R.E. Martin, B. Constantin, R. Noel, and J. Woolard. 2003b. Monitoring predators to optimize their management for marine turtle nest protection. *Biological Conservation* 113:171–178.

Engeman, R. M., R. E. Martin, H.T. Smith, J. Woolard, C. K. Crady, S. A. Shwiff, B. Constantin, M. Stahl, and J. Griner. 2005. Dramatic reduction in predation on marine turtle nests through improved predator monitoring and management. *Oryx* 39:318–326.

Engeman, R. M., M. J. Pipas, K. S. Gruver, and L. Allen. 2000. Monitoring coyote population changes with a passive activity index. *Wildlife Research* 27:553–57.

Engeman R. M., D. V. Rodriquez, M. A. Linnell, and M. E. Pitzler. 1998. A review of the case histories of the brown tree snakes (*Boiga irregularis*) located by detector dogs on Guam. *International Biodeterioration and Biodegradation* 42:161–165.

Engeman, R. M., D. S. Vice, D. York, and K. S. Gruver. 2002. Sustained evaluation of the effectiveness of detector dogs for locating brown tree snakes in cargo outbound from Guam. *International Biodeterioration and Biodegradation* 49:101–106.

Engeman, R. M., D. Whisson, J. Quinn, F. Cano, P. Quiñones, and T. H. White Jr. 2005. Monitoring invasive mammalian predator populations sharing habitat with the Critically Endangered Puerto Rican parrot *Amazona vittata. Oryx* 39:1–8.

Erb, J. 2006a. *Predator/furbearer scent station survey summary, 2005.* Minnesota Department of Natural Resources. Grand Rapids, MN.

———. 2006b. *Predator/furbearer scent station survey summary, 2004.* Minnesota Department of Natural Resources. Grand Rapids, MN. www.files.dnr.state.mn .us/publications/wildlife/populationstatus2005/scent post_track_survey.pdf. (accessed 19 December 2007).

Ernest, H. B., M. C. T. Penedo, B. P. May, M. Syvanen, and W. M. Boyce. 2000. Molecular tracking of mountain lions in the Yosemite Valley region in California: genetic analysis using microsatellites and faccal DNA. *Molecular Ecology* 9:433–441.

Ernest, H. B., E. S. Rubin, and W. M. Boyce. 2002. Fecal DNA analysis and risk assessment of mountain lion predation of bighorn sheep. *Journal of Wildlife Management* 66:75–85.

Evans M. R., and J. L. Burn. 1996. An experimental analysis of mate choice in the wren: a monomorphic, polygynous passerine. *Behavioral Ecology* 7:101–108.

Evans, T. J., A. Fischbach, S. Schliebe, B. Manly, S. Kalxdorff, and G. York. 2003. Polar bear aerial survey in the eastern Chukchi sea: a pilot study. *Arctic* 56:359–366.

Ewen, K. R., M. Bahlo, S. A. Treloar, D. F. Levinson, B. Mowry, J. W. Barlow, and S. J. Foote. 2000. Identification and analysis of error types in high-throughput genotyping. *American Journal of Human Genetics* 67:727–736.

Ezeh P. I., L. J. Myers, L. A. Hanrahan, R. J. Kemppainen, and K. A. Cummins. 1992. Effects of steroids on the olfactory function of the dog. *Physiology and Behavior* 51:1183–1187.

Fagerstone, K. A. 1987. Black-footed ferret, long-tailed weasel, short-tailed weasel, and least weasel. Pages 549–573 *in* M. Novak, J. A. Baker, M. E. Obbard, and B. Malloch, editors. *Wild furbearer management and conservation in North America.* Ontario Ministry of Natural Resources, Toronto, Ontario, Canada.

Fagre, D. B., B. A. Butler, W. E. Howard, and R. Teranishi. 1981. Behavioral responses of coyotes to selected odors and tastes. Pages 966–983 *in* J. A. Chapman and D. Pursley, editors. *Worldwide furbearer conference proceedings.* R. R. Donnelley and Sons, Falls Church, VA.

Falcucci, A., L. Maiorano, and L. Boitani. 2007. Changes in land-use/land-cover patterns in Italy and their implications for biodiversity conservation. *Landscape Ecology* 22:617–631.

Farhadinia, M. S. 2004. The last stronghold: cheetah in Iran. *Cat News* 40:11–14.

Farnsworth, G. L., K. H. Pollock, J. D. Nichols, T. R. Simons, J. E. Hines, and J. R. Sauer. 2002. A removal model for estimating detection probabilities from point-count surveys. *Auk* 119:414–425.

Farrell, L. E., J. Roman, and M. E. Sunquist. 2000. Dietary separation of sympatric carnivores identified by molecular analysis of scats. *Molecular Ecology* 9:1583–1590.

Farrell, R. K., R. W. Leader, and S. D. Johnston. 1973. Differentiation of salmon poisoning disease and Elokomin fluke fever: studies with the black bear (*Ursus americanus*). *American Journal of Veterinary Research* 34:919–922.

FDA (USDHHS, US Food and Drug Administration). 2006. *The bad bug book.* www.cfsan.fda.gov/~mow/ intro.html (accessed 30 November 2007).

Fecske, D. M., J. A. Jenks, and V. J. Smith. 2002. Field evaluation of a habitat-relation model for the American marten. *Wildlife Society Bulletin* 30:775–782.

Fedriani, J. M., T. K. Fuller, and R. M. Sauvajot. 2001. Does availability of anthropogenic food enhance densities of omnivorous animals? An example with coyotes in southern California. *Ecography* 24:325–331.

Felicetti, L. A., C. C. Schwartz, R. O. Rye, K. A. Gunther, J. G. Crock, M. A. Haroldson, L. Waits, and C. T. Robbins. 2004. Use of naturally occurring mercury to determine the importance of cutthroat trout to Yellowstone grizzly bears. *Canadian Journal of Zoology* 82:493–501.

Fernández, G. J., J. C. Corley, and A. F. Capurro. 1997. Identfication of cougar and jaguar feces through bile acid chromatography. *Journal of Wildlife Management* 61:506–510.

Field, S. A., A. J. Tyre, N. Jonzen, J. R. Rhodes, and H. P. Possingham. 2004. Minimizing the cost of environmental management decisions by optimizing statistical thresholds. *Ecology Letters* 7:669–675.

Field, S. A., A. J. Tyre, and H. P. Possingham. 2005. Optimizing allocation of monitoring effort under economic and observational constraints. *Journal of Wildlife Management* 69:473–482.

Finley, R. B., Jr. 1965. Adverse effects on birds of Phosphamidon applied to a Montana forest. *Journal of Wildlife Management* 29:580–591.

Fischer, C. V. 1998. Habitat use by free-ranging felids in an agroecosystem. MS thesis. Texas A&M University, Kingsville.

Fisher, J. T. 2003. *Assessment of wolverine monitoring methodologies in Alberta.* Alberta Research Council, Vegreville, Alberta, Canada.

———. 2004. *Alberta wolverine experimental monitoring project: 2003–2004 Annual Report.* Alberta Research Council, Vegreville, Alberta, Canada.

———. 2005. *Alberta wolverine experimental monitoring project, 2004–2005 Annual Report.* Alberta Research Council, Vegreville, Alberta, Canada.

Fjelline, D. P., and T. M. Mansfield. 1989. Methods to standardize the procedure for measuring mountain lion tracks. Pages 49–51 *in* R. H. Smith, editor. *Proceedings of the third mountain lion workshop.* Arizona Game and Fish Department, Phoenix.

Flagstad, O., E. Hedmark, A. Landa, H. Broseth, J. Persson, R. Andersen, P. Segerstrom, and H. Ellegren. 2004. Colonization history and noninvasive monitoring of a reestablished wolverine population. *Conservation Biology* 18:676–688.

Fleishman, E., R. MacNally, J. P. Fay, and D. D. Murphy. 2001. Modeling and predicting species occurrence using broad-scale environmental variables: an example with butterflies of the Great Basin. *Conservation Biology* 15:1674–1685.

Flinders, J. T., and J. A. Crawford. 1977. Composition and degradation of jackrabbit and cottontail fecal pellets, Texas high plains. *Journal of Range Management* 30:217–220.

Foley, C. A. H., S. Papageorge, and S. K. Wasser. 2001. Stress and reproductive measures of social and ecological pressures in free-ranging African elephants. *Conservation Biology* 15:1134–1142.

Follmann, E. H., D. G. Ritter, and G. M. Baer. 1988. Immunization of arctic foxes (*Alopex lagopus*) with oral rabies vaccine. *Journal of Wildlife Diseases* 24:477–483.

Foran, D. R. 2006. Relative degradation of nuclear and mitochondrial DNA: an experimental approach. *Journal of Forensic Science* 51:766–770.

Foran, D. R., K. R. Crooks, and S. C. Minta. 1997a. Species identification from scat: an unambiguous genetic method. *Wildlife Society Bulletin* 25:835–839.

Foran, D. R., S. C. Minta, and K. S. Heinemeyer. 1997b. DNA-based analysis of hair to identify species and individuals for population research and monitoring. *Wildlife Society Bulletin* 25:840–847.

Foresman, K. R., and D. E. Pearson. 1998. Comparison of proposed survey procedures for detection of forest carnivores. *Journal of Wildlife Management* 62:1217–1226.

Foreyt, W. J., J. R. Gorham, J. S. Green, C. W. Leathers, and B. R. LeaMaster. 1987. Salmon poisoning disease in juvenile coyotes: clinical evaluation and infectivity of metacercariae and rickettsiae. *Journal of Wildlife Diseases* 23:412–417.

Forrest, L. R. 1988. *Field guide to tracking animals in snow.* Stackpole Books, Harrisburg, PA.

Forsey, E. S., and E. M. Baggs. 2001. Winter activity of mammals in riparian zones and adjacent forests prior to and following clear-cutting at Copper Lake, Newfoundland, Canada. *Forest Ecology and Management* 145:163–171.

Foster, D. R., G. Motzkin, D. Bernardos, and J. Cardoza. 2002. Wildlife dynamics in the changing New England landscape. *Journal of Biogeography* 29:1337–1357.

Fowler, C. H. 1995. Techniques for detecting and monitoring martens and fishers in forest habitats of California. MS thesis. Humboldt State University, Arcata, CA.

Fowler, C. H., and R. T. Golightly. 1993. *Marten and fisher survey techniques.* Final report. USDA Forest Service, Pacific Southwest Research Station, Arcata, CA.

———. 1994. *Fisher and marten survey techniques on the Tahoe National Forest.* Final Report to the USDA Forest Service. Agreement No. PSW-90-0034CA. Humboldt State University, Arcata, CA.

Franklin, A. B. 2002. Exploring ecological relationships in survival and estimating rates of population change using program MARK. Pages 350–356 *in* R. Fields, R. J. Warren, and H. Okarma, editors. *Integrating people and wildlife for a sustainable future: Proceedings of the Second International Wildlife Management Congress.* Godollo, Hungary.

Frantz, A. C., L. C. Pope, P. J. Carpenter, T. J. Roper, G. J. Wilson, and R. J. Delahay. 2003. Reliable microsatellite genotyping of the Eurasian badger (*Meles meles*) using faecal DNA. *Molecular Ecology* 23:1649–1661.

Frantz, A. C., M. Schaul, L. C. Pope, F. Fack, L. Schley, C. P. Muller, and T. J. Roper. 2004. Estimating population size by genotyping remotely plucked hair: the Eurasian badger. *Journal of Applied Ecology* 41:985–995.

Frantzen, M. A. J., J. B. Silk, J. W. H. Ferguson, R. K. Wayne, and M. H. Kohn. 1998. Empirical evaluation of preservation methods for faecal DNA. *Molecular Ecology* 7:1423–1428.

French, J. A., K. L. Bales, A. J. Baker, and J. M. Dietz. 2003. Endocrine monitoring of wild dominant and subordinate female *Leontopithecus rosalia. International Journal of Primatology* 24:1281–1300.

Fujita, S., F. Mitsunaga, H. Sugiura, and K. Shimizu. 2001. Measurement of urinary and fecal steroid metabolites during the ovarian cycle in captive and wild Japanese macaques, *Macaca fuscata*. *American Journal of Primatology* 53:167–176.

Fuller, A. 2006. Multi-scalar responses of forest carnivores to habitat and spatial pattern: case studies with Canada lynx and American martens. PhD dissertation. University of Maine, Orono.

Fuller, T. K. 1978. Variable home range sizes in female gray foxes. *Journal of Mammalogy* 59:446–449.

Fuller, T. K., E. C. York, S. M. Powell, T. A. Decker, and R. M. DeGraaf. 2001. An evaluation of territory mapping to estimate fisher density. *Canadian Journal of Zoology* 79:1691–1696.

Funk, S. M., C. V. Fiorello, S. Cleaveland, and M. E. Gompper. 2001. The role of disease in carnivore ecology and conservation. Pages 443–446 *in* J. L. Gittleman, S. M. Funk, D. Macdonald, and R. K. Wayne, editors. *Carnivore conservation*. Cambridge University Press, Cambridge, UK.

Funston, P. J., E. Herrmann, P. Babupi, H. Kruiper, H. Jaggers, K. Masule, and Kruiper. 2001. Spoor frequency estimates as a method of determining lion and other large mammal densities in the Kgalagadi Transfrontier Park. Pages 36–52 *in* J. L. Gittleman, S. M. Funk, D. Macdonald, and R. K. Wayne, editors. *Carnivore conservation*. Cambridge University Press, Cambridge, UK.

Furgal C. M., S. Innes, and K. M. Kovacs. 1996. Characteristics of ringed seal, *Phoca hispida*, subnivean structures and breeding habitat and their effects on predation. *Canadian Journal of Zoology* 74:858–874.

Gagneux, P., C. Boesch, and D. S. Woodruff. 1997. Microsatellite scoring errors associated with noninvasive genotyping based on nuclear DNA amplified from shed hair. *Molecular Ecology* 6:861–868.

Gagnon, J. W., R. E. Schweinsburg, N. L. Dodd, and A. L. Manzano. 2005. Use of video surveillance to assess wildlife behavior and use of wildlife underpasses in Arizona. Pages 534–544 *in* C. L. Irwin, P. Garrett, and K. P. McDermott, editors. *Proceedings of the 2005 International Conference on Ecology and Transportation*. Center for Transportation and the Environment, North Carolina State University, Raleigh.

Gaines, W. L. 2001. Large carnivore surveys in the North Karakorum Mountains, Pakistan. *Natural Areas Journal* 21:168–171.

Galama, W. T., L. H. Graham, and A. Savage. 2004. Comparison of fecal storage methods for steroid analysis in black rhinoceros (*Diceros bicornis*). *Zoo Biology* 23:291–300.

Gardner, C. 2005. *Landscape-scale wolverine distribution and habitat use in interior Alaska: identification of key habitat parameters*. Alaska Department of Fish and Game, Juneau, AK.

Garnier, J. N., D. I. Green, A. R. Pickard, H. J. Shaw, and W. V. Holt. 1998. Diagnosis of pregnancy in wild black rhinoceros (*Diceros bicornis minor*) by faecal steroid analysis. *Reproduction, Fertility and Development* 10:451–458.

Garrott, R. A., and L. E. Eberhardt. 1987. Arctic fox. Pages 395–406 *in* M. Novak, J. A. Baker, M. E. Obbard, and B. Malloch, editors. *Wild furbearer management and conservation in North America*. Ontario Ministry of Natural Resources, Toronto, Ontario, Canada.

Garrott, R. A., S. L. Monfort, P. J. White, K. L. Mashburn, and J. G. Cook. 1998. One-sample pregnancy diagnosis in elk using fecal steroid metabolites. *Journal of Wildlife Diseases* 34:126–131.

Garshelis, D. L. 2006. On the allure of noninvasive genetic sampling—putting a face to the name. *Ursus* 17:109–123.

Gaston, K. J. 1991. How large is a species' geographic range? *Oikos* 61:434–37.

———. 1994. Measuring geographic range sizes. *Ecography* 17:198–205.

Gau, R. J., R. Case, D. F. Penner, and P. D. McLoughlin. 2002. Feeding patterns of barren-ground grizzly bears in the central Canadian Arctic. *Arctic* 55:339–344.

Geary, S. M. 1984. *Fur trapping in North America*. Winchester Press, Piscataway, NJ.

Gehring, T. M., and R. K. Swihart. 2003. Body size, niche breadth, and ecologically scaled responses to habitat fragmentation: mammalian predators in an agricultural landscape. *Biological Conservation* 109:283–295.

———. 2004. Home range and movements of long-tailed weasels in a landscape fragmented by agriculture. *Journal of Mammalogy* 85:79–86.

Gehrt, S. D. 2003. Raccoon (*Procyon lotor* and allies). Pages 611–634 *in* G. A. Feldhamer, B. C. Thompson, and J. A. Chapman, editors. *Wild mammals of North America: biology, management, and conservation*. 2nd ed. John Hopkins University Press, Baltimore, MD.

Gehrt, S. D. and S. Prange. 2007. Interference competition

between coyotes and raccoons: test of the mesopredator release hypothesis. *Behavioral Ecology* 18:204–214.

Gelok, P. A. 2005. Seasonal habitat associations of American martens (*Martes americana*) in central Ontario. MS thesis. University of Toronto, Toronto, Ontario, Canada.

Gerard, P. D., D. R. Smith, and G. Weerakkody. 1998. Limits of retrospective power analysis. *Journal of Wildlife Management* 62:801–807.

Gerrodette, T. 1987. A power analysis for detecting trends. *Ecology* 68:1364–1372.

———. 1993. TRENDS: software for a power analysis of linear regression. *Wildlife Society Bulletin* 21:515–516.

Gese, E. M. 2001. Monitoring of terrestrial carnivore populations. Pages 372–396 *in* J. L Gittleman, S. M. Funk, D.W. Macdonald, and R. K Wayne, editors. *Carnivore conservation*. Cambridge University Press, Cambridge, UK.

Gibbs, J. P., S. Droege, and P. Eagle. 1998. Monitoring populations of plants and animals. *BioScience* 48:935–940.

Gibbs, J. P., H. L. Snell, and C. E. Causton. 1999. Effective monitoring for adaptive wildlife management: lessons from the Galapagos Islands. *Journal of Wildlife Management* 63:1055–1065.

Gibson, L. A., B. A. Wilson, D. M. Cahill, and J. Hill. 2004. Spatial prediction of rufous bristlebird habitat in a coastal heathland: a GIS-based approach. *Journal of Applied Ecology* 41:213–223.

Gilks, W. R., A. Thomas, and D. J. Spiegelhalter. 1994. A language and program for complex Bayesian modelling. *Statistician* 43:169–178.

Gillette, R. L. 2004. Optimizing the scenting ability of the dog. *Athletic and Working Dog Newsletter*, May.

Ginsberg, J. R. 2001. Setting priorities for carnivore conservation: what makes carnivores different? Pages 498–523 *in* J. L. Gittleman, S. M. Funk, D. Macdonald, and R. K. Wayne, editors. *Carnivore conservation*. Cambridge University Press, Cambridge, UK.

Glennon, M. J., W. F. Porter, and C. L. Demers. 2002. An alternative field technique for estimating diversity of small-mammal populations. *Journal of Mammalogy* 83:734–742.

Godbois, I. A., L. M. Conner, B. D. Leopold, and R. J. Warren. 2005. Effect of diet on mass loss of bobcat scat after exposure to field conditions. *Wildlife Society Bulletin* 33:149–153.

Goldsmith, F. B. 1991. *Monitoring for conservation and ecology*. Chapman and Hall, London, UK.

Gompper, M. E., and H. M. Hackett. 2005. The long-term, range-wide decline of a once common carnivore: the eastern spotted skunk (*Spilogale putorius*). *Animal Conservation* 8:195–201.

Gompper, M. E., R. W. Kays, J. C. Ray, S. D. LaPoint, D. A. Bogan, and J. R. Cryan. 2006. A comparison of noninvasive techniques to survey carnivore communities in Northeastern North America. *Wildlife Society Bulletin* 34:1142–1151.

Gompper, M. E., and A. N. Wright. 2005. Altered prevalence of raccoon roundworm (*Baylisascaris procyonis*) owing to manipulated contact rates of hosts. *Journal of Zoology* 266:215–219.

Gonzales, A. G. 1997. Weasels, skunks, ringtail, raccoon, and badger. Pages 55–66 *in* J. E. Harris and C. V. Ogan, editors. *Mesocarnivores of northern California: biology, management, and survey techniques*. Wildlife Society, California North Coast Chapter, Arcata, CA.

González-Esteban, J., I. Villate, and I. Irizar. 2004. Assessing camera traps for surveying the European mink, *Mustela lutreola* (Linnaeus, 1761), distribution. *European Journal of Wildlife Research* 50:33–36.

Goosens, B., L. P. Waits, and P. Taberlet. 1998. Plucked hair samples as a source of DNA: reliability of dinucleotide microsatellite genotyping. *Molecular Ecology* 7:1237–1241.

Gorman, M. L., and B. J. Trowbridge. 1989. The role of odor in the social lives of carnivores. Pages 57–88 *in* J. L. Gittleman, editor. *Carnivore behavior, ecology, and evolution*. Cornell University Press, Ithaca, NY.

Gosse, J. W., R. Cox, and S. W. Avery. 2005. Home range characteristics and habitat use by American martens in eastern Newfoundland. *Journal of Mammalogy* 86:1156–1163.

Gould, L., T. E. Ziegler, and D. J. Wittwer. 2005. Effects of reproductive and social variables on fecal glucocorticoid levels in a sample of adult male ring-tailed lemurs (*Lemur catta*) at the Beza Mahafaly reserve, Madagascar. *American Journal of Primatology* 67:5–23.

Goymann, W. 2005. Monitoring of hormones in bird droppings. *Annals of the New York Academy of Sciences* 1046:35–53.

Goymann, W., E. Möstl, T. Van't Hof, M. L. East, and H. Hofer. 1999. Fecal monitoring of glucocorticoids in spotted hyenas (*Crocuta crocuta*). *General and Comparative Endocrinology* 114:340–348.

Goymann, W., M. L. East, B. Wachter, O. P. Honer, E. Möstl, T. J. Van't Hof, and H. Hofer. 2001. Social, state-dependent and environmental modulation of faecal corticosteroid levels in free-ranging female spotted hyenas. *Proceedings of the Royal Society, London B. Biological Sciences* 268:2453–2459.

Graham, I. M. 2002. Estimating weasel *Mustela nivalis* abundance from tunnel tracking indices at fluctuating field vole *Microtus agrestis* density. *Wildlife Biology* 8:279–287.

Graves, G. E., and M. L. Boddicker. 1987. Field evaluation of olfactory attractants and strategies used to capture depredating coyotes. Eighth Great Plains Wildlife Damage Control Workshop. Rapid City, SD.

Green, G. I., and D. J. Mattson. 2003. Tree rubbing by Yellowstone grizzly bears *Ursus arctos*. *Wildlife Biology* 9:1–9.

Green, G. I., D. J. Mattson, and J. M. Peek. 1997. Spring feeding on ungulate carcasses by grizzly bears in Yellowstone National Park. *Journal of Wildlife Management* 61:1040–1055.

Green, J. S., and J. T. Flinders. 1981. Diameter and pH comparisons of coyote and red fox scats. *Journal of Wildlife Management* 45:765–767.

Green, R. 2006. Distribution and habitat associations of forest carnivores and an evaluation of the California Wildlife Habitat relationship model for American marten in Sequoia and Kings Canyon National Parks. MS thesis. Humboldt State University, Arcata, CA.

Greenwood, R. J., W. E. Newton, G. L. Pearson, and G. J. Schamber. 1997. Population and movement characteristics of radio-collared striped skunks in North Dakota during an epizootic of rabies. *Journal of Wildlife Diseases* 33:226–241.

Gregory, T. 1929. Camera trapping in the snow. *Journal of Mammalogy* 10:142–148.

Grenier, M., editor. 2003. Swift fox conservation team 2002 annual report. *Wyoming Game and Fish Department*, Lander.

Grigione, M. M., P. Burman, V. C. Bleich, and B. M. Pierce. 1999. Identifying individual mountain lions *Felis concolor* by their tracks: refinement of an innovative technique. *Biological Conservation* 88:25–32.

Grinnell, G. B. 1876. Report of a reconnaissance from Carroll, Montana, to Yellowstone National Park and return. US Govt. Print. Office, Washington, DC.

Grinnell, J., J. S. Dixon, and J. M. Linsdale. 1937. *Fur-bearing mammals of California*. University of California Press, Berkeley.

Gu, W. D., and R. K. Swihart. 2004. Absent or undetected? Effects of non-detection of species occurrence on wildlife-habitat models. *Biological Conservation* 116:195–203.

Guggisberg, C. A. W. 1977. *Early wildlife photographers*. Taplinger, New York.

Gusset, M., and N. Burgener. 2005. Estimating larger carnivore numbers from track counts and measurements. *African Journal of Ecology* 43:320–324.

Gutzwiller, K. J. 1990. Minimizing dog-induced biases in game bird research. *Wildlife Society Bulletin* 18:351–356.

Halfpenny, J. C. 1986. *A field guide to mammal tracking in North America*. Johnson Books, Boulder, CO.

Halfpenny, J. C., R. W. Thompson, S. C. Morse, T. Holden, and P. Rezendes. 1995. Snow tracking. Pages 91–124 *in* W. J. Zielinski, and T. E. Kucera, editors. *American marten, fisher, lynx and wolverine: survey methods for their detection*. USDA Forest Service Pacific Southwest Research Station General Technical Report PSW-GTR-157, Albany, CA.

Hallberg, D. L., and G. R. Trapp. 1984. Gray fox temporal and spatial activity in a riparian-agricultural zone in California's Central Valley. Pages 920–928 *in* R. E. Warner and K. M. Hendrix, editors. *Proceedings of the California Riparian Systems Conference*. University of California Press, Berkeley.

Hamm, K. A., L. W. Diller, R. R. Klug, and T. L. McDonald. 2003. Spatial independence of fisher (*Martes pennanti*) detections at track plates in northwestern California. *American Midland Naturalist* 149:201–210.

Hanson, G. 1989. *Fur bearers and their glands*. Glen Hanson, Whitehall, MT.

Haroldson, M. A., K. A. Gunther, D. P. Reinhart, S. R. Podruzny, C. Cegelski, L. Waits, T. Wyman, and J. Smith. 2005. Changing number of spawning cutthroat trout in tributary streams of Yellowstone Lake and estimates of grizzly bears visiting stream from DNA. *Ursus* 16:167–180.

Harper, J. M., and S. N. Austad. 2000. Fecal glucocorticoids: a method of measuring adrenal activity in wild and captive rodents. *Physiological and Biochemical Zoology* 73:12–22.

———. 2004. Fecal corticosteroid levels in free-living populations of deer mice (*Peromyscus maniculatus*) and southern red-backed voles (Clethrionomys gapperi). *American Midland Naturalist* 152:400–409.

Harris, C. E., and F. F. Knowlton. 2001. Differential

responses of coyotes to novel stimuli in familiar and unfamiliar settings. *Canadian Journal of Zoology* 79:2005–2013.

Harrison, R. L. 1997. Chemical attractants for Central American felids. *Wildlife Society Bulletin* 25:93–97.

———. 2003. Swift fox demography, movements, denning, and diet in New Mexico. *Southwestern Naturalist* 48:261–273.

———. 2006. A comparison of survey methods for detecting bobcats. *Wildlife Society Bulletin* 34:548–552.

Harrison, R. L., D. J. Barr, and J. W. Dragoo. 2002. A comparison of population survey techniques for swift foxes (*Vulpes velox*) in New Mexico. *American Midland Naturalist* 148:320–337.

Harrison, R. L., P. G. S. Clarke, and C. M. Clarke. 2004. Indexing swift fox populations in New Mexico using scats. *American Midland Naturalist* 151:42–49.

Harrison, R. L., M. J. Patrick, and C. G. Schmitt. 2003. Foxes, fleas, and plague in New Mexico. *Southwestern Naturalist* 48:720–722

Harrison, T. M., S. H. Harrison, W. K. Rumbeiha, J. Sikarskie, and M. McClean. 2006. Surveillance for selected bacterial and toxicological contaminants in donated carcass meat fed to carnivores. *Journal of Zoo and Wildlife* Medicine 37:102–107.

Harris, T. R., and S. L. Monfort. 2003. Behavioral and endocrine dynamics associated with infanticide in a black and white colobus monkey (*Colobus guereza*). *American Journal of Primatology* 61:135–142.

———. 2005. Mating behavior and endocrine profiles in wild black and white colobus monkeys (*Colobus guereza*): toward an understanding of their life history and mating system. *American Journal of Primatology* 68:383–396.

Hartup, B. K., G. H. Olsen, and N. M. Czekala. 2005. Fecal corticoid monitoring in whooping cranes (*Grus americana*) undergoing reintroduction. *Zoo Biology* 24:15–28.

Harveson, P. M., M. E. Tewes, G. L. Anderson, and L. L. Laack. 2004. Habitat use by ocelots in south Texas: implications for restoration. *Wildlife Society Bulletin* 32:948–954.

Hash, H. S., and M. G. Hornocker. 1980. Immobilizing wolverines with ketamine hydrochloride. *Journal of Wildlife Management* 44:713–715.

Hayes, R. D., and Λ. S. Harestad. 2000. Demography of a recovering wolf population in the Yukon. *Canadian Journal of Zoology* 78:36–48.

Hayward, G. D., D. G. Miquelle, E. N. Smirnov, and C. Nations. 2002. Monitoring Amur tiger populations: characteristics of track surveys in snow. *Wildlife Society Bulletin* 30:1150–1159.

He, F., and K. J. Gaston. 2003. Occupancy, spatial variance, and the abundance of species. *American Naturalist* 162:366–375.

Hebblewhite, M., P. C. Paquet, D. H. Pletscher, R. B. Lessard, and C. J. Callaghan. 2003. Development and application of a ratio estimator to estimate wolf kill rates and variance in a multiple-prey system. *Wildlife Society Bulletin* 31:933–946.

Hedmark, E., Ø. Flagstad, P. Segerström, J. Persson, A. Landa, and H. Ellegren. 2004. DNA-based individual and sex identification from wolverine (*Gulo gulo*) faeces and urine. *Conservation Genetics* 5:405–410.

Hegglin, D., F. Bontadina, S. Gloor, J. Romer, U. Mueller, U. Breitenmoser, and P. Deplazes. 2004. Baiting red foxes in an urban area: a camera trap study. *Journal of Wildlife Management* 68:1010–1017.

Heilbrun, R. D., N. J. Silvy, M. J. Peterson, and M. E. Tewes. 2006. Estimating bobcat abundance using automatically triggered cameras. *Wildlife Society Bulletin* 34:69–73.

Heilbrun, R. D., N. J. Silvy, M. E. Tewes, and M. J. Peterson. 2003. Using automatically triggered cameras to individually identify bobcats. *Wildlife Society Bulletin* 31:748–755.

Hein, E. W., and W. F. Andelt. 1994. Evaluation of coyote attractants and an oral-delivery device for chemical agents. *Wildlife Society Bulletin* 22:651–655.

Heinemeyer, K. S. 2002. Translating individual movements into population patterns: American marten in fragmented forested landscapes. PhD dissertation. University of California, Santa Cruz.

Heinemeyer, K. S., B. C. Aber, and D. Doak. 2000. Aerial surveys for wolverine presence and potential winter recreation impacts to predicted wolverine denning habitat in the southwestern Yellowstone Ecosystem. Unpublished report. Department of Environmental Studies, University of California, Santa Cruz.

Heinemeyer, K. S., and J. P. Copeland. 1999. Wolverine denning habitat and surveys on the Targhee National Forest, 1998–1999 Annual Report. Unpublished report. Department of Environmental Studies, University of California, Santa Cruz.

Heistermann, M., M. Agil, A. Buthe, and J. K. Hodges.

1998. Metabolism and excretion of oestradiol-17 beta and progesterone in the Sumatran rhinoceros (*Dicerorhinus sumatrensis*). *Animal Reproduction Science* 53:157–172.

Heistermann, M., R. Palme, and A. Ganswindt. 2006. Comparison of different enzyme immunoassays for assessment of adrenocortical activity in primates based on fecal analysis. *American Journal of Primatology* 68: 257–273.

Helle, E., P. Helle, H. Linden, and M. Wikman. 1996. Wildlife populations in Finland during 1990–1995, based on wildlife triangle data. *Finnish Game Research* 49:12–17.

Helle, P., and A. Nikula. 1996. Usage of geographic information systems (GIS) in analyses of wildlife triangle data. *Finnish Game Research* 49:26–36.

Hellstedt, P., J. Sundell, P. Helle, and H. Henttonen. 2006. Large-scale spatial and temporal patterns in population dynamics of the stoat, *Mustela erminea*, and the least weasel, *M. nivalis*, in Finland. *Oikos* 115:286–298.

Henderson, F. R. 1994. Weasels. Pages C119–C122 *in* S. E. Hygnstrom, R. M. Timm, and G. E. Larson, editors. *Prevention and control of wildlife damage.* University of Nebraska, Lincoln.

Henschel, P., and J. C. Ray. 2003. *Leopards in African rainforests: survey and monitoring techniques.* Wildlife Conservation Society Global Carnivore Program, Bronx, NY.

Hepper, P. G., and D. L. Wells. 2005. How many footsteps do dogs need to determine the direction of an odour trail? *Chemical Senses* 30:291–298.

Herbert, P. D. N., M. Y. Stoeckle, T. S. Zemlak, and C. M. Francis. 2004. Identification of birds through DNA barcodes. *PLoS Biology* 2:1657–1663.

Herrick, J. R., G. Agoramoorthy, R. Rudran, and J. D. Harder. 2000. Urinary progesterone in free-ranging red howler monkeys (*Alouatta seniculus*): preliminary observations of the estrous cycle and gestation. *American Journal of Primatology* 51:257–263.

Herzog, C. J., R. W. Kays, J. C. Ray, M. E. Gompper, W. J. Zielinksi, R. Higgins, and M. Tymeson. 2007. Using patterns in track plate footprints to identify individual fishers. *Journal of Wildlife Management* 71:955–963.

Heske, E. J. 1995. Mammalian abundances on forest-farm edges vs. forest interiors in southern Illinois: is there an edge effect? *Journal of Mammalogy* 76:562–568.

Heske, E. J., S. K. Robinson, and J. D. Brawn. 1999. Predator activity and predation on songbird nests on forest-field edges in east-central Illinois. *Landscape Ecology* 14:345–354.

Hicks, N. G., M. A. Menzel, and J. Laerm. 1998. Bias in the determination of temporal activity patterns of syntopic *Peromyscus* in the southern Appalachians. *Journal of Mammalogy* 79:1016–1020.

Hill, S., and J. Hill. 1987. Richard Henry of Resolution Island. John McIndoe, Dunedin, New Zealand.

Hilty, J. A., and A. M. Merenlender. 2000. A comparison of covered track-plates and remotely-triggered cameras. *Transactions of the Western Section of the Wildlife Society* 36:27–31.

Hines, J. E., and J. D. Nichols. 2002. Investigations of potential bias in the estimate of lambda using Pradel's (1996) model for capture-recapture data. *Journal of Applied Statistics* 29:573–587.

Hirth, D. H., J. M. A. Petty, and C. W. Kilpatrick. 2002. Black bear, *Ursus americanus*, hair and apple trees, *Malus pumila*, in northeast North America. *Canadian Field Naturalist* 116:305–307.

Hobbs, G. I., E. A. Chadwick, F. M. Slater, and M. W. Bruford. 2006. Landscape genetics applied to a recovering otter (*Lutra lutra*) population in the UK: preliminary results and potential methodologies. *Hystrix Italian Journal of Mammalogy* 17:47–63.

Hodges, J. K., N. M. Czekala, and B. L. Lasley. 1979. Estrogen and luteinizing hormone secretion in diverse primate species from simplified urinary analysis. *Journal of Medical Primatology* 8:349–364.

Hodges, J. K., and M. Heistermann. 2003. Field endocrinology: monitoring hormonal changes in free-ranging primates. Pages 282–294 *in* J. M. Setchell and D. J. Curtis, editors. *Field and laboratory methods in primatology.* Cambridge University Press, Cambridge, UK.

Hoelzel, A. R., and A. Green. 1992. Analysis of population-level variation by sequencing PCR-amplified DNA. Pages 159–187 *in* A. R. Hoelzel, editor. *Molecular genetic analysis of populations: a practical approach.* IRL Press, Oxford, UK.

Hofreiter, M., D. Serre, H. N. Poinar, M. Kuch, and S. Paabo. 2001. Ancient DNA. *Nature Reviews Genetics* 2:353–359.

Hogmander, H., and A. Penttinen. 1996. Some statistical aspects of Finnish wildlife triangles. *Finnish Game Research* 49:37–43.

Holling, C. S. 1978. *Adaptive environmental assessment and management.* John Wiley & Sons, New York.

Homan, H. J., G. Linz, and B. D. Peer. 2001. Dogs increase recovery of passerine carcasses in dense vegetation. *Wildlife Society Bulletin* 29:292–296.

Horne, J. S. 1998. Habitat partitioning of sympatric ocelot and bobcat in southern Texas. MS thesis. Texas A&M University, Kingsville.

Hornocker, M. G., J. J. Craighead, and E. W. Pfeiffer. 1965. Immobilizing mountain lions with succinylcholine chloride and pentobarbital sodium. *Journal of Wildlife Management* 29:880–883.

Hornocker, M. G., and H. S. Hash. 1981. Ecology of the wolverine in northwestern Montana. *Canadian Journal of Zoology* 59:1286–1301.

Hoving, C. L., D. J. Harrison, W. B. Krohn, W. J. Jakubas, and M. A. McCollough. 2004. Canada lynx *Lynx canadensis* habitat and forest succession in northern Maine. *Wildlife Biology* 10:285–294.

Hoving, C. L., D. J. Harrison, W. B. Krohn, R. A. Joseph, and M. O'Brien. 2005. Broad-scale predictors of Canada lynx occurrence in eastern North America. *Journal of Wildlife Management* 69:739–751.

Huggins, R. M. 1989. On the statistical analysis of capture experiments. *Biometrika* 76:133–140.

———. 1991. Some practical aspects of a conditional likelihood approach to capture experiments. *Biometrics* 47:725–732.

Hung, C. M., S. H. Li, and L. L. Lee. 2004. Faecal DNA typing to determine the abundance and spatial organisation of otters (*Lutra lutra*) along two stream systems in Kinmen. *Animal Conservation* 7:301–311.

Hunt, K. E., R. M. Rolland, S. D. Kraus, and S. K. Wasser. 2006. Analysis of fecal glucocorticoids in the North Atlantic right whale (*Eubalaena glacialis*). *General and Comparative Endocrinology* 148:260–272.

Hunt, K. E., and S. K. Wasser. 2003. Effect of long-term preservation methods on fecal glucocorticoid concentrations of grizzly bear and African elephant. *Physiological and Biochemical Zoology* 76:918–928.

Hurt, A., B., Davenport, and E. Greene. 2000. Training dogs to distinguish between black bear (*Ursus americanus*) and grizzly bear (*Ursus arctos*) feces. *University of Montana Under-Graduate Biology Journal.* ibscore.dbs.umt.edu/journal/Articles_all/2000/Hurt.htm (accessed 7 December 2007).

Hutto, R. L., and J. S. Young. 2003. On the design of monitoring programs and the use of population indices: a reply to Ellingson and Lukacs. *Wildlife Society Bulletin* 31:903–910.

Huxley, A. 1932. *Brave new world.* HarperCollins, New York.

IACUC (Institutional Animal Care and Use Committee). 2006. *Research policies and procedure guidelines.* www.iacuc.org/ (accessed 30 November 2007).

Ihaka, R., and R. Gentleman. 1996. R: a language for data analysis and graphics. *Journal of Computational and Graphical Statistics* 5:299–314.

Immell, D. A., D. H. Jackson, and R. G. Anthony. 2004. Black bear population estimation in small study areas—an alternative method of hair collection. Page 35 in *Proceedings of 15th International Conference on Bear Research and Management.* San Diego, CA.

Inman, R. M., R. R. Wigglesworth, K. H. Inman, M. K. Schwartz, B. L. Brock, and J. D. Rieck. 2004. Wolverine makes extensive movements in the Greater Yellowstone Ecosystem. *Northwest Science* 78:261–266.

Ivan, J. S. 2000. Effectiveness of covered track plates for detecting American marten. MS thesis. University of Montana, Missoula.

Iyengar, A., V. N. Babu, S. Hedges, A. B. Venkataraman, N. MacLean, and P. A. Morin. 2005. Phylogeography, genetic structure, and diversity in the dhole (*Cuon alpinus*). *Molecular Ecology* 14:2281–2297.

Jacobson, H. A., J. C. Kroll, R. W. Browning, B. H. Koerth, and M. H. Conway. 1997. Infrared-triggered cameras for censusing white-tailed deer. *Wildlife Society Bulletin* 25:547–556.

Jansen, P. A., and J. Den Ouden. 2005. Observing seed removal: remote video monitoring of seed selection, predation and dispersal. Pages 363–378 in P. M. Forget, J. E. Lambert, P. E. Hulme, and S. B. Vander Wall, editors. *Seed fate: predation, dispersal and seedling establishment.* CAB International, Wallingford, UK.

Janzen, D. H. 1977. Why fruits rot, seeds mold, and meat spoils. *American Naturalist* 111:691–713.

Jefferson, T. A. 2000. Population biology of the Indo-Pacific hump-backed dolphin in Hong Kong waters. *Wildlife Monographs* 144.

Jeffreys, A. J., V. Wilson, and S. L. Thein. 1985a. Individual-specific fingerprints of human DNA. *Nature* 316:76–79.

———. 1985b. Hypervariable "minisatellite" regions in human DNA. *Nature* 314:67–73.

Jenkins, D., A. Watson, and G. R. Miller. 1963. Population studies on red grouse, *Lagopus lagopus scoticus* (Lath.) in north-east Scotland. *Journal of Animal Ecology* 32:317–376.

Jennelle, C. S., M. C. Runge, and D. I. MacKenzie. 2002. The use of photographic rates to estimate densities of tigers and other cryptic mammals: a comment on misleading conclusions. *Animal Conservation* 5:119–120.

Jesse's Hunting and Outdoors. 2006a. Jesse's hunting and outdoors, game trail cameras and camcorders. www .jesseshunting.com/forums/index.php?showforum=50 (accessed 19 December 2007).

———. 2006b. Jesse's hunting and outdoors, game trail cameras and camcorders. One year camera, topic 76232. www.jesseshunting.com/forums/index .php?showtopic=76232 (accessed 19 December 2007).

Jewell, Z. C., S. K. Alibhai, and P. R. Law. 2001. Censusing and monitoring black rhino (*Diceros bicornis*) using an objective spoor (footprint) identification technique. *Journal of Zoology* 254:1–16.

Johnson, C. J., D. R. Seip, and M. S. Boyce. 2004. A quantitative approach to conservation planning: using resource selection functions to map the distribution of mountain caribou at multiple spatial scales. *Journal of Applied Ecology* 41:238–251.

Johnson, D. H. 1980. The comparison of usage and availability measurements for evaluating resource preference. *Ecology* 61:65–71.

Johnson, D. H., and G. A. Sargeant. 2002. Toward better atlases: improving presence-absence information. Pages 391–397 *in* J. M. Scott, P. J. Heglund, M. L. Morrison, J. B. Haufler, M. G. Raphael, W. A. Wall, and F. B. Samson, editors. *Predicting species occurrences*. Island Press, Washington, DC.

Johnson, M. K., and R. C. Belden. 1984. Differentiating mountain lion and bobcat scats. *Journal of Wildlife Management* 48:239–244.

Johnston, J. M. 1999. *Canine detection capabilities: operational implications of recent R & D findings*. Institute for Biological Detection Systems, Auburn University, Auburn, AL.

Jones, C., H. Moller, and W. Hamilton 2004. A review of potential techniques for identifying individual stoats (*Mustela erminea*) visiting control or monitoring stations. *New Zealand Journal of Zoology* 31:193–203.

Jones, J. L., and E. O. Garton. 1994. Selection of successional stages by fishers in north-central Idaho. Pages 377–387 *in* S. W. Buskirk, A. S. Harestad, M. G. Raphael, and R. A. Powell, editors. *Martens, sables, and fishers*. Cornell University Press, Ithaca, NY.

Jones, L. L. C., and M. G. Raphael. 1993. *Inexpensive camera systems for detecting martens, fishers, and other animals: guidelines for use and standardization*. USDA Forest Service Pacific Northwest Research Station General Technical Report PNW-GTR-306, Portland, OR.

Jones, L. L. C., M. G. Raphael, J. T. Forbes, and L. A. Clark. 1997. Using remotely activated cameras to monitor maternal dens of martens. Pages 329–349 *in* G. Proulx, H. N. Bryant, and P. M. Woodard, editors. *Martes taxonomy, ecology, techniques and management*. Provincial Museum of Alberta, Edmonton, Alberta, Canada.

Jonzén, N., J. R. Rhodes, and H. P. Possingham. 2005. Trend detection in source-sink systems: when should sink habitats be monitored? *Ecological Applications* 15:326–334.

Jordan, M. J. 2007. Fisher ecology in the Sierra National Forest, California. PhD dissertation. University of California, Berkeley.

Joseph, L. N., S. A. Field, C. Wilcox, and H. P. Possingham. 2006. Presence-absence versus abundance data for monitoring threatened species. *Conservation Biology* 20:1679–1687.

Kalinowski, S. T. 2006. HW-QUICKCHECK: an easy-to-use computer program for checking genotypes for agreement with Hardy-Weinberg expectations. *Molecular Ecology Notes* 6:974–979.

Kamler, J. F., W. B. Ballard, R. L. Gilliland, and K. Mote. 2002. Improved trapping methods for swift foxes and sympatric coyotes. *Wildlife Society Bulletin* 30:1262–1266.

Karanth, K. U. 1995. Estimating tiger (*Panthera tigris*) populations from camera-trap data using capture-recapture models. *Biological Conservation* 71:333–338.

Karanth, K. U., R. S. Chundawat, J. D. Nichols, and N. S. Kumar. 2004a. Estimation of tiger densities in the tropical dry forests of Panna, Central India, using photographic capture-recapture sampling. *Animal Conservation* 7:285–290.

Karanth, K. U., and J. D. Nichols. 1998. Estimation of tiger densities in India using photographic captures and recaptures. *Ecology* 79:2852–2862.

———. 2002. *Monitoring tigers and their prey*. Centre for Wildlife Studies, Bangalore, India.

Karanth, K. U., J. D. Nichols, N. S. Kumar, and J. E. Hines. 2006. Assessing tiger population dynamics using photographic capture-recapture sampling. *Ecology* 87:2925–2937.

Karanth, K. U., J. D. Nichols, N. S. Kumar, W. A. Link, and J. E. Hines. 2004b. Tigers and their prey: predicting carnivore densities from prey abundance. *Proceedings of the National Academy of Sciences of the United States of America* 101:4854–4858.

Karanth, K. U., J. D. Nichols, J. Seidensticker, E. Dinerstein, J. L. D. Smith, C. McDougal, A. J. T. Johnsingh, R. S. Chundawat, and V. Thapar. 2003. Science deficiency in conservation practice: the monitoring of tiger populations in India. *Animal Conservation* 6: 141–146.

Kauffman, M. J., M. Sanjayan, J. Lowenstein, A. Nelson, R. M. Jeo, and K. R. Crooks. 2007. Remote camera-trap methods and analyses reveal impacts of rangeland management on Namibian carnivore communities. *Oryx* 41:70–78.

Kaufmann, J. H. 1987. Ringtail and coati. Pages 500–509 *in* M. Novak, J. A. Baker, M. E. Obbard, and B. Malloch, editors. *Wild furbearer management and conservation in North America*. Ontario Ministry of Natural Resources, Toronto, Ontario, Canada.

Kauhala, K., and P. Helle. 2000. The interactions of predator and hare populations in Finland—A study based on wildlife monitoring counts. *Annales Zoologici Fennici* 37:151–160.

Kawanishi, K. 2002. Population status of tigers (*Panthera tigris*) in a primary rainforest of peninsular Malaysia. PhD dissertation. University of Florida, Gainesville.

Kays, R. W., and D. E. Wilson. 2002. *Mammals of North America*. Princeton University Press, Princeton, NJ.

Kelly, B. T., and E. O. Garton. 1997. Effects of prey size, meal size, meal composition, and daily frequency of feeding on the recovery of rodent remains from carnivore scats. *Canadian Journal of Zoology* 75:1811–1817.

Kelly, M. J. 2001. Computer-aided photograph matching in studies using individual identification: an example from Serengeti cheetahs. *Journal of Mammalogy* 82:440–449.

———. 2003. Jaguar monitoring in the Chiquibul forest, Belize. *Caribbean Geography* 13:19–32.

Kendall, K., J. Stets, and A. Macleod. 2004. *Opportunities for passive collection of bear hair*. Poster. Wildlife Society Annual Conference, 18–22 September 2004, Calgary, Alberta, Canada. www.nrmsc.usgs.gov/research/NCDE_TWSposter.htm (accessed 30 November 2007).

Kendall, K. C., L. H. Metzgar, D. A. Patterson, and B. M. Steele. 1992. Power of sign surveys to monitor population trends. *Ecological Applications* 2:422–430.

Kendall, K. C., and L. P. Waits. 2003. *Monitoring grizzly bear populations using DNA*. www.nrmsc.usgs.gov/research/glac_beardna.htm (accessed 30 November 2007).

Kendall, W. L., J. D. Nichols, and J. E. Hines. 1997. Estimating temporary emigration using capture-recapture data with Pollock's robust design. *Ecology* 78:563–578.

Kerley, L. L., and G. P. Salkina. 2007. Using scent-matching dogs to identify individual Amur tigers from scat. *Wildlife Society Bulletin* 71:1341–1356.

Kery, M., J. A. Royle, and H. Schmid. 2005. Modeling avian abundance from replicated counts using binomial mixture models. *Ecological Applications* 15:1450–1461.

Khan, M. Z., J. Altmann, S. S. Isani, and J. Yu. 2002. A matter of time: evaluating the storage of fecal samples for steroid analysis. *General and Comparative Endocrinology* 128:57–64.

Kimball, B. A., J. R. Mason, F. S. Blom, J. J. Johnston, and D. E. Zemlicka. 2000. Development and testing of seven new synthetic coyote attractants. *Journal of Agricultural and Food Chemistry* 48:1892–1897.

King, C. M., and R. L. Edgar. 1977. Techniques for trapping and tracking stoats (*Mustela erminea*): a review, and a new system. *New Zealand Journal of Zoology* 4: 193–212.

King, C. M., R. M. McDonald, R. D. Martin, G. W. Tempero, and S. J. Holmes. 2007. Long-term automated monitoring of the distribution of small carnivores. *Wildlife Research* 34:140–148

King, C. M., C. F. J. O'Donnell, and S. M. Phillipson. 1994. Monitoring and control of mustelids on conservation lands. Part 2: field and workshop guide. *Department of Conservation Technical Series 4*. Department of Conservation, Wellington, New Zealand.

Kirk, T. 2007. Landscape-scale habitat associations of the Amercian marten (*Martes americana*) in northeastern

California. MS thesis. Humboldt State University, Arcata, CA.

Kirkpatrick, J. F., D. F. Gudermuth, F. L. Flagan, J. C. McCarthy, and B. L. Lasley. 1993. Remote monitoring of ovulation and pregnancy of Yellowstone bison. *Journal of Wildlife Management* 57:407–412.

Kirkpatrick, J. F., V. Kincy, K. Bancroft, S. E. Shideler, and B. L. Lasley. 1991a. Oestrous cycle of the North American bison (*Bison bison*) characterized by urinary pregnanediol-3-glucuronide. *Journal of Reproduction and Fertility* 93:541–547.

Kirkpatrick, J. F., J. C. McCarthy, D. F. Gudermuth, S. E. Shideler, and B. L. Lasley. 1996. An assessment of the reproductive biology of Yellowstone bison (*Bison bison*) subpopulations using noncapture methods. *Canadian Journal of Zoology* 74:8–14.

Kirkpatrick, J. F., S. E. Shideler, and B. L. Lasley. 1991b. Pregnancy determination in uncaptured feral horses by means of fecal steroid conjugates. *Theriogenology* 35:753–760.

Kistner, T., D. Wyse, and J. A. Schmitz. 1979. Pathogenicity attributed to massive infection of *Nanophyetus salmincola* in a cougar. *Journal of Wildlife Diseases* 15:419–420.

Kitchener, A. 1991. *The natural history of the wild cats.* Comstock Publishing Associates, Ithaca, NY.

Kitchings, J. T., and J. D. Story. 1979. Home range and diet of bobcats in eastern Tennessee. *National Wildlife Federation Scientific and Technical Series* 6:47–52.

Klute, D. S., M. J. Lovallo, and W. M. Tzilkowski. 2002. Autologistic regression modeling of American woodcock habitat use with spatially dependent data. Pages 335–343 *in* J. M. Scott, P. J. Heglund, M. L. Morrison, J. B. Haufler, M. G. Raphael, W. A. Wall, and F. B. Samson, editors. *Predicting species occurrences.* Island Press, Washington, DC.

Knorr, L. 2004. Black bear mark-recapture study using remote cameras. MS thesis. Central Connecticut State University, New Britain.

Knowlton, F. F. 1984. Feasibility of assessing coyote abundance on small areas. Unpublished report of Work Unit 909: 01, US Fish and Wildlife Service, Denver Wildlife Research Center, Denver, CO.

Knowlton, F. F., and L. C. Stoddart. 1984. Feasibility of assessing coyote abundance on small areas. US Fish and Wildlife Service, Washington, DC.

Kocher, T. W., W. Thomas, A. Meyer, S. Edwards, S. Pääbo, F. Villablanca, and A. Wilson. 1989. Dynamics of mitochondrial DNA evolution in animals: amplification and sequencing with conserved primers. *Proceedings of the National Academy of Sciences* 86:6196–6200.

Kohlmann, S. G., G. A. Schmidt, and D. K. Garcelon. 2005. A population viability analysis for the island fox on Santa Catalina Island, California. *Ecological Modeling* 183:77–94.

Kohn, B. E., N. F. Payne, J. E. Ashbrenner, and W. A. Creed. 1993. *The fisher in Wisconsin.* Technical Bulletin No. 183, Wisconsin Department of Natural Resources, Madison.

Kohn, M. H., and R. K. Wayne. 1997. Facts from feces revisited. *Trends in Ecology and Evolution* 12:223–227.

Kohn, M. H., W. J. Murphy, E. A. Ostrander, and R. K. Wayne. 2006. Genomics and conservation genetics. *Trends in Ecology and Evolution* 21:629–637.

Kohn, M. H., E. C. York, D. A. Kamradt, G. Haught, R. M. Sauvajot, and R. K. Wayne. 1999. Estimating population size by genotyping faeces. *Proceedings of the Royal Society of London B* 266:657–663.

Kolb, H. H., and R. Hewson. 1980. A study of fox populations in Scotland from 1971 to 1976. *Journal of Applied Ecology* 17:7–19.

Koopman, M. E., B. L. Cypher, and J. H. Scrivner. 2000. Dispersal patterns of San Joaquin kit foxes (*Vulpes macrotis mutica*). *Journal of Mammalogy* 81:213–222.

Kraemer, H. C., and S. Thiemann. 1987. *How many subjects? Statistical power analysis in research.* Sage, Newbury Park, CA.

Krebs, C. J. 1966. Demographic changes in fluctuating populations of *Microtus californicus. Ecological Monographs* 36:239–273.

———. 1985. *Ecology: the experimental analysis of distribution and abundance.* 3rd edition. Harper and Row, New York.

———. 1991. The experimental paradigm and long-term population studies. *Ibis* 133: suppl. 1:3–8.

———. 1998. *Ecological methodology.* Addison Wesley Longman, Menlo Park, CA.

Kruse, C. D., K. F. Higgins, and B. A. Vander-Lee. 2001. Influence of predation on piping plover, *Charadrius melodus*, and least tern, *Sterna antillarum*, productivity along the Missouri River in South Dakota. *Canadian Field-Naturalist* 115:480–486.

Krutova, I. V. 2001. Use of the method of odor identification by dogs in studies of large carnivorous mammals.

*Bulletin of Moscow Society of Naturalists* (in Russian) 106:3–12.

Kruuk, H., and J. W. H. Conroy. 1987. Surveying otter *Lutra lutra* populations: a discussion of problems with spraints. *Biological Conservation* 41:179–183.

Kruuk, H., J. W. H. Conroy, U. Glimmerveen, and E. J. Ouwerkerk. 1986. The use of spraints to survey populations of otters *Lutra lutra*. *Biological Conservation* 35:187–194.

Kucera, T. E., and R. H. Barrett. 1993. The Trailmaster camera system for detecting wildlife. *Wildlife Society Bulletin* 21:505–508.

———. 1995. Trailmaster camera system: response. *Wildlife Society Bulletin* 23:110–113.

Kucera, T. E., A. M. Soukkala, and W. J. Zielinski. 1995a. Photographic bait stations. Pages 25–65 *in* W. J. Zielinski and T. E. Kucera, editors. *American marten, fisher, lynx, and wolverine: survey methods for their detection*. USDA Forest Service, Pacific Southwest Research Station General Technical Report PSW-GTR-157, Albany, CA.

Kucera, T. E., W. J. Zielinski, and R. H. Barrett. 1995b. Current distribution of the American marten, *Martes americana*, in California. *California Fish and Game* 81:96–103.

Kurki, S., A. Nikula, P. Helle, and H. Linden. 1998. Abundances of red fox and pine marten in relation to the composition of boreal forest landscapes. *Journal of Animal Ecology* 67:874–886.

Kuznetsov, V. A., A. V. Tchabovsky, I. E. Kolosova, and M. P. Moshkin. 2004. Effect of habitat type and population density on the stress level of midday gerbils (*Meriones meridianus* Pall.) in free-living populations. *Biology Bulletin* 31:628–632.

Laack, L. L. 1991. Ecology of the ocelot (*Felis pardalis*) in south Texas. MS thesis. Texas A&M University, Kingsville.

Laliberte, A. S., and W. J. Ripple. 2004. Range contractions of North American carnivores and ungulates. *Bioscience* 54:123–138.

Lancia, R. A., W. L. Kendall, K. H. Pollock, and J. D. Nichols. 2005. Estimating the number of animals in wildlife populations. Pages 106–153 in C. E. Braun, editor. *Techniques for wildlife investigations and management*. Port City Press, Baltimore, MD.

LaPoint, S. C., R. W. Kays, and J. C. Ray. 2003. Do mammals use culverts to cross Interstate 87 within the Adirondack Park? *Adirondack Journal of Environmental Studies* 10:11–17.

Larrucea, E. S., P. F. Brussard, M. M. Jaeger, and R. H. Barrett. 2007. Cameras, coyotes, and the assumption of equal detectability. *Journal of Wildlife Management* 71:1682–1689.

Lasley, B. L., and J. F. Kirkpatrick. 1991. Monitoring ovarian function in captive and free-ranging wildlife by means of urinary and fecal steroids. *Journal of Zoo and Wildlife Medicine* 22:23–31.

Leberg, P. L., and M. L. Kennedy. 1987. Use of scent-station methodology to assess raccoon abundance. *Proceedings of the Annual Conference of the Southeastern Association of Fish and Wildlife Agencies* 41:394–403.

Leonard, J. A., C. Vila, and R. Wayne. 2005 Legacy lost: genetic variability and population size of extirpated US grey wolves (*Canis lupus*). *Molecular Ecology* 14:9–17.

Leonard, J. A., R. K. Wayne, and A. Cooper. 2000. Population genetics of Ice Age brown bears. *Proceedings of the National Academy of Sciences USA* 97:1651–1654.

Lewis, J. C., and D. W. Stinson. 1998. *Washington state status report for the fisher*. Washington Department of Fish and Wildlife, Olympia.

Lewison, R., E. L. Fitzhugh, and S. P. Galentine. 2001. Validation of a rigorous track classification technique: identifying individual mountain lions. *Biological Conservation* 99:313–321.

Liebenberg, L. 1990. *The art of tracking: the origin of science*. David Philip Publishers, Cape Town, South Africa.

Linden, H., E. Helle, P. Helle, and M. Wikman. 1996. Wildlife triangle scheme in Finland: methods and aims for monitoring wildlife populations. *Finnish Game Research* 49:4–11.

Lindsay, S. 1999. *Handbook of applied dog behavior and training*, Vol. 1. Iowa State University Press, Ames. NetLibrary e-book.

Linhart, S. B., G. J. Dasch, J. D. Roberts, and P. J. Savarie. 1977. Test methods for determining the efficacy of coyote attractants and repellents. Pages 114–122 *in* W. B. Jackson and R. E. Marsh, editors. *Test methods for vertebrate pest control and management materials*. American Society for Testing and Materials, STP 625, Sacramento, CA.

Linhart, S. B., and F. F. Knowlton. 1975. Determining the relative abundance of coyotes by scent station lines. *Wildlife Society Bulletin* 3:119–124.

Link, W.A., and J. Hatfield. 1990. Power calculations and model selection for trend analysis: a comment. *Ecology* 71:1217–1220.

Link, W. A., and J. R. Sauer. 1998a. Estimating population change from count data: application to the North American Breeding Bird Survey. *Ecological Applications* 8:258–268.

———. 1998b. Estimating relative abundance from count data. *Austrian Journal of Statistics* 27:83–97.

———. 1999. Controlling for varying effort in count surveys: an analysis of Christmas Bird Count data. *Journal of Agricultural, Biological, and Environmental Statistics* 4:116–125.

———. 2002. A hierarchical analysis of population change with application to Cerulean Warblers. Ecology 83:2832–2840.

Linkie, M., G. Chapron, D. J. Martyr, J. Holden, and N. Leader-Williams. 2006. Assessing the viability of tiger subpopulations in a fragmented landscape. *Journal of Applied Ecology* 43:576–586.

Linnell, J. D. C., J. E. Swenson, and R. Andersen. 2000. Conservation of biodiversity in Scandinavian boreal forests: large carnivores as flagships, umbrellas, indicators, or keystones. *Biodiversity and Conservation* 9:857–868.

Livingston, T. R., P. S. Gipson, W. B. Ballard, D. M. Sanchez, and P. R. Krausman. 2005. Scat removal: a source of bias in feces-related studies. *Wildlife Society Bulletin* 33:172–178.

Lomolino, M. V., and D. R. Perault. 2001. Island biogeography and landscape ecology of mammals inhabiting fragmented, temperate rain forests. *Global Ecology & Biogeography* 10:113–132.

Long, R. A. 2006. Developing predictive occurrence models for carnivores in Vermont using data collected with multiple noninvasive methods. PhD dissertation. University of Vermont, Burlington.

Long, R. A., T. M. Donovan, P. MacKay, W. J. Zielinski, and J. S. Buzas. 2007a. Effectiveness of scat detection dogs for detecting forest carnivores. *Journal of Wildlife Management* 71:2007–2017.

———. 2007b. Comparing scat detection dogs, cameras, and hair snares for surveying carnivores. *Journal of Wildlife Management* 71:2018–2025.

Loukmas J. J., and R. S. Halbrook. 2001. A test of the mink habitat suitability index model for riverine systems. *Wildlife Society Bulletin* 29:821–826.

Loukmas, J. J., D. T. Mayack, and M. E. Richmond. 2003. Track plate enclosures: box designs affecting attractiveness to riparian mammals. *American Midland Naturalist* 149:219–224.

Lovallo, M. J., and E. M. Anderson. 1996. Bobcat movements and home ranges relative to roads in Wisconsin. *Wildlife Society Bulletin* 24:71–76.

Lowery, J. C. 2006. *The tracker's field guide: a comprehensive handbook for animal tracking in the United States.* Falcon Press Publishing, Helena, MT.

Lozano, J., E. Virgos, A. F. Malo, D. L. Huertas, and J. G. Casanovas. 2003. Importance of scrub-pastureland mosaics for wild-living cats occurrence in a Mediterranean area: implications for the conservation of the wildcat (*Felis silvestris*). *Biodiversity and Conservation* 12:921–935.

Lucchini, V., E. Fabbri, F. Marucco, S. Ricci, L. Boitani, and E. Randi. 2002. Noninvasive molecular tracking of colonizing wolf (*Canis lupus*) packs in the western Italian Alps. *Molecular Ecology* 11:857–868.

Luikart G., and J.-M. Cornuet. 1998. Empirical evaluation of a test for identifying recently bottlenecked populations from allele frequency data. *Conservation Biology* 12:228–237.

Luikart, G., P. R. England, D. Tallmon, S. Jordan, and P. Taberlet. 2003. The power and promise of population genomics: from genotyping to genome typing. *Nature Reviews Genetics* 4:981–994.

Lukacs, P. M. 2005. Statistical aspects of using genetic markers for individual identification in capture-recapture studies. PhD dissertation. Colorado State University, Fort Collins.

Lukacs, P. M., and K. P. Burnham. 2005a. Estimating population size from DNA-based closed capture-recapture data incorporating genotyping error. *Journal of Wildlife Management* 69:396–403.

———. 2005b. Review of capture-recapture methods applicable to noninvasive genetic sampling. *Molecular Ecology* 14:3909–3919.

Lukacs, P. M., K. P. Burnham, B. P. Dreher, K. T. Scribner, and S. R. Winterstein. 2008a. Extending the robust design for DNA-based capture-recapture data incorporating genotyping error and multiple data sources. *Environmental and Ecological Statistics* 15:in press.

Lukacs, P. M., L. S. Eggert, and K. P. Burnham. 2008b. Estimating population size from multiple detections with non-invasive genetic data. *Wildlife Biology in Practice* 4:in press.

Lukins, W. J., S. Creel, B. Erbes, and G. Spong. 2004. An assessment of the Tobacco Root Mountain Range in southwestern Montana as a linkage zone for grizzly bears. *Northwest Science* 78:168–172.

Lynch, J. W., M. Z. Khan, J. Altmann, M. N. Njahira, and N. Rubenstein. 2003. Concentrations of four fecal steroids in wild baboons: short-term storage conditions and consequences for data interpretation. *General and Comparative Endocrinology* 132:264–271.

Lynch, J. W., T. E. Ziegler, and K. B. Strier. 2002. Individual and seasonal variation in fecal testosterone and cortisol levels of wild male tufted capuchin monkeys, *Cebus apella nigritus*. *Hormones and Behavior* 41:275–287.

Lynch, M. 1988. Estimation of relatedness by DNA fingerprinting. *Molecular Biology and Evolution* 5:584–589.

Lyren, L. M. 2001. Movement patterns of coyotes and bobcats relative to roads and underpasses in the Chino Hills area of Southern California. MS thesis. California State Polytechnic University, Pomona.

MacCarthy, K. A., T. C. Carter, B. J. Steffen, and G. A. Feldhamer. 2006. Efficacy of the mist-net protocol for Indiana bats: a video analysis. *Northeastern Naturalist* 13:25–28.

MacDonald, D. W. 2001. Postscript—Carnivore conservation: science, compromise and tough choices. Pages 524–538 *in* J. L. Gittleman, S. M. Funk, D. MacDonald, and R. K. Wayne, editors. *Carnivore conservation.* Cambridge University Press, Cambridge, UK.

Macdonald, D. W., C. D. Buesching, P. Stopka, J. Henderson, S. A. Ellwood, and S. E. Baker. 2004. Encounters between two sympatric carnivores: red foxes (*Vulpes vulpes*) and European badgers (*Meles meles*). *Journal of Zoology* 263:385–396.

Macdonald, D. W., G. Mace, and S. Rushton. 1998. *Proposals for future monitoring of British mammals.* Department of the Environment, Transport and the Regions: London, Bressenden Place, London, UK.

Macdonald, D. W., P. D. Stewart, P. J. Johnson, J. Porkert, and C. Buesching. 2002. No evidence of social hierarchy amongst feeding badgers, *Meles meles*. *Ethology* 108:613–628.

Mace, R. D., S. A. Minta, T. L. Manley, and K. E. Aune. 1994. Estimating grizzly bear population size using camera sightings. *Wildlife Society Bulletin* 22:74–83.

MacKenzie, D. I., L. L. Bailey, and J. D. Nichols. 2004. Investigating species co-occurrence patterns when species are detected imperfectly. *Journal of Animal Ecology* 73:546–555.

MacKenzie, D. I., and J. D. Nichols. 2004. Occupancy as a surrogate for abundance estimation. *Animal Biodiversity and Conservation* 27:461–467.

MacKenzie, D. I., J. D. Nichols, J. E. Hines, M. G. Knutson, and A. B. Franklin. 2003. Estimating site occupancy, colonization, and local extinction when a species is detected imperfectly. *Ecology* 84:2200–2207.

MacKenzie, D. I., J. D. Nichols, G. B. Lachman, S. Droege, J. A. Royle, and C. A. Langtimm. 2002. Estimating site occupancy rates when detection probabilities are less than one. *Ecology* 83:2248–2255.

MacKenzie, D. I., J. D. Nichols, J. A. Royle, K. H. Pollock, L. L. Bailey, and J. E. Hines. 2006. *Occupancy estimation and modeling: inferring patterns and dynamics of species occurrence.* Academic Press, Burlington, MA.

MacKenzie, D. I., and J. A. Royle. 2005. Designing occupancy studies: general advice and allocating survey effort *Journal of Applied Ecology* 42:1105–1114.

Maffei, L., E. Cuéllar, and A. Noss. 2002. Uso de trampas-cámara para la evaluación de mamíferos en el ecotono Chaco-Chiquitanía. *Revista Boliviana de Ecología y Conservación Ambiental* 11:55–65.

———. 2004. One thousand jaguars (*Panthera onca*) in Bolivia's Chaco? Camera trapping in the Kaa-Iya National Park. *Journal of Zoology* 262:295–304.

Maffei, L., A. J. Noss, E. Cuéllar, and D. Rumiz. 2005. Ocelot (*Felis pardalis*) population densities, activity, and ranging behavior in the dry forests of eastern Bolivia: data from camera trapping. *Journal of Tropical Ecology* 21:349–353.

Magoun, A., N. Dawson, R. Klafki, J. Bowman, and J. C. Ray. 2005. A method of identifying individual wolverines using remote cameras. *1st International Symposium on Wolverine Research and Management.* Jokkmokk, Sweden.

Magoun, A. J., J. C. Ray, D. S. Johnson, P. Valkenburg, F. N. Dawson, and J. Bowman. 2007. Modeling wolverine occurrence using aerial surveys of tracks in snow. *Journal of Wildlife Management* 71:2221–2229.

Mahon, P. S., P. B. Banks, and C. R. Dickman. 1998. Population indices for wild carnivores: a critical study in sand-dune habitat, south-western Queensland. *Wildlife Research* 25:11–22.

Main, M. B., P. B. Walsh, K. M. Portier, and S. F. Coates. 1999. Monitoring the expanding range of coyotes in Florida: results of the 1997–98 statewide scent station surveys. *Florida Field Naturalist* 27:150–162.

Major, M., M. K. Johnson, W. S. Davis, and T. F. Kellogg. 1980. Identifying scats by recovery of bile acids. *Journal of Wildlife Management* 44:290–293.

Mammato, B. 1997. *Pet first aid: cats and dogs.* C. V. Mosby, St. Louis, MO.

Manel, S., O. E. Gaggiotti, and R. S. Waples. 2005. Assignment methods: matching biological questions with appropriate techniques. *Trends in Ecology and Evolution* 20:136–142.

Manel, S., M. K. Schwartz, G. Luikart, and P. Taberlet. 2003. Landscape genetics: combining landscape ecology and population genetics. *Trends in Ecology and Evolution* 18:189–197.

Manley, P. N., B. Van Horne, J. K. Roth, W. J. Zielinski, M. M. McKenzie, T. J. Weller, F. W. Weckerly, and C. Vojta. 2006. *Multiple species inventory and monitoring technical guide.* USDA Forest Service General Technical Report GTR-WO-73, Washington, DC.

Manley, P. N., W. J. Zielinski, M. D. Schlesinger, and S. R. Mori. 2004. Evaluation of a multiple-species approach to monitoring species at the ecoregional scale. *Ecological Applications* 14:296–310.

Manly, B. F. J. 1994. Multivariate statistical methods: a primer. Chapman and Hall, London, UK.

Manly, B. F. J., L. L. McDonald, and S. C. Amstrup. 2005. Introduction to the handbook. Pages 1–21 *in* S. C. Amstrup, T. L. McDonald, and B. F. J. Manly, editors. *Handbook of capture-recapture analysis.* Princeton University Press, Princeton, NJ.

Manly B. F. J., L. L. McDonald, and D. L. Thomas. 1993. *Resource selection by animals: statistical design and analysis for field studies.* Chapman & Hall, New York.

Marsh, R. E., W. E. Howard, S. M. McKenna, B. Butler, and D. A. Barnum. 1982. A new system for delivery of predacides or other active ingredients for coyote management. Pages 229–233 *in Proceedings of the 10th Vertebrate Pest Conference.* Monterey, CA.

Martin, D. J., and D. B. Fagre. 1988. Field evaluation of a synthetic coyote attractant. *Wildlife Society Bulletin* 16:390–396.

Mason, J. R., J. Belant, A. E. Barras, and J. W. Guthrie. 1999. Effectiveness of color as an M-44 attractant for coyotes. *Wildlife Society Bulletin* 27:86–90.

Massolo, A., and A. Meriggi. 1998. Factors affecting habitat occupancy by wolves in northern Apennines (northern Italy): a model of habitat suitability. *Ecography* 21:97–107.

Mata, C., I. Hervas, J. Herranz, F. Suarez, and J.E. Malo. 2005. Complementary use by vertebrates of crossing structures along a fenced Spanish motorway. *Biological Conservation* 124:397–405.

Mateo, J. M., and S. A. Cavigelli. 2005. A validation of extraction methods for noninvasive sampling of glucocorticoids in free-living ground squirrels. *Physiological and Biochemical Zoology* 78:1069–1084.

Matos, H., and M. Santos-Reis. 2006. Distribution and status of pine marten *Martes martes* in Portugal. Pages 47–62 *in Martes in carnivore communities.* Alpha Wildlife Publications, Alberta, Canada.

Matyushkin, E. N. 2000. Tracks and tracking techniques in studies of large carnivorous mammals. *Zoologichesky Zhurnal* 79:412–429.

Maudet, C., G. Luikart, D. Dubray, A. von Hardenberg, and P. Taberlet. 2004. Low genotyping error rates in wild ungulate faeces sampled in winter. *Molecular Ecology Notes* 4:772–775.

Mayer, M. V. 1957. A method for determining the activity of burrowing mammals. *Journal of Mammalogy* 38:531.

McCleery, R. A., G. W. Foster, R. R. Lopez, M. J. Peterson, D. J. Forrester, and N. J. Silvy. 2005. Survey of raccoons on Key Largo, Florida, for *Baylisascaris procyonis. Journal of Wildlife Diseases* 41:250–252.

McCullagh, P., and J. A. Nelder. 1989. *Generalized linear models.* Chapman and Hall. London, UK.

McDaniel, G. W., K. S. McKelvey, J. R. Squires, and L. F. Ruggiero. 2000. Efficacy of lures and hair snares to detect lynx. *Wildlife Society Bulletin* 28:119–123.

McDonald, T. 2003. Review of environmental monitoring methods: survey designs. *Environmental Monitoring and Assessment* 85:277–292.

McEwen, B. S. 1998. Stress, adaptation, and disease: allostasis and allostatic load. *Annals of the New York Academy of Science* 840:33–44.

———. 2000. Stress, definitions and concepts of. Pages 508–509 *in* G. Fink, editor. *Encyclopedia of stress.* Academic Press, San Diego, CA.

McKelvey, K. S., J. J. Claar, G. W. McDaniel, and G. Hanvey. 1999. *National Lynx Detection Protocol.* USDA Forest Service Rocky Mountain Research Station, Missoula, MT.

McKelvey, K. S., J. von Kienast, K. B. Aubry, G. M. Koehler, B. T. Maletzke, J. R. Squires, E. L. Lindquist, S. Loch, and M. K. Schwartz. 2006. DNA analysis of hair and scat collected along snow tracks to document the presence of Canada lynx (*Lynx canadensis*). *Wildlife Society Bulletin* 34:451–455.

McKelvey, K. S., and M. K. Schwartz. 2004a. Providing reliable and accurate genetic capture-mark-recapture estimates in a cost-effective way. *Journal of Wildlife Management* 68:453–456.

———. 2004b. Genetic errors associated with population estimation using molecular tagging: problems and new solutions. *Journal of Wildlife Management* 68:439–448.

———. 2005. DROPOUT: a program to identify problem loci and samples for genetic samples in a capture-mark-recapture framework. *Molecular Ecology Notes* 5:716–718.

McQuillen, H. L., and L. W. Brewer. 2000. Methodological considerations for monitoring wild bird nests using video technology. *Journal of Field Ornithology* 71:167–172.

Meadows, R. 2002. Scat-sniffing dogs. *Zoogoer*, September/October.

Mech, L. D., and S. M. Barber. 2002. *A critique of wildlife radio-tracking and its use in national parks: report to the US National Park Service.* United States National Park Service, Fort Collins, CO.

Melquist, W. E., and A. E. Dronkert. 1987. River otter. Pages 626–641 *in* M. Novak, J. A. Baker, M. E. Obbard, and B. Malloch, editors. *Wild furbearer management and conservation in North America.* Ontario Ministry of Natural Resources, Toronto, Ontario, Canada.

Miller, C. R., J. R. Adams, and L. P. Waits. 2003. Pedigree-based assignment tests for reversing coyote (*Canis latrans*) introgression into the wild red wolf (*Canis rufus*) population. *Molecular Ecology* 12:3287–3301.

Miller, C. R., and L. P. Waits. 2003. The history of effective population size and genetic diversity in the Yellowstone grizzly (*Ursus arctos*): implications for conservation. *Proceedings of the National Academy of Science* 100:4334–4339.

Miller, C. R., P. Joyce, and L. P. Waits. 2002. Assessing allelic dropout and genotype reliability using maximum likelihood. *Genetics* 160:357–366.

———. 2005. A new method for estimating the size of small populations from genetic mark-recapture data. *Molecular Ecology* 14:1991–2005.

Miller, M. W., N. T. Hobbs, and M. C. Sousa. 1991. Detecting stress responses in Rocky Mountain bighorn sheep (*Ovis canadensis canadensis*): reliability of cortisol concentrations in urine and feces. *Canadian Journal of Zoology* 69:15–24.

Mills, L. S., J. J. Citta, K. P. Lair, M. K. Schwartz, and D. A. Tallmon. 2000a. Estimating animal abundance using noninvasive DNA sampling: promise and pitfalls. *Ecological Applications* 10:283–294.

Mills, L. S., K. Pilgrim, M. K. Schwartz, and K. McKelvey. 2000b. Identifying lynx and other North American felids based on MtDNA analysis. *Conservation Genetics* 1:285–288.

Millspaugh, J. J., and B. E. Washburn. 2004. Use of fecal glucocorticoid metabolite measures in conservation biology research: considerations for application and interpretation. *General and Comparative Endocrinology* 138:189–199.

Millspaugh, J. J., B. E. Washburn, M. A. Milanick, R. Slotow, and G. van Dyk. 2003. Effects of heat and chemical treatments on fecal glucocorticoid measurements: implications for sample transport. *Wildlife Society Bulletin* 31:399–406.

Ministry of Environment, Lands and Parks. 1999. *Inventory methods for medium-sized territorial carnivores: coyote, red fox, lynx, bobcat, wolverine, fisher and badger. Standards for components of British Columbia's Biodiversity* No. 25. Resources Inventory Committee, Province of British Columbia, Canada.

Minser III, W. G. 1984. Comments "On Scent-station method for monitoring furbearers." *Wildlife Society Bulletin* 12:328.

Minta, S. C., P. M. Karieva, and A. P. Curlee. 1999. Carnivore research and conservation: learning from history and theory. Pages 323–404 *in* T. W. Clark, A. P. Curlee, S. C. Minta, and P. M Karieva, editors. *Carnivores in ecosystems: the Yellowstone experience.* Yale University Press. New Haven, CT.

Minta, S. C., and R. E. Marsh. 1988. Badgers (*Taxidea taxus*) as occasional pests in agriculture. Pages 199–208 *in Proceedings of the 13th Vertebrate Pest Conference.* University of California, Davis.

Miura, S., M. Yasuda, and L. C. Ratnam, 1997. Who steals the fruits? Monitoring frugivory of mammals in a tropical rain forest. *Malayan Nature Journal* 50:183–193.

Mladenoff, D. J., T. A. Sickley, R. G. Haight, and A. P. Wydevan. 1995. A regional landscape analysis and prediction of favorable gray wolf habitat in the northern Great Lakes region. *Conservation Biology* 9:279–294.

Moberg, G. P. 1985. Influence of stress on reproduction: measure of well-being. Pages 245–267 *in* G. P. Moberg, editor. *Animal stress.* American Physiological Society, Bethesda, MD.

Moilanen, A. 2002. Implications of empirical data quality to metapopulation model parameter estimation and application. *Oikos* 96:516–530.

Monfort, S. L. 2003. Endocrine measures of reproduction and stress in wild populations. Pages 147–165 *in* D. E. Wildt, W. Holt, and A. Pickard, editors. *Reproduction and integrated conservation science.* Cambridge University Press, Cambridge, UK.

Monfort, S. L., N. P. Arthur, and D. E. Wildt. 1990. Monitoring ovarian function and pregnancy by evaluating excretion of urinary oestrogen conjugates in semi-free-ranging Przewalski's horses (*Equus przewalskii*). *Journal of Reproduction and Fertility* 91:155–164.

Monfort, S. L., K. L. Mashburn, B. A. Brewer, and S. R. Creel. 1998. Evaluating adrenal activity in African wild dogs (*Lycaon pictus*) by fecal corticosteroid analysis. *Journal of Zoo and Wildlife Medicine* 29:129–133.

Monfort, S. L., C. C. Schwartz, and S. K. Wasser. 1993. Monitoring reproduction in moose using urinary and fecal steroid metabolites. *Journal of Wildlife Management* 57:400–407.

Monfort, S. L., S. K. Wasser, K. L. Mashburn, M. Burke, B. A. Brewer, and S. R. Creel. 1997. Steroid metabolism and validation of non invasive endocrine monitoring in the African wild dog (*Lycaon pictus*). *Zoo Biology* 16:533–548.

Moore, T. D., L. E. Spence, C. E. Dugnolle, and W. G. Hepworth. 1974. *Identification of the dorsal guard hairs of some mammals of Wyoming.* Wyoming Game and Fish Department Bulletin No. 14, Cheyenne.

Morin, P. A., K. E. Chambers, C. Boesch, and L. Vigilant. 2001. Quantitative polymerase chain reaction analysis of DNA from samples for accurate microsatellite genotyping of wild chimpanzees (*Pan troglodytes verus*). *Molecular Ecology* 10:1835–1844.

Morin, P. A., G. Luikart, R. K. Wayne, and the SNP workshop group. 2004. SNPs in ecology, evolution and conservation. *Trends in Ecology and Evolution* 19:208–216.

Morin, P. A. and M. McCarthy. 2007. Highly accurate SNP genotyping from historical and low-quality samples. *Molecular Ecology Notes* 7:937–946.

Morin, P. A., J. J. Moore, R. Chakraborty, L. Jin, J. Goodall, and D. S. Woodruff. 1994. Kin selection, social structure, gene flow, and the evolution of chimpanzees. *Science* 265:1193–1201.

Morin, P. A., and D. S. Woodruff. 1992. Paternity exclusion using multiple hypervariable microsatellite loci amplified from nuclear DNA of hair cells. Pages 63–81 *in* R. D. Martin, A. F. Dixson, and E. J. Wickings, editors. *Paternity in primates: genetic tests and theories.* Karger, Basel, Switzerland.

Morrison, D. W., R. M. Edmunds, G. Linscombe, and J. W. Goertz. 1981. Evaluation of specific scent station variables in northcentral Louisiana. *Proceedings of the Annual Conference of the Southeastern Association of Fish and Wildlife Agencies* 35:281–291.

Morrison, M. L., W. M. Block, M. D. Strickland, and W. L. Kendall. 2001. *Wildlife study design.* Springer-Verlag, New York.

Morrow, C. J., and S. L. Monfort. 1998. Ovarian activity in the scimitar-horned oryx (*Oryx dammah*) determined by fecal steroid analysis. *Animal Reproduction Science* 53:191–207.

Moruzzi, T. L., T. K. Fuller, R. M. DeGraaf, R. T. Brooks, and W. Li. 2002. Assessing remotely triggered cameras for surveying carnivore distribution. *Wildlife Society Bulletin* 30:380–386.

Moruzzi, T. L., K. J. Royar, C. Grove, R. T. Brooks, C. Bernier, F. L. J. Thompson, R. M. DeGraaf, and T. K. Fuller. 2003. Assessing an American marten, *Martes americana*, reintroduction in Vermont. *Canadian Field Naturalist* 117:190–195.

Moss, A. M., T. H. Clutton-Brock, and S. L. Monfort. 2001. Longitudinal gonadal steroid excretion in free-living male and female meerkats (*Suricata suricatta*). *General and Comparative Endocrinology* 122:158–171.

Möstl, E., H. S. Choi, W. Wurm, N. Ismail, and E. Bamberg. 1984. Pregnancy diagnosis in cows and heifers by determination of oestradiol-17 in faeces. *British Veterinary Journal* 140:287–291.

Mowat, G. 2001. *Measuring wolverine distribution and abundance in Alberta.* Alberta Species at Risk Report Number 32. Alberta Sustainable Resource Development, Fish and Wildlife Division, Edmonton, Alberta, Canada.

———. 2006. Winter habitat associations of American martens *Martes americana* in interior wet-belt forests. *Wildlife Biology* 12:17–27.

Mowat, G., and D. C. Heard. 2006. Major components of grizzly bear diet across North America. *Canadian Journal of Zoology* 84:473–489.

Mowat, G., D. C. Heard, D. R. Seip, K. G. Poole, G. Stenhouse, and D. W. Paetkau. 2005. Grizzly *Ursus arctos* and black bear *U. americanus* densities in the interior mountains of North America. *Wildlife Biology* 11:31–48

Mowat, G., C. Kyle, and D. Paetkau. 2003. *Testing methods for detecting wolverine.* Alberta Sustainable Resource Development, Fish and Wildlife Division, Alberta Species at Risk Report No. 71. Edmonton, Alberta, Canada.

Mowat, G., and D. Paetkau. 2002. Estimating marten *Martes americana* population size using hair capture and genetic tagging. *Wildlife Biology* 8:201–209.

Mowat, G., K. G. Poole, and M. O'Donoghue. 1999. Ecology of lynx in northern Canada and Alaska. Pages 265–306 *in* L. F. Ruggiero, K. B. Aubry, S. W. Buskirk, G. M. Koehler, C. J. Krebs, K. S. McKelvey, and J. R. Squires, editors. *Ecology and conservation of lynx in the United States.* University Press of Colorado, Boulder. www.fs.fed.us/rm/pubs/rmrs_gtr030.html (accessed 19 December 2007).

Mowat, G., C. Shurgot, and K. G. Poole. 2000. Using track plates and remote cameras to detect marten and short-tailed weasels in coastal cedar hemlock forests. *Northwestern Naturalist* 81:113–121.

Mowat, G., and C. Strobeck 2000. Estimating population size of grizzly bears using hair capture, DNA profiling, and mark-recapture analysis. *Journal of Wildlife Management* 64:183–193.

Mulders, R., J. Boulanger, and D. Paetkau. 2005. *Using genetic analysis to estimate wolverine abundance in northern Canada.* Abstract of talk presented at the First International Symposium on Wolverine Research and Management, Jokkmokk, Sweden.

Muller, M. N., and R. W. Wrangham. 2004a. Dominance, aggression and testosterone in wild chimpanzees: a test of the "challenge hypothesis." *Animal Behavior* 67:113–123.

———. 2004b. Dominance, cortisol and stress in wild chimpanzees (*Pan troglodytes schweinfurthii*). *Behavioral Ecology and Sociobiology* 55:332–340.

Murie, A. 1940. *The wolves of Mount McKinley.* US Government Printing Office, Washington, DC.

Murie, O. J. 1954. *A field guide to animal tracks.* Peterson Field Guide Series, Houghton Mifflin, Boston, MA.

Murphy, E. C., B. K. Clapperton, and P. M. F. Bradfield. 1999. Secondary poisoning of stoats after an aerial 1080 poison operation in Pureora forest, New Zealand. *New Zealand Journal of Ecology* 23:175–182.

Murphy, M. A., L. P. Waits, and K. C. Kendall. 2000. Quantitative evaluation of fecal drying methods for brown bear DNA analysis. *Wildlife Society Bulletin* 28:951–957.

Murphy, M. A., L. P. Waits, K. C. Kendall, S. K. Wasser, J. A. Higbee, and R. Bogden. 2002. An evaluation of long-term preservation methods for brown bear (*Ursus arctos*) faecal samples. *Conservation Genetics* 3:347–359.

Murray, D. L., S. Boutin, and M. O'Donoghue. 1994. Winter habitat selection by lynx and coyotes in relation to snowshoe hare abundance. *Canadian Journal of Zoology* 72:1444–1451.

Murray, D. L., S. Boutin, M. O'Donoghue, and V. O. Nams. 1995. Hunting behaviour of a sympatric felid and canid in relation to vegetative cover. *Animal Behaviour* 50:1203–1210.

Myers L. J., L. A. Hanrahan, L. J. Swango, and K. E. Nusbaum. 1988a. Anosmia associated with canine distemper. *American Journal of Veterinary Research* 49:1295–1297.

Myers, L. J., K. E. Nusbaum, L. J. Swango, L. N. Hanrahan, and E. Sartin. 1988b. Dysfunction of sense of smell caused by canine parainfluenza virus infection in dogs. *American Journal of Veterinary Research* 49:188–190.

Nakagawa, S., N. J. Gemmell, and T. Burke. 2004. Measuring vertebrate telomeres: applications and limitations. *Molecular Ecology* 34:2523–2534.

Nams, V. O., and E. A. Gillis. 2003. Changes in tracking tube use by small mammals over time. *Journal of Mammalogy* 84:1374–1380.

National Center for Biotechnology Information. 2007. GenBank. www.ncbi.nlm.nih.gov/Genbank/index.html (accessed 19 December 2007).

National Research Council. 2001. *Grand Challenges in the Environmental Sciences.* National Academy Press, Washington, DC.

NDFA (North Dakota Furtakers Association). 1997. *North Dakota furtakers educational manual.* North Dakota Game and Fish Department, Bismarck, ND. www.npwrc.usgs.gov/resource/mammals/furtake/mink.htm (accessed 30 November 2007).

Newhouse, N. J., and T. A. Kinley. 2000. *Ecology of American badgers near their range limit in southeastern British Columbia.* Columbia Basin Fish and Wildlife

Compensation Program, Nelson, British Columbia, Canada.

Ng, S. J., J. W. Dole, R. M. Sauvajot, S. P. D. Riley, and T. J. Valone. 2004. Use of highway undercrossings by wildlife in southern California. *Biological Conservation* 115:499–507.

Nichols, J. D., and C. R. Dickman. 1996. Capture-recapture methods. Pages 217–225 *in* D. Wilson, F. R. Cole, J. D. Nichols, R. Rudran, and M. S. Foster, editors. *Measuring and monitoring biological diversity: standard methods for mammals.* Smithsonian Institution Press, Washington, DC.

Nichols, J. D., J. E. Hines, J. R. Sauer, F. W. Fallon, J. E. Fallon, and P. J. Heglund. 2000. A double-observer approach for estimating detection probability and abundance from point counts. *Auk* 117:393–408.

Nichols, J. D., and K. U. Karanth. 2002. Statistical concepts: assessing spatial distributions. Pages 29–38 *in* K. U. Karanth and J. D. Nichols, editors. *Monitoring tigers and their prey: a manual for researchers, managers and conservationists in tropical Asia.* Centre for Wildlife Studies, Bangalore, India.

Nichols, N., and G. Ward. 1998. *Year of the tiger.* National Geographic, Washington, DC.

Niswender, G. D., A. M. Akbar, and T. M. Nett. 1975. Use of specific antibodies for quantification of steroid hormones. Pages 119–142 *in* B. W. O'Malley and J. G. Hardman, editors. *Methods in enzymology.* Academic Press, New York.

Norris, J. L., and K. H. Pollock. 1996. Nonparametric MLE under two closed capture-recapture models with heterogeneity. *Biometrics* 52:639–649.

Noss, R. 1990. Indicators for monitoring biodiversity: a hierarchical approach. *Conservation Biology* 4:355–364.

Noss, R. F., C. Carroll, K. Vance-Borland, and G. Wuerthner. 2002. A multicriteria assessment of the irreplaceability and vulnerability of sites in the greater Yellowstone ecosystem. *Conservation Biology* 16:895–908.

Nottingham, B. G., Jr., K. G. Johnson, and M. R. Pelton. 1989. Evaluation of scent-station surveys to monitor raccoon density. *Wildlife Society Bulletin* 17:29–35.

Nova Scotia American Marten Recovery Team. 2006. Recovery strategy for American marten (*Martes americana*) on Cape Breton Island, Nova Scotia in Canada. Nova Scotia, Canada.

Novaro, A. J., M. C. Funes, and R. S. Walker. 2005. An empirical test of source-sink dynamics induced by hunting. *Journal of Applied Ecology* 42:910–920.

Nsubuga, A. M., M. M. Robbins, A. D. Roeder, P. A. Morin, C. Boesch, and L. Vigilant. 2004. Factors affecting the amount of genomic DNA extracted from ape faeces and the identification of an improved sample storage method. *Molecular Ecology* 13:2089–2094.

O'Connell, A. F., Jr., N. W. Talancy, L. L. Bailey, J. R. Sauer, R. Cook, and A. T. Gilbert. 2006. Estimating site occupancy and detection probability parameters for meso- and large mammals in a coastal ecosystem. *Journal of Wildlife Management* 70:1625–1633.

O'Donnell, C. F. J. 1996. Predators and the decline of New Zealand forest birds: an introduction to the hole-nesting bird and predator programme. *New Zealand Journal of Ecology* 23:213–219.

O'Farrell, T. P. 1987. Kit fox. Pages 423–431 *in* M. Novak, J. A. Baker, M. E. Obbard, and B. Malloch, editors. *Wild furbearer management and conservation in North America.* Ontario Ministry of Natural Resources, Toronto, Ontario, Canada.

Oehler, J. D., and J. A. Litvaitis. 1996. The role of spatial scale in understanding responses of medium sized carnivores to forest fragmentation. *Canadian Journal of Zoology* 74:2070–2079.

Olson, C. A., and P. A. Werner. 1999. Oral rabies vaccine contact by raccoons and nontarget species in a field trial in Florida. *Journal of Wildlife Diseases* 4:687–695.

Orloff, S. G., A. W. Flannery, and K. C. Belt. 1993. Identification of San Joaquin kit fox (*Vulpes macrotis mutica*) tracks on aluminum tracking plates. *California Fish and Game* 79:45–53.

OSPAR (Convention for the Protection of the Marine Environment of the North-East Atlantic). 2004. *Musk xylene and other musks.* Hazardous Substances Series No. 200. www.ospar.org/documents/dbase/publications/p00200_BD%20on%20musk%20xylene.pdf (accessed 30 November 2007).

Osterberg, D. M. 1962. Activity of small mammals as recorded by a photographic device. *Journal of Mammalogy* 43:219–229.

Østergaard, S., M. M. Hansen, V. Loeschcke, and E. E. Nielson. 2003. Long-term temporal changes of genetic composition in brown trout (*Salmo trutta L.*) populations inhabiting an unstable environment. *Molecular Ecology* 12:3123–3135.

Otani, T. 2001. Measuring fig foraging frequency of the

Yakushima macaque by using automatic cameras. *Ecological Research* 16:49–54.

Otis, D. L., K. P. Burnham, G. C. White, and D. R. Anderson. 1978. Statistical inference from capture data on closed animal populations. *Wildlife Monographs* 62:1–135.

Overton, W. S., and S. V. Stehman. 1996. Desirable design characteristics for long-term monitoring of ecological variables. *Environmental and Ecological Statistics* 3:349–361.

Paetkau, D. 2003. An empirical exploration of data quality in DNA-based population inventories. *Molecular Ecology* 12:1375–1387.

Page, K. L., R. K. Swihart, and K. R. Kazacos. 2001. Seed preferences and foraging by granivores at raccoon latrines in the transmission dynamics of the raccoon roundworm (*Baylisascaris procyonis*). *Canadian Journal of Zoology* 76:617–622.

Palme, R. 2005. Measuring fecal steroids: guidelines for practical applications. *Annals of the New York Academy of Sciences* 1046:75–80.

Palme, R., S. Rettenbacher, C. Touma, S. M. El-Bahr, and E. Möstl. 2005. Stress hormones in mammals and birds—comparative aspects regarding metabolism, excretion, and measurement in fecal samples. *Annals of the New York Academy of Sciences* 1040:162–171.

Palomares, F., and T. M. Caro. 1999. Interspecific killing among mammalian carnivores. *American Naturalist* 153:492–508.

Palomares, F., J. A. Godoy, A. Piriz, S. J. O'Brien, and W. E. Johnson 2002. Faecal genetic analysis to determine the presence and distribution of elusive carnivores: design and feasibility for the Iberian lynx. *Molecular Ecology* 11:2171–2182.

Palsbøll, P. J., J. Allen, M. Bérubé, P. J. Clapham, T. P. Feddersen, P. S. Hammond, R. R. Hudson, H. Jørgensen, S. Katona, A. H. Larsen, F. Larsen, et al. 1997. Genetic tagging of humpback whales. *Nature* 388:676–679.

Palsbøll, P. J., M. Bérubé, and F. W. Allendorf. 2007. Identification of management units using population genetic data. *Trends in Ecology and Evolution* 22:11–16.

Parker, G. 1995. *Eastern coyote: the story of its success*. Nimbus Publishing, Halifax, Nova Scotia, Canada.

Patterson, B. R., N. W. S. Quinn, E. F. Becker, and D. B. Meier. 2004. Estimating wolf densities in forested areas using network sampling of tracks in snow. *Wildlife Society Bulletin* 32:938–947.

Paxinos, E., C. McIntosh, K. Ralls, and R. Fleischer. 1997. A noninvasive method for distinguishing among canid species: amplification and enzyme restriction of DNA from dung. *Molecular Ecology* 6:483–486.

Pearson, O. P. 1959. A traffic survey of Microtus-Reithrodontomys runways. *Journal of Mammalogy* 40:169–180.

Pedlar, J. H., L. Fahrig, and H. G. Merriam. 1997. Raccoon habitat use at two spatial scales. *Journal of Wildlife Management*. 61:102–112.

Pellet, J., and B. R. Schmidt. 2005. Monitoring distributions using call surveys: estimating site occupancy, detection probabilities and inferring absence. *Biological Conservation* 123:27–35.

Pelletier, F., J. Bauman, and M. Festa-Bianchet. 2003. Fecal testosterone in bighorn sheep (*Ovis canadensis*): behavioural and endocrine correlates. *Canadian Journal of Zoology* 81:1678–1684.

Pellikka, J., H. Rita, and H. Linden. 2005. Monitoring wildlife richness—Finnish applications based on wildlife triangle censuses. *Annales Zoologici Fennici* 42:123–134.

Peterjohn, B. B. 1994. The North American Breeding Bird Survey. *Birding* 26:386–398.

Peterjohn, B. G., J. R. Sauer, and W. A. Link. 1994. The 1992 and 1993 summary of the North American Breeding Bird Survey. *Bird Populations* 2:46–51.

Petit, E., and N. Valiere. 2006. Estimating population size with noninvasive capture-mark-recapture data. *Conservation Biology* 20:1062–1073.

Pettitt, B. A., C. J. Wheaton, and J. M. Waterman. 2007. Effects of storage treatment on fecal steroid hormone concentrations of a rodent, the Cape ground squirrel (*Xerus inauris*). *General and Comparative Endocrinology* 150:1–11.

PHAC (Public Health Agency of Canada). 2006. *Material safety data sheets—Infectious substances*. www.phac-aspc.gc.ca/msds-ftss/ (accessed 30 November 2007).

Pierce, B. M., V. C. Bleich, C. B. Cetkiewicz, and J. D. Wehausen. 1998. Timing of feeding bouts of mountain lions. *Journal of Mammalogy* 79:222–226.

Pietz, P. J., and D. A. Granfors. 2000. Identifying predators and fates of grassland passerine nests using miniature video cameras. *Journal of Wildlife Management* 64:71–87.

Piggott, M. P. 2004. Effect of sample age and season of collection on the reliability of microsatellite genotyping of faecal DNA. *Wildlife Research* 31:485–493.

Piggott, M. P., S. C. Banks, N. Stone, C. Banffy, and A. C. Taylor. 2006. Estimating population size of endangered brush-tailed rock-wallaby (*Petrogale penicillata*) colonies using faecal DNA. *Molecular Ecology* 15:81–91.

Piggott, M. P., and A. C. Taylor. 2003a. Remote collection of animal DNA and its applications in conservation management and understanding the population biology of rare and cryptic species. *Wildlife Research* 30:1–13.

———. 2003b. Extensive evaluation of faecal preservation and DNA extraction methods in Australian native and introduced species. *Australian Journal of Zoology* 51:341–355.

Pilgrim, K. L., K. S. McKelvey, A. E. Riddle, and M. K. Schwartz. 2005. Felid sex identification based on genetic samples. *Molecular Ecology Notes* 5:60–61.

Pilgrim, K. L., W. J. Zielinski, M. J. Mazurek, F. V. Schlexer, and M. K. Schwartz. 2006. Development and characterization of microsatellite markers in Point Arena mountain beaver *Aplodontia rufa nigra*. *Molecular Ecology Notes* 6:800–802.

Pina, G. P. L., R. A. C. Gamez, and C. A. L. Gonzalez. 2004. Distribution, habitat association, and activity patterns of medium and large-sized mammals of Sonora, Mexico. *Natural Areas Journal* 24:354–357.

Pittaway, R. 1983. Fisher and red fox interactions over food. *Ontario Field Biologist* 37:88–90.

Pledger, S. 2000. Unified maximum likelihood estimates for closed capture-recapture models using mixtures. *Biometrics* 56:434–442.

Pocock, M. J. O., A. C. Frantz, D. P. Cowan, P. C. L. White, and J. B. Searle. 2004. Tapering bias inherent in minimum number alive (MNA) population indices. *Journal of Mammalogy* 85:959–962.

Poglayen-Neuwall, I., and D. E. Toweill. 1988. *Bassariscus astutus*. *Mammalian Species* 327:1–8.

Pollock, K. H. 1982. A capture-recapture design robust to unequal capture probability. *Journal of Wildlife Management* 46:752–757.

Pollock, K. H., J. D. Nichols, J. E. Hines, and C. Brownie. 1990. Statistical inference for capture-recapture experiments. *Wildlife Monographs* 107.

Pollock, K. H., J. D. Nichols, T. R. Simons, G. L. Farnsworth, L. L. Bailey, and J. R. Sauer. 2002. Large scale wildlife monitoring studies: statistical methods for design and analysis. *Environmetrics* 13:105–119.

Poole, J. H., L. H. Kasman, E. C. Ramsay, and B. L. Lasley. 1984. Musth and urinary testosterone in the African elephant (*Loxodonta africana*). *Journal of Reproduction and Fertility* 70:255–290.

Poole, K. G., G. Mowat, and D. A. Fear. 2001. DNA-based population estimate for grizzly bears *Ursus arctos* in northeastern British Columbia, Canada. *Wildlife Biology* 7:105–115.

Porter, J., P. Arzberger, H. Braun, P. Bryant, S. Gage, T. Hansen, P. Hanson, C. Lin, F. Lin, T. Kratz, W. Michener, S. Shapiro, and T. Williams. 2005. Wireless sensor networks for ecology. *BioScience* 55:561–572.

Poszig, D., C. D. Apps, and A. Dibb. 2004. Predation on two mule deer, *Odocoileus hemionus*, by a Canada lynx, *Lynx canadensis*, in the Southern Canadian Rocky Mountains. *Canadian Field Naturalist* 118:191–194.

Powell, R. A., and W. J. Zielinski. 1983. Competition and coexistence in mustelid communities. *Acta Zoologica Fennica* 174:223–227.

Powell, R. A., S. W. Buskirk, and W. J. Zielinski. 2003. Fisher and marten (*Martes pennanti* and *Martes americana*). Pages 635–649 in G. A. Feldhamer, B. C. Thompson, and J. A. Chapman, editors. *Wild mammals of North America: biology, management, and conservation*. 2nd ed. John Hopkins University Press, Baltimore, MD.

Powell, R. A., and G. Proulx. 2003. Trapping and marking terrestrial mammals for research: integrating ethics, performance criteria, techniques, and common sense. *Institute of Laboratory Animal Resources Journal* 44:259–276.

Pradel, R. 1996. Utilization of mark-recapture for the study of recruitment and population growth rate. *Biometrics* 52:703–709.

Proctor, M., and J. Boulanger. 2006. *Pilot studies for grizzly bear population trend monitoring using scat-DNA methods*. Prepared for Parks Canada; Banff National Park, Lake Louise, Alberta, Canada.

Proctor, M., J. Boulanger, S. Nielsen, C. Servheen, W. Kasworm, T. Radandt, and D. Paetkau. 2007. *Abundance and density of Central Purcell, South Purcell, Yahk, and South Selkirk Grizzly Bear Population Units in southeast British Columbia*. Report submitted to British Columbia Ministry of Environment, Canada.

Proctor, M. F. 2003. Genetic analysis of movement,

dispersal and population fragmentation of grizzly bears in southwestern Canada. PhD dissertation. University of Calgary, Alberta, Canada.

Proctor, M. F., B. McLellan, and C. Strobeck. 2002. Population fragmentation of grizzly bears in southeastern British Columbia, Canada. *Ursus* 13:153–160.

Proctor, M. F., B. N. McLellan, C. Strobeck, and R. M. R. Barclay. 2004. Gender-specific dispersal distances of grizzly bears estimated by genetic analysis. *Canadian Journal of Zoology* 82:1108–1118.

———. 2005. Genetic analysis reveals demographic fragmentation of grizzly bears yielding vulnerably small populations. *Proceedings of the Royal Society Biology* 272(1579):2409–2416.

Prugh, L. R., and C. J. Krebs. 2004. Snowshoe hare pellet-decay rates and aging in different habitats. *Wildlife Society Bulletin* 32:386–393.

Prugh, L. R., and C. E. Ritland. 2005. Molecular testing of observer identification of carnivore feces in the field. *Wildlife Society Bulletin* 33:189–194.

Prugh, L. R., C. E. Ritland, S. M. Arthur, and C. J. Krebs. 2005. Monitoring coyote population dynamics by genotyping faeces. *Molecular Ecology* 14:1585–1596.

Pulliainen, E. 1981. A transect survey of small land carnivore and red fox populations on a subarctic fell in Finnish forest Lapland over 13 winters. *Annales Zoologici Fennici* 18:270–278.

Putnam, R. J. 1984. Facts from faeces. *Mammal Review* 14:79–97.

Quinn, G. P., and M. J. Keough. 2002. *Experimental design and data analysis for biologists.* Cambridge University Press, Cambridge, UK.

R Development Core Team. 2005. *R: a language and environment for statistical computing.* R Foundation for Statistical Computing, Vienna, Austria. ISBN 3-900051-07-0. www.R-project.org (accessed 19 December 2007).

Ragg, J. R., and H. Moller. 2000. Microhabitat selection by feral ferrets (*Mustela furo*) in a pastoral habitat, East Otago, New Zealand. *New Zealand Journal of Ecology* 24:39–46.

Raine, R. M. 1983. Winter habitat use and responses to snow cover of fisher (*Martes pennanti*) and marten (*Martes americana*) in southeastern Manitoba. *Canadian Journal of Zoology* 61:25–34.

Ralls, K., and D. A. Smith. 2004. Latrine use by San Joaquin kit foxes (*Vulpes macrotis mutica*) and coyotes

(*Canis latrans*). *Western North American Naturalist* 64:544–547.

Randa, L. A., and J. A. Yunger. 2004. The influence of prey availability and vegetation characteristics on scent station visitation rates of coyotes, *Canis latrans*, in a heterogeneous environment. *Canadian Field Naturalist* 118:341–353.

———. 2006. Carnivore occurrence along an urban-rural gradient: a landscape-level analysis. *Journal of Mammalogy* 87:1154–1164.

Raphael, M. G. 1994. Techniques for monitoring populations of fishers and American martens. Pages 224–240 *in* S. W. Buskirk, A. S. Harestad, M. G. Raphael, and R. A. Powell, editors. *Martens, sables, and fishers.* Cornell University Press, Ithaca, NY.

Ray, J. C. 2005. Large carnivorous animals as tools for conserving biodiversity: assumptions and uncertainties. Pages 34–56 *in* J. C. Ray, K. H. Redford, R. S. Steneck, and J. Berger, editors. *Large carnivores and the conservation of biodiversity.* Island Press, Washington, DC.

Ray, J. C., K. H. Redford, J. Berger, and R. S. Steneck. 2005. Conclusion: is large carnivore conservation equivalent to biodiversity conservation and how can we achieve both? Pages 400–427 *in* J. C. Ray, K. H. Redford, R. S. Steneck, and J. Berger, editors. *Large carnivores and the conservation of biodiversity.* Island Press, Washington, DC.

Ray, J. C., and M. E. Sunquist. 2001. Trophic relations in a community of African rainforest carnivores. *Oecologia* 127:395–408.

Rebmann, A., E. David, and M. H. Sorg. 2000. *Cadaver dog handbook: forensic training and tactics for the recovery of human remains.* CRC Press LLC, Boca Raton, FL.

Reed, J. E., R. J. Baker, W. B. Ballard, and B. T. Kelly. 2004. Differentiating Mexican gray wolf and coyote scats using DNA analysis. *Wildlife Society Bulletin* 32:685–692.

Reed, S. E., and E. F. Leslie. 2005. Patterns of carnivore co-occurrence on the North Rim, Grand Canyon National Park. Pages 309–316 *in* C. van Riper and D. J. Mattson, editors. *The Colorado Plateau II: biophysical, socioeconomic and cultural research.* University of Arizona Press, Tucson.

Reid, D. G., M. B. Bayer, T. E. Code, and B. McLean. 1987. A possible method for estimating river otter, *Lutra canadensis*, populations using snow tracks. *Canadian Field Naturalist* 101:576–580.

Reimers, T. J., B. A. Cowan, H. P. Davidson, and E. D. Colby. 1981. Validation of radioimmunoassays for tri-iodothyronine, thyroxine, and hydrocortisone (cortisol) in canine, feline, and equine sera. *American Journal of Veterinary Research* 42:2016–2021.

Reindl-Thompson, S. A., J. A. Shivik, A. Whitelaw, A. Hurt, K. F. Higgins. 2006. Efficacy of scent dogs in detecting black-footed ferrets at a reintroduction site in South Dakota. *Wildlife Society Bulletin* 34:1435–1439.

Resource Inventory Committee 1999. *Inventory methods for marten and weasels: standards for components of British Columbia's biodiversity.* No. 24. BC Ministry of Environment, Lands, and Parks, Resource Inventory Branch for the Terrestrial Ecosystems Task Force, Resource Information Standards Committee, British Columbia, Canada.

Rexstad, E., and K. P. Burnham. 1991. *User's guide for interactive program CAPTURE: abundance estimation of closed animal populations.* Colorado State University, Fort Collins. www.mbr-pwrc.usgs.gov/software/doc/capturemanual.pdf (accessed 19 December 2007).

Reynolds, J. C., and N. J. Aebischer. 1991. Comparison and quantification of carnivore diet by faecal analysis: a critique, with recommendations, based on a study of the fox *Vulpes vulpes*. *Mammal Review* 21:97–122.

Reynolds, J. C., M. J. Short, and R. J. Leigh. 2004. Development of population control strategies for mink *Mustela vison*, using floating rafts as monitors and trap sites. *Biological Conservation* 120:533–543.

Reynolds, S. A., F. Levy, and E. S. Walker. 2006. Hand sanitizer alert. *Emerging Infectious Diseases* 12:527–29. www.cdc.gov/ncidod/eid/vol12no03/05-0955.htm#cit (accessed 30 November 2007).

Rezendes, P. 1992. *Tracking and the art of seeing: how to read animal tracks and sign.* Camden House Publishing, Charlotte, VT.

Rhodes, J. R., A. J. Tyre, N. Jonzen, C. A. McAlpine, and H. P. Possingham. 2006. Optimizing presence-absence surveys for detecting population trends. *Journal of Wildlife Management* 70:8–18.

Rice, C. G. 1995. Trailmaster camera system: response. *Wildlife Society Bulletin* 23:110–111.

Rice, C. G., J. Rohlman, J. Beecham, and S. Pozzanghera. 2001. Power analysis of bait station surveys in Idaho and Washington. *Ursus* 12:227–236.

Riddle, A. E., K. L. Pilgrim, L. S. Mills, K. S. McKelvey, and L. F. Ruggiero. 2003. Identification of mustelids using mitochondrial DNA and non-invasive sampling. *Conservation Genetics* 4:241–243.

Ritland, K. 1996. Estimators for pairwise relatedness and individual inbreeding coefficients. *Genetical Research* 67:175–185.

Rizzo, W. 2005. Return of the jaguar? *Smithsonian.* www.smithsonianmagazine.com/issues/2005/december/phenomena.php (accessed 19 December 2007).

Robbins, C. S., D. Bustrak, and P. H. Geissler. 1986. *The breeding bird survey: its first fifteen years.* 1965–1979. US Fish and Wildlife Service, Resource Publication 157, Patuxent Wildlife Research Center, Laurel, MD.

Robbins, C. S., S. Droege, and J. R. Sauer. 1989. Monitoring bird populations with breeding bird survey and atlas data. *Annales Zoologici Fennici* 26:297–304.

Robbins, M. M., and N. M. Czekala. 1997. A preliminary investigation of urinary testosterone and cortisol levels in wild male mountain gorillas. *American Journal of Primatology* 43:51–64.

Robinson, W. B. 1962. Methods of controlling coyotes, bobcats, and foxes. Pages 32–56 *in Proceedings of the 1st Vertebrate Pest Conference.* University of Nebraska, Lincoln.

Robitaille, J.-F., and K. Aubry. 2000. Occurrence and activity of American martens *Martes americana* in relation to roads and other routes. *Acta Theriologica* 45:137–143.

Robson, M. S., and S. R. Humphrey. 1985. Inefficacy of scent-station methodology for assessing trends in carnivore populations. *Wildlife Society Bulletin* 13:558–561.

Roeder, A. D., F. I. Archer, H. N. Poinar, and P. A. Morin. 2004. A novel method for collection and preservation of faeces for genetic studies. *Molecular Ecology Notes* 4:761–764.

Rolland, R. M., P. K. Hamilton, S. D. Kraus, B. Davenport, R. M. Gillett, and S. K. Wasser. 2006. Faecal sampling using detection dogs to study reproduction and health in North Atlantic right whales (*Eubalaena glacialis*). *Journal of Cetacean Research and Management* 8:121–125.

Romain-Bondi, K. A., R. B. Wielgus, L. Waits, W. F. Kasworm, M. Austin, and W. Wakkinen. 2004. Density and population size estimates for North Cascade grizzly bears using DNA hair-sampling techniques. *Biological Conservation* 117:417–428.

Roon, D. A., M. E. Thomas, K. C. Kendall, and L. P. Waits. 2005. Evaluating mixed samples as a source of error in non-invasive genetic studies using microsatellites. *Molecular Ecology* 14:195–201.

Roon, D. A., L. P. Waits, and K. C. Kendall. 2003. A quantitative evaluation of two methods for preserving hair samples. *Molecular Ecology Notes* 3:163–166.

Rosatte, R. C. 1987. Striped, spotted, hooded, and hognosed skunk. Pages 599–613 *in* M. Novak, J. A. Baker, M. E. Obbard, and B. Malloch, editors. *Wild furbearer management and conservation in North America.* Ontario Ministry of Natural Resources, Toronto, Ontario, Canada.

Rosatte, R., and S. Lariviére. 2003. Skunks (Genera *Mephitis, Spilogale, and Conepatus*). Pages 692–707 *in* G .A. Feldhamer, B. C. Thompson, and J. A. Chapman, editors. *Wild mammals of North America: biology, management, and conservation.* 2nd ed. John Hopkins University Press, Baltimore, MD.

Rotenberry, J. T., S. T. Knick, and J. Dunn, E. 2002. A minimalist approach to mapping species' habitat: Pearson's planes of closest fit. Pages 281–289 *in* J. M. Scott, P. J. Heglund, M. L. Morrison, J. B. Haufler, M. G. Raphael, W. A. Wall, and F. B. Samson, editors. *Predicting species occurrences.* Island Press, Washington, DC.

Roughton, R. D. 1979. Developments in scent station technology. Pages 17–44 *in Proceedings of the Midwest Furbearer Conference.* Kansas State University, Cooperative Extension Service, Manhattan.

———. 1982. A synthetic alternative to fermented egg as a canid attractant. *Journal of Wildlife Management* 46:230–234.

Roughton, R. D., and D. C. Bowden. 1979. Experimental design for field evaluation of odor attractants for predators. Pages 249–254 *in* J. R. Beck, editor. *Vertebrate pest control and management materials.* American Society for Testing and Materials, STP 680, Sacramento, CA.

Roughton, R. D., and M. W. Sweeny. 1982. Refinements in scent-station methodology for assessing trends in carnivore populations. *Journal of Wildlife Management* 46:217–229.

Routledge, R. G. 2000. Use of track plates to detect changes in American marten (*Martes americana*) abundance. MS thesis. Laurentian University, Sudbury, Ontario, Canada.

Rovero, F., T. Jones, and J. Sanderson. 2005. Notes on Abbott's duiker (*Cephalophus spadix*, True 1890) and other forest antelopes of Mwanihana forest, Udzungwa Mountains, Tanzania, as revealed by camera-trapping and direct observations. *Tropical Zoology* 18:13–23.

Roy, C. C. 1999. 1998 Swift fox conservation team annual report. Kansas Department of Wildlife and Parks, Emporia.

Roy, C. C., M. A. Sovada, and G. A. Sargeant. 1999. An improved method for determining the distribution of swift foxes in Kansas. Pages 4–17 *in* C. C. Roy, editor. *1998 Swift Fox Conservation Team annual report.* Kansas Department of Wildllife and Parks, Emporia.

Royle, J. A. 2004a. Modeling abundance index data from anuran calling surveys. *Conservation Biology* 18:1378–1385.

———. 2004b. N-Mixture models for estimating population size from spatially replicated counts. *Biometrics* 60:108–115.

———. 2006. Site occupancy models with heterogeneous detection probabilities. *Biometrics* 62:97–102.

Royle, J. A., D. K. Dawson, and S. Bates. 2004. Modeling abundance effects in distance sampling. *Ecology* 85:1591–1597.

Royle, J. A., and M. Kéry. 2006. A Bayesian state-space representation of dynamic occupancy models. *Ecology* 88:1813–1823.

Royle, J. A., and W. A. Link. 2006. Generalized site occupancy models allowing for false positive and false negative errors. *Ecology* 87:835–841.

Royle, J. A., and J. D. Nichols. 2003. Estimating abundance from repeated presence-absence data or point counts. *Ecology* 84:777–790.

Royle, J. A., J. D. Nichols, and M. Kéry. 2005. Modelling occurrence and abundance of species when detection is imperfect. *Oikos* 110:353–359.

Rucker, R. A. 1983. An assessment of scent stations as a means of monitoring raccoon populations in middle Tennessee. MS thesis. Tennessee Technological University, Cookeville.

Ruell, E. W., and K. R. Crooks. 2007. Evaluation of noninvasive genetic sampling methods for felid and canid populations. *Journal of Wildlife Management* 71:1690–1694.

Ruggiero, L. F., K. B. Aubry, S. W. Buskirk, L. J. Lyon, and W. J. Zielinski. 1994. *The scientific basis for conserving forest carnivores: American marten, fisher, lynx, and wolverine in the Western United States*. USDA Forest Service Rocky Mountain Forest and Experiment Station General Technical Report RM-254. Fort Collins, CO.

Ruiz-Olmoa, J., D. Saavedra, and J. Jiminez. 2001. Testing the surveys and visual and track censuses of Eurasian otters (*Lutra lutra*). *Journal of Zoology* 253:359–369.

Sadlier, L. M. J., C. C. Webbon, P. J. Baker, and S. Harris. 2004. Methods of monitoring red foxes *Vulpes vulpes* and badgers *Meles meles*: are field signs the answer? *Mammal Review* 34:75–98.

Saiki, R. K., D. H. Gelfand, S. Stoffel, S. J. Scharf, R. Higuchi, G. T. Horn, K. B. Mullis, and H. A. Erlich. 1988. Primer-directed enzymatic amplification of DNA with a thermostable DNA polymerase. *Science* 239:487–491.

Sala, E. 2007. Top predators provide insurance against climate change. *Trends in Ecology and Evolution* 21:479–480.

Sanchez, D. M., P. R. Krausman, T. R. Livingston, and P. S. Gipson. 2004. Persistence of carnivore scat in the Sonoran Desert. *Wildlife Society Bulletin* 32:366–372.

Sanderson, G. C. 1987. Raccoon. Pages 486–499 *in* M. Novak, J. A. Baker, M. E. Obbard, and B. Malloch, editors. *Wild furbearer management and conservation in North America*. Ontario Ministry of Natural Resources, Toronto, Ontario, Canada.

Sanderson, J. G., and M. Trolle. 2005. Monitoring elusive mammals. *American Scientist* 93:148–155.

Sands, J., and S. Creel. 2004. Social dominance, aggression and fecal glucocorticoid levels in a wild population of wolves, *Canis lupus*. *Animal Behavior* 67:387–396.

Sargeant, A. B., W. K. Pfeifer, and S. H. Allen. 1975. Spring aerial census of red foxes in North Dakota. *Journal of Wildlife Management* 39:30–39.

Sargeant, G. A., W. E. Berg, and D. H. Johnson. 1996. *Minnesota carnivore population index scent-station survey results, 1986–95*. Jamestown, ND: Northern Prairie Wildlife Research Center Online. www.npwrc.usgs.gov/resource/mammals/mncarns/index.htm (Accessed 19 December 2007).

Sargeant, G. A., D. H. Johnson, and W. E. Berg. 1998. Interpreting carnivore scent station surveys. *Journal of Wildlife Management* 62:1235–1245.

———. 2003a. Sampling designs for carnivore scent-station surveys. *Journal of Wildlife Management* 67:289–298.

Sargeant, G. A., M. A. Sovada, C. C. Slivinski, and D. H. Johnson. 2005. Markov chain Monte Carlo estimation of species distributions: a case study of the swift fox in western Kansas. *Journal of Wildlife Management* 69:483–497.

Sargeant, G. A., P. J. White, M. A. Sovada, and B. L. Cypher. 2003b. Scent station survey techniques for swift and kit foxes. Pages 99–105 *in* M. A. Sovada and L. Carbyn, editors. *Ecology and conservation of swift foxes in a changing world*. Canadian Plains Research Center, University of Regina, Saskatchewan, Canada.

Sato, Y., T. Aoi, K. Kaji, and S. Takasuki. 2004. Temporal changes in the population density and diet of brown bears in eastern Hokkaido, Japan. *Mammal Study* 29:47–53.

Sauer J. R., and S. Droege. 1990. Survey designs and statistical methods for the estimation of avian population trends. US Fish and Wildlife Service, Biological Report 90(1).

Sauer, J. R., J. E. Fallon, and R. Johnson. 2003. Use of North American Breeding Bird Survey data to estimate population change for bird conservation regions. *Journal of Wildlife Management* 67:372–389.

Savolainen, V., R. S. Cowan, A. P. Vogler, G. K. Roderick, and R. Lane. 2005. Towards writing the encyclopaedia of life: an introduction to DNA barcoding. *Philosophical Transactions of the Royal Society of London, Series B*. 360:1805–1811.

Sawaya, M., and T. Ruth. 2006. *Development and testing of noninvasive genetic sampling techniques for cougars in Yellowstone National Park*. Wildlife Conservation Society, Bronx, NY.

Sawaya, M., T. Ruth, S. Kalinowski, and S. Creel. 2005. *Development and testing of noninvasive genetic sampling techniques for cougars in Yellowstone National Park*. Poster presented at the 8th Mountain Lion Workshop, Leavenworth, WA.

Scandura, M. 2005. Individual sexing and genotyping from blood spots on the snow: a reliable source of DNA for non-invasive genetic surveys. *Conservation Genetics* 6:871–874.

Schaeffer, R. L., W. Mendenhall, and L. Ott. 1990. *Elementary survey sampling*. PWS-Kent, Boston, MA.

Schauster, E. R., E. M. Gese, and A. M. Kitchen. 2002. An evaluation of survey methods for monitoring swift fox abundance. *Wildlife Society Bulletin* 30:464–477.

Schemnitz, S. D. 1996. Capturing and handling wild animals. Pages 106–124 *in* T. A. Bookhout, editor. *Research and management techniques for wildlife and habitats.* Wildlife Society, Bethesda, MD.

Schipper, J. 2007. Camera-trap avoidance by kinkajous *Potos flavus*: rethinking the "non-invasive" paradigm. *Small Carnivore Conservation* 36:38–41.

Schmeiser, H. H., R. Gminski, and V. Mersch-Sundermann. 2001. Evaluation of health risks caused by musk ketone. *International Journal of Hygiene and Environmental Health* 203:293–299.

Schmitt, C. G. 2000. *Swift fox conservation team annual report, 1999.* New Mexico Department of Game and Fish, Santa Fe, NM.

Schoenecker, K. A., R. O. Lyda, and J. Kirkpatrick. 2004. Comparison of three fecal steroid metabolites for pregnancy detection used with single sampling in bighorn sheep (*Ovis canadensis*). *Journal of Wildlife Diseases* 40:273–281.

Schoon, G. A. 1996. Scent identification lineups by dogs (*Canis familiaris*): experimental design and forensic application. *Applied Animal Behaviour Science* 49:257–267.

Schoon, G. A., and R. Haak. 2002. K9 suspect discrimination: training and practicing scent identification lineups. Detselig Enterprises, Calgary, Alberta, Canada.

Schwartz, C. C., S. L. Monfort, P. Dennis, and K. J. Hundertmark. 1995. Fecal progesterone concentration as an indicator of the estrous cycle and pregnancy in moose. *Journal of Wildlife Management* 59:590–583.

Schwartz, E. R., and C. W. Schwartz. 1991. A quarter-century study of survivorship in a population of three-toed box turtles in Missouri. *Copeia* 1991:1120–1123.

Schwartz, M. K., K. B. Aubry, K. S. McKelvey, K. L. Pilgrim, J. P. Copeland, J. R. Squires, R. M. Inman, S. M. Wisely, and L. F. Ruggiero. 2007. Inferring geographic isolation of wolverine in California using historical DNA. *Journal of Wildlife Management* 71:2170–2179.

Schwartz, M. K., S. A. Cushman, K. S. McKelvey, J. Hayden, and C. Engkjer. 2006. Detecting genotyping errors and describing black bear movement in North Idaho. *Ursus* 17:138–148.

Schwartz, M. K., G. Luikart, R. S. Waples. 2007. Genetic monitoring as a promising tool for conservation and management. *Trends in Ecology and Evolution* 22:25–33.

Schwartz, M. K., K. L. Pilgrim, K. S. McKelvey, E. L. Lindquist, J. J. Claar, S. Loch, and L. F. Ruggiero. 2004. Hybridization between Canada lynx and bobcats: genetic results and management implications. *Conservation Genetics* 5:349–355.

Schwartz, M. K., D. Tallmon, and G. Luikart. 1998. Using genetic sampling methods and new analytical tools to detect population declines and minimize extinctions. *Animal Conservation* 1:293–299.

Schwarzenberger, F., E. Möstl, R. Palme, and E. Bamberg. 1996. Faecal steroid analysis for non-invasive monitoring of reproductive status in farm, wild and zoo animals. *Animal Reproduction Science* 42:515–526.

Scott, J. M., P. J. Heglund, M. L. Morrison, J. B. Haufler, M. G. Raphael, W. A. Wall, and F. B. Samson, editors. 2002. *Predicting species occurrences.* Island Press, Washington, DC.

Scott-Brown, J. M., S. Herrero, and J. Reynolds. 1987. Swift fox. Pages 433–441 *in* M. Novak, J. A. Baker, M. E. Obbard, and B. Malloch, editors. *Wild furbearer management and conservation in North America.* Ontario Ministry of Natural Resources, Toronto, Ontario, Canada.

Seal, B. S., C. W. Lutze, L. C. Kreutz, T. Sapp, G. C. Dulac, and J. D. Neill. 1995. Isolation of calciviruses from skunks that are antigenically and genotypically related to San Miguel sea lion virus. *Virus Research* 37:1–12.

Seber, G. A. F. 1982. *The estimation of animal abundance and related parameters.* 2nd edition. Macmillan, New York.

Seddon, J. M., H. G. Parker, E. A. Ostrander, and H. Ellegren. 2005. SNPs in ecological and conservation studies: a test in the Scandinavian wolf population. *Molecular Ecology* 14:503–511.

Sequin, E. S., M. M. Jaeger, P. F. Brussard, and R. H. Barrett. 2003. Wariness of coyotes to camera traps relative to social status and territory boundaries. *Canadian Journal of Zoology* 81:2015–2025.

Serfass, T. L., R. P. Brooks, and W. M. Tzilkowski. 2001. *Fisher reintroduction in Pennsylvania, final project report.* University of Maryland, Frostberg.

Servheen, C., J. S. Waller, and P. Sandstrom. 2001. *Identification and management of linkage zones for grizzly bears between large blocks of public land in the Northern Rocky Mountains.* US Fish and Wildlife Service. Missoula, MT.

Servin, J. I., J. R. Rau, and M. Delibes. 1987. Use of radio tracking to improve the estimation by track counts of the relative abundance of red fox. *Acta Theriologica* 32:489–492.

Seton, E. T. 1937. *Biography of an arctic fox.* Appleton-Century. New York, NY.

Seutin, G., B. N. White, and P. T. Boag. 1991. Preservation of avian blood and tissue samples for DNA analysis. *Canadian Journal of Zoology* 69:82–90.

Sharma, S., Y. Jhala, and V. B. Sawarkar. 2005. Identification of individual tigers (*Panthera tigris*) from their pugmarks. *Journal of Zoology* 267:9–18.

Shenk, T. M. 2001. *Post-release monitoring of lynx reintroduced to Colorado.* Colorado Division of Wildlife, Annual Progress Report for the US Fish and Wildlife Service, December 2001, Denver, CO.

Shinn, K. J. 2002. Ocelot distribution in the Lower Rio Grande Valley National Wildlife Refuge. MS thesis. University of Texas–Pan American, Edinburg, TX.

Shivik, J. A. 2006. Tools for the edge: what's new for conserving carnivores. *BioScience* 53:253–259.

Shivik, J. A., D. J. Martin, M. J. Pipas, J. Turnan, and T. J. DeLiberto. 2005. Initial comparison: jaws, cables, and cage-traps to capture coyotes. *Wildlife Society Bulletin* 33:1375–1383.

Sidorovich, V. E., D. A. Krasko, and A. A. Dyman. 2005. Landscape-related differences in diet, food supply and distribution pattern of the pine marten, *Martes martes,* in the transitional mixed forest of northern Belarus. *Folia Zoologica* 54:39–52.

Silveira, L., A. T. A. Jacomo, and J. A. F. Diniz-Filho. 2003. Camera trap, line transect census and track surveys: a comparative evaluation. *Biological Conservation* 114:351–355.

Silver, S. 2004. *Assessing jaguar abundance using remotely triggered cameras.* Wildlife Conservation Society, Bronx, NY

Silver, S. C., L. E. T. Ostro, L. K. Marsh, L. Maffei, A. J. Noss, M. J. Kelly, R. B. Wallace, H. Gomez, and G. Ayala. 2004. The use of camera traps for estimating jaguar *Panthera onca* abundance and density using capture/recapture analysis. *Oryx* 38:148–154.

Silvy, A. J., R. R. Lopez, and M. J. Peterson. 2005. Wildlife marking techniques. Pages 339–375 *in* C. E. Braun, ed-

itor. *Techniques for wildlife investigations and management.* Port City Press, Baltimore, MD.

Sinclair, K. E., G. R. Hess, C. E. Moorman, and J. H. Mason. 2005. Mammalian nest predators respond to greenway width, landscape context and habitat structure. *Landscape and Urban Planning* 71:277–292.

Skalski, J. R., and D. S. Robson. 1992. *Techniques for wildlife investigations: design and analysis of capture data.* Academic Press, San Diego, CA.

Slauson, K. M., R. L. Truex, and W. J. Zielinski. 2008. Determining the sexes of American martens and fishers at track plates. *Northwest Science* 82:in press.

Slauson, K. M., W. J. Zielinski, and J. P. Hayes. 2007. Habitat selection by American martens in coastal California. *Journal of Wildlife Management* 71:458–468.

Sloane, M. A., P. Sunnicks, D. Alpers, L. B. Beheregaray, and A. C. Taylor. 2000. Highly reliable genetic identification of individual northern hairy-nosed wombats from single remotely collected hairs: a feasible censusing method. *Molecular Ecology* 9:1233–1240.

Smallwood, K. S. 1994. Trends in California mountain lion populations. *Southwestern Naturalist* 39:67–72.

Smallwood, K. S., and E. L. Fitzhugh. 1993. A rigorous technique for identifying individual mountain lions *Felis concolor* by their tracks. *Biological Conservation* 65:51–59.

———. 1995. A track count for estimating mountain lion *Felis concolor californica* population trend. *Biological Conservation* 71:251–259.

Smith, D. A., K. Ralls, B. L. Cypher, H. O. Clark Jr., P. A. Kelly, D. F. Williams, and J. E. Maldonado. 2006a. Relative abundance of endangered San Joaquin kit foxes (*Vulpes macrotis mutica*) based on scat-detection dog surveys. *Southwestern Naturalist* 51:210–219.

Smith, D. A., K. Ralls, B. L. Cypher, and J. E. Maldonado. 2005. Assessment of scat-detection dog surveys to determine kit fox distribution. *Wildlife Society Bulletin* 33:897–904.

Smith D. A., K. Ralls, B. Davenport, B. Adams, and J. E. Maldonado. 2001. Canine assistants for conservationists. *Science* 291:435.

Smith, D. A., K. Ralls, A. Hurt, B. Adams, M. Parker, B. Davenport, M. C. Smith, and J. E. Maldonado. 2003. Detection and accuracy rates of dogs trained to find scats of San Joaquin kit foxes (*Vulpes macrotis mutica*). *Animal Conservation* 6:339–346.

Smith, D. A., K. Ralls, A. Hurt, B. Adams, M. Parker, and J. E. Maldonado. 2006b. Assessing reliability of microsatellite genotypes from kit fox faecal samples using genetic and GIS analyses. *Molecular Ecology* 15:387–406.

Smith, D. W., T. D. Drummer, K. M. Murphy, D. S. Guernsey, and S. B. Evans. 2004. Winter prey selection and estimation of wolf kill rates in Yellowstone National Park, 1995–2000. *Journal of Wildlife Management* 68:153–166.

Smith, J. B., J. A. Jenks, and R. W. Klaver. 2007. Evaluating detection probabilities for American marten in the Black Hills, South Dakota. *Journal of Wildlife Management* 71:2412–2416.

Smith, L. M., and I. L. Brisbin. 1984. An evaluation of total trapline captures as estimates of furbearer abundance. *Journal of Wildlife Management* 48:1452–1455.

Smith, T. G., and I. Stirling. 1975. The breeding habitat of the ringed seal (*Phoca hispida*). The birth lair and associated structures. *Canadian Journal of Zoology* 53:1297–1305.

Smith, W. P., D. L. Borden, and K. M. Endres. 1994. Scent-station visits as an index to abundance of raccoons: an experimental manipulation. *Journal of Mammalogy* 75:637–747.

Soisalo, M. K., and S. M. C. Cavalcanti. 2006. Estimating the density of a jaguar population in the Brazilian Pantanal using camera-traps and capture-recapture sampling in combination with GPS radio-telemetry. *Biological Conservation* 129:487–496.

Sovada, M. A., and C. C. Roy. 1996. Summary of swift fox research activities conducted in western Kansas: annual report. Pages 64–68 *in* B. Luce and F. Lindzey, editors. *Annual report of the Swift Fox Conservation Team.* Lander, WY.

Squires, J. R., J. Copeland, T. J. Ulizio, M. K. Schwartz, and L. F. Ruggiero. 2007. Sources and patterns of wolverine mortality in western Montana. *Journal of Wildlife Management* 71:2213–2220.

Squires, J. R., K. S. McKelvey, and L. F. Ruggiero. 2004. A snow-tracking protocol used to delineate local lynx, *Lynx canadensis*, distributions. *Canadian Field-Naturalist* 118:583–589.

Srbek-Araujo, A. C., and A. G. Chiarello. 2005. Is camera-trapping an efficient method for surveying mammals in neotropical forests? A case study in south-eastern Brazil. *Journal of Tropical Ecology* 21:121–125.

Stager, K. E. 1964. The role of olfaction in food location by the turkey vulture (*Cathartes aura*). *Los Angeles County Museum Contributions to Science.* Los Angeles, CA.

Stake, M. M., and D. A. Cimprich. 2003. Using video to monitor predation at black-capped vireo nests. *Condor* 105:348–357.

Stander, P. E. 1998. Spoor counts as indices of large carnivore populations: the relationship between spoor frequency, sampling effort and true density. *Journal of Applied Ecology* 35:378–385.

Stanley, T. R., and J. Bart. 1991. Effects of roadside habitat and fox density on a snow track survey for foxes in Ohio. *Ohio Journal of Science* 91:186–190.

Stanley, T. R., and K. P. Burnham. 1999. A closure test for time-specific capture-recapture data.. *Environmental and Ecological Statistics* 6:197–209.

Stanley, T. R., and J. D. Richards. 2005. CloseTest: a program for testing capture-recapture data for closure. *Wildlife Society Bulletin* 33:782–785.

Stanley, T. R., and J. A. Royle. 2005. Estimating site occupancy and abundance using indirect detection indices. *Journal of Wildlife Management* 69:874–883.

Stapper, R. J., D. L. Rakestraw, D. B. Fagre, and N. J. Silvy. 1992. Evaluation of two lures for furbearer scent-station surveys. *Proceedings of the Annual Conference of the Southeastern Association of Fish and Wildlife Agencies* 46:75–78.

Stauffer, H. B., C. J. Ralph, and S. L. Miller. 2002. Incorporating uncertainty of detection into presence-absence survey design and analysis, with application to the marbled murrelet: a component of accuracy assessment. Pages 357–366 *in* J. M. Scott, P. J. Heglund, M. L. Morrison, J. B. Haufler, M. G. Raphael, W. A. Wall, and F. B. Samson, editors. *Predicting species occurrences: issues of scale and accuracy.* Island Press, Washington, DC.

Steelman, H. G., S. E. Henke, and G. M. Moore. 1998. Gray fox response to baits and attractants for oral rabies vaccination. *Journal of Wildlife Diseases* 34:764–770.

Stephens, P. A., O. Y. Zaumyslova, D. G. Miquelle, A. I. Myslenkov, and G. D. Hayward. 2006. Estimating population density from indirect sign: track counts and the Formozov-Malyshev-Pereleshin formula. *Animal Conservation* 9:339–348.

Stevens, S. S., and T. L. Serfass. 2005. Sliding behavior in nearctic river otters: locomotion or play? *Northeast Naturalist* 12:241–244.

Stevick, P. T., P. J. Palsboll, T. D. Smith, M. V. Bravington, and P. S. Hammond. 2001. Errors in identification using natural markings: rates, sources, and effects on capture-recapture estimates of abundance. *Canadian Journal of Fisheries and Aquatic Science* 58:1861–1870.

St-Georges, M., S. Nadeau, D. Lambert, and R. Decarie. 1995. Winter habitat use by ptarmigan, snowshoe hares, red foxes, and river otters in the boreal forest-tundra transition zone of western Quebec. *Canadian Journal of Zoology* 73:755–764.

Stirling, I., N. J. Lunn, J. Iacozza, C. Elliott, and M. Obbard. 2004. Polar bear distribution and abundance on the Southwestern Hudson Bay Coast during open water season, in relation to population trends and annual ice patterns. *Arctic* 57:15–26.

Stoops, M. A., G. B. Anderson, B. L. Lasley, and S. E. Shideler. 1999. Use of fecal steroid metabolites to estimate the pregnancy rate of a free-ranging herd of tule elk. *Journal of Wildlife Management* 63:661–665.

Strickland, M. A., and C. W. Douglas. 1987. Marten. Pages 530–547 *in* M. Novak, J. A. Baker, M. E. Obbard, and B. Malloch, editors. *Wild furbearer management and conservation in North America.* Ontario Ministry of Natural Resources, Toronto, Ontario, Canada.

Strier, K. B., and T. E. Ziegler. 1997. Behavioral and endocrine characteristics of the reproductive cycle in wild muriqui monkeys, *Brachyteles arachnoides. American Journal of Primatology* 42:299–310.

Stuart, J. N. 2006. Swift fox research in New Mexico: 2004 update. Pages 11–13 *in* J. N. Stuart and S. Wilson, editors. *Swift fox conservation team: annual report for 2004.* New Mexico Department of Game and Fish, Santa Fe.

Sundqvist, A.-K., H. Ellegren, and C. Vilà. 2008. Wolf or dog? Genetic identification of predators from saliva collected around bite wounds on prey. *Conservation Genetics* 9:in press.

Swann, D. E., C. C. Hass, D. C. Dalton, and S. A. Wolf. 2004. Infrared-triggered cameras for detecting wildlife: an evaluation and review. *Wildlife Society Bulletin* 32:357–365.

Swanson, B. J., B. P. Kelly, C. K. Maddox, and J. R. Moran. 2006. Shed skin as a source of DNA for genotyping seals. *Molecular Ecology Notes* 6:1006–9.

Swihart, R. K., and N. A. Slade. 1985. Testing for independence of observations in animal movements. *Ecology* 66:1176–1184.

Swimley, T. J., T. L. Serfass, R. P. Brooks, and W. M. Tzilkowski. 1998. Predicting river otter latrine sites in Pennsylvania. *Wildlife Society Bulletin* 26:836–845.

Syrotuck, W. G. 1972. *Scent and the scenting dog.* 4th reprint ed. (January 2000). Barkleigh Productions. Mechanicsburg, PA.

Szetei, V., Á. Miklósi, J. Topál, and V. Csáynyi. 2003. When dogs seem to lose their nose: an investigation on the use of visual and olfactory cues in communicative context between dog and owner. *Applied Animal Behaviour Science* 83:141–152.

Taberlet, P., J. J. Camarra, S. Griffin, E. Uhres, O. Hanotte, L. P. Waits, C. Dubois-Paganon, T. Burke, and J. Bouvet. 1997. Noninvasive genetic tracking of the endangered Pyrenean brown bear population. *Molecular Ecology* 6:869–876.

Taberlet, P., S. Griffin, B. Goossens, S. Questiau, V. Manceau, N. Escaravage, L. P. Waits, and J. Bouvet. 1996. Reliable genotyping of samples with very low DNA quantities using PCR. *Nucleic Acids Research* 26:3189–3194.

Taberlet, P., L. Waits, and G. Luikart. 1999. Noninvasive genetic sampling: look before you leap. *Trends in Ecology and Evolution* 14:323–27.

Talancy, N. W. 2005. Effects of habitat fragmentation and landscape context on medium-sized mammalian predators in northeastern national parks. MS thesis. University of Rhode Island, Providence.

Taussky, H. H. 1954. A microcolorimetric determination of creatine in urine by the Jaffe reaction. *Journal of Biological Chemistry* 208:853–861.

Tautz, D. 1989. Hypervariability of simple sequences as a general source for polymorphic DNA markers. *Nucleic Acids Research* 17:6463–6471.

Taylor, A. C., A. Horsup, C. N. Johnson, P. Sunnucks, and W. B. Sherwin. 1997. Relatedness structure detected by microsatellite analysis and attempted pedigree reconstruction in an endangered marsupial, the northern hairy-nosed wombat, *Lasiorhinus krefftii. Molecular Ecology* 6:9–19.

Taylor, C. A., and M. G. Raphael. 1988. Identification of mammal tracks from sooted track stations in the Pacific Northwest. *California Fish and Game* 74:217–229.

Taylor, W. 1971. The excretion of steroid hormone metabolites in bile and feces. *Vitamins and Hormones* 29:201–285.

Tempel, D. J., and R. J. Gutierrez. 2003. Fecal corticosterone levels in California spotted owls exposed to low-intensity chainsaw sound. *Wildlife Society Bulletin* 31:698–702.

———. 2004. Factors related to fecal corticosterone levels in California spotted owls: implications for assessing chronic stress. *Conservation Biology* 18:538–547.

Temple, S. A., and A. J. Temple. 1986. Geographic distributions and patterns of relative abundance of Wisconsin birds: a WSO research project. *Passenger Pigeon* 48:58–68.

Terio, K. A., J. L. Brown, R. Moreland, and L. Munson. 2002. Comparison of different drying and storage methods on quantiiable concentrations of fecal steroids in the cheetah. *Zoo Biology* 21:215–222.

Tewes, M. E. 1986. Ecological and behavioral correlates of ocelot spatial patterns. PhD dissertation. University of Idaho, Moscow.

Thacker, R. K., J. T. Flinders, B. H. Blackwell, and H. D. Smith. 1995. *Comparison and use of four techniques for censusing three sub-species of kit fox*. Final Report. Utah Division of Wildlife Resources, Salt Lake City.

Theberge, J. B., and C. H. R. Wedeles. 1989. Prey selection and habitat partitioning in sympatric coyote and red fox populations, southwest Yukon. *Canadian Journal of Zoology* 67:1285–1290.

Thogmartin, W. E., J. R. Sauer, and M. G. Knutson. 2004. A hierarchical spatial model of avian abundance with application to cerulean warblers. *Ecological Applications* 14:1766–1779.

Thomas, L., and C. J. Krebs. 1997. A review of statistical power analysis software. *Bulletin of the Ecological Society of America* 78:126–139.

Thompson, I. D., I. J. Davidson, S. O'Donnell, and F. Brazeau. 1989. Use of track transects to measure the relative occurrence of some boreal mammals in uncut forest and regeneration stands. *Canadian Journal of Zoology* 67:1816–1823.

Thompson, J., R. Golightly, and L. Diller. 2004. Preliminary data: the abundance and density of fisher (*Martes pennanti*) on managed timberlands in north coastal California. *Annual Conference of the Western Section of the Wildlife Society*. Rohnert Park, CA.

Thompson, J. L. 2007. Abundance and density of fisher on managed timberlands in north coastal California. MS thesis. Humboldt State University, Arcata, CA.

Thompson, S. K. 1992. *Sampling*. John Wiley & Sons, New York.

Thompson, W. L. 2004a. Future directions in estimating abundance of rare or elusive species. Pages 389–399 *in* W. L. Thompson, editor. *Sampling rare or elusive species: concepts, designs, and techniques for estimating population parameters*. Island Press, Washington, DC.

———, editor. 2004b. *Sampling rare or elusive species: concepts, designs, and techniques for estimating population parameters*. Island Press, Washington, DC.

Thompson, W. L., G. C. White, and C. Cowan. 1998. *Monitoring vertebrate populations*. Academic Press, San Diego, CA.

Touma, C., and R. Palme. 2005. Measuring fecal glucocorticoid metabolites in mammals and birds: the importance of validation. *Annals of the New York Academy of Sciences* 1046:54–74.

Touma, C., N. Sachser, E. Möstl, and R. Palme. 2003. Effects of sex and time of day on metabolism and excretion of corticosterone in urine and feces of mice. *General and Comparative Endocrinology* 130: 267–278.

Trapp, G. R. 1978. Comparative behavioral ecology of the ringtail (*Bassariscus astutus*) and gray fox (*Urocyon cinereoargenteus*) in southwestern Utah. *Carnivore* 1:3–32.

Travaini, A., R. Laffitte, and M. Delibes. 1996. Determining the relative abundance of European red foxes by scent-station methodology. *Wildlife Society Bulletin* 24:500–504.

Tredick, C. A. 2005. Population abundance and genetic structure of black bears in coastal North Carolina and Virginia using noninvasive genetic techniques. MS thesis. Virginia Polytechnic Institute and State University, Blacksburg.

Triant, D. A., R. M. Pace, and M. Stine. 2004. Abundance, genetic diversity and conservation of Louisiana black bears (*Ursus americanus luteolus*) as detected through noninvasive sampling. *Conservation Genetics* 5:647–659.

Trolle, M. 2003. Mammal survey in the southeastern Pantanal, Brazil. *Biodiversity and Conservation* 12:823–836.

Trolle, M., and M. Kery. 2003. Estimation of ocelot density in the Pantanal using campture-recapture analysis of camera-trapping data. *Journal of Mammalogy* 84:607–614.

Truex, R. L. 2003. *Sierra Nevada forest plan amendment adaptive management strategy: forest carnivore moni-*

*toring protocol.* USDA Forest Service Pacific Southwest Region, Vallejo, CA.

Tucker, A. O., and S. S. Tucker. 1988. Catnip and the catnip response. *Economic Botany* 42:214–231.

Turkowski, F. J., M. L. Popelka, B. B. Green, and R. W. Bullard. 1979. Testing the responses of coyotes and other predators to odor attractants. Pages 255–269 *in* J. R. Beck, editor. *Vertebrate pest control and management materials.* American Society for Testing and Materials, STP 680, Sacramento, CA.

Tyre, A. J., B. Tenhumberg, S. A. Field, D. Niejalke, K. Parris, and H. P. Possingham. 2003. Improving precision and reducing bias in biological surveys: estimating false-negative error rates. *Ecological Applications* 13: 1790–1801.

Ulizio, T. J. 2005. A noninvasive survey method for detecting wolverine. MS thesis. University of Montana Wildlife Biology Program, Missoula.

Ulizio, T., J. R. Squires, D. H. Pletscher, M. Schwartz, J. Claar, and L. F. Ruggiero. 2006. The efficacy of obtaining genetic-based identification from putative wolverine snow tracks. *Wildlife Society Bulletin* 34:1326–1332.

Unger, K. 2006. Follow the footprints. *Science* 313:784–785.

Uresk, D. W., K. E. Severson, and J. Javersak. 2003. *Detecting swift fox: smoked-plate scent stations versus spotlighting.* USDA Forest Service, Rocky Mountain Research Station, Research Paper RMRS-RP-39, Ogden, UT. www.treesearch.fs.fed.us/pubs/5448 (accessed 30 November 2007).

Urquhart, N. S., and T. M. Kincaid. 1999. Trend detection in repeated surveys of ecological responses. *Journal of Agricultural, Biological, and Environmental Statistics* 4:404–414.

———. 1999. Designs for detecting trend from repeated surveys of ecological resources. *Journal of Agricultural, Biological, and Environmental Statistics* 4:404–414.

Urquhart, N. S., S. G. Paulsen, and D. P. Larsen. 1998. Monitoring for policy-relevant regional trends over time. *Ecological Applications* 8:246–257.

USDA. 2001. *Sierra Nevada forest plan amendment: final environmental impact statement.* USDA Forest Service, Pacific Southwest Region, Sacramento, CA.

USDA (Food Safety and Inspection Service). 2005. *Foodborne illness: what consumers need to know.* www.fsis.usda.gov/Fact_Sheets/Foodborne_Illness_What_Consumers_Need_to_Know/index.asp (accessed 30 November 2007).

USDA Forest Service and USDI Bureau of Land Management. 2000. *Final supplemental environmental impact statement for amendment to the Survey and Manage, protection buffer, and other mitigation measures standards and guidelines.* Vol. 2. USDI Bureau of Land Management and USDA Forest Service, Portland, OR.

USDI Geological Survey. 2003. *Lure testing to test the effectiveness of a variety of scent lures.* Northern Divide Grizzly Bear Project. www.nrmsc.usgs.gov/research/glac_scentlure.htm (accessed 30 November 2007).

USDI Fish and Wildlife Service. 2004. 12-month finding for a petition to list the west coast distinct population segment of the fisher (*Martes pennanti*). *Federal Register* 69:18770–18792.

Usher, M. B. 1991. Scientific requirements of a monitoring programme. Pages 15–32 *in* F. B. Goldsmith, editor. *Monitoring for conservation and ecology.* Chapman and Hall, London, UK.

Utami, S. S., B. Goossens, M. W. Bruford, J. De Ruiter, and J. A. R. Van Hoof. 2002. Male bimaturism and reproductive success in Sumatran orangutans. *Behavioral Ecology* 13:643–652.

Valenzuela, D., and G. Ceballos. 2000. Habitat selection, home range, and activity of the white-nosed coati (*Nasua narica*) in a Mexican tropical dry forest. *Journal of Mammalogy* 81:810–819.

Valiere, N., L. Fumagalli, L. Gielly, C. Miquel, B. Lequette, M. L. Poulle, J. M. Weber, R. Arlettaz, and P. Taberlet. 2003. Long-distance wolf recolonization of France and Switzerland inferred from non-invasive genetic sampling over a period of 10 years. *Animal Conservation* 6:83–92.

Valiere, N., P. Berthier, D. Mouchiroud, and D. Pontier. 2002. GEMINI: software for testing the effects of genotyping errors and multitubes approach for individual identification. *Molecular Ecology Notes* 2:83–86.

Van Ballenberghe, V. 1984. Injuries to wolves sustained during live-capture. *Journal of Wildlife Management* 48:1425–1429.

Van Dyke, F. G., R. H. Brocke, and H. G. Shaw. 1986. Use of road track counts as indices of mountain lion presence. *Journal of Wildlife Management* 50:102–9.

Van Oosterhout, C., W. F. Hutchinson, D. P. M. Willis and P. Shipley. 2004. MICRO-CHECKER: software for identifying and correcting genotyping errors in microsatellite data. *Molecular Ecology Notes* 4:535–438.

Van Sickle, W. D., and F. G. Lindzey. 1992. Evaluation of road track surveys for cougars (*Felis concolor*). *Great Basin Naturalist* 52:232–236.

Van Wieren, S. E., and P. B. Worm. 2001. The use of a motorway wildlife overpass by large mammals. *Netherlands Journal of Zoology* 51:97–105.

Vanschaik, C. P., and M. Griffiths. 1996. Activity Periods of Indonesian Rain Forest Mammals. *Biotropica* 28:105–112.

Vickers, N. J. 2000. Mechanisms of animal navigation in odor plumes. *Biological Bulletin* 198:203–212.

Vinkey, R., M. K. Schwartz, K. S. McKelvey, K. R. Foresman, K. L. Pilgrim, B. J. Giddings, and E. C. LoFroth. 2006. When reintroduction efforts are augmentations: the genetic legacy of fisher (*Martes pennanti*) in Montana. *Journal of Mammology* 87:265–271.

von der Ohe, C. G., and C. Servheen. 2002. Measuring stress in mammals using fecal glucocorticoids: opportunities and challenges. *Wildlife Society Bulletin* 30:1215–1225.

von der Ohe, C. G., S. K. Wasser, K. E. Hunt, and C. Servheen. 2004. Factors associated with fecal glucocorticoids in Alaskan brown bears (*Ursus arctos horribilis*). *Physiological and Biochemical Zoology* 77:313–320.

Waits, L. 2004. Using noninvasive genetic sampling to detect and estimate abundance of rare wildlife species. Pages 211–228 *in* W. L. Thompson, editor. *Sampling rare or elusive species*. Island Press, Washington, DC.

Waits, J. L., and P. L. Leberg. 2000. Biases associated with population estimation using molecular tagging. *Animal Conservation* 3:191–199.

Waits, L.P., G. Luikart, and P. Taberlet. 2001. Estimating the probability of identity among genotypes in natural populations: cautions and guidelines. *Molecular Ecology* 10:249–256.

Waits, L. P., and D. Paetkau. 2005. Noninvasive sampling tools for wildlife biologists: a review of applications and recommendations for accurate data collection. *Journal of Wildlife Management* 64:1419–1433.

Wallace, R. B., H. Gomez, G. Ayala, and F. Espinoza. 2003. Camera trapping for jaguar (*Panthera onca*) in the Tuichi Valley, Bolivia. *Mastozoología Neotropical* 10:133–139.

Wallmo, O. C., A. W. Jackson, T. L. Hailey, and R. L. Carlisle. 1962. Influence of rain of the count of deer pellet groups. *Journal of Wildlife Management* 26:50–55.

Wan, Q.-H., S.-G. Fang, G.-F. Chen, Z.-M. Wang, P. Ding, M.-Y. Zhu, K.-S. Chen, J.-H. Yu, and Y.-P. Zhao. 2003. Use of oligonucleotide fingerprinting and faecal DNA in identifying the distribution of the Chinese tiger (*Panthera tigris amoyensis* Hilzheimer). *Biodiversity and Conservation* 12:1641–1648.

Warrick, G. D., and C. E. Harris. 2001. Evaluation of spotlight and scent-station surveys to monitor kit fox abundance. *Wildlife Society Bulletin* 29:827–832.

Washburn, B. E., and J. J. Millspaugh. 2002. Effects of simulated environmental conditions on glucocorticoid metabolite measurements in white-tailed deer feces. *General and Comparative Endocrinology* 127:217–222.

Wasser, S. K. 1996. Reproductive control in wild baboons measured by fecal steroids. *Biology of Reproduction* 55:393–399.

Wasser, S. K., K. Bevis, G. King, and E. Hanson. 1997. Noninvasive physiological measures of disturbance in the northern spotted owl. *Conservation Biology* 11:1019–1022.

Wasser, S. K., B. Davenport, E. R. Ramage, K. E. Hunt, M. Parker, C. Clarke, and G. Stenhouse. 2004. Scat detection dogs in wildlife research and management: application to grizzly and black bears in the Yellowhead ecosystem, Alberta, Canada. *Canadian Journal of Zoology* 82:475–492.

Wasser, S. K., C. S. Houston, G. M. Koehler, G. G. Cadd, and S. R. Fain. 1997. Techniques for application of faecal DNA methods to field studies of Ursids. *Molecular Ecology* 6:1091–1097.

Wasser, S. K., K. E. Hunt, J. L. Brown, K. Cooper, C. M. Crockett, U. Bechert, J. J. Millspaugh, S. Larson, and S. L. Monfort. 2000. A generalized fecal glucocorticoid assay for use in a diverse array of nondomestic mammalian and avian species. *General and Comparative Endocrinology* 120:260–275.

Wasser, S. K., S. L. Monfort, J. Southers, and D. E. Wildt. 1994. Excretion rates and metabolites of oestradiol and progesterone in baboon (*Papio cynocephalus cynocephalus*) faeces. *Journal of Reproduction and Fertility* 101:213–220.

Wasser, S. K., S. Papageorge, C. Foley, and J. L. Brown. 1996. Excretory fate of estradiol and progesterone in the African elephant (*Loxodonta africana*) and pattern of fecal steroid concentrations throughout the estrous cycle. *General and Comparative Endocrinology* 102: 255–262.

Wassmer, D. A. 1982. Demography, movements, activity, habitat utilization and marking behavior of a bobcat (*Lynx rufus*) population in south-central Florida. MA thesis. University of South Florida, Tampa.

Way, J. G., I. M. Ortega, P. J. Auger, and E. G. Strauss. 2002. Box-trapping eastern coyotes in southeastern Massachusetts. *Wildlife Society Bulletin* 30:695–702.

Weaver, J. L., P. Wood, and D. Paetkau. 2003. *A new noninvasive technique to survey ocelots.* Wildlife Conservation Society, New York.

Weaver, J. L., P. Wood, D. Paetkau, and L. L. Laack. 2005. Use of scented hair snares to detect ocelots. *Wildlife Society Bulletin* 33:1384–1391.

Weber, J. L., and P. E. May. 1989. Abundant class of human DNA polymporphisms which can be typed using the polymerase chain reaction. *American Journal of Human Genetics* 44:388–396.

Webster's Ninth New Collegiate Dictionary. 1988. Merriam-Webster, Springfield, MA.

Weckel, M., W. Giuliano, and S. Silver. 2006. Jaguar (*Panthera onca*) feeding ecology: distribution of predator and prey through time and space. *Journal of Zoology* 270:25–30.

Wegge, P., C. Pokheral, and S. R. Jnawali. 2004. Effects of trapping effort and trap shyness on estimates of tiger abundance from camera trap studies. *Animal Conservation* 7:251–256.

Weir, L. A., J. A. Royle, P. Nanjappa and R. E. Jung. 2005. Modeling anuran detection and site occupancy on North American Amphibian Monitoring Program (NAAMP) routes in Maryland. *Journal of Herpetology* 39:627–639.

Weston, J. L., and I. L. Brisbin Jr. 2003. Demographics of a protected population of gray foxes (*Urocyon cinereoargenteus*) in South Carolina. *Journal of Mammalogy* 84:996–1005.

Wetton, J. H., R. E. Carter, D. T. Parkin, and D. Walters. 1987. Demographic study of a wild house sparrow population by DNA fingerprinting. *Nature* 327:147–149.

White, G. C., D. R. Anderson, K. P. Burnham, and D. L. Otis. 1982. *Capture-recapture and removal methods for sampling closed populations.* Los Alamos Nat. Lab. Rep. LA-8787-NERP, Los Alamos, NM.

White, G. C., and K. P. Burnham. 1999. Program MARK: survival estimation from populations of marked animals. *Bird Study* 46:120–138.

Whitten, P. J., D. K. Brockman, and R. C. Stavisky. 1998. Recent advances in noninvasive techniques to monitor hormone-behavior interactions. *American Journal of Physical Anthropology* 27 (suppl.):1–23.

Whittington, J., C. C. St. Clair, and G. Mercer. 2004. Path tortuosity and the permeability of roads and trails to wolf movement. *Ecology and Society* 9:4. www.ecologyandsociety.org/vol9/iss1/art4 (accessed 19 December 2007).

———. 2005. Spatial responses of wolves to roads and trails in mountain valleys. *Ecological Applications* 15:543–553.

Wikramanayake, E. D., E. Dinerstein, J. G. Robinson, U. Karanth, A. Rabinowitz, D. Olson, T. Mathew, P. Hedao, M. Conner, G. Hemley, and D. Bolze. 1998. An ecology-based method for defining priorities for large mammal conservation: the tiger as a case study. *Conservation Biology* 12:865–878.

Wildt, D. E., M. Bush, K. L. Goodrowe, C. Packer, A. E. Pusey, J. L. Brown, P. Joslin, and S. J. O'Brien. 1987. Reproductive and genetic consequences of founding isolated lion populations. *Nature* 329:328–331.

Williams, B. K., J. D. Nichols, and M. J. Conroy. 2002. *Analysis and management of animal populations.* Academic Press, New York.

Williams, C. L., K. Blejwas, J. J. Johnson, and M. M. Jaeger. 2003. A coyote in sheep's clothing: predator identification from saliva. *Wildlife Society Bulletin* 31:926–932.

Williams, G. E., and P. B. Wood. 2002. Are traditional methods of determining nest predators and nest fates reliable? An experiment with wood thrushes (*Hylocichla mustelina*) using miniature video cameras. *Auk* 119:1126–1132.

Williams, M., and J. M. Johnston. 2002. Training and maintaining the performance of dogs (*Canis familiaris*) on an increasing number of odor discriminations in a controlled setting. *Applied Animal Behaviour Science* 78:55–65.

Wilson, D., and D. Reeder. 2005. *Mammal species of the world: a taxonomic and geographic reference.* Johns Hopkins University Press, Baltimore, MD.

Wilson, G. J., and R. J. Delahay. 2001. A review of methods to estimate the abundance of terrestrial carnivores using field sign and observation. *Wildlife Research* 28: 151–164.

Wilson, K. R., and D. R. Anderson. 1985. Evaluation of two density estimators of small mammal population size. *Journal of Mammalogy* 66:13–21.

Windberg, L. A., and F. F. Knowlton. 1990. Relative vulnerability of coyotes to some capture methods. *Wildlife Society Bulletin* 18:282–290.

Winter, M., D. H. Johnson, and J. Faaborg. 2000. Evidence for edge effects on multiple levels in tallgrass prairie. *Condor* 102:256–266.

Winter, W. 1981. *Black-footed ferret search dogs.* Southwestern Research Institute, Santa Fe, NM.

Wintle, B. A., M. A. McCarthy, K. M. Parris, and M. A. Burgman. 2004. Precision and bias of methods for estimating point survey detection probabilities. *Ecological Applications* 14:703–712.

Wise, K. K., M. R. Conover, and F. F Knowlton. 1999. Response of coyotes to avian distress calls: testing the startle-predator and predator-attraction hypotheses. *Behaviour* 136:935–949.

Wisely, S. M., S. W. Buskirk, G. A. Russell, K. B. Aubry, and W. J. Zielinski. 2004. Genetic diversity and structure of the fisher (*Martes pennanti*) in a peninsular and peripheral metapopulation. *Journal of Mammalogy* 85:640–648.

Wolf, K. N., F. Elvinger, and J. L. Pilcicki. 2003. Infrared-triggered photography and tracking plates to monitor oral rabies vaccine bait contact by raccoons in culverts. *Wildlife Society Bulletin* 31:387–391.

Wong, S. T., C. Servheen, C., L. Ambu, and A. Norhayati. 2005. Impacts of fruit production cycles on Malayan sun bears and bearded pigs in lowland tropical forest of Sabah, Malaysian Borneo. *Journal of Tropical Ecology* 21:627–639.

Wood, J. E. 1959. Relative estimates of fox population levels. *Journal of Wildlife Management* 23:53–63.

Wood, W. F., M. N. Terwilliger, and J. P. Copeland. 2005. Volatile compounds from the anal glands of the wolverine, *Gulo gulo. Journal of Chemical Ecology* 12: 2111–2117.

Woods, J. G., D. Paetkau, D. Lewis, B. N. McLellan, M. Proctor, and C. Strobeck. 1999. Genetic tagging free-ranging black and brown bears. *Wildlife Society Bulletin* 27:616–627.

Wright, S. 1969. *Evolution and the genetics of populations. Vol. 2. The theory of gene frequencies.* University of Chicago Press, Chicago, IL.

Wydeven, A. P., R. N. Schultz, and R. P. Thiel. 1995. Monitoring a recovering gray wolf population in Wisconsin, 1979–1995. Pages 147–156 *in* L. N. Carbyn, S. H. Fritts, and D. R. Seip, editors. *Ecology and conservation of wolves in a changing world.* Canadian Circumpolar Institute, Edmonton, Alberta, Canada.

Wyshinski, N. 2001. *Formulating and compounding animal baits and lures.* 3rd ed. Nick Wyshinski, Berwick, PA.

Yahoo Camera Trap Group. 2007. uk.groups.yahoo.com/group/cameratraps/ (accessed 30 November 2007).

Yalow, R. S., and S. A. Berson. 1959. Assay of plasma insulin in human subjects by immunological methods. *Nature* 184:1648–1649.

Yanes, M., and F. Suarez. 1996. Incidental nest predation and lark conservation in an Iberian semiarid shrubsteppe. *Conservation Biology* 10:881–887.

Yang, H. C., and A. Chao. 2005. Modeling animals' behavioral response by Markov chain models for capture-recapture experiments. *Biometrics* 61:1010–1017.

Yanosky, A. A., and C. Mercolli. 1992. Habitat preferences and activity in the ring-tailed coati (*Nasua nasua*) at El Bagual Ecological Reserve (Argentina). *Miscellania Zoologia* 16:179–182.

Yasuda, M. 2004. Monitoring diversity and abundance of mammals with camera traps: a case study on Mount Tsukuba, central Japan. *Mammal Study* 29:37–46.

Yasuda, M., and K. Kawakami. 2002. New method of monitoring remote wildlife via the Internet. *Ecological Research* 17:119–124.

Yoccoz, N. G., J. D. Nichols, and T. Boulinier. 2001. Monitoring of biological diversity in space and time. *Trends in Ecology and Evolution* 16:446–453.

York, E. C., T. L. Moruzzi, T. K. Fuller, J. F. Organ, R. M. Sauvajot, and R. M. DeGraaf. 2001. Description and evaluation of a remote camera and triggering system to monitor carnivores. *Wildlife Society Bulletin* 29:1228–1237.

Young, A. J., A. A. Carlson, S. L. Monfort, A. F. Russell, N. C. Bennett, and T. H. Clutton-Brock. 2006. Stress and the suppression of subordinate reproduction in cooperatively breeding meerkats. *Proceedings of the National Academy of Sciences* 103:12005–12010.

Young, K. M., S. L. Walker, C. Lanthier, W. T. Waddell, S. L. Monfort, and J. L. Brown. 2004. Monitoring of adrenocortical activity in carnivores by fecal glucocorticoid analyses. *General and Comparative Endocrinology* 137: 148–165.

Young, S. P. 1958. *The bobcat of North America.* Wildlife Management Institute, Washington, DC.

Zabala, J., and I. Zuberogoitia. 2003. Badger, *Meles meles* (Mustelidae, Carnivora), diet assessed through scat-analysis: a comparison and critique of different methods. *Folia Zoologica* 52:23–30.

Zaniewski, A. E., A. Lehmann, and J. M. Overton. 2002. Predicting species spatial distributions using presence-only data: a case of native New Zealand ferns. *Ecological Modelling* 157:261–280.

Zezulak, D. S., and R. G. Schwab. 1979. A comparison of density, home range, and habitat utilization of bobcat populations at Lava Beds and Joshua Tree National Monuments. *National Wildlife Federation Scientific and Technical Series* 6:74–79.

Ziegler, C., and C. Carroll. 2005. Following the stealth hunter. *National Geographic* (November):66–77.

Ziegler, T., J. K. Hodges, P. Winkler, and M. Heistermann. 2000. Hormonal correlates of reproductive seasonality in wild female Hanuman langurs (*Presbytis entellus*). *American Journal of Primatology* 51:119–134.

Ziegler, T. E., C. V. Santos, A. Pissinatti, and K. B. Strier. 1997. Steroid excretion during the ovarian cycle in captive and wild muriquis, *Brachyteles arachnoides*. *American Journal of Primatology* 42:311–321.

Ziegler, T. E., and D. J. Wittwer. 2005. Fecal steroid research in the field and laboratory: improved methods for storage, transport, processing, and analysis. *American Journal of Primatology* 67:159–174.

Zielinski, W. J. 1995. Track plates. Pages 67–89 *in* W. J. Zielinski and T. E. Kucera, editors. *American marten, fisher, lynx, and wolverine: survey methods for their detection.* USDA Forest Service Pacific Southwest Research Station General Technical Report PSW-GTR-157, Albany, CA.

Zielinski, W. J., and T. E. Kucera. 1995a. *American marten, fisher, lynx, and wolverine: survey methods for their detection.* USDA Forest Service, Pacific Southwest Research Station General Technical Report PSW-GTR-157, Albany, CA.

———. 1995b. Introduction to detection and survey methods. Pages 1–16 *in* W. J. Zielinski and T. E. Kucera, editors. *American marten, fisher, lynx, and wolverine: survey methods for their detection.* USDA Forest Ser-vice, Pacific Southwest Research Station General Technical Report PSW-GTR-157, Albany, CA.

Zielinski, W. J., T. E. Kucera, and R. H. Barrett. 1995a. The current distribution of fisher, *Martes pennanti*, in California. *California Fish and Game* 81:104–112.

Zielinski, W. J., T. E. Kucera, and J. C. Halfpenny. 1995b. Definition and distribution of sample units. Pages 17–24 *in* W. J. Zielinski and T. E. Kucera, editors. *American marten, fisher, lynx, and wolverine: survey methods for their detection.* USDA Forest Service Pacific Southwest Research Station General Technical Report PSW-GTR-157, Albany, CA.

Zielinski, W. J., and S. Mori. 2001. What is the status and change in the geographic distribution and relative abundance of fisher? Study plan, adaptive management strategy, Sierra Nevada Framework. USDA Forest Service, Pacific Southwest Research Station, unpublished manuscript, Albany, CA.

Zielinski, W. J., F. V. Schlexer, K. L. Pilgrim, and M. K. Schwartz. 2006. The efficacy of wire and glue hair snares in identifying mesocarnivores. *Wildlife Society Bulletin* 34:1152–1161.

Zielinski, W. J., K. M. Slauson, A. E. Bowles. 2007. *The effect of off-highway vehicle use on the American marten in California.* Final report submitted to USDA Forest Service, Pacific Southwest Region. Redwood Sciences Laboratory, Arcata, CA.

Zielinski, W. J., K. M. Slauson, C. R. Carroll, C. J. Kent, and D. G. Kudrna. 2001. Status of American martens in coastal forests of the Pacific states. *Journal of Mammalogy* 82:478–490.

Zielinski, W. J., and H. B. Stauffer. 1996. Monitoring Martes populations in California: survey design and power analysis. *Ecological Applications* 6:1254–1267.

Zielinski, W. J., and R. L. Truex. 1995. Distinguishing tracks of marten and fisher at track plate stations. *Journal of Wildlife Management* 59:571–579.

Zielinski, W. J., R. L. Truex, C. V. Ogan, and K. Busse. 1997. Detection surveys for fishers and American martens in California, 1989–1994: summary and interpretations. Pages 372–392 *in* G. Proulx, H. N. Bryant, and P. M. Woodard, editors. *Martes: taxonomy, ecology, techniques, and management.* Provincial Museum of Alberta, Edmonton, Alberta, Canada.

Zielinski, W. J., R. L. Truex, F. V. Schlexer, L. A. Campbell, and C. Carroll. 2005. Historical and contemporary distributions of carnivores in forests of the Sierra Nevada, CA. *Journal of Biogeography* 32:1385–1407.

Zoellick, B. W., and N. S. Smith. 1992. Size and spatial organization of home ranges of kit foxes in Arizona. *Journal of Mammalogy* 73:83–88.

Zoellick, B. W., H. M. Ulmschneider, B. S. Cade, and A. W. Stanley. 2004. Isolation of Snake River islands and mammalian predation of waterfowl nests. *Journal of Wildlife Management* 68:650–662.

Zoellick, B. W., H. M. Ulmschnieder, and A. W. Stanley 2005. Distribution and composition of mammalian predators along the Snake River in southwestern Idaho. *Northwest Science* 79:265–272.

Zuercher, G. L., P. S. Gipson, and G. C. Stewart. 2003. Identification of carnivore feces by local peoples and molecular analyses. *Wildlife Society Bulletin* 31:961–970.

Zwickel, F. C. 1980. Use of dogs in wildlife biology. Pages 531–536 *in* S. D. Schemnitz, editor. *Wildlife techniques manual.* 4th ed. Wildlife Society, Washington, DC.

# Contributors

**Jon P. Beckmann**
Wildlife Conservation Society
2023 Stadium Drive, Suite 1A
Bozeman, MT 59715
jbeckmann@wcs.org

**Lori A. Campbell**
USDA Forest Service
Sierra Nevada Research Center, Pacific Southwest
    Research Station
1731 Research Park Drive
Davis, CA 95618
lcampbell@fs.fed.us

**Samuel Cushman**
USDA Forest Service
Forestry Sciences Laboratory, Rocky Mountain
    Research Station
800 East Beckwith
Missoula, MT 59801
scushman@fs.fed.us

**John Erb**
Minnesota Department of Natural Resources
Forest Wildlife Populations and Research Group
1201 E. Highway 2
Grand Rapids, MN 55744
john.erb@dnr.state.mn.us

**Robert L. Harrison**
Department of Biology

University of New Mexico
Albuquerque, NM 87131–1091
rharison@unm.edu

**Stephen W. R. Harrison** (illustrator)
California State University, Stanislaus
Endangered Species Recovery Program
PO Box 9622
Bakersfield CA 93389
sharrison@esrp.csustan.edu

**Kimberly S. Heinemeyer**\*
Round River Conservation Studies
2591 Kid Curry Drive
Bozeman, MT 59718
kim@roundriver.org

**Roland W. Kays**\*
New York State Museum
CEC 3140
Albany, NY 12230
rkays@mail.nysed.gov

**Katherine C. Kendall**\*
USGS Glacier Field Station
Glacier National Park
West Glacier, MT 59936-0128
kkendall@usgs.gov

**Patricia Kernan** (illustrator)
New York State Museum

CEC 3140
Albany, NY 12230
pkernan@mail.nysed.gov

**Robert A. Long** (editor)*
Western Transportation Institute, Montana State
    University
P.O. Box 1654
Ellensburg, WA 98926
robert.long@coe.montana.edu

**Paul M. Lukacs**
Colorado Division of Wildlife
317 W. Prospect Road
Fort Collins, CO 80526
paul.lukacs@state.co.us

**Paula MacKay** (managing editor)*
204 East 10th Avenue
Ellensburg, WA 98926
paularob@gmavt.net

**Kevin S. McKelvey**
USDA Forest Service
Forestry Sciences Laboratory, Rocky Mountain
    Research Station
800 East Beckwith
Missoula, MT 59801
kmckelvey@fs.fed.us

**Steven L. Monfort**
Smithsonian Institution
National Zoological Park, Conservation and
    Research Center
1500 Remount Road
Front Royal, VA 22630
monforts@si.edu

**Megan Parker**
Conservation and Ecosystem Sciences
University of Montana
Missoula, MT 59812
mnparkcr@igc.org

**Justina C. Ray** (editor)
Wildlife Conservation Society Canada
720 Spadina Avenue, Suite 600
Toronto, Ontario M5S_2T9
Canada
jray@wcs.org

**J. Andrew Royle**
USGS Patuxent Wildlife Research Center
12100 Beech Forest Road
Laurel, MD 20708
aroyle@usgs.gov

**Fredrick V. Schlexer**
USDA Forest Service
Pacific Southwest Research Station
1700 Bayview Drive
Arcata, CA 95521
rschlexer@fs.fed.us

**Michael K. Schwartz**
USDA Forest Service
Forestry Sciences Laboratory, Rocky Mountain
    Research Station
800 East Beckwith
Missoula, MT 59801
mkschwartz@fs.fed.us

**Keith M. Slauson**
USDA Forest Service
Pacific Southwest Research Station
1700 Bayview Drive
Arcata, CA 95521
kslauson@fs.fed.us

**Deborah A. Smith**
Working Dogs for Conservation Foundation
52 Eustis Road
Three Forks, MT 59752
info@workingdogsforconservation.org

**Thomas R. Stanley**
USGS Fort Collins Science Center

2150 Centre Avenue, Building C
Fort Collins, CO 80526
stanleyt@usgs.gov

**Jeffrey B. Stetz**
University of Montana
USGS Glacier Field Station
Glacier National Park
West Glacier, MT 59936-0128
jstetz@usgs.gov

**Richard L. Truex**
USDA Forest Service, Pacific Southwest Region
Sequoia National Forest
1839 South Newcomb Street
Porterville, CA 93257
rtruex@fs.fed.us

**Todd J. Ulizio**
USDA Forest Service
Rocky Mountain Research Station
800 East Beckwith
Missoula, MT 59802
todd_ulizio@yahoo.com

**William J. Zielinski** (editor)*
USDA Forest Service
Pacific Southwest Research Station
1700 Bayview Drive
Arcata, CA 95521
bzielinski@fs.fed.us

*Attended planning workshop in Essex, Massachusetts, June 1–3, 2005.

# About the Editors

**Robert A. Long** is a research ecologist with the Road Ecology Program of the Western Transportation Institute at Montana State University. He currently coordinates wildlife monitoring in the Cascades of Washington and holds an adjunct faculty position at Central Washington University. His research interests include carnivore ecology and conservation, landscape permeability for wildlife, and wildlife survey and monitoring.

**Paula MacKay** is a research associate with the Road Ecology Program of the Western Transportation Institute at Montana State University and a freelance writer/editor. She has assisted with numerous wildlife surveys on land and at sea, and is currently conducting wildlife monitoring in the Cascades of Washington. Her writings have been published in magazines, newspapers, and books, and she performs communications work for various conservation organizations.

**Justina C. Ray** is director of Wildlife Conservation Society Canada and holds adjunct faculty positions at University of Toronto and Trent University. While her research has ranged from tropical rainforests to subarctic taiga, the ecology and conservation of forest carnivores have remained common themes. Her current emphasis is on research and policy activities associated with conservation planning in large intact landscapes of Canada's northern boreal forests. She is senior editor of *Large Carnivores and the Conservation of Biodiversity* (Island Press 2005).

**William J. Zielinski** is a research ecologist with the USDA Forest Service Pacific Southwest Research Station who specializes in the ecology of forest mammals. He also holds adjunct faculty positions at Humboldt State University and Oregon State University. He has a long-standing research interest in the field of wildlife survey, inventory, and monitoring, and in the transfer of research developments to the wildlife management community.

# Index